Texts and Monographs in Physics

Springer-Verlag Berlin Heidelberg GmbH

Texts and Monographs in Physics

From Microphysics to Macrophysics I + II Methods and Applications of Statistical Physics By R. Balian

Variational Methods in Mathematical Physics A Unified Approach By P. Blanchard and E. Brüning

Quantum Mechanics: Foundations and Applications 3rd enlarged edition By A. Bohm

The Early Universe Facts and Fiction 3rd corrected and enlarged edition By G. Börner

Geometry of the Standard Model of Elementary Particles By A. Derdzinski

Random Walks, Critical Phenomena, and Triviality in Quantum Field Theory By R. Fernández, J. Fröhlich and A. D. Sokal

Quantum Relativity A Synthesis of the Ideas of Einstein and Heisenberg By D. Finkelstein

Quantum Mechanics I + II By A. Galindo and P. Pascual

The Elements of Mechanics By G. Gallavotti

Local Quantum Physics Fields, Particles, Algebras Corrected 2nd printing By R. Haag

Elementary Particle Physics Concepts and Phenomena By O. Nachtmann

Inverse Schrödinger Scattering in Three Dimensions By R. G. Newton

Scattering Theory of Waves and Particles 2nd edition By R. G. Newton

Quantum Entropy and Its Use By M. Ohya and D. Petz

Generalized Coherent States and Their Applications By A. Perelomov

Essential Relativity Special, General, and Cosmological Revised 2nd edition By W. Rindler

Path Integral Approach to Quantum Physics An Introduction By G. Roepstorff

Advanced Quantum Theory and Its Applications Through Feynman Diagrams 2nd edition By M. D. Scadron

Finite Quantum Electrodynamics The Causal Approach 2nd edition By G. Scharf

From Electrostatics to Optics A Concise Electrodynamics Course By G. Scharf

The Mechanics and Thermodynamics of Continuous Media By M. Silhavy

Large Scale Dynamics of Interacting Particles By H. Spohn

General Relativity and Relativistic Astrophysics By N. Straumann

The Dirac Equation By B. Thaller

Relativistic Quantum Mechanics By F. J. Ynduráin

The Theory of Quark and Gluon Interactions 2nd completely revised and enlarged edition By F. J. Ynduráin

G. Scharf

Finite Quantum Electrodynamics

The Causal Approach

Second Edition
With 17 Figures

Springer

Professor Dr. G. Scharf

Institut für Theoretische Physik
Universität Zürich, Büro 36-K-70
Winterthurer Strasse 190
CH-8057 Zürich, Switzerland

ISBN 978-3-642-63345-4 ISBN 978-3-642-57750-5 (eBook)
DOI 10.1007/978-3-642-57750-5

Library of Congress Cataloging-in-Publication Data.
Scharf, G. (Günter), 1938-
Finite quantum electrodynamics: the causal approach / G. Scharf. – 2nd ed.
p. cm. – (Texts and monographs in physics)
Includes bibliographical references and index.
ISBN 978-3-642-63345-4
1. Quantum electrodynamics. 2. Quantum field theory. 3. Perturbation (Quantum dynamics) I. Title.
II. Series. QC680.S32 1995 537.6'7—dc20 95-35562

Originally published by Springer-Verlag Berlin Heidelberg New York in 1995
Softcover reprint of the hardcover 2nd edition

Typesetting: Data conversion by K. Mattes, Heidelberg
Cover design: Springer-Verlag, Design & Production

SPIN: 10508490 55/3144-543210 - Printed on acid-free paper

Preface

Quantum field theory as it is usually formulated is full of problems with ultraviolet and infrared divergences. This is somewhat surprising, because there is a simple way out which one learns in mathematics. One must only adopt the following two rules. First, use well-defined quantities only, for example free fields. Second, make justified operations only in the calculations; in particular do not multiply certain distributions by discontinuous step functions. If one really follows these rules, then no infinity can appear and life is beautiful. The question then is how to construct the standard theory according to these rules. This one can learn from an old paper by Epstein and Glaser (*Annales de l'Institut Poincaré A 19, p. 211 (1973)*). The main tool in this method is causality.

The causal method was developed by Stückelberg and Bogoliubov in the 1950s. One reason for the limited resonance it found was perhaps the highly non-trivial nature of the causality condition. We therefore start slowly. After a chapter on the classical Dirac theory of electrons and positrons and the quantization of free fields, we study the external field problem in some detail. We will find that the (second quantized) scattering matrix (S-matrix) for this problem is uniquely determined up to a phase. This phase contains physical effects, namely the so-called vacuum polarization which is produced by the external field. Therefore, it is needed to complete the construction of the S-matrix, and here is the place where the causality condition comes in for the first time. With this experience we are then able to construct the S-matrix of full QED by causal perturbation theory in Chap. 3. The important point is that this directly leads to the finite ("renormalized") perturbation series. In fact, no divergent Feynman integral and no ultraviolet cutoff will appear in this book, explaining why the title "Finite QED" was chosen.

It is a common belief that QED with a cutoff or scale parameter should be considered as part of a more fundamental theory where the scale parameter disappears, and that the theory is only mathematically well defined in this bigger framework. We will see that there is no scale parameter in QED in the causal approach if the electrons are massive. If one considers massless fermions, then a scale parameter appears in a natural way, because the central splitting solution (Sect. 3.2) no longer exists. This suggests that it seems indeed necessary to study a bigger theory if one wants to attack the mass problem. But if we take the electron mass as a given finite parameter, QED still has a good chance of being well defined. In fact, the perfect agreement of the perturbative results with experiment cannot be an accident; there must exist well-defined objects (perhaps the adiabatically switched S-matrix $S(g)$ of Sect. 3.1) which are approximated by perturbation theory.

The fact that the causal theory is perturbative has not only a technical but also a deep physical reason. In any realistic quantum field theory one must draw a sharp distinction between the fundamental fields that appear in the elementary interaction and the asymptotic states describing the real incoming and outgoing particles. This is well known today from the theory of strong interactions (quantum chromodynamics, QCD) where the quark fields are the fundamental Fermi fields, while the mesons and nucleons are complicated bound states of them. But even the electron is complicated because it carries the Coulomb field, so it must be regarded as a bound state where (scalar) photons are confined to a Dirac field. Compositeness is the normal case. Only the photon and the neutrini seem to be elementary in the sense that they can simply be generated by fundamental fields. In the causal theory the very hard problem of the asymptotic states is clearly separated from the rest of the theory by the method of adiabatic switching: the interaction is multiplied by a test function $g(x)$ and one performs the adiabatic limit $g \to 1$ at the very end in observable quantities. This means that the confinement is switched off in the asymptotic region in a "gedanken-experiment", so that free fields are coming out, instead of the complicated real physical particles. The switching is then removed in the adiabatic limit. From the study of this limit one can learn something about the structure of the real asymptotic states. It turns out that the limit does not always exist. Only if the right inclusive cross sections are considered does the limit come out finite and unique. In this way the S-matrix itself dictates the structure of the physical particles, as it must be. This highly important fact, which is even more important in non-abelian theories, can already be seen in perturbation theory, as we will discuss at the end of Chap. 3. But it seems to be rather hopeless to jump by some non-perturbative guess directly to the correct description of the asymptotic states.

The inductive construction of the S-matrix enables us to give simple inductive proofs of the various properties of the theory, in particular gauge invariance and unitarity. These themes are described in Chap. 4. The discussion of other electromagnetic couplings in Chap. 5. brings in new features which are important for preparing the extension of the causal method to non-abelean gauge theories. One might regret that this subject is not yet included, but the Epilogue gives a short account of the present status of this field. In the historical introduction the various lines of development in quantum field theory are discussed. From the beginning in the 1920s until today this was a fascinating sequence of successes and failures, where each attempt contained its piece of truth.

The book differs considerably from its first edition: Chapter 3 was completely rewritten and the Chaps. 4 and 5 are new. The bibliographical notes give some hints for further reading.

Acknowledgements. I wish to thank M. Dütsch and F. Krahe for many important comments and for their help in correcting the manuscript. I am also grateful to W. Beiglböck and to the staff at Springer-Verlag for the excellent collaboration.

Zürich, May 1995 *G. Scharf*

Contents

0. Preliminaries 1

0.0 Historical Introduction 1

0.1 Minkowski Space and the Lorentz Group 6

0.2 Tensors in Minkowski Space 11

0.3 Some Topics of Scattering Theory 14

0.4 Problems 19

1. Relativistic Quantum Mechanics 21

1.1 Spinor Representations of the Lorentz Group 21

1.2 Invariant Field Equations 26

1.3 Algebraic Properties of the Dirac Equation 32

1.4 Discussion of the Free Dirac Equation 36

1.5 Gauge Invariance and Electromagnetic Fields 44

1.6 The Hydrogen Atom 54

1.7 Problems 62

2. Field Quantization 66

2.1 Second Quantization in Fock Space 67

2.2 Quantization of the Dirac Field 78

2.3 Discussion of the Commutation Functions 87

2.4 The Scattering Operator (S-Matrix) in Fock Space 93

2.5 Perturbation Theory 105

2.6 Electron Scattering 111

2.7 Pair Production 118

2.8 The Causal Phase of the S-Matrix 124

2.9 Non-Perturbative Construction of the Causal Phase 134

2.10 Vacuum Polarization 141

2.11 Quantization of the Radiation Field 146

2.12 Problems 156

3. Causal Perturbation Theory 159

3.1 The Method of Epstein and Glaser 160

3.2 Splitting of Causal Distributions 170

3.3 Application to QED 183

3.4 Electron Scattering (Moeller Scattering) 186

3.5 Electron-Photon Scattering (Compton Scattering) 195

3.6 Vacuum Polarization 202
3.7 Self-Energy 208
3.8 Vertex Function: Causal Distribution 214
3.9 Vertex Function: Retarded Distribution 228
3.10 Form Factors 236
3.11 Adiabatic Limit 239
3.12 Charged Particles in Perturbative QED 248
3.13 Charge Normalization 258
3.14 Problems 261

4. Properties of the S-Matrix 263
4.1 Vacuum Graphs 263
4.2 Operator Character of the S-Matrix 268
4.3 Normalizability of QED 271
4.4 Discrete Symmetries 275
4.5 Poincaré Covariance 282
4.6 Gauge Invariance and Ward Identities 289
4.7 Unitarity 300
4.8 Renormalization Group 308
4.9 Interacting Fields and Operator Products 314
4.10 Field Equations 323
4.11 Problems 333

5. Other Electromagnetic Couplings 335
5.1 Scalar QED: Basic Properties 335
5.2 Scalar QED: Gauge Invariance 344
5.3 Axial Anomalies 351
5.4 (2+1)-Dimensional QED: Vacuum Polarization 362
5.5 (2+1)-Dimensional QED: Mass Generation 368
5.6 Problems 375

6. Epilogue: Non-Abelian Gauge Theories 376

Appendices
A: The Hydrogen Atom
According to the Schrödinger Equation 381
B: Regularly Varying Functions 384
C: Spence Functions 390
D: Grassmann Test Functions 392

Bibliographical Notes 397

Subject Index 403

0. Preliminaries

We start the numbering with zero because this chapter is preparatory. At the beginning of each chapter we want to make some general introductory remarks because, we think, the reader has a right to know in advance why the material that follows is presented to him. We begin with an introduction into the history of quantum field theory. To understand the striking success of this theory, it is helpful and clarifying to remember how the fundamental ideas have been introduced in the past and how they got modified in the course of time. After this historical introduction of those concepts we start with their physical introduction.

The object of physics is the description of observable phenomena in space and time and the investigation of the mathematical structure behind these phenomena. Therefore in the first section the 4-dimensional space of space-time points and the corresponding transformation group of the reference systems is described. The tensor calculus, which is briefly discussed in Sect. 0.2, is a tool to write the equations in a form independent of the reference system. The third section is concerned with some basic concepts of scattering theory. As we shall see much later, it is difficult, in general, to formulate the time-evolution of a system in quantum field theory, contrary to non-relativistic quantum mechanics. In this situation, scattering theory becomes of central importance. We show how the scattering matrix can be constructed using causality instead of dynamical equations. This is precisely what we will do in the case of full QED in Chap. 3. Causality will be the cornerstone in the book.

0.0 Historical Introduction

The dawn of quantum field theory coincides with the development of quantum mechanics in the 1920's. When M. Born and P. Jordan (*Zeitschrift f. Physik 34, 886 (1925)*) clarified the structure of Heisenberg's matrix mechanics, they added a chapter IV with the title "Remarks on Electrodynamics". They pointed out that the quantum mechanical treatment of the harmonic oscillator, which was of crucial importance for the discovery of the theory, is also relevant for the electromagnetic field: Although the latter is a system of infinitely many degrees of freedom, the theory of the one-dimensional oscillator is sufficient for its treatment, because the radiation field can be regarded as a system of uncoupled oscillators. Then the electric and magnetic field strength $\boldsymbol{E}, \boldsymbol{H}$ with periodic time dependence become matrices. The authors, therefore, used the notion "matrix electrodynamics". But they only considered the free electromagnetic field.

The name quantum electrodynamics (QED) was introduced by P.A.M. Dirac (*Proc. Roy. Soc. London A 114, 243 (1927)*) in his paper on "The Quantum Theory of Emission and Absorption of Radiation" after Schrödinger's formulation of quantum mechanics in 1926. Dirac had the time-dependent perturbation theory at his disposal, therefore, he was able to treat the radiation field in interaction with an atom. He observed that light quanta must obey Bose-Einstein statistics and calculated Einstein's A- and B-coefficients for the emission and absorption rates. Here spontaneous emission was explained for the first time. The procedure of quantizing the radiation field still remained somewhat unclear. This point was further considered by P. Jordan and W. Pauli (*Z. Phys. 47, 151 (1928)*) in their paper "On Quantum Electrodynamics of Fields without Charges". They gave a Lorentz invariant quantization of the electromagnetic field and introduced the invariant D-function which was later called Jordan-Pauli distribution. They arrived directly at the commutation relations for the electric and magnetic fields $\boldsymbol{E}, \boldsymbol{H}$ and noticed that there exist no simple invariant commutation relation for the vector potential. They also noticed the difficulty of the infinite zero-point energy. Jordan continued this line of research together with E. Wigner (*Z. Phys. 47, 631 (1928)*) in the paper "On Pauli's Exclusion Principle", where they showed that Pauli's principle implies field quantization with anticommutators. This led them to an elegant theory of the Fermi gas.

At the same time Dirac established the second pillar of QED, namely the relativistic equation for the electron in his paper "The Quantum Theory of the Electron" (*Proc. Roy. Soc. London 117, 610 and 118, 351 (1928)*). This famous equation immediately explained the spin of the electron and its magnetic moment $e\hbar/2mc$, as well as the fine-structure of the spectrum of the hydrogen atom. Despite these brilliant successes, there was a serious difficulty in the theory which was realized by Dirac: The equation has solutions with unbounded negative energy. This problem occupied Dirac for almost two years. At the beginning of 1930 (*Proc. Roy. Soc. London 126, 360 (1930)*) he gave a solution in his paper "A Theory of Electrons and Protons" (originally he thought the negative energy states to be protons). He interpreted the theory as a multiparticle theory and used the exclusion principle for the electrons. He did not put all pieces together, because he was not using Jordan and Wigner's method for quantization of Fermi fields which would be the appropriate tool, but developed a picture of his own in his "hole theory". It rests on the assumption that all states with negative energy are filled up with electrons, so that no electron can jump into one of these occupied states according to the exclusion principle. This new picture of the vacuum state has observable consequences in electron-photon scattering, and it predicts new effects: A hole in the sea of negative states appears as a particle with opposite (positive) charge. Dirac first thought that this must be the proton, because no other particle with positive charge was known. But then the two particles would annihilate in a hydrogen atom. Finally (*Proc. Roy. Soc. London 133, 60 (1931)*) he assumed that the holes are new, yet unknown "anti-electrons" with the same mass as electrons but charge $+e$. By analogy he also thought that anti-protons might exist. When the anti-electron (positron) was indeed found by C.D. Anderson in the cosmic rays in 1932, this was the first particle correctly predicted by theory. The anti-proton was observed much later in 1955.

As already said, Dirac with his hole theory did not follow the ideas of quantum field theory. This direction was further pursued by W. Heisenberg and W. Pauli in their paper "On Quantum Dynamics of Wave Fields" (*Z. Phys. 56, 1 (1929)*). Here the general method of canonical quantization was systematically developed. The problem was reduced to quantum mechanics by dividing the 3-space into cells and treating the field variables in these cells like the mechanical coordinates and momenta. Pauli has sometimes used this old method in later years for basic reasoning. When the method was applied to electrodynamics, some difficulties appeared, because the time-component of the vector potential has no conjugate momentum. This problem was brilliantly circumvented by introducing a gauge-fixing term, as we call it today. However, for the electron field satisfying the Dirac equation the two possibilities with commutation or anticommutation relations were treated upon the same footing. Obviously, the connection of spin and statistics was not yet understood. For Pauli this was a theme for a long time (*Phys. Rev. 58, 716 (1940)*). The problem of the negative energy states was still not solved, as well as the zero-point energy of the radiation field and the infinite self-energy of the electron.

That the zero-point energy of the electromagnetic field in infinite space has no physical meaning was clear to many authors. But the radiation field poses more problems. To treat the interaction with matter, it is necessary to use potentials. Then, however, it is difficult to perform the quantization in a manifest Lorentz covariant form. Dirac in the second edition of his book on "Quantum Mechanics" (*Oxford 1935*) gave an elegant solution to the problem using results of E. Fermi (*Rev. Mod. Phys. 4, 125 (1932)*). The positron problem was even harder because there is a polarization of the vacuum (*W. Heisenberg, Z. Phys. 90, 209 (1934)*). Heisenberg found a pragmatic solution: he quantized the free electron-positron field in accordance with Dirac's hole theory and then developed perturbation theory. At the same time W.H. Furry and J.R. Oppenheimer wrote a paper "On the Theory of Electron and Positive" (*Phys. Rev. 45, 245 (1934)*) where they discuss (second) quantization of the Dirac field in the modern way. When Pauli summarized the status of the theory in his review article "Relativistic Field Theory of Elementary Particles" (*Rev. Mod. Phys. 13, 203 (1941)*), he quantized all interesting fields in a completely satisfactory manner, apart from a small reservation in case of the Dirac field. This article was called the "New Testament" by the younger collaborators of Pauli in contrast to his work of 1933 on quantum mechanics (*Handbuch der Physik, 2. Aufl., Bd. 24/1*), which was the "Old Testament".

However, the situation with respect to the other infinities that are due to interaction could not be improved until after the Second World War. The key point was to formulate QED in a manifest relativistically covariant form. This was independently achieved by S. Tomonaga and collaborators, J. Schwinger and R.P. Feynman in different manners. They won the Nobel prize together in 1965. Tomonaga's work (*Progr. Theoret. Phys. Kyoto 2, 101 (1947)*) was closest to the older quantum field theory, because he started from the Schrödinger picture, went over to the Heisenberg picture and established perturbation theory. Schwinger (*Phys. Rev. 74, 1439 (1948), 75, 651 (1949)*) worked in the intermediate interaction representation which Tomonaga had implicitly also used, and constructed the Lorentz invariant collision operator (S-matrix). He calculated mostly in x-space which required ingenious formal tricks, because most objects are much more singular here than in momentum space. Feynman worked in a totally different way. In his paper "Space-Time Structure of Quantum Electro Dynamics" (*Phys. Rev. 76, 769 (1949)*) he avoided quantized fields altogether, using a quantum mechanical propagator theory instead. But the field quantization is hidden in the rules for many-body processes and in the choice of the propagator functions. F.J. Dyson (*Phys. Rev. 75, 486 (1949)*) showed the equivalence of this theory with Tomonaga's and Schwinger's and derived the Feynman rules by means of of quantum field theory. Feynman's formulation in momentum space was of greatest importance for the further development of field theory and particle physics, because it gives by far the simplest scheme for the explicit calculations.

Unfortunately, the Feynman rules still lead to ill-defined integrals which are ultraviolet and partially also infrared divergent. But in the covariant theory it was possible to calculate unique finite results which are in perfect agreement with experiments. This was achieved by regularization of the integrals and absorption of the infinities into the mass and charge terms, the well-known method of renormalization (F.J. Dyson *Phys. Rev. 75, 1736 (1949)*). Although the final results of the theory were certainly correct, it was clear that this was not yet the right formulation. Tomonaga said in his Nobel lecture: "It is a very pleasant thing that no divergence is involved in the theory except for the two infinities of electronic mass and charge. We cannot say that we have no divergences in the theory, since the mass and charge are in fact infinite." And Feynman in his Nobel lecture (*Science 153, 699 (1966)*) was even more critical of his own work: "I think that the renormalization theory is simply a way to sweep the difficulties of the divergences of electrodynamics under the rug. I am, of course, not sure of that." Twenty years later in his popular book with the remarkable title "The Strange Theory of Light and Matter" (*Princeton N.J. 1985*) he still wrote: "What is certain is that we do not have a good mathematical way to describe the theory of quantum electrodynamics." Another critic was Dirac. He called the theory "an ugly and incomplete one" (*Proc. Roy. Soc. A 209, 291 (1951)*). In his book "Dreams of a Final Theory" (*London 1993, p. 91*) S. Weinberg reported on discussions with Dirac and wrote: "I did not see what was so terrible about an infinity in the bare mass and charge as long as the final answers for physical quantities turn out to be finite and unambiguous and in agreement with experiment. It seemed to me that a theory that is as spectacularly successful as quantum electrodynamics has to be more or less correct, although we may not be formulating it in just the right way. But Dirac was unmoved by these arguments. I do not agree with his attitude towards quantum electrodynamics, but I do not think that he was just being stubborn; the demand for a completely finite theory is similar to a host of other aesthetic judgements that theoretical physicists always need to make." Dirac's point, perhaps, was that mathematical consistency is more fundamental than aesthetic judgements.

The third Nobel laureate of 1965 said nothing about the divergence problems, instead Schwinger made the following introductory remark: "I shall begin by describing to you the logical foundations of relativistic quantum field theory. No dry recital of lifeless "axioms" is intended ..." What are the lifeless axioms? In the 1950's A.S. Wightman and others (*R.F. Streater and A.S. Wightman, PCT, Spin and Statistics, and All That, New York 1964*) started to analyse the general structure which underlies all quantum field theories. From the well understood theory of free fields they extracted general properties (formulated as axioms) and studied the relations between them with rigorous mathematical methods. The resulting "general theory of quantized fields" (this better name is the title of a book by *R. Jost, Providence, Rhode Island 1965*) supplied various important results. But the main question whether the basic notions apply to realistic theories remained open. Only in lower dimensions non-trivial models satisfying the Wightman axioms have been constructed (*J. Glimm, A. Jaffe, Quantum Physics, Springer-Verlag 1981*). The failure of some constructive methods in four dimensions has given rise to speculations that a non-perturbative definition of QED might not exist. One must be careful with such statements, because one can only prove that *a particular construction* does not work.

There exists another more pragmatic approach which is the basis of this book. It goes back to Heisenberg (*Z. Phys. 120, 513 (1943)*) and takes the scattering operator (S-matrix) as the basic quantity. The S-matrix maps the asymptotically incoming, free fields on the outgoing ones and, hence, it should be possible to express it completely by the well-defined free fields. E.C.G. Stückelberg and collaborators (*Helv. Phys. Acta 23, 215 (1950), 24, 153 (1951)*) showed that this is possible in perturbation theory if one uses a causality condition in addition to unitarity of the S-matrix. Later on N.N. Bogoliubov and D.V. Shirkov (*Introduction to the Theory of Quantized Fields, New York 1959*) simplified the causality condition by using the important tool of adiabatic switching with a test function. This tool must be used for mathematical reasons because the S-matrix is an operator-valued functional and not an operator, and also for physical reasons since the real asymptotic states are not simply generated by free fields, as briefly discussed in the preface.

Unfortunately, these authors did not solve the divergence problems because they arrived at the usual defective expression for the S-matrix involving naively defined time-ordered products. As mentioned in the preface, the program was successfully carried through for scalar theories by H. Epstein and V. Glaser in 1973 (*Annales de l'Institut Poincaré A 19, 211 (1973)*). In their method the perturbation series for the S-matrix was constructed inductively, order by order, by means of causality and translation invariance; unitarity was not used. The most delicate step in this construction is the decomposition of distributions with causal support into retarded and advanced parts. If this distribution splitting is carried out without care by multiplication with step functions, then the usual ultraviolet divergences appear. But if it is carefully done by first multiplying with a C^∞ function and then performing the limit to the step function, everything is finite and well-defined. In this way the ultraviolet problem which has plagued field theorists for more than fifty years does not arise at all. Unfortunately, it is still not clear how the perturbation series can be summed up. Therefore, problems occurring in partial resummation, like the Landau pole (*M. Gell-Mann, F. Low, Phys. Rev. 95, 1300 (1954)*), cannot be treated yet. One should notice that this problem does not arise, if one considers the adiabatically switched S-matrix $S(g)$ (Sect. 3.1).

Summing up, we have looked at the history of quantum electrodynamics like a doctor examining the course of a disease. In fact, the force driving this history was mainly the attempt to "cure" the illness of the various divergences. The infinities were present in QED from the very beginning and their slow disappearance indicates our progress in understanding. Sometimes the disease has been considered so grave that radical treatment was recommended. But until now quantum field theory has always survived and we hope that it will be completely healthy one fine day.

0.1 Minkowski Space and the Lorentz Group

The framework of a physical description is the four-dimensional real space $\mathbb{R}^4$ of space-time points $x = (x^0, x^1, x^2, x^3) = (x^\mu)$, $x^0 = ct$. The velocity of light c has been introduced into the time component in order to have the same dimension in all four components of x. Throughout we use the convention that greek indices assume the values 0,1,2,3, whereas latin indices are used for the spatial values 1,2,3. Specifying the position $\boldsymbol{x}$ of a physical object as a function of time t, defines a curve in $\mathbb{R}^4$. The light rays outgoing from the origin move on the light-cone

$$c^2t^2 - |\boldsymbol{x}|^2 = 0. \tag{0.1.1}$$

This double-cone consists of the past-cone $t < 0$ and the future-cone $t > 0$. A change of the frame of reference is described by a linear transformation

$$x \longrightarrow x' = \Lambda x, \tag{0.1.2}$$

where Λ is a real 4×4-matrix. Introducing components with respect to a basis $e_\mu, \mu = 0, 1, 2, 3$

$$x = x^\mu e_\mu,$$

the transformation (0.1.2) is written as follows

$$x'^\mu = \Lambda^\mu{}_\nu x^\nu \tag{0.1.3}$$

where the convention of summing over double upper and lower indices is always assumed. The reason for using upper and lower indices will be explained in the following section.

The basis of relativity is the principle of constant velocity of light. In view of (0.1.1) it can be expressed as follows: If

$$(x^0)^2 - \boldsymbol{x}^2 = 0$$

in one frame of reference then this also holds in another frame

$$(x'^0)^2 - \boldsymbol{x}'^2 = 0.$$

It is convenient to write the quadratic forms appearing here as

$$Q(x) = x^T g\, x \tag{0.1.4}$$

$$Q'(x) = Q(\Lambda x) = x^T \Lambda^T g\, \Lambda\, x, \tag{0.1.5}$$

where

$$g = (g_{\mu\nu}) = \begin{pmatrix} 1 & 0 & 0 & 0 \\ 0 & -1 & 0 & 0 \\ 0 & 0 & -1 & 0 \\ 0 & 0 & 0 & -1 \end{pmatrix} \tag{0.1.6}$$

is the fundamental metric tensor. Both forms (0.1.4, 5) vanish for fixed $\boldsymbol{x}$ if $x^0 = \pm|\boldsymbol{x}|$, therefore

$$Q'(x) = \lambda(x^0 - |\boldsymbol{x}|)(x^0 + |\boldsymbol{x}|) = \lambda((x^0)^2 - \boldsymbol{x}^2) = \lambda\, Q(x).$$

The case $\lambda \neq 1$ corresponds to a change of units which we disregard. Then we arrive at

$$x^T \Lambda^T g\, \Lambda\, x = x^T g\, x$$

for all $x \in \mathbb{R}^4$, or

$$\Lambda^T g\, \Lambda = g. \tag{0.1.7}$$

We emphasize that we have used the condition of constant $x^2 = x'^2$ only for light rays ($x^2 = 0$). All transformations satisfying (0.1.7) are called Lorentz transformations. They obviously form a group, the Lorentz group $\mathcal{L}$. Equation (0.1.7) suggests the introduction of the indefinite scalar product

$$(x,y) = x^T g\, y = x^0y^0 - \boldsymbol{x}\cdot\boldsymbol{y} = x^0y^0 - x^1y^1 - x^2y^2 - x^3y^3. \qquad (0.1.8)$$

It is invariant under Lorentz transformations

$$(x',y') = (\Lambda x, \Lambda y) = (\Lambda x)^T g\,\Lambda\, y = x^T \Lambda^T g\,\Lambda\, y = x^T g\, y = (x,y).$$

The four-dimensional real vector space with scalar product (0.1.8) is called Minkowski space $\mathbb{M}$. Lorentz transformations are the congruency transformations of $\mathbb{M}$. The elements of $\mathbb{M}$ are called points or (four) vectors in the following.

There are three classes of vectors in $\mathbb{M}$: (i) time-like vectors x with $x^2 > 0$, (ii) space-like vectors y with $y^2 < 0$ and (iii) light-like vectors z with $z^2 = 0$. Each class is mapped into itself under Lorentz transformations because x^2 remains constant. We shall often find that functions of a four-vector x behave differently for time-like or space-like x. A three-dimensional surface S in $\mathbb{M}$ is called time-like or space-like if any tangent vector to S is time-like or space-like, respectively. Two disjoint sets X, Y of points are space-like separated if every vector $x - y, x \in X, y \in Y$ is space-like. Then it is impossible to connect the points x, y in a causal way, for instants by light signals. If $x - y$ is time-like, then the two points are causally connected. This causal structure of Minkowski space will be of crucial importance later.

Equation (0.1.7) implies det $\Lambda = \pm 1$ for all $\Lambda \in \mathcal{L}$. Examples of determinant $= -1$ are time-reflection T and space-reflection P (parity transformation)

$$T = \begin{pmatrix} -1 & 0 & 0 & 0 \\ 0 & 1 & 0 & 0 \\ 0 & 0 & 1 & 0 \\ 0 & 0 & 0 & 1 \end{pmatrix}, \qquad P = \begin{pmatrix} 1 & 0 & 0 & 0 \\ 0 & -1 & 0 & 0 \\ 0 & 0 & -1 & 0 \\ 0 & 0 & 0 & -1 \end{pmatrix}. \qquad (0.1.9)$$

The Lorentz transformations Λ with det $\Lambda = +1$ form the subgroup

$$\mathcal{L}_+ = SO(1,3)$$

of $\mathcal{L}$. It is a special pseudo-orthogonal group. The defining Eq. (0.1.7) means that the rows and columns of a Lorentz matrix $\Lambda^\mu{}_\nu$ are orthogonal with respect to the Minkowski scalar product (0.1.8), for example

$$\Lambda^0{}_\mu \Lambda^0{}_\nu - \sum_{j=1}^{3} \Lambda^j{}_\mu \Lambda^j{}_\nu = \begin{cases} 0, & \text{for } \mu \neq \nu \\ 1, & \text{for } \mu = \nu = 0\,. \\ -1 & \text{for } \mu = \nu \neq 0 \end{cases} \qquad (0.1.10)$$

Taking $\mu = \nu = 0$,we have

$$(\Lambda^0{}_0)^2 - \sum_{j=1}^{3} (\Lambda^j{}_0)^2 = 1$$

and therefore

$$(\Lambda^0{}_0)^2 \geq 1 \quad \text{i.e.} \quad \Lambda^0{}_0 \geq 1 \quad \text{or} \quad \Lambda^0{}_0 \leq -1.$$

For $\Lambda^0{}_0 \geq 1$, the direction of time is not reversed. The subgroup

$$\mathcal{L}_+^\uparrow = \{\Lambda \in \mathcal{L}_+ | \Lambda^0{}_0 \geq 1\} \tag{0.1.11}$$

is the proper Lorentz group. Only this group is an exact symmetry group of physics (neglecting gravitation), because parity and time-reversal (0.1.9) are not conserved in weak interactions.

We now consider subgroups of the proper Lorentz group $\mathcal{L}_+^\uparrow$. First we assume that a Lorentz matrix $\Lambda \in \mathcal{L}_+^\uparrow$ has the following structure

$$\Lambda(R) = \begin{pmatrix} 1 & 0 \\ 0 & R_3 \end{pmatrix}, \tag{0.1.12}$$

where R_3 is a real 3×3 matrix. Equation (0.1.7) implies $R_3^T R_3 = 1$, which means that R_3 is a 3-dimensional rotation $\in SO(3)$. Another subgroup is constituted by the Lorentz boosts, for example

$$\Lambda(\chi) = \begin{pmatrix} \cosh\chi & 0 & 0 & \sinh\chi \\ 0 & 1 & 0 & 0 \\ 0 & 0 & 1 & 0 \\ \sinh\chi & 0 & 0 & \cosh\chi \end{pmatrix}. \tag{0.1.13}$$

This is a special Lorentz transformation along the 3-axis

$$x'^0 = x^0 \cosh\chi + x^3 \sinh\chi$$

$$x'^3 = x^0 \sinh\chi + x^3 \cosh\chi.$$

Every $\Lambda \in \mathcal{L}_+^\uparrow$ can be generated by means of these special transformations (0.1.12) and (0.1.13):

Theorem 1.1. Every proper Lorentz transformation $\Lambda \in \mathcal{L}_+^\uparrow$ can be expressed in the following form

$$\Lambda = \Lambda(R_1)\,\Lambda(\chi)\,\Lambda(R_2), \tag{0.1.14}$$

where $R_1, R_2 \in SO(3)$, $\Lambda(R)$ is given by (0.1.12) and $\Lambda(\chi)$ is the boost (0.1.13).

Proof. From the given Lorentz matrix $\Lambda^\mu{}_\nu$ we form the three-vector $\boldsymbol{f} = (\Lambda^1{}_0, \Lambda^2{}_0, \Lambda^3{}_0) \neq 0$ and normalize it

$$\boldsymbol{e}_3 = \lambda \boldsymbol{f} = (c_1, c_2, c_3), \quad c_1^2 + c_2^2 + c_3^2 = 1. \tag{0.1.15}$$

We choose two normalized three-vectors $\boldsymbol{e}_1 = (a_1, a_2, a_3)$ and $\boldsymbol{e}_2 = (b_1, b_2, b_3)$ orthogonal in three-space such that $\boldsymbol{e}_1, \boldsymbol{e}_2, \boldsymbol{f}$ is a basis of three-space with positive orientation, in particular

$$(\boldsymbol{e}_j, \boldsymbol{f}) = 0, \quad j = 1, 2. \tag{0.1.16}$$

Then the matrix

$$R_1^{-1} = \begin{pmatrix} 1 & 0 & 0 & 0 \\ 0 & a_1 & a_2 & a_3 \\ 0 & b_1 & b_2 & b_3 \\ 0 & c_1 & c_2 & c_3 \end{pmatrix}$$

is a rotation $\in SO(3)$ and therefore

$$R_1^{-1} \Lambda = \begin{pmatrix} \Lambda^0{}_0 & \Lambda^0{}_1 & \Lambda^0{}_2 & \Lambda^0{}_3 \\ 0 & d_{11} & d_{12} & d_{13} \\ 0 & d_{21} & d_{22} & d_{23} \\ d_{30} & d_{31} & d_{32} & d_{33} \end{pmatrix} \tag{0.1.17}$$

is in $\mathcal{L}_+^\uparrow$. The two zeros in the first column follow from (0.1.16). Now we consider the two three-vectors $\boldsymbol{f}_1 = (d_{11}, d_{12}, d_{13})$ and $\boldsymbol{f}_2 = (d_{21}, d_{22}, d_{23})$ which are orthonormal

$$(\boldsymbol{f}_1, \boldsymbol{f}_2) = 0, \quad \boldsymbol{f}_1^2 = 1 = \boldsymbol{f}_2^2,$$

because the rows in (0.1.17) are orthonormal (0.1.10). We add a third orthonormal vector $\boldsymbol{f}_3 = (g_1, g_2, g_3)$ such that again a three-basis with positive orientation is obtained. Now the matrix

$$R_2^{-1} = \begin{pmatrix} 1 & 0 & 0 & 0 \\ 0 & d_{11} & d_{21} & g_1 \\ 0 & d_{12} & d_{22} & g_2 \\ 0 & d_{13} & d_{23} & g_3 \end{pmatrix}$$

is in $SO(3)$. Then the product

$$R_1^{-1} \Lambda R_2^{-1} = \begin{pmatrix} \Lambda^0{}_0 & 0 & 0 & h_{03} \\ 0 & 1 & 0 & 0 \\ 0 & 0 & 1 & 0 \\ d_{30} & 0 & 0 & h_{33} \end{pmatrix} = \Lambda(\chi)$$

is just the Lorentz boost (0.1.13) which proves the theorem. In the special case where the vector $\boldsymbol{f}$, we started with, is zero, it follows from (0.1.10) that $\Lambda \in SO(3)$. □

It is easily seen from this proof that the representation (0.1.14) is not unique. Since the rotations R_1, R_2 can be continuously deformed into the identity and the boost $\Lambda(\chi)$ as well ($\chi \to 0$), the proper Lorentz group is connected.

The transition from one frame of reference to another can also been made by translations $x \longrightarrow x + a, a \in \mathbb{R}^4$. This leads to inhomogeneous Lorentz transformations (a, Λ)

$$x' = \Lambda x + a, \quad \Lambda \in \mathcal{L}_+^\uparrow, \quad a \in \mathbb{M}. \tag{0.1.18}$$

These transformations form the proper Poincaré group $\mathcal{P}_+^\uparrow$ which is the most important symmetry group of physics. To write equations in a Poincaré invariant form, one needs the tensor calculus which is briefly described in the next section.

0.2 Tensors in Minkowski Space

Lorentz tensors are linear forms over Minkowski space. The real linear forms A' on $\mathbb{M}$

$$\mathbb{M} \ni A \longrightarrow \langle A', A\rangle \in \mathbb{R} \tag{0.2.1}$$

form the dual space $\mathbb{M}'$ of $\mathbb{M}$. Every linear form is a scalar product with some element of $\mathbb{M}$. However, in tensor calculus it is convenient to distinguish between $\mathbb{M}'$ and $\mathbb{M}$. Let e_ν be a basis of $\mathbb{M}$ and e^μ the corresponding dual basis in $\mathbb{M}'$

$$\langle e^\mu, e_\nu\rangle = \delta^\mu{}_\nu . \tag{0.2.2}$$

Then an element

$$\mathbb{M}' \ni A' = A'_\mu e^\mu$$

operates on

$$\mathbb{M} \ni B = B^\nu e_\nu$$

as follows

$$\langle A', B\rangle = A'_\mu B^\mu = A'^\nu g_{\nu\mu} B^\mu \tag{0.2.3}$$

because of (0.2.2). This leads to the definition of covariant (A'_μ) and contravariant (A'^ν) components

$$A'_\mu = A'^\nu g_{\nu\mu} \tag{0.2.4}$$

and to the lowering of indices by means of the metric tensor g. If upper and lower indices are contracted in couples as in (0.2.3), we get a number. Writing the inverse matrix of g as

$$g_{\mu\nu} g^{\nu\lambda} = \delta_\mu{}^\lambda , \tag{0.2.5}$$

we find

$$g^{\mu\nu} = g_{\mu\nu} = g_{\nu\mu} .$$

Multiplying (0.2.4) with the inverse g^{-1}, we have lifted an index

$$A^\mu = A_\nu g^{\nu\mu} = g^{\mu\nu} A_\nu .$$

A bilinear form T over $\mathbb{M}$

$$T(A, B) = T_{\mu\nu} A^\mu B^\nu$$

is a covariant tensor of second rank. By lifting one index, we obtain a mixed tensor $T^\mu{}_\nu$. An example of this is the Lorentz transformation

$$\tilde{A}^\mu = \Lambda^\mu{}_\nu A^\nu. \tag{0.2.6}$$

Since the covariant components transform with the inverse transposed matrix Λ^{-1T}

$$\tilde{B}_\mu = \Lambda_\mu{}^\lambda B_\lambda \quad , \quad (\Lambda^{-1})^\lambda{}_\mu = \Lambda_\mu{}^\lambda, \tag{0.2.7}$$

it follows that by contracting an upper with a lower index, we get a Lorentz invariant

$$\tilde{A}_\mu \tilde{B}^\mu = \Lambda^\mu{}_\nu \Lambda_\mu{}^\lambda A^\nu B_\lambda = A^\nu B_\nu.$$

Next we consider vectors and tensors which are space and time-dependent, like $A^\mu(x)$, $T_{\mu\nu}(x)$. These objects are called vector and tensor fields. They are differentiable with respect to x, if the increments can be linearly approximated:

$$\lim_{\varepsilon\to 0} \frac{1}{\varepsilon}[T(x+\varepsilon y) - T(x)] = \langle DT(x), y\rangle \quad , \quad y \in \mathbb{M}.$$

Since this is a linear form on $\mathbb{M}$, differentiation increases the covariant degree of a tensor field by one. We write in components

$$(DT)^{\mu\nu\cdots}_\lambda = \frac{\partial T^{\mu\nu\cdots}(x)}{\partial x^\lambda} = T^{\mu\nu\cdots}{}_{,\lambda}\,.$$

We give some important examples:

1) A scalar field $\phi(x)$ is a tensor field of rank 0. Then

$$(D\phi(x))_\mu = \frac{\partial\phi(x)}{\partial x^\mu} = \partial_\mu \phi \tag{0.2.8}$$

is a covariant vector field, the gradient.

2) Let $A^\mu(x)$ be a contravariant vector field. Differentiating it, we obtain the mixed second rank tensor

$$(DA(x))^\mu{}_\nu = \frac{\partial A^\mu(x)}{\partial x^\nu} = A^\mu{}_{,\nu}\,. \tag{0.2.9}$$

If this is contracted, we have the scalar field

$$A^\mu{}_{,\mu} = \operatorname{div} A(x), \tag{0.2.10}$$

which is the divergence of $A(x)$.

3) If we differentiate a covariant vector field $A_\mu(x)$

$$(DA(x))_{\mu\nu} = \frac{\partial A_\mu(x)}{\partial x^\nu} = A_{\mu,\nu} \tag{0.2.11}$$

and form the antisymmetric combination

$$A_{\nu,\mu} - A_{\mu,\nu} = (\operatorname{curl} A)_{\mu\nu}, \tag{0.2.12}$$

we get the curl of $A(x)$. It is an antisymmetric second rank tensor.

4) We now take the contravariant components of grad ϕ

$$(\mathrm{grad}\,\phi)^\mu = g^{\mu\nu}\frac{\partial\phi}{\partial x^\nu} \tag{0.2.13}$$

and form the divergence according to 2) above:

$$g^{\mu\nu}\frac{\partial^2\phi}{\partial x^\mu\partial x^\nu} = \frac{\partial^2\phi}{\partial(x^0)^2} - \frac{\partial^2\phi}{\partial(x^1)^2} - \frac{\partial^2\phi}{\partial(x^2)^2} - \frac{\partial^2\phi}{\partial(x^3)^2} = \partial_\mu\partial^\mu\phi. \tag{0.2.14}$$

This gives the wave operator which obviously is Lorentz invariant.

Finally we mention the integral theorems which we have to use later. The Lebesgue measure on $\mathbb{R}^4$

$$d^4x = dx^0dx^1dx^2dx^3 \tag{0.2.15}$$

is invariant under Lorentz transformation Λ because $|\det\Lambda| = 1$. We only need the following simple form of Gauss' theorem: Let $A^\mu(x)$ be a continuously differentiable contravariant vector field defined on a region G in Minkowski space with smooth boundary ∂G, and let A^μ vanish on ∂G. Then we have

$$\int_G \mathrm{div}\,A(x)d^4x = \int_G \frac{\partial A^\mu(x)}{\partial x^\mu}d^4x = 0. \tag{0.2.16}$$

This theorem immediately extends to tensor fields: Let $a \in \mathbb{M}$ be an arbitrary constant vector. Then, given a differentiable tensor field $T_{\mu\nu}(x)$ in G vanishing on ∂G, Gauss' theorem applied to the vector field $T_{\mu\nu}(x)a^\nu$ leads to

$$\int_G \partial^\mu T_{\mu\nu}(x)d^4x\,a^\nu = 0$$

for arbitrary $a \in \mathbb{M}$. Therefore

$$\int_G \partial^\mu T_{\mu\nu}(x)d^4x = 0. \tag{0.2.17}$$

Partial integration is another consequence of Gauss' theorem:

$$\int_G d^4x\partial_\mu(A^\mu(x)g(x)) = 0 = \int_G d^4xA^\mu(x)\partial_\mu g(x) + \int_G d^4x(\partial_\mu A^\mu)g. \tag{0.2.18}$$

0.3 Some Topics of Scattering Theory

We consider a quantum mechanical system described by a (time independent) Hamiltonian H which is a selfadjoint operator on a Hilbert space $\mathcal{H}$. The time evolution of the system is then given by the unitary transformation

$$\psi(t) = e^{-iHt}\psi_0. \tag{0.3.1}$$

We assume H to be of the following form

$$H = H_0 + V, \tag{0.3.2}$$

where H_0 is the free Hamiltonian and the interaction V has short range. The latter means that the so-called wave operators

$$W_{\substack{\text{in}\\ \text{out}}} = s - \lim_{t\to\mp\infty} e^{iHt}e^{-iH_0t} \tag{0.3.3}$$

exist as strong limits on $\mathcal{H}$.

In the case of a time-dependent interaction $V = V(t)$, the time evolution is given by a unitary propagator

$$\psi(t) = U(t,s)\psi(s) \quad , \quad U(t,s)^+ = U(s,t) \tag{0.3.4}$$

$$U(t,s)U(s,r) = U(t,r) \tag{0.3.5}$$

instead of (0.3.1). The wave operators then are defined as follows

$$\begin{aligned} W_{\substack{\text{in}\\ \text{out}}} &= s - \lim_{t\to\mp\infty} U(t,0)^+e^{-iH_0t} \\ &= s - \lim_{t\to\mp\infty} U(0,t)e^{-iH_0t}. \end{aligned} \tag{0.3.6}$$

The plus in the exponent always means the adjoint in Hilbert space, the asterisk $*$ is reserved for complex conjugation, while the bar is used later for Dirac conjugation.

The central object of scattering theory is the scattering matrix (S-matrix)

$$S = W_{\text{out}}^+ W_{\text{in}} = \lim_{\substack{s\to-\infty\\ t\to+\infty}} e^{iH_0t}U(t,s)e^{-iH_0s}. \tag{0.3.7}$$

From this definition the physical meaning of the S-matrix can be read of: A normalized initial asymptotic state φ considered at time $t = 0$, say, is first transformed to $s = -\infty$ by the free dynamics, then it is evolved from $-\infty$ to $t = +\infty$ by the full interacting dynamics and finally it is transformed back from $+\infty$ to $t = 0$ by the free dynamics. The resulting state $S\varphi$ is therefore the outgoing scattering state, transformed to $t = 0$ by the free time evolution. It can then be compared with an arbitrary normalized outgoing asymptotic state ψ by calculating the scalar product $(\psi, S\varphi)$. The absolute square of this is the probability for a transition from φ to ψ

$$P(\varphi \to \psi) = |(\psi, S\varphi)|^2. \tag{0.3.8}$$

The state $\psi(t)$ (0.3.4) is the solution of the Schrödinger equation

$$i\frac{d}{dt}\psi(t) = (H_0 + V(t))\psi(t) \quad , \quad \hbar = 1. \tag{0.3.9}$$

We go over to the so-called interaction picture by the substitution

$$\psi(t) = e^{-iH_0 t}\varphi(t). \tag{0.3.10}$$

$\varphi(t)$ then satisfies the simple equation

$$i\frac{d}{dt}\varphi(t) = \tilde{V}(t)\varphi(t), \tag{0.3.11}$$

where

$$\tilde{V}(t) = e^{iH_0 t}V(t)e^{-iH_0 t} \tag{0.3.12}$$

is the operator $V(t)$ in the interaction picture. Note that the S-matrix (0.3.7) is just the limit of the time evolution in the interaction picture:

$$S = \lim_{\substack{s\to-\infty \\ t\to+\infty}} \tilde{U}(t,s). \tag{0.3.13}$$

Equation (0.3.11) can be written as an integral equation

$$\varphi(t) = \varphi(s) - i\int_s^t dt_1 \tilde{V}(t_1)\varphi(t_1). \tag{0.3.14}$$

If the interaction $V(t)$ is a bounded operator, Eq. (0.3.14) can be iterated, leading to the so-called Dyson series

$$\varphi(t) = \left[1 + \sum_{n=1}^{\infty}(-i)^n \int_s^t dt_1 \int_s^{t_1} \cdots \int_s^{t_{n-1}} dt_n \tilde{V}(t_1)\ldots\tilde{V}(t_n)\right]\varphi(s). \tag{0.3.15}$$

This series converges in operator norm, because the n-th order term U_n can be estimated as follows:

$$\|U_n\| \le \int_s^t dt_1 \ldots \int_s^{t_{n-1}} dt_n \|\tilde{V}(t_1)\| \ldots \|\tilde{V}(t_n)\|$$

$$= \frac{1}{n!}\left[\int_s^t d\tau \|\tilde{V}(\tau)\|\right]^n . \tag{0.3.16}$$

The Dyson series (0.3.15) can be taken as definition of the unitary propagator (0.3.4).

If the time dependent interaction $V(t)$ decreases for large times such that

$$\int_{-\infty}^{+\infty} ds\|V(s)\| < \infty, \tag{0.3.17}$$

then the Dyson series for the S-matrix

$$S = \sum_{n=0}^{\infty}(-i)^n \int_{-\infty}^{+\infty} dt_1 \int_{-\infty}^{t_1} dt_2 \ldots \int_{-\infty}^{t_{n-1}} dt_n \tilde{V}(t_1)\ldots\tilde{V}(t_n) \tag{0.3.18}$$

is also norm-convergent. It then defines a unitary operator in $\mathcal{H}$. Consequently, asymptotic completeness is no problem here. This is in sharp contrast to scattering theory for static potentials, where unitarity of the S-matrix is difficult to prove. At first sight, it seems to be very restrictive to suppose the interaction $V(t)$ to be bounded in norm for almost all t (cf. (0.3.17)). However, as we shall discuss later, unbounded external fields are unphysical in quantum electrodynamics, because they do not allow quantization of the Dirac field (cf. below (2.5.29) in Chap. 2).

The domain of integration in the iterated integral (0.3.18) is a simplex in $\mathbb{R}^n$. Such an integral can be extended to an integral over a cube. However, since the factors $\tilde{V}(t_j)$ do not commute, it is necessary to maintain the time ordering:

$$S = \sum_{n=0}^{\infty} \frac{(-i)^n}{n!} \int_{-\infty}^{+\infty} dt_1 \int_{-\infty}^{+\infty} dt_2 \ldots \int_{-\infty}^{+\infty} dt_n\, T[\tilde{V}(t_1)\ldots\tilde{V}(t_n)] \tag{0.3.19}$$

$$\stackrel{\text{def}}{=} T \exp -i \int_{-\infty}^{+\infty} dt\, \tilde{V}(t). \tag{0.3.20}$$

The symbol T before a product means that the factors are arranged with decreasing time coordinate. The order of the factors in such a time-ordered product is immaterial, because the time-ordering T ensures the correct order. Hence, the integrand in (0.3.19) is symmetric in $t_1, \ldots, t_n$; every simplex gives the same contribution. The time-ordered exponential (0.3.20) is a compact notation for the series (0.3.18, 19).

We now give a second derivation of the perturbation series (0.3.19) *without using the Schrödinger equation.* Since in quantum field theory one does not have well-defined dynamical evolution equations, in general, it is this derivation which will be taken over to QED in Chap. 3 (cf. Sect. 3.1). In view of this later application, let us multiply the interaction by a switching function $g(t)$, which vanishes rapidly for $t \to \pm\infty$. We want to construct the S-matrix as a power series in g of the form

$$S(g) = 1 + \sum_{n=1}^{\infty} \frac{1}{n!} \int dt_1 \ldots dt_n\, T_n(t_1, \ldots, t_n) g(t_1) \ldots g(t_n) \tag{0.3.21}$$

$$\stackrel{\text{def}}{=} 1 + T.$$

By definition, $T_n(t_1, \ldots, t_n)$ must be symmetric with respect to permutations of the t_j, otherwise the contribution to (0.3.21) would be zero. Then it is appropriate to consider the disordered set of time points $X = \{t_1, \ldots, t_n\}$ as argument of T_n.

Simultaneously with $S(g)$, the inverse $S(g)^{-1}$ can be expanded as follows

$$S(g)^{-1} = 1 + \sum_{n=1}^{\infty} \frac{1}{n!} \int dt_1 \ldots dt_n \, \tilde{T}_n(t_1, \ldots, t_n) g(t_1) \ldots g(t_n) \tag{0.3.22}$$

$$= (1 + T)^{-1} = 1 + \sum_{r=1}^{\infty} (-T)^r. \tag{0.3.23}$$

The $\tilde{T}_n$ follow from (0.3.23) by expanding the r-th power of $-T$

$$\tilde{T}_n(X) = \sum_{r=1}^{n} (-)^r \sum_{P_r} T_{n_1}(X_1) \ldots T_{n_r}(X_r), \tag{0.3.24}$$

where the second sum runs over all partitions P_r of X into r disjoint subsets

$$X = X_1 \cup \ldots \cup X_r \, , \; X_r \neq \emptyset \, , \; |X_j| = n_j .$$

In particular we have

$$\tilde{T}_1(t) = -T_1(t). \tag{0.3.25}$$

Instead of the Schrödinger equation, we will use a causality property of the S-matrix which follows from (0.3.13). Let us consider two switching functions g_1, g_2 with disjoint support in time

$$\operatorname{supp} g_1 \subset (-\infty, s) \, , \; \operatorname{supp} g_2 \subset (s, +\infty). \tag{0.3.26}$$

Then, in virtue of (0.3.13) we have

$$S(g_1 + g_2) = U_0(0, \infty) U(+\infty, -\infty) U_0(-\infty, 0)$$

$$= U_0(0, \infty) U(+\infty, s) U_0(s, 0) U_0(0, s) U(s, -\infty) U_0(-\infty, 0) = S(g_2) S(g_1), \tag{0.3.27}$$

where the arguments $\pm\infty$ stand for the corresponding strong limits in (0.3.7) and U_0 for the free time evolution. Equation (0.3.27) expresses causality in the sense that, what happens for $t < s$ (described by $S(g_1)$) is not influenced by what happens for $t > s$ (described by $S(g_2)$).

We now want to find a perturbative formulation of the causality condition (0.3.27). Substituting (0.3.21) into the left hand side of (0.3.27), we get

$$S(g_1 + g_2) = \sum_n \frac{1}{n!} \int dt_1 \ldots dt_n \, T_n(t_1, \ldots, t_n)$$

$$\times \, (g_1(t_1) + g_2(t_1)) \ldots (g_1(t_n) + g_2(t_n)). \tag{0.3.28}$$

By permutation of the integration variables t_j in the 2^n terms, we may arrange the switching functions in the form $g_2(t_1)\dots g_2(t_m)g_1(t_{m+1})\dots g_1(t_n)$. Since there are

$$\frac{n!}{m!(n-m)!} \text{ permutations, } \sum_{m=0}^{n} \frac{n!}{m!(n-m)!} = 2^n,$$

we arrive at

$$S(g_1+g_2) = \sum_{n=0}^{\infty}\sum_{m=0}^{n} \frac{1}{m!(n-m)!}\int dt_1\dots dt_n\, T_n(t_1,\dots,t_n)$$

$$\times\, g_2(t_1)\dots g_2(t_m)g_1(t_{m+1})\dots g_1(t_n)$$

$$= S(g_1)S(g_2) = \sum_{n=0}^{\infty}\sum_{m=0}^{n} \frac{1}{m!(n-m)!}\int dt_1\dots dt_n$$

$$T_m(t_1,\dots,t_m)T_{n-m}(t_{m+1},\dots,t_n)g_2(t_1)\dots g_2(t_m)g_1(t_{m+1})\dots g_1(t_n). \tag{0.3.29}$$

This leads to the desired perturbative causality condition

$$T_n(t_1,\dots,t_n) = T_m(t_1,\dots,t_m)T_{n-m}(t_{m+1},\dots,t_n), \tag{0.3.30}$$

if all $\{t_1,\dots,t_m\}$ are greater than all $\{t_{m+1},\dots,t_n\}$.

We now claim that all orders T_n in (0.3.21) can be inductively determined by means of (0.3.30), provided the first order $T_1(t)$ is given. We proceed as follows: We form the two functions

$$A_2'(t_1,t_2) = \tilde{T}_1(t_1)T_1(t_2) = -T_1(t_1)T_1(t_2) \tag{0.3.31}$$

$$R_2'(t_1,t_2) = T_1(t_2)\tilde{T}_1(t_1) = -T_1(t_2)T_1(t_1), \tag{0.3.32}$$

and consider the so-called advanced and retarded functions

$$A_2(t_1,t_2) = A_2'(t_1,t_2) + T_2(t_1,t_2)$$

$$R_2(t_1,t_2) = R_2'(t_1,t_2) + T_2(t_2,t_1). \tag{0.3.33}$$

It follows from (0.3.30) that A_2 vanishes for $t_1 > t_2$ and R_2 vanishes for $t_1 < t_2$. The combination

$$D_2 = R_2 - A_2 = R_2' - A_2' \tag{0.3.34}$$

is known to us because T_2 drops out due to the symmetry $T_2(t_1,t_2) = T_2(t_2,t_1)$. Then, R_2 and A_2 can be identified in (0.3.34) by their support properties, for example

$$R_2(t_1,t_2) = \Theta(t_1-t_2)D_2(t_1,t_2)$$

$$= \Theta(t_1-t_2)(T_1(t_1)T_1(t_2) - T_1(t_2)T_1(t_1)). \tag{0.3.35}$$

R_2 is uniquely determined up to its value at $t_1 = t_2$. Finally, $T_2(t_1, t_2)$ can be determined from (0.3.33)

$$T_2(t_1,t_2) = R_2(t_1,t_2) - R_2'(t_1,t_2)$$

$$= \Theta(t_1 - t_2)T_1(t_1)T_1(t_2) + \Theta(t_2 - t_1)T_1(t_2)T_1(t_1)$$

$$= T\{T_1(t_1)T_1(t_2)\}. \tag{0.3.36}$$

This is exactly the second order time-ordered product in agreement with (0.3.19). For the induction step from $n-1$ to n we refer to Sect. 3.1 in Chap. 3. Although the induction is carried out there for the case of field theory, it is an easy exercise to adapt it to quantum mechanics.

The two constructions of the time-ordered exponential (0.3.20), discussed here, are analogous to the different definitions of the exponential function in elementary calculus: either by a differential equation (0.3.9) or by a functional equation (0.3.27). In field theory the second construction is superior.

0.4 Problems

0.1. Consider the special Lorentz transformation

$$x'^0 = x^0 \cosh\chi + x^3 \sinh\chi$$

$$x'^3 = x^0 \sinh\chi + x^3 \cosh\chi \tag{0.4.1}$$

$$x'^2 = x^2 \quad , \quad x'^1 = x^1 .$$

Study the motion of the spatial origin of the primed system ($x'^3 = 0$) and interpret the parameter χ by means of the relative velocity v.

0.2. a) For the Lorentz boost (1) compare the time t^0 measured by a clock at $x^3 = 0$ with the time in the primed system (time dilatation).

b) A rod at rest in the primed system has its endpoints at $x'^3 = 0, l'$. What is its length l at time $x^0 = 0$ in the other system (length contraction)?

0.3. Calculate the product of two boosts with relative velocities v_1, v_2 in the 3-direction. What is the relative velocity v of the resulting transformation (velocity addition law). Show that $v < c$ always holds.

0.4. A carriage moves in the 1-direction with velocity v relative to the ground. A second carriage rolls on top of the first in the same direction with velocity v relative to the first; a third carriage rolls on top of the second etc. What is the velocity of the n-th carriage relative to the ground? What comes out in the limit $n \to \infty$?

0.5. Write down the equation $x' = \Lambda x$ for a Lorentz boost with relative velocity $\boldsymbol{v}$ in an arbitrary direction.

0.6. Compute the transformation matrix for a boost in 1-direction followed by a boost in 2-direction. Do the two transformations commute? Show that the result is not a Lorentz boost.

0.7. Write the transformation of problem 0.6 with boost parameters χ_1, χ_2 in the form

$$\Lambda = R_1 \, \Lambda(\chi) \, R_2,$$

where R_1 and R_2 are rotations around the 3-axis and $\Lambda(\chi)$ a Lorentz boost. Compute the rotation angles of R_1, R_2 in terms of χ_1, χ_2.

0.8. Determine the multiplication law of the Poincaré group $\mathcal{P}$.

0.9. Show that the translation group $\mathbb{R}^4$ is an invariant subgroup of $\mathcal{P}$ such that $\mathcal{P}$ is the semi-direct product of $\mathbb{R}^4$ and $\mathcal{L} = O(1,3)$:

$$\mathcal{P} \ni g = g_1 g_2 \quad , \quad g_1 \in \mathbb{R}^4, \quad g_2 \in \mathcal{L}.$$

Proof also $\mathcal{P}/\mathbb{R}^4 = O(1,3)$.

1. Relativistic Quantum Mechanics

This chapter is concerned with the one-particle Dirac theory. It is not entirely correct to use the notion "particle" because, as we will see, there is no consistent physical interpretation of this sometimes called "first quantized" theory. Nevertheless, we must study it in great detail since later, after transforming it into a quantum field theory, it will lead us to the correct second quantized theory.

1.1 Spinor Representations of the Lorentz Group

There exists a well-known correspondence between the rotation group $SO(3)$ and the unimodular group $SU(2)$ of complex 2×2 matrices with determinant equal to 1. This correspondence is most explicitly described by means of the Pauli matrices

$$\sigma_1 = \begin{pmatrix} 0 & 1 \\ 1 & 0 \end{pmatrix} \quad \sigma_2 = \begin{pmatrix} 0 & -i \\ i & 0 \end{pmatrix} \quad \sigma_3 = \begin{pmatrix} 1 & 0 \\ 0 & -1 \end{pmatrix}. \tag{1.1.1}$$

They are self-adjoint matrices which obey the following important multiplication rule:

$$(\boldsymbol{x}\cdot\boldsymbol{\sigma})(\boldsymbol{y}\cdot\boldsymbol{\sigma}) = \boldsymbol{x}\cdot\boldsymbol{y} + i\boldsymbol{\sigma}\cdot(\boldsymbol{x}\wedge\boldsymbol{y}), \tag{1.1.2}$$

where $\boldsymbol{\sigma} = (\sigma_1, \sigma_2, \sigma_3)$ and dot and wedge mean the euclidian scalar and vector products, in particular

$$\boldsymbol{x}\cdot\boldsymbol{\sigma} = x_1\sigma_1 + x_2\sigma_2 + x_3\sigma_3. \tag{1.1.3}$$

A rotation $R(\boldsymbol{\alpha}) \in SO(3)$ is specified by the direction $\boldsymbol{n}$ of the rotation axis ($\|\boldsymbol{n}\| = 1$) and an angle φ between 0 and π. We set

$$\boldsymbol{\alpha} = \boldsymbol{n}\varphi \quad , \quad \varphi = \alpha \in [0, \pi), \tag{1.1.4}$$

to represent the elements $R(\boldsymbol{\alpha}) \in SO(3)$. Let us now consider

$$U(\boldsymbol{\alpha}) = \exp -\frac{i}{2}\boldsymbol{\alpha}\cdot\boldsymbol{\sigma} = \cos\frac{\alpha}{2} - i\sin\frac{\alpha}{2}\frac{\boldsymbol{\alpha}\cdot\boldsymbol{\sigma}}{\alpha}. \tag{1.1.5}$$

One immediately sees from the exponential expression that $U(\boldsymbol{\alpha})$ is unitary and from the trigonometric expression that $\det U = 1$, hence $U(\boldsymbol{\alpha}) \in SU(2)$.

Furthermore, it can be shown that every element of $SU(2)$ can be written in the form (1.1.5), with

$$0 \le \alpha < 2\pi. \tag{1.1.6}$$

With help of (1.1.2), one easily computes

$$U(\boldsymbol{\alpha})\boldsymbol{x}\cdot\boldsymbol{\sigma}U(\boldsymbol{\alpha})^+ = [\cos\alpha\boldsymbol{x} + \sin\alpha\,\boldsymbol{n}\wedge\boldsymbol{x} + (1-\cos\alpha)\boldsymbol{n}\cdot\boldsymbol{x}\boldsymbol{n}]\cdot\boldsymbol{\sigma} = [R(\boldsymbol{\alpha})\boldsymbol{x}]\cdot\boldsymbol{\sigma}, \tag{1.1.7}$$

where in the bracket Euler's formula for 3-dimensional rotations is obtained. If R_1, R_2 are two rotations, then it follows from (1.1.7) that

$$U(R_1R_2) = U(R_1)U(R_2). \tag{1.1.8}$$

This means that the mapping

$$SO(3) \ni R(\boldsymbol{\alpha}) \longrightarrow U(\boldsymbol{\alpha}) \in SU(2),$$

defined by (1.1.7) is a representation of $SO(3)$. However this mapping is not 1:1, but 1:2, because we have $0 \le \alpha < \pi$ for rotations in contrast to (1.1.6). In fact, the elements $U(\boldsymbol{\alpha})$ and $-U(\boldsymbol{\alpha})$ of $SU(2)$ correspond to the same rotation $R(\boldsymbol{\alpha}) \in SO(3)$.

This so-called spinor representation of $SO(3)$ will now be extended to four dimensions. We add the unit matrix

$$\sigma_0 = \begin{pmatrix} 1 & 0 \\ 0 & 1 \end{pmatrix} \tag{1.1.9}$$

to the three Pauli matrices (1.1.1). For $x = (x^\mu) \in \mathbb{M}$ we consider

$$\hat{x} = \sum_{\mu=0}^{3} x^\mu \sigma_\mu = \begin{pmatrix} x^0 + x^3 & x^1 - ix^2 \\ x^1 + ix^2 & x^0 - x^3 \end{pmatrix}, \tag{1.1.10}$$

instead of (1.1.3). The matrix (1.1.10) is self-adjoint and any self-adjoint 2×2 matrix is of this form with $x \in \mathbb{M}$. We compute

$$\det \hat{x} = x^{02} - \boldsymbol{x}^2 = x^2. \tag{1.1.11}$$

Instead of (1.1.7) we now form

$$A\hat{x}A^+ = \widehat{x'}, \tag{1.1.12}$$

for any non-singular complex 2×2 matrix A. Since (1.1.12) is self-adjoint, x' must be a real vector in $\mathbb{M}$. From

$$\det A \det \hat{x} \det A^+ = \det \widehat{x'}, \tag{1.1.13}$$

we find for

$$\det A = 1, \quad \text{i.e.} \quad A \in SL(2,\mathbb{C}), \tag{1.1.14}$$

that $x^2 = x'^2$, therefore, x and x' are connected by a Lorentz transformation Λ

$$A(\Lambda)\hat{x}A(\Lambda)^+ = \widehat{(\Lambda x)}. \tag{1.1.15}$$

As above (1.1.8), this equation shows that $A(\Lambda)$ is a representation

$$A(\Lambda_1\Lambda_2) = A(\Lambda_1)A(\Lambda_2). \tag{1.1.16}$$

Let us compute by means of (1.1.2), what comes out for self adjoint A:

$$A = a^0 + \boldsymbol{a}\cdot\boldsymbol{\sigma}, \quad (a^0, \boldsymbol{a}) \quad \text{real},$$

$$(a^0 + \boldsymbol{a}\cdot\boldsymbol{\sigma})(x^0 + \boldsymbol{x}\cdot\boldsymbol{\sigma})(a^0 + \boldsymbol{a}\cdot\boldsymbol{\sigma}) = x^0(a^{02} + \boldsymbol{a}^2) + 2a^0\,\boldsymbol{a}\cdot\boldsymbol{x}$$
$$+(2a^0x^0\boldsymbol{a} + a^{02} - \boldsymbol{a}^2\boldsymbol{x} + 2\boldsymbol{a}\cdot\boldsymbol{x}\boldsymbol{a})\cdot\boldsymbol{\sigma}. \tag{1.1.17}$$

Since $A \in SL(2,\mathbb{C})$, we may write

$$a^{02} = \cosh^2\frac{\chi}{2}, \quad \boldsymbol{a}^2 = \sinh^2\frac{\chi}{2}, \quad \boldsymbol{a} = \boldsymbol{n}|\boldsymbol{a}|, \tag{1.1.18}$$

then

$$x'^0 = x^0\cosh\chi + \boldsymbol{x}\cdot\boldsymbol{n}\sinh\chi$$
$$\boldsymbol{x}' = x^0\sinh\chi\,\boldsymbol{n} + \boldsymbol{n}\cdot\boldsymbol{x}\,\boldsymbol{n}\cosh\chi + \boldsymbol{x} - \boldsymbol{n}\cdot\boldsymbol{x}\,\boldsymbol{n}. \tag{1.1.19}$$

This is just a Lorentz boost in the direction $-\boldsymbol{n}$. Since rotations are already represented by unitary matrices $U \in SU(2)$, we have obtained a representation of the whole proper Lorentz group according to Theorem 1.1.

This representation is again not unique. Let us assume that two elements $A_1, A_2 \in SL(2,\mathbb{C})$ correspond to the same Lorentz transformation, then

$$A_1\hat{x}A_1^+ = A_2\hat{x}A_2^+, \quad \text{for all} \quad x \in \mathbb{M}.$$

It follows

$$B\hat{x}B^+ = \hat{x}, \quad \text{with} \quad B = A_1^{-1}A_2. \tag{1.1.20}$$

In the special case $\hat{x} = 1$, we find $BB^+ = 1$, that means B is unitary $\in SU(2)$. Then we know already that B and $-B$ correspond to the same rotation. Consequently

$$A_2 = A_1B \quad \text{and} \quad A_2' = -A_1B = -A_2 \tag{1.1.21}$$

correspond to the same Lorentz transformation. Therefore, the two elements $A(\Lambda)$ and $-A(\Lambda)$ of $SL(2,\mathbb{C})$ are mapped on the same Λ, which means that $SL(2,\mathbb{C})$ is a two-valued representation of $\mathcal{L}_+^\uparrow$, the so-called spinor representation $D^{(1/2,0)}$.

Other representations of the Lorentz group can be constructed by the spinor calculus of van der Waerden (*B.L. van der Waerden, Group Theory and Quantum Mechanics, Springer 1974*). The representation $D^{(1/2,0)}$, constructed above, is realized in the space

$$\mathbb{C}^2 \ni u = \begin{pmatrix} u_1 \\ u_2 \end{pmatrix} = (u_\alpha) \tag{1.1.22}$$

as follows

$$u' = A\,u = (A_\alpha{}^\beta u_\beta) \quad , \quad A \in SL(2,\mathbb{C}). \tag{1.1.23}$$

We now consider the symplectic bilinear form

$$\langle u, v\rangle = u_1 v_2 - u_2 v_1 = \det(u, v). \tag{1.1.24}$$

It is invariant under $SL(2,\mathbb{C})$ because

$$\langle u', v'\rangle = \det(Au, Av) = \det[A(u,v)A^T] = (\det A)^2 \det(u,v) = \langle u, v\rangle.$$

We use the notation

$$\langle u, v\rangle = u_\alpha \varepsilon^{\alpha\beta} v_\beta \tag{1.1.25}$$

with

$$\varepsilon^{\alpha\beta} = \begin{pmatrix} 0 & 1 \\ -1 & 0 \end{pmatrix}. \tag{1.1.26}$$

It is convenient to introduce tensor notation

$$v^\alpha = \varepsilon^{\alpha\beta} v_\beta \quad , \quad \langle u, v\rangle = u_\alpha v^\alpha. \tag{1.1.27}$$

It expresses the fact that summation over a pair of upper and lower indices gives an invariant under $SL(2,\mathbb{C})$.

Now we introduce a second representation space

$$\bar{\mathbb{C}}^2 \ni \bar{u} = \begin{pmatrix} u_{\bar{1}} \\ u_{\bar{2}} \end{pmatrix} \tag{1.1.28}$$

and therein the representation

$$\Lambda \longrightarrow A(\Lambda)^*, \tag{1.1.29}$$

where the asterisk denotes the complex conjugate matrix. This is indeed a representation, denoted by $D^{(0,1/2)}$, because

$$A(\Lambda_1\Lambda_2)^* = A(\Lambda_1)^* A(\Lambda_2)^*.$$

We are going to show that it is inequivalent to $D^{(1/2,0)}$. We have for every $A \in SL(2,\mathbb{C})$

$$A^T \varepsilon A = \varepsilon, \tag{1.1.30}$$

where T means the transposed matrix. This follows simply by multiplying the 2×2-matrices. Multiplying (1.1.30) by $A^{-1}\varepsilon^{-1}$ and taking the adjoint, we get

$$A^* = \varepsilon^{-1}(A^{-1})^+\varepsilon. \tag{1.1.31}$$

For unitary A, A^* is equivalent to A

$$A^* = \varepsilon^{-1} A \varepsilon,$$

but for selfadjoint A this is not true:

$$A^* = \varepsilon^{-1} A^{-1} \varepsilon.$$

We have therefore obtained a new inequivalent representation $D^{(0,1/2)}$ of $\mathcal{L}_+^\uparrow$ or $SL(2,\mathbb{C})$, in contrast to the case of $SU(2)$, where there is only one spinor representation $D^{1/2}$.

Further representations are constructed as tensor products. Let us consider the tensor product space

$$(\mathbb{C}^2)^{\otimes n} \ni u_{\alpha_1 \cdots \alpha_n} = u,$$

carrying the following tensor representation

$$(D^{(1/2,0)})^{\otimes n} : u \longrightarrow A_{\alpha_1}{}^{\beta_1}(\Lambda) \ldots A_{\alpha_n}{}^{\beta_n}(\Lambda) u_{\beta_1 \cdots \beta_n}. \tag{1.1.32}$$

However, this representation is reducible. This is easily seen by projecting onto the following anti-symmetric subspace

$$v_{\alpha_3 \cdots \alpha_n} = \varepsilon^{\alpha_1 \alpha_2} u_{\alpha_1 \alpha_2 \cdots \alpha_n} = u_{12\alpha_3 \cdots \alpha_n} - u_{21\alpha_3 \cdots \alpha_n}. \tag{1.1.33}$$

Since the contraction with the ε-tensor leads to an invariant in the first two indices, (1.1.33) is indeed a subrepresentation of (1.1.32). Only the symmetric part of (1.1.32) can be irreducible, and in fact, it is (see for example *van der Waerden*, cited above). The same is true for the complex conjugate tensor representation $(D^{(0,1/2)})^{\otimes n}$. The spinors of rank (n,m) are the elements of

$$\mathrm{sym}(\mathbb{C}^2)^{\otimes n} \times \mathrm{sym}(\bar{\mathbb{C}}^2)^{\otimes m} \ni u_{\alpha_1 \cdots \alpha_n \bar{\beta}_1 \cdots \bar{\beta}_m}, \tag{1.1.34}$$

which are symmetric in $(\alpha_1 \ldots \alpha_n)$ and $(\bar{\beta}_1 \ldots \bar{\beta}_m)$ separately. They generate the irreducible representation $D^{(n/2,m/2)}$:

$$u \longrightarrow A_{\alpha_1}{}^{\gamma_1} \ldots A_{\alpha_n}{}^{\gamma_n} A^*_{\bar{\beta}_1}{}^{\bar{\delta}_1} \ldots A^*_{\bar{\beta}_m}{}^{\bar{\delta}_m} u_{\gamma_1 \cdots \gamma_n \bar{\delta}_1 \cdots \bar{\delta}_m}. \tag{1.1.35}$$

All finite dimensional irreducible representations are obtained in this way. Their dimensions are given by the dimensions of the representations of the SU(2) subgroup

$$\dim D^{(n/2,m/2)} = \left(2\frac{n}{2}+1\right)\left(2\frac{m}{2}+1\right) = (n+1)(m+1). \tag{1.1.36}$$

1.2 Invariant Field Equations

We now turn to spinor fields. These are spinors depending on $x = (x^0, \boldsymbol{x})$, which transform under Lorentz transformations

$$x \longrightarrow x' = \Lambda x \tag{1.2.1}$$

according to a representation $D^{(n/2,m/2)}$ as discussed in the last section. Let us go through the representations of lowest dimensions, which are the most important ones:

1) $n = 0$, $m = 0$: The corresponding field is a scalar field $\psi(x)$ which transforms trivially under the change of the reference system (1.2.1)

$$\psi'(x') = \psi(x) = \psi(\Lambda^{-1}x'). \tag{1.2.2}$$

2) $n = 1/2$, $m = 0$: This is the two-dimensional self-representation of $SL(2,\mathbb{C})$ which is realized by a spinor field $\psi_\alpha(x)$. It is given by

$$\psi'_\alpha(x') = A_\alpha{}^\beta(\Lambda)\psi_\beta(x) = A_\alpha{}^\beta(\Lambda)\psi_\beta(\Lambda^{-1}x'). \tag{1.2.3}$$

3) $n = 0$, $m = 1/2$: This is the case of a spinor field $\psi_{\bar{\alpha}}(x)$ which transforms with the complex conjugate representation of (1.2.3).

4) $n = 1/2$, $m = 1/2$: The corresponding field $\psi_{\alpha\bar{\beta}}(x)$ has four components and transforms as follows:

$$\psi'_{\alpha\bar{\beta}}(x') = A_\alpha{}^\gamma(\Lambda)A^*_{\bar{\beta}}{}^{\bar{\delta}}(\Lambda)\psi_{\gamma\bar{\delta}}(x). \tag{1.2.4}$$

One suggests that this four-component field might have something to do with a vector field. In fact, introducing the 2×2-matrix

$$\hat{\psi}_{\gamma\bar{\delta}} = \sum_{\mu=0}^{3} \psi^\mu(\sigma_\mu)_{\gamma\bar{\delta}}\,, \tag{1.2.5}$$

completely analogous to (1.1.10), the transformation (1.2.4) leads to

$$\widehat{\psi'}_{\alpha\bar{\beta}} = A_\alpha{}^\gamma(\Lambda)\hat{\psi}_{\gamma\bar{\delta}}A^*_{\bar{\beta}}{}^{\bar{\delta}}(\Lambda) = (A(\Lambda)\hat{\psi}A(\Lambda)^+)_{\alpha\bar{\beta}} = (\widehat{\Lambda\psi})_{\alpha\bar{\beta}}, \tag{1.2.6}$$

that means

$$\psi'^\mu(x') = \Lambda^\mu{}_\nu\psi^\nu(x). \tag{1.2.7}$$

Consequently, Eq. (1.2.5) indeed defines a one-to-one correspondence between four-vectors ψ^μ and spinors $\psi_{\alpha\bar{\beta}}$ of type 1/2, 1/2.

The correspondence (1.2.5) can immediately be taken over to the differential operators

$$\partial_\mu = \frac{\partial}{\partial x^\mu} \quad \text{and} \quad \partial^\mu = g^{\mu\nu}\partial_\nu,$$

as follows

$$\hat{\partial} = (\partial_{\alpha\bar{\beta}}) = \sigma_\mu\partial^\mu = \sigma_0\partial_0 - \boldsymbol{\sigma}\cdot\boldsymbol{\partial}, \quad \boldsymbol{\partial} = \text{grad}. \tag{1.2.8}$$

Lifting the indices with help of the ε-tensor leads to

$$\partial^{\alpha\bar{\beta}} = \varepsilon^{\alpha\gamma}\varepsilon^{\bar{\beta}\bar{\delta}}\partial_{\gamma\bar{\delta}} = (\varepsilon\hat{\partial}\varepsilon^T)^{\alpha\bar{\beta}} = (\sigma_0\partial_0 + \boldsymbol{\sigma}^T\cdot\boldsymbol{\partial})^{\alpha\bar{\beta}}. \tag{1.2.9}$$

Using (1.2.8, 9) and the rules of the tensor calculus, it is straight-forward to write down invariant differential equations for the various fields. For a scalar field we need a scalar differential operator, which exists only in second order:

$$\partial_{\alpha\bar{\beta}}\partial^{\alpha\bar{\beta}}\psi(x) + a\psi(x) = 0, \tag{1.2.10}$$

where a is a constant. The invariant second order differential operator

$$\partial_{\gamma\bar{\beta}}\partial^{\alpha\bar{\beta}} = (\sigma_0\partial_0 - \boldsymbol{\sigma}\cdot\boldsymbol{\partial})_{\gamma\bar{\beta}}(\sigma_0\partial_0 + \boldsymbol{\sigma}^T\cdot\boldsymbol{\partial})^{\alpha\bar{\beta}}$$

$$= (\partial_0 - \boldsymbol{\sigma}\cdot\boldsymbol{\partial})(\partial_0 + \boldsymbol{\sigma}\cdot\boldsymbol{\partial})\delta_\gamma{}^\alpha = ({\partial_0}^2 - \boldsymbol{\partial}^2)\delta_\gamma{}^\alpha = \Box\,\delta_\gamma{}^\alpha, \tag{1.2.11}$$

is just the wave operator. Equation (1.2.10) now reads

$$[2\Box + a]\psi(x) = 0. \tag{1.2.12}$$

The constant a has dimension 1/length^2. Introducing the Compton wave length of a particle of mass m

$$\lambda = \frac{\hbar}{mc},$$

Eq. (1.2.12) becomes the Klein-Gordon equation

$$\Box\,\psi(x) + \left(\frac{mc}{\hbar}\right)^2\psi(x) = 0. \tag{1.2.13}$$

For $m = 0$ one gets the wave equation, in accordance with the fact that mass 0 particles move with the velocity of light. Planck's constant $\hbar$ then drops out in this case. There is no first quantized mass zero theory.

Now we turn to a spinor field $\psi_\alpha(x)$. In this case a first order equation is possible:

$$\partial^{\alpha\bar{\beta}}\psi_\alpha(x) = 0. \tag{1.2.14}$$

We sum over the first index because of the transposed Pauli matrices in (1.2.9). In matrix notation the equation can be written as follows

$$(\partial_0 + \boldsymbol{\sigma}\cdot\boldsymbol{\partial})\psi(x) = 0, \quad \text{where} \quad \psi = \begin{pmatrix}\psi_1\\ \psi_2\end{pmatrix} \tag{1.2.15}$$

is a two component (Weyl) spinor. This is the Weyl equation. It has two defects which for a long time have questioned its usefulness: First, there is no mass term $a\psi_\alpha$ possible in (1.2.14). Consequently, the Weyl equation describes mass 0 particles. In fact, we have

$$\partial_{\gamma\bar{\beta}}\partial^{\alpha\bar{\beta}}\psi_\alpha = \delta_\gamma{}^\alpha\Box\psi_\alpha = \Box\psi_\gamma = 0. \tag{1.2.16}$$

The second defect of (1.2.15) is that it is not invariant under space reflection (parity). Considering the parity transformation P

$$P : \boldsymbol{x} \longrightarrow -\boldsymbol{x}, \quad \tilde{\psi}(x) = \psi(x^0, -\boldsymbol{x}), \tag{1.2.17}$$

the transformed Eq. (1.2.15) reads

$$\partial_0 \tilde{\psi} - \boldsymbol{\sigma} \cdot \boldsymbol{\partial} \tilde{\psi} = 0. \tag{1.2.18}$$

This equation would be equivalent to (1.2.15), if there exists a linear transformation S with

$$\partial_0 S \tilde{\psi} - S \boldsymbol{\sigma} \cdot \boldsymbol{\partial} \tilde{\psi} = \partial_0 S \tilde{\psi} + \boldsymbol{\sigma} \cdot \boldsymbol{\partial} S \tilde{\psi},$$

for all solutions $\tilde{\psi}$, which implies

$$\sigma_j S = -S \sigma_j \quad \text{for} \quad j = 1, 2, 3. \tag{1.2.19}$$

Hence, S anticommutes with all σ_j. This is only possible if $S = 0$, which shows that $\tilde{\psi}(x)$ obeys a different Eq. (1.2.18) and, consequently, the Weyl equation is not invariant under space reflection. For this reason the equation has long been considered as useless for physics. However, since parity is not conserved in weak interactions, the equation can be used there for the description of mass 0 neutrinos.

We are now looking for an equation without the two defects just discussed, that means, we seek a parity-invariant equation which allows $m \neq 0$. Then we need a second spinor field to write the coupled equations

$$\partial^{\alpha\bar{\beta}} \varphi_\alpha = a \chi^{\bar{\beta}}$$
$$\partial_{\alpha\bar{\beta}} \chi^{\bar{\beta}} = b \varphi_\alpha, \tag{1.2.20}$$

or

$$(\sigma_0 \partial_0 + \boldsymbol{\sigma}^T \cdot \boldsymbol{\partial})^T \varphi = a \chi \tag{1.2.21}$$
$$(\sigma_0 \partial_0 - \boldsymbol{\sigma} \cdot \boldsymbol{\partial}) \chi = b \varphi. \tag{1.2.22}$$

The two matrix transpositions in (1.2.21) can be dropped. Denoting the space-reflected spinors by

$$\tilde{\varphi}(x) = \varphi(Px) \quad , \quad \tilde{\chi}(x) = \chi(Px), \tag{1.2.23}$$

the transformed equations

$$(\sigma_0 \partial_0 - \boldsymbol{\sigma} \cdot \boldsymbol{\partial}) \tilde{\varphi} = a \tilde{\chi} \quad , \quad (\sigma_0 \partial_0 + \boldsymbol{\sigma} \cdot \boldsymbol{\partial}) \tilde{\chi} = b \tilde{\varphi}$$

are identical with (1.2.21, 22), if the two equations are interchanged, that means

$$a = b \quad , \quad \tilde{\varphi} = \text{const}\, \chi \quad , \quad \tilde{\chi} = \text{const}\, \varphi. \tag{1.2.24}$$

Choosing

$$a = -\frac{i}{\hbar}mc,$$

we arrive at the Dirac equation for a particle with mass m

$$i\hbar(\sigma_0\partial_0 + \boldsymbol{\sigma}\cdot\boldsymbol{\partial})\varphi = mc\chi \quad , \quad i\hbar(\sigma_0\partial_0 - \boldsymbol{\sigma}\cdot\boldsymbol{\partial})\chi = mc\varphi. \tag{1.2.25}$$

The two coupled two-component equations (1.2.25) can be written in the usual four-component form by introducing a Dirac spinor

$$\psi = \begin{pmatrix} \varphi_\alpha \\ \chi^\beta \end{pmatrix} \tag{1.2.26}$$

and instead of the Pauli matrices

$$\sigma_\mu = (\sigma_0, \boldsymbol{\sigma}) \quad , \quad \sigma^\mu = (\sigma_0, -\boldsymbol{\sigma}) \tag{1.2.27}$$

the 4×4 γ-matrices

$$\gamma^\mu = \begin{pmatrix} 0 & \sigma^\mu \\ \sigma_\mu & 0 \end{pmatrix} \quad \text{or}$$

$$\gamma^0 = \begin{pmatrix} 0_2 & 1_2 \\ 1_2 & 0_2 \end{pmatrix} \quad , \quad \gamma^k = \begin{pmatrix} 0 & -\sigma_k \\ \sigma_k & 0 \end{pmatrix}. \tag{1.2.28}$$

One calls (1.2.28) the chiral or Weyl representation of the γ's. Then the Dirac equation (1.2.25) assumes the usual form

$$i\hbar\gamma^\mu\partial_\mu\psi(x) = mc\psi(x). \tag{1.2.29}$$

After construction, the Dirac equation transforms under $\mathcal{L}_+^\uparrow$ according to the reducible representation $D^{(1/2,0)} \oplus D^{(0,1/2)}$. It is irreducible, however, with respect to $\mathcal{L}^\uparrow$. The so-called chiral form (1.2.26) of the Dirac spinors is particular important in electro-weak theory where the fundamental particles are "chiral fermions". From the definition (1.2.28) of the γ-matrices, we get the anti- commutation relations

$$\gamma^\mu\gamma^\nu + \gamma^\nu\gamma^\mu = 2g^{\mu\nu}, \tag{1.2.30}$$

because

$$\sigma_j\sigma_k + \sigma_k\sigma_j = 2\delta_{jk}. \tag{1.2.31}$$

We finally observe that each component of the Dirac spinor fulfills the Klein-Gordon equation (1.2.13). Indeed, operating with $i\hbar\gamma^\nu\partial_\nu$ on (1.2.29) and adding the equation with μ and ν interchanged, we get by means of (1.2.30)

$$-2\hbar^2 g^{\mu\nu}\partial_\nu\partial_\mu\psi = 2mci\hbar\gamma^\nu\partial_\nu\psi = 2m^2c^2\psi$$

or

$$\hbar^2\partial^\mu\partial_\mu\psi = -m^2c^2\psi.$$

The advantage of the described approach to the Dirac equation lies in the fact that we already know the transformation properties of ψ (1.2.26) under Lorentz transformations

$$x \longrightarrow x' = \Lambda x \quad , \quad \text{namely}$$

$$\psi = \begin{pmatrix} \varphi_\alpha \\ \chi^{\bar\beta} \end{pmatrix} \longrightarrow \psi'(x') = S(\Lambda)\psi(x), \tag{1.2.32}$$

where

$$S(\Lambda) = \begin{pmatrix} A(\Lambda) & 0 \\ 0 & \varepsilon^T A(\Lambda)^* \varepsilon \end{pmatrix}, \tag{1.2.33}$$

the ε appears in the lower corner of $S(\Lambda)$, because the index $\bar\beta$ in (1.2.32) is lifted. The matrix $A(\Lambda) \in SL(2,\mathbb{C})$ was introduced in the previous section (1.1.15). Since the transformed spinor (1.2.32) satisfies the same equation

$$i\hbar\gamma^\mu \partial'_\mu \psi'(x') = mc\psi'(x'),$$

we conclude

$$i\hbar\gamma^\mu \partial'_\mu S(\Lambda)\psi(x) = mcS(\Lambda)\psi(x).$$

Using

$$\partial'_\mu = \Lambda_\mu{}^\nu \partial_\nu,$$

we find

$$i\hbar\Lambda_\mu{}^\nu S(\Lambda)^{-1}\gamma^\mu S(\Lambda)\partial_\nu\psi(x) = mc\psi(x)$$

for all $\psi(x)$, consequently

$$\Lambda_\mu{}^\nu S(\Lambda)^{-1}\gamma^\mu S(\Lambda) = \gamma^\nu$$

or after applying the inverse Lorentz transformation $\Lambda^\mu{}_\nu$

$$S(\Lambda)^{-1}\gamma^\mu S(\Lambda) = \Lambda^\mu{}_\nu\gamma^\nu. \tag{1.2.34}$$

The matrix on the left side of (1.2.34) transforms like a four-vector. We also know the transformation properties of the Dirac spinor under space reflection

$$x \longrightarrow x' = Px = (x^0, -\boldsymbol{x}).$$

According to Eq. (1.2.24), the two Weyl-spinors in (1.2.32) are interchanged:

$$\psi'(x') = \begin{pmatrix} 0_2 & 1_2 \\ 1_2 & 0_2 \end{pmatrix} \psi(x), \tag{1.2.35}$$

hence

$$\psi'(x') = \pm\gamma^0\psi(x). \tag{1.2.36}$$

In order to understand what kind of particles the Dirac equation describes, we look at the behaviour of a Dirac spinor under rotations $\Lambda = R(\boldsymbol{\alpha})$. Then

$$A(R(\boldsymbol{\alpha})) = U(\boldsymbol{\alpha}) \in SU(2)$$

and from (1.1.31) we get

$$\varepsilon^T A(R)^*\varepsilon = \varepsilon^T\varepsilon^{-1}(U^{-1})^+\varepsilon\varepsilon = U(\boldsymbol{\alpha}).$$

The transformation law (1.2.32, 33) now reads

$$\psi'(x') = \begin{pmatrix} U(\boldsymbol{\alpha}) & 0 \\ 0 & U(\boldsymbol{\alpha}) \end{pmatrix} \psi(R(\boldsymbol{\alpha})^{-1}x'). \tag{1.2.37}$$

The unitary matrix in front realizes a spinor representation $D^{1/2} \oplus D^{1/2}$ of the rotation group. The Dirac equation, therefore, describes particles with spin $\frac{1}{2}$. Differentiating (1.2.37) with respect to α_j at $\boldsymbol{\alpha} = 0$, we find the infinitesimal generators

$$\boldsymbol{J} = \boldsymbol{S} + \boldsymbol{L} \quad \text{with}$$

$$S^j = \frac{1}{2}\begin{pmatrix} \sigma_j & 0 \\ 0 & \sigma_j \end{pmatrix} \quad , \quad L^j = \frac{1}{i}\left(\boldsymbol{x}' \wedge \frac{\partial}{\partial \boldsymbol{x}'}\right)_j . \tag{1.2.38}$$

$\boldsymbol{S}$ is the spin and $\boldsymbol{L}$ the orbit angular momentum, such that $\boldsymbol{J}$ is the total angular momentum.

We finally go one step further in our discussion of invariant wave equations, namely to the representation $D^{(1/2,1/2)}$ considered as case 4) at the beginning of this section. The corresponding field is related to a vector field $A^\mu(x)$ by (1.2.5)

$$(\hat{A})_{\alpha\bar{\beta}} = A^\mu(\sigma_\mu)_{\alpha\bar{\beta}}.$$

The simplest invariant differential equation for this field is

$$\partial^{\alpha\bar{\beta}}\hat{A}_{\alpha\bar{\beta}} = 0, \tag{1.2.39}$$

or with (1.2.9)

$$\mathrm{Tr}(\partial_0 + \boldsymbol{\sigma}\cdot\boldsymbol{\partial})(A^0 - \boldsymbol{\sigma}\cdot\boldsymbol{A}) = \partial_0 A^0 - \boldsymbol{\partial}\cdot\boldsymbol{A} = 0, \tag{1.2.40}$$

since the Pauli matrices have trace 0. This is the Lorentz gauge condition. It is obviously not an equation of motion to specify the space-time behaviour of a vector field. The next more complicated equation is of second order

$$\partial_{\alpha\bar{\beta}}\partial^{\gamma\bar{\beta}}A_{\gamma\bar{\delta}} + m^2 A_{\alpha\bar{\delta}} = 0. \tag{1.2.41}$$

Using (1.2.11), we see that this is just the Klein-Gordon equation for every component A^μ

$$\Box A^\mu + m^2 A^\mu = 0. \tag{1.2.42}$$

We want to determine the spin content of this theory. Restricting the representation

$$D^{(1/2,1/2)}|_{SU(2)} = D^{1/2} \otimes D^{1/2} = D^1 + D^0 \tag{1.2.43}$$

to the rotation group, we see that $A^\mu(x)$ simultaneously describes a spin 0 and a spin 1 field. The former can be eliminated by the Lorentz condition (1.2.40). The Eqs. (1.2.40) and (1.2.42) together define the so-called Proca theory. For $m = 0$ one gets the Maxwell field in the Lorentz gauge.

1.3 Algebraic Properties of the Dirac Equation

We first study the algebra C_4 of the γ-matrices $\gamma^\mu, \mu = 0,1,2,3$. For doing so, we do not use the explicit representation (1.2.28), but consider C_4 as an abstract algebra with the basic anticommutation relations (1.2.30)

$$\gamma^\mu \gamma^\nu + \gamma^\nu \gamma^\mu = 2g^{\mu\nu}. \tag{1.3.1}$$

This is a special Clifford algebra. A basis of C_4 is formed by the following 16 elements:

$$\mathbf{1},\, \gamma^\mu,\, \gamma^\mu\gamma^\nu\, (\mu < \nu),\, \gamma^\mu\gamma^\nu\gamma^\lambda\, (\mu < \nu < \lambda), \tag{1.3.2}$$

$$\gamma^0\gamma^1\gamma^2\gamma^3 = -i\gamma^5. \tag{1.3.3}$$

The last equation is the definition of γ^5. These 16 elements are also a basis for all 4×4-matrices.

Besides C_4 we consider the multiplicative group G_4 generated by the γ^μ. It is a finite group, consisting of the 32 elements

$$\pm\mathbf{1}, \pm\gamma^\mu, \pm\gamma^\mu\gamma^\nu (\mu < \nu)\,, \pm\gamma^\mu\gamma^\nu\gamma^\lambda (\mu < \nu < \lambda), \pm i\gamma^5. \tag{1.3.4}$$

It follows easily from (1.3.1) that G_4 has the following equivalence classes of conjugated elements

$$[\mathbf{1}], [-\mathbf{1}], [\pm\gamma^\mu], [\pm\gamma^\mu\gamma^\nu], [\pm\gamma^\mu\gamma^\nu\gamma^\lambda], [\pm i\gamma^5]. \tag{1.3.5}$$

Since there are 17 classes, we know from the representation theory of finite groups (see e.g. *C. W. Curtis, I. Reiner, Representation Theory of Finite Groups and Associative Algebras, Interscience Publ. 1962*) that there must exist 17 irreducible representations of G_4. Since

$$\gamma^\mu\gamma^\nu = \pm\gamma^\nu\gamma^\mu,$$

the factor group

$$G_4/[\mathbf{1}, -\mathbf{1}] = G_4'$$

is abelian. G_4' has 16 classes (the $-\mathbf{1}$ in (1.3.5) is missing) and, hence, has 16 one-dimensional representations. These 16 representations are also representations of G_4 but not of C_4, because both $+\mathbf{1}$ and $-\mathbf{1}$ are represented by 1. There remains one irreducible representation of G_4 of dimension $d \neq 1$ to be considered. Since, according to the general theory, the sum of the squares of the dimensions is equal to the number of elements of the group, we have

$$16 \cdot 1^2 + d^2 = 32.$$

The solution $d = 4$ shows that the self-representation is the only non-trivial representation of C_4. We have thus proved the following

Theorem of Pauli. Let $\tilde{\gamma}^\mu, \mu = 0,1,2,3$ be four arbitrary matrices (not necessarily 4×4) with

$$\tilde{\gamma}^\mu \tilde{\gamma}^\nu + \tilde{\gamma}^\nu \tilde{\gamma}^\mu = 2g^{\mu\nu}. \tag{1.3.6}$$

If the multiplicative group generated by them is irreducible, the $\tilde{\gamma}^\mu$ are 4×4 matrices and there exists a non-singular matrix C such that

$$\tilde{\gamma}^\mu = C\gamma^\mu C^{-1}, \tag{1.3.7}$$

where γ^μ are the γ-matrices in the chiral representation (1.2.28).

It follows from this theorem that the Dirac equation is valid in an arbitrary representation of the γ-matrices. In fact, the equation

$$i\hbar C\gamma^\mu C^{-1}\partial_\mu \tilde{\psi}(x) = m\tilde{\psi}(x), \quad c = 1$$

is a consequence of

$$i\hbar\gamma^\mu \partial_\mu C^{-1}\tilde{\psi}(x) = mC^{-1}\tilde{\psi}(x). \tag{1.3.8}$$

This shows that the change of the representation is simply a linear transformation of the spinor field. Besides the chiral representation (1.2.28), the so-called standard representation is very useful. It is obtained by taking the unitary matrix

$$C = \frac{1}{\sqrt{2}}\begin{pmatrix} 1_2 & 1_2 \\ 1_2 & -1_2 \end{pmatrix} = C^{-1} = C^+ \tag{1.3.9}$$

in (1.3.7). Then

$$\tilde{\gamma}^0 = \begin{pmatrix} 1 & 0 \\ 0 & -1 \end{pmatrix} \quad , \quad \tilde{\gamma}^j = \begin{pmatrix} 0 & \sigma_j \\ \sigma_j & 0 \end{pmatrix}. \tag{1.3.10}$$

This representation of the γ's is particularly useful for the discussion of the non-relativistic limit. We note that the spin matrices (1.2.38) remain unchanged:

$$\tilde{S}_j = \frac{1}{4}\begin{pmatrix} 1 & 1 \\ 1 & -1 \end{pmatrix}\begin{pmatrix} \sigma_j & 0 \\ 0 & \sigma_j \end{pmatrix}\begin{pmatrix} 1 & 1 \\ 1 & -1 \end{pmatrix} = \frac{1}{2}\begin{pmatrix} \sigma_j & 0 \\ 0 & \sigma_j \end{pmatrix} = S_j. \tag{1.3.11}$$

Now we consider the adjoint matrices. In the chiral representation (1.2.28) we have

$$\gamma^{0+} = \gamma^0 \quad , \quad \gamma^{j+} = -\gamma^j,$$

because the Pauli matrices σ_j are selfadjoint. In an arbitrary representation, we then find

$$\tilde{\gamma}^{\mu+} = C^{-1+}\gamma^{\mu+}C^+ = \pm C^{+-1}\gamma^\mu C^+ = \pm C\gamma^\mu C^{-1} = \pm\tilde{\gamma}^\mu,$$

if C is unitary. That means, the hermitian structure is unchanged for unitary equivalent representations. Therefore, all representations used in the following are assumed to be unitary equivalent to the chiral representation. Then the adjoint γ-matrices are given by

$$\gamma^{\mu +} = \gamma^0 \gamma^\mu \gamma^0. \tag{1.3.12}$$

The chiral representation is most useful for the analysis of transformation properties. We now want to form bilinear expressions from the spinor (1.2.32)

$$\psi = \begin{pmatrix} \varphi_\alpha \\ \chi^{\bar{\beta}} \end{pmatrix} \quad \text{and} \quad \psi^+ = (\varphi_\alpha^+, \chi^{\bar{\beta}+})$$

with simple transformation properties under Lorentz transformations

$$x \longrightarrow x' = \Lambda x.$$

According to (1.2.32, 33), we have

$$\psi'(x') = S(\Lambda)\psi(x) \tag{1.3.13}$$

and

$$\psi'(x')^+ = \psi(x)^+ S(\Lambda)^+, \tag{1.3.14}$$

with

$$S(\Lambda)^+ = \begin{pmatrix} A(\Lambda)^+ & 0 \\ 0 & \varepsilon^T A(\Lambda)^T \varepsilon \end{pmatrix}. \tag{1.3.15}$$

Since $A \in SL(2, \mathbb{C})$, we get from (1.1.30)

$$A^+ \varepsilon A^* = \varepsilon \quad \text{and} \quad A^+ = \varepsilon A^{*-1} \varepsilon^{-1}.$$

Therefore, we can write (1.3.15) as follows

$$S(\Lambda)^+ = \begin{pmatrix} \varepsilon A^{*-1} \varepsilon^T & 0 \\ 0 & A^{-1} \end{pmatrix}.$$

Multiplying with

$$\gamma^0 = \begin{pmatrix} 0 & 1_2 \\ 1_2 & 0 \end{pmatrix},$$

from both sides, we finally obtain

$$S(\Lambda)^+ = \gamma^0 \begin{pmatrix} A^{-1} & 0 \\ 0 & \varepsilon^T A^{*-1} \varepsilon \end{pmatrix} \gamma^0 = \gamma^0 S(\Lambda)^{-1} \gamma^0. \tag{1.3.16}$$

For the transformed adjoint spinor (1.3.14), we now have

$$\psi'(x')^+ \gamma^0 = \psi(x)^+ \gamma^0 S(\Lambda)^{-1} = \overline{\psi}(x) S(\Lambda)^{-1}, \tag{1.3.17}$$

where the Dirac adjoint spinor

$$\overline{\psi}(x) = \psi(x)^+ \gamma^0 \tag{1.3.18}$$

has been introduced. It transforms as follows

$$\overline{\psi'}(x') = \overline{\psi}(x) S(\Lambda)^{-1}. \tag{1.3.19}$$

Its transformation under space reflection P follows from (1.2.36):

$$\psi'(Px)^+ = \pm\psi^+(x)\gamma^0,$$
$$\overline{\psi'}(Px) = \pm\psi^+(x). \tag{1.3.20}$$

Now let us form bilinear combinations of ψ and $\overline{\psi}$:

1) The product

$$\overline{\psi'}(x')\psi'(x') = \overline{\psi}(x)S(\Lambda)^{-1}S(\Lambda)\psi(x) \tag{1.3.21}$$

is invariant under proper Lorentz transformations. It is also invariant under space reflections

$$\overline{\psi'}(Px)\psi'(Px) = \psi^+(x)\gamma^0\psi(x) = \overline{\psi}(x)\psi(x)$$

and, therefore, is a scalar.

2) A vector is obtained by

$$\overline{\psi'}(x')\gamma^\mu\psi'(x') = \overline{\psi}(x)S(\Lambda)^{-1}\gamma^\mu S(\Lambda)\psi(x) = \Lambda^\mu{}_\nu\overline{\psi}(x)\gamma^\nu\psi(x), \tag{1.3.22}$$

where (1.2.34) has been used. The transformation law under space reflections is easily verified:

$$\overline{\psi'}(Px)\gamma^\mu\psi'(Px) = \psi^+(x)\gamma^\mu\gamma^0\psi(x) = \begin{cases} \overline{\psi}\gamma^0\psi \\ -\overline{\psi}\gamma^j\psi \end{cases}.$$

3) To get a tensor, we use the antisymmetric combination

$$\sigma^{\mu\nu} = \frac{i}{2}[\gamma^\mu, \gamma^\nu] \qquad \mu < \nu \tag{1.3.23}$$

the square brackets always denote the commutator. As in (1.3.22), we find

$$\overline{\psi'}(x')\sigma^{\mu\nu}\psi'(x') = \Lambda^\mu{}_{\mu'}\Lambda^\nu{}_{\nu'}\overline{\psi}(x)\sigma^{\mu'\nu'}\psi(x).$$

For spatial indices, σ^{jl} is connected with the spin vector (1.2.38), because

$$\sigma^{jl} = \varepsilon^{jlm}\begin{pmatrix} \sigma_m & 0 \\ 0 & \sigma_m \end{pmatrix}. \tag{1.3.24}$$

Here ε^{jlm} is the 3-dimensional totally antisymmetric tensor

$$\varepsilon^{jlm} = \begin{cases} 1, & \text{if } j,l,m \text{ is an even permutation of } 1,2,3 \\ -1, & \text{if } j,l,m \text{ is an odd permutation of } 1,2,3 \\ 0 & \text{otherwise.} \end{cases}$$

4) The matrix γ^5 (1.3.3) anticommutes with all γ^μ

$$\gamma^5\gamma^\mu = -\gamma^\mu\gamma^5, \mu = 0,1,2,3. \tag{1.3.25}$$

Since it is diagonal in the chiral representation,

$$\gamma^5 = \begin{pmatrix} 1_2 & 0 \\ 0 & -1_2 \end{pmatrix}, \tag{1.3.26}$$

we find from (1.2.33):

$$S(\Lambda^{-1})\gamma^5 S(\Lambda) = \gamma^5. \tag{1.3.27}$$

Then,

$$\overline{\psi'}(x')\gamma^5\psi'(x') = \overline{\psi}(x)\gamma^5\psi(x) \tag{1.3.28}$$

and

$$\overline{\psi'}(Px)\gamma^5\psi'(Px) = \psi^+(x)\gamma^5\gamma^0\psi(x) = -\overline{\psi}(x)\gamma^5\psi(x).$$

This is a pseudo-scalar.

5) A pseudo-vector is given by

$$\begin{aligned}\overline{\psi'}(x')\gamma^\mu\gamma^5\psi'(x') &= \overline{\psi}(x)S(\Lambda)^{-1}\gamma^\mu\gamma^5 S(\Lambda)\psi(x)\\ &= \Lambda^\mu{}_\nu\overline{\psi}(x)\gamma^\nu\gamma^5\psi(x),\end{aligned} \tag{1.3.29}$$

with

$$\overline{\psi'}(Px)\gamma^\mu\gamma^5\psi'(Px) = \begin{cases} -\overline{\psi}(x)\gamma^0\gamma^5\psi(x) \\ \overline{\psi}(x)\gamma^j\gamma^5\psi(x)\end{cases}. \tag{1.3.30}$$

Note that the signs here are different from 2).

Totally, we have obtained 16 linear independent combinations. Since there are 16 linear independent 4×4 matrices, these are all bilinear combinations.

Finally, we write down the Dirac equation for the adjoint spinor (1.3.18). Taking the adjoint of (1.2.29), multiplying by γ^0 from the right, we obtain

$$\psi^+\gamma^{\mu+}\gamma^0 = \psi^+\gamma^0\gamma^\mu = \overline{\psi}\gamma^\mu$$

in virtue of (1.3.12), and

$$-i\hbar\partial_\mu\overline{\psi}\gamma^\mu = mc\overline{\psi}. \tag{1.3.31}$$

This is the so-called adjoint Dirac equation.

1.4 Discussion of the Free Dirac Equation

In this section we take $c = \hbar = 1$. We want to cast the Dirac equation

$$i(\gamma^0\partial_0 + \gamma^j\partial_j)\psi(x) = m\psi(x) \tag{1.4.1}$$

into a hamiltonian form. Multiplying (1.4.1) by γ^0 and introducing the new matrices

$$\alpha^j = \gamma^0\gamma^j \quad , \quad \beta = \gamma^0, \tag{1.4.2}$$

one obtains

$$i\frac{\partial\psi}{\partial t} = \left[m\beta - \boldsymbol{\alpha}\, i\frac{\partial}{\partial \boldsymbol{x}}\right]\psi(x). \tag{1.4.3}$$

The new matrices (1.4.2) satisfy

$$\{\alpha^j,\alpha^k\} \stackrel{\text{def}}{=} \alpha^j\alpha^k + \alpha^k\alpha^j = 2\delta_{jk} \quad , \quad \{\alpha^j,\beta\} = 0 \tag{1.4.4}$$

and

$$(\alpha^j)^2 = \mathbf{1}_4 = \beta^2. \tag{1.4.5}$$

The reason for introducing the new matrices is that they are selfadjoint. In fact, β is selfadjoint in the standard, chiral or any unitary equivalent representation, and so is

$$\alpha^{j+} = \gamma^{j+}\gamma^{0+} = -\gamma^j\gamma^0 = \gamma^0\gamma^j = \alpha^j.$$

Consequently, (1.4.3) assumes the usual quantum mechanical form

$$i\frac{\partial\psi}{\partial t} = H_0\,\psi. \tag{1.4.6}$$

The hamiltonian H_0 is selfadjoint in the Hilbert space $\mathcal{H} = (L^2(\mathbb{R}^3))^4$ of elements

$$f = \begin{pmatrix} f_1 \\ f_2 \\ f_3 \\ f_4 \end{pmatrix} \tag{1.4.7}$$

with scalar product

$$(f,\, g) = \sum_j \int f_j(\boldsymbol{x})^* g_j(\boldsymbol{x}) d^3x = \int f^+(\boldsymbol{x}) g(\boldsymbol{x}) d^3x, \tag{1.4.8}$$

where the asterisk means complex conjugation and f^+ is the row matrix $(f_1^*,\ f_2^*,\ f_3^*,\ f_4^*)$. In fact, we have

$$(\beta f,\, g) = (f,\, \beta g) \quad \text{and}$$

$$\left(\alpha^j i\frac{\partial f}{\partial x_j},\, g\right) = \left(i\frac{\partial f}{\partial x_j},\, \alpha^j g\right) = \left(f,\, \alpha^j i\frac{\partial g}{\partial x_j}\right).$$

The solution of the Dirac equation (1.4.3) is then given by a unitary group in $\mathcal{H}$

$$\psi(t) = e^{-iH_0t}\psi(0), \tag{1.4.9}$$

which leaves the scalar product (1.4.8) invariant:

$$\int \psi(t,\boldsymbol{x})^+\psi(t,\boldsymbol{x}) d^3x = \int \psi(0,\boldsymbol{x})^+\psi(0,\boldsymbol{x}) d^3x. \tag{1.4.10}$$

The hamiltonian H_0 is best discussed in momentum space. Let the Fourier transformation on $\mathcal{H}$ be defined by

$$\hat{f}(\boldsymbol{p}) = (2\pi)^{-3/2} \int f(\boldsymbol{x}) e^{-i\boldsymbol{p}\cdot\boldsymbol{x}} d^3x. \tag{1.4.11}$$

Then H_0 becomes a multiplication operator

$$\widehat{(H_0\psi)}(\boldsymbol{p}) = [m\beta + \boldsymbol{\alpha}\cdot\boldsymbol{p}]\hat{\psi}(\boldsymbol{p}), \tag{1.4.12}$$

with domain

$$D(H_0) = \{\psi(\boldsymbol{x}) \in \mathcal{H} \mid p_j\hat{\psi}(\boldsymbol{p}) \in \mathcal{H}\}.$$

There remains to diagonalize the matrix

$$H_0(\boldsymbol{p}) = \beta m + \boldsymbol{\alpha} \cdot \boldsymbol{p}. \tag{1.4.14}$$

Working in the standard representation

$$\beta = \begin{pmatrix} 1 & 0 \\ 0 & -1 \end{pmatrix} \quad , \quad \alpha^j = \begin{pmatrix} 0 & \sigma_j \\ \sigma_j & 0 \end{pmatrix} , \tag{1.4.15}$$

the corresponding eigenvalue problem

$$H_0\psi = \begin{pmatrix} m & \boldsymbol{\sigma} \cdot \boldsymbol{p} \\ \boldsymbol{\sigma} \cdot \boldsymbol{p} & -m \end{pmatrix} \psi = E(\boldsymbol{p})\psi \tag{1.4.16}$$

is solved with the help of the Ansatz

$$\psi = \begin{pmatrix} a_0\chi_s \\ \boldsymbol{b} \cdot \boldsymbol{\sigma}\chi_s \end{pmatrix} , \tag{1.4.17}$$

where

$$\begin{aligned} \chi_s &= \begin{pmatrix} 1 \\ 0 \end{pmatrix} \quad \text{for} \quad s = 1 \\ &= \begin{pmatrix} 0 \\ 1 \end{pmatrix} \quad \text{for} \quad s = -1. \end{aligned} \tag{1.4.18}$$

Then it follows from (1.4.16) that

$$\boldsymbol{b} = \frac{a_0}{E + m}\boldsymbol{p} \tag{1.4.19}$$

and

$$E(\boldsymbol{p}) = \pm\sqrt{\boldsymbol{p}^2 + m^2} \quad \text{or} \tag{1.4.20}$$

$$= \pm\sqrt{c^2\boldsymbol{p}^2 + m^2c^4}, \tag{1.4.21}$$

with the light velocity inserted. This is just the relation between energy and momentum for a relativistic free particle. However, we have got positive and negative energies! The hamiltonian is unbounded from below. This makes a consistent physical interpretation of this one-particle theory, as it stands, impossible.

Substituting (1.4.19, 20) into (1.4.17), we obtain the following eigenvectors:

for positive energy $E > 0$

$$u_s(\boldsymbol{p}) = a_0 \begin{pmatrix} \chi_s \\ \dfrac{\boldsymbol{\sigma} \cdot \boldsymbol{p}}{E + m}\chi_s \end{pmatrix} \tag{1.4.22}$$

and for negative energy $-E < 0$

$$E = +\sqrt{\boldsymbol{p}^2 + m^2} \tag{1.4.23}$$

$$\tilde{v}_s(\boldsymbol{p}) = \tilde{a}_0 \begin{pmatrix} \chi_s \\ \dfrac{\boldsymbol{\sigma} \cdot \boldsymbol{p}}{-E + m} \chi_s \end{pmatrix} . \tag{1.4.24}$$

Since $s = \pm 1$ (1.4.18), every eigenvalue E is two-fold degenerated. It is therefore possible to choose the eigenvectors in a different way. For this purpose, we introduce the helicity operator

$$h = \begin{pmatrix} \boldsymbol{\sigma} \cdot \boldsymbol{p} & 0 \\ 0 & \boldsymbol{\sigma} \cdot \boldsymbol{p} \end{pmatrix} , \tag{1.4.25}$$

which describes the spin component in the direction of the momentum. One easily verifies

$$[H_0, h] = 0 \quad \text{and} \tag{1.4.26}$$

$$(\boldsymbol{\sigma} \cdot \boldsymbol{p})^2 = \boldsymbol{p}^2 = E^2 - m^2 . \tag{1.4.27}$$

Then, new eigenvectors of H_0 are obtained by

$$v_s(-\boldsymbol{p}) = h \tilde{v}_s(\boldsymbol{p}) = \tilde{a}_0 \begin{pmatrix} \boldsymbol{\sigma} \cdot \boldsymbol{p} \chi_s \\ -(E + m) \chi_s \end{pmatrix} .$$

Choosing the new factor $a_0 = -\tilde{a}_0(E + m)$, one gets the eigenvectors

$$v_s(-\boldsymbol{p}) = a_0 \begin{pmatrix} -\dfrac{\boldsymbol{\sigma} \cdot \boldsymbol{p}}{E + m} \chi_s \\ \chi_s \end{pmatrix} \tag{1.4.28}$$

for negative energy

$$H_0 v_s(-\boldsymbol{p}) = -E v_s(-\boldsymbol{p}) \tag{1.4.29}$$

instead of (1.4.24). The vectors are normalized in $\mathbb{C}^4$

$$u_s^+(\boldsymbol{p}) u_{s'}(\boldsymbol{p}) = |a_0|^2 \left(1 + \frac{E^2 - m^2}{(E + m)^2} \right) \delta_{ss'} = \delta_{ss'}$$

$$v_s^+(-\boldsymbol{p}) v_{s'}(-\boldsymbol{p}) = \delta_{ss'} , \tag{1.4.30}$$

if we take

$$a_0 = \sqrt{\frac{E + m}{2E}} . \tag{1.4.31}$$

They are also orthogonal

$$u_s^+(\boldsymbol{p}) v_{s'}(-\boldsymbol{p}) = 0 . \tag{1.4.32}$$

The total eigenfunctions in x-space are given by

$$\psi_{s+}(\boldsymbol{p}, \boldsymbol{x}) = u_s(\boldsymbol{p}) e^{i\boldsymbol{p}\boldsymbol{x}} \quad , \quad \psi_{s-}(\boldsymbol{p}, \boldsymbol{x}) = v_s(-\boldsymbol{p}) e^{i\boldsymbol{p}\boldsymbol{x}} , \tag{1.4.33}$$

satisfying

$$H_0 \psi_{s+} = E \psi_{s+}$$

$$H_0\psi_{s-} = -E\psi_{s-} \quad , \quad s = \pm 1. \tag{1.4.34}$$

The general solution of the Dirac equation can now be written down

$$\Phi(t, \boldsymbol{x}) = (2\pi)^{-3/2} \int d^3p \left[\Phi_{s+}(\boldsymbol{p})u_s(\boldsymbol{p})e^{-i(Et-\boldsymbol{px})} + \Phi_{s-}(\boldsymbol{p})v_s(\boldsymbol{p})e^{i(Et-\boldsymbol{px})}\right]. \tag{1.4.35}$$

The corresponding initial wave packet at $t = 0$ is

$$\Phi(\boldsymbol{x}) = (2\pi)^{-3/2} \int d^3p \left[\Phi_{s+}(\boldsymbol{p})u_s(\boldsymbol{p})e^{i\boldsymbol{px}} + \Phi_{s-}(\boldsymbol{p})v_s(\boldsymbol{p})e^{-i\boldsymbol{px}}\right]. \tag{1.4.36}$$

This formula is the analog of the Fourier transformation for the Dirac equation. For later purpose, we want to invert it. Transforming $\boldsymbol{p}$ into $-\boldsymbol{p}$ in the last term in (1.4.36) and using the inverse theorem of ordinary Fourier transformation, we get

$$\Phi_{s+}(\boldsymbol{p})u_s(\boldsymbol{p}) + \Phi_{s-}(-\boldsymbol{p})v_s(-\boldsymbol{p}) = (2\pi)^{-3/2} \int d^3x\, \Phi(\boldsymbol{x})e^{-i\boldsymbol{px}} = \hat{\Phi}(\boldsymbol{p}), \tag{1.4.37}$$

where the hat always means the ordinary Fourier transform in L^2. This can be solved for the Dirac-Fourier transforms by means of the orthonormality (1.4.30, 32) of the spinors $u_s(\boldsymbol{p})$ and $v_s(-\boldsymbol{p})$:

$$\Phi_{s+}(\boldsymbol{p}) = u_s^+(\boldsymbol{p})\hat{\Phi}(\boldsymbol{p}) \quad , \quad \Phi_{s-}(\boldsymbol{p}) = v_s^+(\boldsymbol{p})\hat{\Phi}(-\boldsymbol{p}). \tag{1.4.38}$$

The unitarity of the transformation is expressed by

$$(\Phi, \Psi) = \sum_s \int d^3p\, [\Phi_{s+}(\boldsymbol{p})^*\Psi_{s+}(\boldsymbol{p}) + \Phi_{s-}(\boldsymbol{p})^*\Psi_{s-}(\boldsymbol{p})]. \tag{1.4.39}$$

The Dirac-Fourier transformation gives the spectral decomposition of H_0. Accordingly, the $\pm$ signs in the foregoing equations mean that the corresponding quantities belong to the positive $[m, +\infty)$ and negative $(-\infty, -m]$ part of the spectrum of H_0. The corresponding projection operators are

$$P_+(\boldsymbol{p}) = \sum_s u_s(\boldsymbol{p})u_s(\boldsymbol{p})^+ = \frac{E+m}{2E}\begin{pmatrix} 1 & \dfrac{\boldsymbol{\sigma}\cdot\boldsymbol{p}}{E+m} \\ \dfrac{\boldsymbol{\sigma}\cdot\boldsymbol{p}}{E+m} & \dfrac{E^2-m^2}{(E+m)^2} \end{pmatrix}$$

$$= \frac{1}{2E}(E + \boldsymbol{\alpha}\cdot\boldsymbol{p} + \beta m) \tag{1.4.40}$$

and similarly

$$P_-(\boldsymbol{p}) = \sum_s v_s(-\boldsymbol{p})v_s(-\boldsymbol{p})^+ = \frac{1}{2E}(E - \boldsymbol{\alpha}\cdot\boldsymbol{p} - \beta m). \tag{1.4.41}$$

They are orthogonal projection operators in $\mathbb{C}^4$, satisfying

$$P_+(\boldsymbol{p}) + P_-(\boldsymbol{p}) = \mathbf{1}$$

$$P_+(\boldsymbol{p})P_-(\boldsymbol{p}) = 0$$

$$P_\pm(\boldsymbol{p})^2 = P_\pm(\boldsymbol{p})$$

$$P_\pm(\boldsymbol{p})^+ = P_\pm(\boldsymbol{p}). \tag{1.4.42}$$

Multiplying (1.4.40) and (1.4.41) by γ^0, one gets the covariant projection operators

$$\sum_s u_s(\boldsymbol{p})\overline{u}_s(\boldsymbol{p}) = \frac{\not{p}+m}{2E} \tag{1.4.43}$$

and

$$\sum_s v_s(-\boldsymbol{p})\overline{v_s}(-\boldsymbol{p}) = \frac{1}{2E}(E\gamma^0 + \not{\boldsymbol{p}}\cdot\boldsymbol{\gamma} - m)$$

$$\sum_s v_s(\boldsymbol{p})\overline{v_s}(\boldsymbol{p}) = \frac{\not{p}-m}{2E}, \tag{1.4.44}$$

where the slash always means

$$\not{p} = p^0\gamma^0 - \boldsymbol{p}\cdot\boldsymbol{\gamma} = p_\mu\gamma^\mu. \tag{1.4.45}$$

Finally, we analyse the transformation properties of the solutions. For this purpose, it is again convenient to return to the chiral representation by the linear transformation

$$\tilde{\Phi}(x) = C\Phi(x) \tag{1.4.46}$$

with the matric C (1.3.9). Then the u-spinor in (1.4.36) becomes

$$\tilde{u}_s(\boldsymbol{p}) = \sqrt{\frac{E+m}{4E}}\begin{pmatrix} (1+\dfrac{\boldsymbol{\sigma}\cdot\boldsymbol{p}}{E+m})\chi_s \\ (1-\dfrac{\boldsymbol{\sigma}\cdot\boldsymbol{p}}{E+m})\chi_s \end{pmatrix}$$

$$= \begin{pmatrix} A & 0 \\ 0 & \varepsilon^T A^* \varepsilon \end{pmatrix}\sqrt{\frac{m}{2E}}\begin{pmatrix} \chi_s \\ \chi_s \end{pmatrix}, \tag{1.4.47}$$

with

$$A = \sqrt{\frac{E+m}{2m}}\left(1 + \frac{\boldsymbol{\sigma}\cdot\boldsymbol{p}}{E+m}\right). \tag{1.4.48}$$

This is just a Lorentz transformation (1.2.32, 33) with

$$A = \cosh\frac{\chi}{2} + \frac{\boldsymbol{\sigma}\cdot\boldsymbol{p}}{|\boldsymbol{p}|}\sinh\frac{\chi}{2} \tag{1.4.49}$$

$$\cosh\frac{\chi}{2} = \sqrt{\frac{E+m}{2m}}.$$

According to (1.1.19), this is a Lorentz boost in the direction $-\boldsymbol{p}$. It transforms the particle with momentum $\boldsymbol{p}$ at rest. In the rest system, we take $\boldsymbol{p} = 0$, $A = 1$, $E = m$ in (1.4.47):

$$\tilde{u}_s(\mathbf{0}) = \frac{1}{\sqrt{2}} \begin{pmatrix} \chi_s \\ \chi_s \end{pmatrix}. \tag{1.4.50}$$

Now we consider rotations $R(\varphi, \boldsymbol{n})$ (see (1.1.5))

$$U = \cos\frac{\varphi}{2} + i\boldsymbol{n} \cdot \boldsymbol{\sigma} \sin\frac{\varphi}{2} \in SU(2). \tag{1.4.51}$$

The spinor (1.4.50) is transformed as follows

$$R\tilde{u}_s(\mathbf{0}) = \frac{1}{\sqrt{2}} \begin{pmatrix} U\chi_s \\ \varepsilon^T U^* \varepsilon \chi_s \end{pmatrix} = \frac{1}{\sqrt{2}} \begin{pmatrix} U\chi_s \\ U\chi_s \end{pmatrix}. \tag{1.4.52}$$

This is obviously the representation $D^{1/2} \oplus D^{1/2}$ of $SO(3)$. Since the spin of a particle per definition is the angular momentum in the rest frame, the particles described by the Dirac equation have spin $\frac{1}{2}$. The index $s = \pm 1$ in the forgoing equations is therefore the spin index. Similarly, one gets for the v-spinor

$$\begin{aligned} \tilde{v}_s(-\boldsymbol{p}) &= \sqrt{\frac{E+m}{4E}} \begin{pmatrix} (1 - \dfrac{\boldsymbol{\sigma} \cdot \boldsymbol{p}}{E+m})\chi_s \\ -(1 + \dfrac{\boldsymbol{\sigma} \cdot \boldsymbol{p}}{E+m})\chi_s \end{pmatrix} \\ &= S(-\boldsymbol{p}) \sqrt{\frac{m}{2E}} \begin{pmatrix} \chi_s \\ -\chi_s \end{pmatrix}, \end{aligned} \tag{1.4.53}$$

where S is the same matrix as in (1.4.47) with only $-\boldsymbol{p}$ instead of $\boldsymbol{p}$ in (1.4.48).

Finally we discuss the Poincaré covariance of solutions of the free Dirac equation in x-space. A solution $\Phi(x)$ (1.4.35) transforms under a transformation

$$x \longrightarrow x' = \Lambda x + a \tag{1.4.54}$$

as follows

$$\Phi(x) \longrightarrow \Phi'(x') = S(\Lambda)\Phi(\Lambda^{-1}(x' - a)), \tag{1.4.55}$$

because of (1.3.13). We wish to show that this is a unitary transformation in $\mathcal{H}$ with scalar product (1.4.8)

$$(\Phi_1, \Phi_2) = \int\limits_{t=\text{const}} d^3x\, \Phi_1(t, \boldsymbol{x})^+ \Phi_2(t, \boldsymbol{x}) = \int\limits_{t=0} d^3x\, \overline{\Phi_1}(x)\gamma_0\Phi_2(x). \tag{1.4.56}$$

To do so, we rewrite (1.4.56) in an invariant form by introducing the time-like unit vector

$$(n^\mu) = (1, 0, 0, 0) \tag{1.4.57}$$

and the surface element

$$d\sigma^\mu = n^\mu\, d^3x \quad \text{on the plain} \quad P : n \cdot x = 0. \tag{1.4.58}$$

Then we have

$$(\Phi_1, \Phi_2) = \int_P d\sigma^\mu(x) \overline{\Phi}_1(x)\gamma_\mu\Phi_2(x). \tag{1.4.59}$$

Let now Q be an arbitrary space-like plain and assume that Φ_1, Φ_2 are decreasing sufficiently rapidly to 0 for $|\boldsymbol{x}| \to \infty$. Then according to Gauss' theorem, we find

$$\left(\int_P - \int_Q\right) d\sigma^\mu(x)\overline{\Phi}_1(x)\gamma_\mu\Phi_2(x) = \int d^4x\partial^\mu(\overline{\Phi}_1\gamma_\mu\Phi_2) = 0,$$

where the last integrand vanishes since Φ_2 is a solution of the Dirac equation and $\overline{\Phi}_1$ satisfies the adjoint Dirac equation (1.3.31). Accordingly, the scalar product (1.4.56) assumes the following invariant form

$$(\Phi_1, \Phi_2) = \int_Q d\sigma^\mu(x)\overline{\Phi}_1(x)\gamma_\mu\Phi_2(x), \tag{1.4.60}$$

where the integral can be taken over an arbitrary space-like plain or even a smooth space-like surface in virtue of Gauss' theorem.

Let us now return to the Poincaré transformation (1.4.54, 55). The scalar product between the transformed spinors is

$$(\Phi_1', \Phi_2') = \int_{Q'} d\sigma^\mu(x')\overline{\Phi_1'}\gamma_\mu\Phi_2'(x')$$

$$= \int_{Q'} d\sigma^\mu(x')\overline{\Phi}_1(x)S(\Lambda)^{-1}\gamma_\mu S(\Lambda)\Phi_2(x). \tag{1.4.61}$$

Using (1.2.34) and the Lorentz covariance and translation invariance of the surface element

$$d\sigma^\mu(x')\Lambda_\mu{}^\nu = d\sigma^\nu(\Lambda^{-1}x') = d\sigma^\nu(x),$$

(1.4.61) can be written as follows

$$= \int_Q d\sigma^\nu(x)\overline{\Phi}_1(x)\gamma_\nu\Phi_2(x) = (\Phi_1, \Phi_2). \tag{1.4.62}$$

We have thus obtained a unitary representation of the Poincaré group $\mathcal{P}_+^\uparrow$

$$(U(a, \Lambda)\Phi)(x) = S(\Lambda)\Phi(\Lambda^{-1}(x-a)) \tag{1.4.63}$$

in the one-particle Hilbert space.

For later use, we discuss the so-called Gordon decomposition. Multiplying the equation

$$H_0(\boldsymbol{p})u_s(\boldsymbol{p}) = (\beta m + \boldsymbol{\alpha}\cdot\boldsymbol{p})u_s(\boldsymbol{p}) = E(\boldsymbol{p})u_s(\boldsymbol{p}) \tag{1.4.64}$$

by γ^0, one obviously has

$$m\,u_s(p) = (E\gamma^0 - \boldsymbol{\gamma}\cdot\boldsymbol{p})u_s(p) = \not{p}u_s(p), \tag{1.4.65}$$

and, similarly

$$\overline{u}_s(p)\not{p} = m\,\overline{u}_s(p). \tag{1.4.66}$$

Here p^0 is given by $p^0 = E(\boldsymbol{p})$. It is our aim to decompose the expression

$$\overline{u}_s(p)\gamma^\mu u_\sigma(q) = \frac{1}{2m}[\overline{u}_s(p)\not{p}\gamma^\mu u_\sigma(q) + \overline{u}_s(p)\gamma^\mu \not{q} u(q)]. \tag{1.4.67}$$

Using

$$\begin{aligned}\not{p}\gamma^\mu + \gamma^\mu \not{q} &= (p_\nu - q_\nu)\gamma^\nu\gamma^\mu + 2q^\mu \\ &= (q_\nu - p_\nu)\gamma^\mu\gamma^\nu + 2p^\mu\end{aligned}$$

and forming half the sum of the right-hand sides, one may write (1.4.67) in the form

$$\overline{u}_s(p)\gamma^\mu u_\sigma(q) = \frac{1}{2m}\overline{u}_s(p)[p^\mu + q^\mu + i\sigma^{\mu\nu}(p_\nu - q_\nu)]u_\sigma(q), \tag{1.4.68}$$

which is the desired decomposition. Here

$$\begin{aligned} i\sigma^{\mu\nu} &= \frac{1}{2}(\gamma^\nu\gamma^\mu - \gamma^\mu\gamma^\nu) \quad \text{or} \\ \sigma^{\mu\nu} &= \frac{i}{2}[\gamma^\mu, \gamma^\nu], \end{aligned} \tag{1.4.69}$$

which agrees with (1.3.23).

1.5 Gauge Invariance and Electromagnetic Fields

The Dirac equation

$$i\hbar\gamma^\mu\partial_\mu\psi(x) = mc\psi(x) \tag{1.5.1}$$

is also invariant under phase transformations

$$\psi(x) \longrightarrow e^{i\Lambda}\psi(x) \quad , \quad \Lambda \quad \text{real}. \tag{1.5.2}$$

Since the phase $e^{i\Lambda}$ can be viewed as an element of the unitary group $U(1)$, this symmetry is called global $U(1)$ gauge invariance and the transformation (1.5.2) is a gauge transformation of the first kind. We now consider local gauge transformations or gauge transformations of the second kind, taking Λ depending on x

$$\psi(x) \longrightarrow e^{i\Lambda(x)}\psi(x) = \psi'(x). \tag{1.5.3}$$

This transformation does not leave (1.5.1) invariant because

$$i\hbar\partial_\mu\psi'(x) = e^{i\Lambda(x)}i\hbar\partial_\mu\psi(x) - \hbar\partial_\mu\Lambda(x)\psi'(x). \tag{1.5.4}$$

It is the basic idea of gauge theories to compensate the unwanted term in (1.5.4) by introducing an additional field, called a gauge field.

It is known from classical electrodynamics that the electromagnetic potentials $V, \boldsymbol{A}$ can be transformed as follows

$$\boldsymbol{A}(x) \longrightarrow \boldsymbol{A} + \text{grad}\chi$$

$$V(x) \longrightarrow V - \frac{1}{c}\frac{\partial \chi}{\partial t}, \tag{1.5.5}$$

however, representing the same electromagnetic field

$$\boldsymbol{B} = \operatorname{curl} \boldsymbol{A}$$

$$\boldsymbol{E} = -\operatorname{grad} V - \frac{1}{c}\frac{\partial \boldsymbol{A}}{\partial t}. \tag{1.5.6}$$

In relativistic notation $A^\mu = (V, \boldsymbol{A})$, the gauge transformation (1.5.5) reads

$$A^\mu \longrightarrow A^\mu - \partial^\mu \chi.$$

The last term in here has exactly the same form as the additional term in (1.5.4). Consequently, the equation

$$\gamma^\mu \Big(i\hbar\partial_\mu - \frac{e}{c}A_\mu(x)\Big)\psi(x) = mc\psi(x) \tag{1.5.7}$$

is invariant under local gauge transformations, if the electromagnetic potential transforms as follows

$$A_\mu(x) \longrightarrow A'_\mu(x) = A_\mu(x) - \frac{\hbar c}{e}\partial_\mu \Lambda(x). \tag{1.5.8}$$

Equation (1.5.7) is the Dirac equation in the presence of electromagnetic fields, e is the charge of the particle, for electrons we have $e = -|e|$. The particular combination acting on ψ on the left of (1.5.7) is called a covariant derivative, it transforms simply under gauge transformations (1.5.3, 8) and defines the so-called minimal coupling.

It is our aim now to derive the Dirac equation (1.5.7) from a Lagrange variational principle

$$\int d^4x L(x) = \text{stationary}. \tag{1.5.9}$$

The Lagrange density $L(x)$ must be a Lorentz scalar and gauge invariant. The right expression is

$$L(\psi, \overline{\psi}, A_\mu) = \overline{\psi}\Big(\gamma^\mu\Big(i\hbar\partial_\mu - \frac{e}{c}A_\mu\Big) - mc\Big)\psi = L(\psi', \overline{\psi}', A'_\mu). \tag{1.5.10}$$

In fact, as in classical mechanics, the variational principle (1.5.9) implies the Euler-Lagrange equations

$$\frac{\partial L}{\partial \overline{\psi}} - \frac{\partial}{\partial x^\mu}\frac{\partial L}{\partial \overline{\psi}_{,\mu}} = 0, \tag{1.5.11}$$

and similarly for ψ. The second term in (1.5.11) is zero and the first one gives just the Dirac equation (1.5.7).

It is well known from mechanics that symmetries of the Lagrangian $L(x)$ imply conservation laws (Noether's theorem). From this point of view, we consider the *global* gauge invariance (1.5.2). For any gauge invariant Lagrange density L depending on ψ, $\overline{\psi}$ and their first derivatives, we have

$$\begin{aligned} 0 &= \frac{\partial}{\partial \Lambda} L\Big(\psi e^{i\Lambda}, \overline{\psi} e^{-i\Lambda}, \dots\Big)\Big|_{\Lambda=0} \\ &= i\Big(\frac{\partial L}{\partial \psi}\psi - \overline{\psi}\frac{\partial L}{\partial \overline{\psi}} + \frac{\partial L}{\partial \psi_{,\mu}}\psi_{,\mu} - \overline{\psi}_{,\mu}\frac{\partial L}{\partial \overline{\psi}_{,\mu}}\Big). \end{aligned} \tag{1.5.12}$$

Using the Euler-Lagrange equations (1.5.11), this leads to

$$\frac{\partial}{\partial x^\mu}\left(\frac{\partial L}{\partial \psi_{,\mu}}\psi - \overline{\psi}\frac{\partial L}{\partial \overline{\psi}_{,\mu}}\right) = 0. \tag{1.5.13}$$

Since the second term vanishes for the Dirac Lagrangian (1.5.10), we arrive at the continuity equation

$$\partial_\mu \overline{\psi}\gamma^\mu \psi = 0. \tag{1.5.14}$$

It is not hard to derive this conservation law directly from the Dirac equation and its adjoint. The Eq. (1.5.14) means that

$$j^\mu(x) = e\overline{\psi}\gamma^\mu\psi \tag{1.5.15}$$

is a conserved vector field (Noether current). The 0-component

$$j^0(x) = e\psi^+\psi \ge 0 \tag{1.5.16}$$

is non-negative if $e > 0$. Nevertheless, it would be hasty to interpret it as a probability density as in non-relativistic quantum mechanics.

In order to find the interpretation of $j^\mu(x)$, we consider the variation of L (1.5.10) with respect to A_μ

$$\frac{\partial L}{\partial A_\mu} = -\frac{e}{c}\overline{\psi}\gamma^\mu\psi = -\frac{1}{c}j^\mu. \tag{1.5.17}$$

If the electromagnetic potential is not a given external field but also a dynamical quantity, then the variation of the total Lagrangian

$$L_{\text{total}} = L(\psi, \overline{\psi}, A_\mu) + L_{\text{field}}(A_\mu) \tag{1.5.18}$$

with respect to A_μ must give Maxwell's equations for the electromagnetic fields. Then, j^μ in (1.5.17) must be the source of the electromagnetic field, that is the charge-current density

$$j^\mu(x) = (c\varrho, \boldsymbol{j}). \tag{1.5.19}$$

The continuity equation (1.5.14)

$$\frac{\partial \varrho}{\partial t} + \operatorname{div}\boldsymbol{j} = 0 \tag{1.5.20}$$

now can be interpreted as charge conservation. For current densities $\boldsymbol{j}$ vanishing at infinity, we get by integration over $\mathbb{R}^3$

$$\frac{\partial}{\partial t}\int d^3x\varrho(t,\boldsymbol{x}) = 0, \tag{1.5.21}$$

using Gauss' theorem. Accordingly, the total charge

$$Q = \int d^3x\varrho(t,\boldsymbol{x}) = e\int d^3x\psi(t,\boldsymbol{x})^+\psi(t,\boldsymbol{x}) \tag{1.5.22}$$

is constant in time.

We now write the Dirac equation with potentials in hamiltonian form, as we did for the free equation at the beginning of the last section (1.4.6). Multiplying

$$\gamma^0\Big(i\hbar\partial_0 - \frac{e}{c}V\Big)\psi + \gamma^j\Big(i\hbar\partial_j - \frac{e}{c}A_j\Big)\psi = mc\psi \tag{1.5.23}$$

by $c\gamma^0$ and using the matrices (1.4.2), we arrive at

$$i\hbar\frac{\partial\psi}{\partial t} = \Big[mc^2\beta - c\boldsymbol{\alpha}\Big(i\hbar\frac{\partial}{\partial\boldsymbol{x}} + \frac{e}{c}\boldsymbol{A}\Big) + eV\Big]\psi \stackrel{\text{def}}{=} H\psi. \tag{1.5.24}$$

We write this in the standard representation

$$\beta = \begin{pmatrix} \mathbf{1} & 0 \\ 0 & -\mathbf{1} \end{pmatrix} \quad , \quad \boldsymbol{\alpha} = \begin{pmatrix} o & \boldsymbol{\sigma} \\ \boldsymbol{\sigma} & 0 \end{pmatrix}.$$

Introducing

$$\psi = \begin{pmatrix} \varphi \\ \chi \end{pmatrix} \exp -\frac{i}{\hbar}mc^2t \quad , \quad \boldsymbol{p} = \frac{\hbar}{i}\frac{\partial}{\partial\boldsymbol{x}}, \tag{1.5.25}$$

we get two coupled 2-dimensional equations

$$i\hbar\frac{\partial\varphi}{\partial t} = eV\varphi + c\boldsymbol{\sigma}\cdot\Big(\boldsymbol{p} - \frac{e}{c}\boldsymbol{A}\Big)\chi \tag{1.5.26}$$

$$i\hbar\frac{\partial\chi}{\partial t} = (eV - 2mc^2)\chi + c\boldsymbol{\sigma}\cdot\Big(\boldsymbol{p} - \frac{e}{c}\boldsymbol{A}\Big)\varphi. \tag{1.5.27}$$

We consider the non-relativistic limit of the Dirac equations (1.5.26, 27) in some detail to justify the title "relativistic quantum mechanics" of this part. In the limit $c \to \infty$, the leading terms in powers of c are the last two on the right of (1.5.27), leading to

$$\chi = \frac{1}{2mc}\boldsymbol{\sigma}\cdot\Big(\boldsymbol{p} - \frac{e}{c}\boldsymbol{A}\Big)\varphi + O(c^{-2}). \tag{1.5.28}$$

$\chi = O(c^{-1})$ is the so-called small component which is substituted back into (1.5.26). Then the following equation is obtained for the large component φ:

$$i\hbar\frac{\partial\varphi}{\partial t} = eV\varphi + \frac{1}{2m}\Big[\boldsymbol{\sigma}\cdot\Big(\boldsymbol{p}-\frac{e}{c}\boldsymbol{A}\Big)\Big]^2\varphi. \tag{1.5.29}$$

We compute the square bracket

$$\Big[\boldsymbol{\sigma}\cdot\Big(\boldsymbol{p}-\frac{e}{c}\boldsymbol{A}\Big)\Big]^2 = \Big(\boldsymbol{p}-\frac{e}{c}\boldsymbol{A}\Big)^2 + i\boldsymbol{\sigma}\cdot\Big[\Big(\boldsymbol{p}-\frac{e}{c}\boldsymbol{A}\Big)\wedge\Big(\boldsymbol{p}-\frac{e}{c}\boldsymbol{A}\Big)\Big].$$

Note that the last term does not vanish, because $\boldsymbol{p}$ is the differential operator (1.5.25),

$$= \Big(\boldsymbol{p}-\frac{e}{c}\boldsymbol{A}\Big)^2 + i\boldsymbol{\sigma}\cdot\Big[-\frac{\hbar}{i}\Big(\frac{\partial}{\partial\boldsymbol{x}}\wedge\frac{e}{c}\boldsymbol{A}\Big)\Big]$$

$$= \Big(\boldsymbol{p}-\frac{e}{c}\boldsymbol{A}\Big)^2 - \frac{e\hbar}{c}\boldsymbol{\sigma}\cdot\boldsymbol{B} \quad , \quad \boldsymbol{B} = \mathrm{curl}\,\boldsymbol{A}.$$

Using this in (1.5.29), one obtains the Pauli spin equation

$$i\hbar\frac{\partial\varphi}{\partial t} = \frac{1}{2m}\Big(\boldsymbol{p}-\frac{e}{c}\boldsymbol{A}\Big)^2\varphi + eV\varphi - \frac{e\hbar}{2mc}\boldsymbol{\sigma}\cdot\boldsymbol{B}\varphi. \tag{1.5.30}$$

This is just the Schrödinger equation with an additional magnetic term. We find that due to the spin, the particle has acquired a magnetic moment

$$\boldsymbol{\mu} = \mu_B\boldsymbol{\sigma} \quad , \quad \mu_B = \frac{e\hbar}{2mc}. \tag{1.5.31}$$

The value (1.5.31) for μ_B (Bohr's magneton) is correct for leptons (electron, myon etc.), up to small corrections (cf. Sect. 3.9 of Chap. 3). This was a first triumph of the Dirac theory.

In order to calculate the next relativistic correction, we consider the time independent equation

$$eV\varphi + cP\chi = E\varphi \tag{1.5.32}$$

$$(eV - 2mc^2)\chi + cP\varphi = E\chi, \tag{1.5.33}$$

where

$$P = \boldsymbol{\sigma}\cdot\Big(\boldsymbol{p}-\frac{e}{c}\boldsymbol{A}\Big). \tag{1.5.34}$$

As before, we compute the small component

$$\chi = \frac{c}{2mc^2 + E - eV}P\varphi$$

$$= \frac{1}{2mc}\Big(1 - \frac{E-eV}{2mc^2} + \ldots\Big)P\varphi \tag{1.5.35}$$

and substitute this into (1.5.32)

$$E\varphi = eV\varphi + \frac{1}{2m}P\Big(1 - \frac{E-eV}{2mc^2}\Big)P\varphi$$

$$= eV\varphi + \frac{1}{2m}P^2\varphi - \frac{E}{4m^2c^2}P^2\varphi + \frac{e}{4m^2c^2}PVP\varphi,$$

or

$$E\left(1+\frac{1}{4m^2c^2}P^2\right)\varphi = eV\varphi + \frac{1}{2m}P^2\varphi + \frac{e}{4m^2c^2}PVP\varphi. \tag{1.5.36}$$

Let us abbreviate this equation by

$$EB\varphi = H_1\varphi. \tag{1.5.37}$$

Here B and H_1 are hermitian operators, furthermore, B is positive such that the square root $B^{1/2}$ exists. Then (1.5.37) can be written as an ordinary hermitian eigenvalue equation

$$EB^{1/2}\varphi = B^{-1/2}H_1B^{-1/2}B^{1/2}\varphi \stackrel{\text{def}}{=} H\varphi_S, \tag{1.5.38}$$

where

$$\varphi_S = B^{1/2}\varphi \tag{1.5.39}$$

is the Schrödinger wave function. Using

$$B^{-1/2} = 1 - \frac{1}{8m^2c^2}P^2 + O(c^{-4}),$$

we can compute the hermitian Hamiltonian

$$H = B^{-1/2}H_1B^{-1/2} = \left(1-\frac{1}{8m^2c^2}P^2\right)eV\left(1-\frac{1}{8m^2c^2}P^2\right)+\frac{1}{2m}P^2-\frac{1}{8m^3c^2}P^4$$

$$+\frac{e}{4m^2c^2}PVP = \frac{1}{2m}P^2 + eV - \frac{1}{8m^3c^2}P^4 - \frac{e}{8m^2c^2}D \tag{1.5.40}$$

where D is given in terms of commutators:

$$D = P[P, V] - [P, V]P.$$

With

$$[P, V] = \boldsymbol{\sigma}\cdot\frac{\hbar}{i}\frac{\partial V}{\partial \boldsymbol{x}} = i\hbar\boldsymbol{\sigma}\cdot\boldsymbol{E},$$

we finally obtain

$$D = i\hbar\left\{\frac{\hbar}{i}\operatorname{div}\boldsymbol{E} + \hbar\boldsymbol{\sigma}\cdot\operatorname{curl}\boldsymbol{E} - 2i\boldsymbol{\sigma}\cdot\left(\boldsymbol{E}\wedge\left(\boldsymbol{p}-\frac{e}{c}\boldsymbol{A}\right)\right)\right\}. \tag{1.5.41}$$

The first term in (1.5.41), the so-called Darwin term, is proportional to the charge density $\varrho(\boldsymbol{x})$ which generates the electric field $\boldsymbol{E}(\boldsymbol{x})$. In the case of a point charge, this term is proportional to a δ-distribution and affects only the $L = 0$ states. The second term vanishes for static fields. The last term is the spin-orbit coupling, since for a radial field $\boldsymbol{E} \sim \boldsymbol{x}$ the angular momentum operator $\boldsymbol{L} = \boldsymbol{x}\wedge\boldsymbol{p}$ appears.

We now study a further symmetry of the Dirac equation, the so-called charge conjugation. Let us take the complex conjugate of

$$\gamma^\mu(i\hbar\partial_\mu - eA_\mu)\psi(x) = m\psi(x), \tag{1.5.42}$$

assuming $c = 1$ and remembering that A_μ is real,

$$-\gamma^{\mu *}(i\hbar\partial_\mu + eA_\mu)\psi^* = m\psi^*. \tag{1.5.43}$$

In the chiral representation (1.2.28), we have

$$\gamma^{0*} = \gamma^0 = \gamma^{0T}$$

$$\gamma^{k*} = \begin{pmatrix} 0 & -\sigma_k^* \\ \sigma_k^* & 0 \end{pmatrix} = \begin{pmatrix} 0 & -\sigma_k^T \\ \sigma_k^T & 0 \end{pmatrix} = \begin{pmatrix} 0 & \sigma_k \\ -\sigma_k & 0 \end{pmatrix}^T = -\gamma^{kT}. \tag{1.5.44}$$

These relations remain valid in all unitary equivalent representations:

$$(U\gamma^\mu U^+)^* = U^{T+}\gamma^{\mu *}U^T = \pm U^{+T}\gamma^{\mu T}U^T = \pm(U\gamma^\mu U^+)^T,$$

where the upper sign corresponds to $\mu = 0$, the lower one to $\mu = 1, 2, 3$.

The matrices $-\gamma^{\mu T}$ obey the same algebraic relations as γ^μ, because they satisfy the fundamental anticommutation relations (1.3.1). In addition, they have the same properties under hermitian conjugation. According to the theorem of Pauli (1.3.7), there exists a unitary matrix C such that

$$\gamma^{\mu T} = -C^{-1}\gamma^\mu C. \tag{1.5.45}$$

Taking the transposed equation

$$\gamma^\mu = -C^T\gamma^{\mu T}C^{T-1} = C^T C^{-1}\gamma^\mu C C^{T-1},$$

we get

$$\gamma^\mu C^T C^{-1} = C^T C^{-1}\gamma^\mu \quad , \quad \mu = 0, 1, 2, 3.$$

Since the representation of the γ's is irreducible, it follows

$$C^T C^{-1} = \alpha \mathbb{1}$$

or

$$C = \alpha C^T. \tag{1.5.46}$$

Transposing again, one finds $\alpha^2 = 1$ and

$$\alpha = \pm 1. \tag{1.5.47}$$

Suppose $\alpha = 1$, then C is a symmetric matrix. Consequently

$$(C^{-1}\gamma^\mu)^T = \gamma^{\mu T}C^{T-1} = \gamma^{\mu T}C^{-1} = -C^{-1}\gamma^\mu \tag{1.5.48}$$

is antisymmetric. The matrices

$$C^{-1}\gamma^\mu\gamma^\nu = C^{-1}\gamma^\mu C C^{-1}\gamma^\nu = \gamma^{\mu T}\gamma^{\nu T}C^{-1} \tag{1.5.49}$$

are also antisymmetric:

$$(C^{-1}\gamma^\mu\gamma^\nu)^T = C^{-1}\gamma^\nu\gamma^\mu = -C^{-1}\gamma^\mu\gamma^\nu \quad , \quad \mu > \nu.$$

Altogether we have constructed 10 antisymmetric matrices. This is impossible, because there are only 6 linear independent antisymmetric 4×4 matrices. Therefore, we must have $\alpha = -1$ in (1.5.46)

$$C^T = -C. \tag{1.5.50}$$

Now we return to (1.5.43) and multiply by $C\gamma^0$ from the left. Since by (1.5.44)

$$C\gamma^0\gamma^{\mu *} = C\gamma^{\mu T}\gamma^0 = -\gamma^\mu C\gamma^0,$$

we obtain

$$\gamma^\mu(i\hbar\partial_\mu + eA_\mu)C\gamma^0\psi^* = mC\gamma^0\psi^*. \tag{1.5.51}$$

This is the Dirac equation for the wave function

$$\psi_C = U_C\psi \stackrel{\text{def}}{=} C\gamma^0\psi^*, \tag{1.5.52}$$

but with a different sign of the electric charge e. One therefore calls (1.5.52) the charge-conjugate spinor and

$$U_C = C\gamma^0 K \tag{1.5.53}$$

the operator of charge conjugation. K denotes complex conjugation. U_C is an anti-unitary operator

$$(U_C f, U_C g) = (Kf, Kg) = (f, g)^*, \tag{1.5.54}$$

and it is an involution

$$U_C^2 = \mathbb{1}. \tag{1.5.55}$$

Applying charge conjugation U_C to the hamiltonian form of the Dirac equation

$$i\hbar\partial_t\psi = H(e)\psi,$$

one finds

$$-i\hbar\partial_t\psi_C = U_C H(e)U_C^{-1}\psi_C = -H(-e)\psi_C, \tag{1.5.56}$$

where (1.5.51) has been used. This implies

$$U_C H(e)U_C^{-1} = -H(-e). \tag{1.5.57}$$

Accordingly, the expectation values of the two hamiltonian operators in (1.5.57) are related as follows

$$\begin{aligned}(\psi_C, H(-e)\psi_C) &= -(\psi_C, U_C H(e)U_C^{-1}\psi_C) \\ &= -(\psi, H(e)\psi)^* = -(\psi, H(e)\psi).\end{aligned}$$

Since the Dirac operator with potential is unbounded from above, it is also unbounded from below. This unpleasant feature is therefore a necessary consequence of charge symmetry (C-invariance).

It is now easy to show the C-invariance of the S-matrix. Starting from the expression

$$S(e) = U_0(0, +\infty)U(+\infty, -\infty)U_0(-\infty, 0),$$

where $\pm\infty$ stands for the corresponding strong limits (see (0.3.7)) and taking

$$U(t, -t) = e^{-2itH(e)}$$

$$U_C U(t, -t) U_C^{-1} = e^{-2itH(-e)}$$

into account, one arrives at

$$U_C S(e) U_C^{-1} = S(-e). \tag{1.5.58}$$

Obviously, this equation also holds for time-dependent electromagnetic potentials.

In addition, we consider the gauge invariance of the S-matrix. Let us perform a gauge transformation (1.5.3, 8)

$$A_\mu(x) \longrightarrow A'_\mu(x) = A_\mu(x) - \frac{\hbar}{e}\partial_\mu \Lambda(x)$$

$$\psi(x) \longrightarrow \psi'(x) = e^{i\Lambda(x)}\psi(x) \tag{1.5.59}$$

with $\Lambda(x) \to 0$ for $t \to \pm\infty$. Then it follows

$$\psi'(x) = e^{i\Lambda(t,\boldsymbol{x})}U(t, -t)\psi(-t, \boldsymbol{x}')$$

$$= e^{i\Lambda(t,\boldsymbol{x})}U(t, -t)e^{-i\Lambda(-t,\boldsymbol{x}')}\psi'(-t, \boldsymbol{x}').$$

Accordingly, the transformed time evolution operator is

$$U'(t, -t) = e^{i\Lambda(t,\boldsymbol{x})}U(t, -t)e^{-i\Lambda(t,\boldsymbol{x}')}.$$

Since the phase factors vanish for $t \to \pm\infty$, one obtains

$$S[A'] = \text{s}-\lim_{t\to\infty} U_0(0, t)U'(t, -t)U_0(-t, 0) = S[A], \tag{1.5.60}$$

which is the gauge invariance of the S-matrix. The conclusion remains true if $\Lambda(x)$ goes to one and the same constant for $t \to -\infty$ and $t \to \infty$.

For the sake of completeness we finally discuss the remaining discrete symmetries of the Dirac equation with electromagnetic potentials, namely time reversal T and parity P. It follows from classical electrodynamics (*G. Scharf, From Electrostatics to Optics, Springer-Verlag 1994*) that the electric and magnetic fields transform as follows

$$T: \quad \boldsymbol{E}(t, \boldsymbol{x}) \longrightarrow \boldsymbol{E}(-t, \boldsymbol{x}), \quad \boldsymbol{B}(t, \boldsymbol{x}) \longrightarrow -\boldsymbol{B}(-t, \boldsymbol{x}) \tag{1.5.61}$$

$$P: \quad \boldsymbol{E}(t, \boldsymbol{x}) \longrightarrow -\boldsymbol{E}(t, -\boldsymbol{x}), \quad \boldsymbol{B}(t, \boldsymbol{x}) \longrightarrow \boldsymbol{B}(t, -\boldsymbol{x}). \tag{1.5.62}$$

This implies the following transformation properties of the potentials (1.5.6):

$$T: \quad V(t,\boldsymbol{x}) \longrightarrow V(-t,\boldsymbol{x}), \quad \boldsymbol{A}(t,\boldsymbol{x}) \longrightarrow -\boldsymbol{A}(-t,\boldsymbol{x})$$

$$P: \quad V(t,\boldsymbol{x}) \longrightarrow V(t,-\boldsymbol{x}), \quad \boldsymbol{A}(t,\boldsymbol{x}) \longrightarrow -\boldsymbol{A}(t,-\boldsymbol{x}). \tag{1.5.63}$$

Note that the four-potential transforms contrary to the space-time four-vector under time reversal. This forces us to use complex conjugation again to get a symmetry transformation of the Dirac equation with potentials. Since the zeroth component A^0 remains unchanged and the spatial components A^k change sign, we can use (1.5.44) to compensate this. Therefore, we write the Dirac operator in (1.5.43) in terms of time-reversed quantities and take (1.5.44) into account. Denoting the transformed quantities by primes, we have

$$\gamma^{\mu T}(i\hbar\partial'_\mu - eA'_\mu)\psi^* = m\psi^*. \tag{1.5.64}$$

By (1.5.44) this is equivalent to

$$-\gamma^\mu(i\hbar\partial'_\mu - eA'_\mu)C\psi^* = mC\psi^*.$$

To recover the original form (1.5.42) of the Dirac equation, the minus sign must disappear. This can be achieved by multiplication with γ^5:

$$\gamma^\mu(i\hbar\partial'_\mu - eA'_\mu)\gamma^5 C\psi^* = m\gamma^5 C\psi^* \tag{1.5.65}$$

This leads to the following form of the time-reversed Dirac spinor

$$\psi_T(Tx) = \gamma^5 C\psi^*(x), \quad \psi_T(x) = \gamma^5 C\psi^*(T^{-1}x). \tag{1.5.66}$$

The calculation for space reflection P is simpler. Denoting the space-reflected quantities by primes again, we now get from (1.5.42)

$$\gamma^0(i\hbar\partial_0 - eA_0)\psi - \gamma^k(i\hbar\partial'_k - eA'_k)\psi = m\psi. \tag{1.5.67}$$

Multiplication by γ^0 brings us back to the original Dirac equation

$$\gamma^\mu(i\hbar\partial'_\mu - eA'_\mu)\gamma^0\psi = m\gamma^0\psi. \tag{1.5.68}$$

This yields the following form of the space-reflected Dirac spinor

$$\psi_P(Px) = \gamma^0\psi(x), \quad \psi_P(x) = \gamma^0\psi(P^{-1}x), \tag{1.5.69}$$

in agreement with (1.2.36).

To formulate the corresponding symmetries for the S-matrix, we introduce the anti-unitary operators

$$U_P = \gamma^0, \quad U_T = \gamma^5 CK. \tag{1.5.70}$$

Then parity invariance of the time-evolution

$$U_P U(t,-t) U_P^{-1} = U'(t,-t)$$

immediately implies

$$U_P S[A] U_P^{-1} = S[A']. \tag{1.5.71}$$

But for time-reversal we obtain

$$U_T U(t,-t) U_T^{-1} = U'(-t,t),$$

so that

$$U_T S[A] U_T^{-1} = U_0(0,-\infty) U'(-\infty,+\infty) U_0(\infty,0) = S^{-1}[A'] = S^{+}[A']. \tag{1.5.72}$$

The S-matrix elements then satisfy

$$(\Phi_f, S[A]\Phi_i) = (U_T S[A']\Phi_i, U_T\Phi_f) = (S^{+}[A']U_T\Phi_i, U_T\Phi_f)$$

$$= (U_T\Phi_i, S[A']U_T\Phi_f), \tag{1.5.73}$$

which expresses the interchange of initial and final states under time reversal.

Note that the discrete symmetry transformations do not commute:

$$U_C U_P = -U_P U_C, \quad U_T U_P = -U_P U_T, \quad U_C U_T = -U_T U_C. \tag{1.5.74}$$

This is not a defect in the one-particle theory, because only bilinear expressions have a physical meaning.

1.6 The Hydrogen Atom

The most important application of the Dirac equation with potential is the calculation of the bound states in an attractive Coulomb potential, the spectrum of the hydrogen atom. As in non-relativistic quantum mechanics, the Coulomb problem for the Dirac equation can be solved exactly. In fact, there is an exact correspondence between the two problems.

First of all we separate the angular dependence using spinor spherical harmonics. Let us consider the addition of angular momentum and spin

$$\boldsymbol{j} = \boldsymbol{L} + \boldsymbol{s}, \quad \boldsymbol{s} = \tfrac{1}{2}\boldsymbol{\sigma}, \tag{1.6.1}$$

where $\boldsymbol{\sigma} = (\sigma_1, \sigma_2, \sigma_3)$ are the Pauli matrices. A basis in spin space is given by

$$\chi_{1/2} = \begin{pmatrix} 1 \\ 0 \end{pmatrix}, \quad \chi_{-1/2} = \begin{pmatrix} 0 \\ 1 \end{pmatrix}, \tag{1.6.2}$$

whereas the angular momentum basis is yielded by the usual scalar spherical harmonics $Y_l^m(\vartheta, \varphi)$. A basis for the total angular momentum $\boldsymbol{j}$ is now obtained as a linear combination

$$\Omega_{jl}^m = \sum_{\substack{m'=-l \\ s=\pm\frac{1}{2}}}^{l} Y_l^{m'} \chi_s \, (l\, m' \, \tfrac{1}{2} \, s \,|\, l \, \tfrac{1}{2} \, j \, m). \tag{1.6.3}$$

The coefficients in (1.6.3) are the Clebsch-Gordan coefficients. Substituting their values, we arrive at the following expressions for the spinor spherical harmonics:

$$\Omega^m_{jj-\frac{1}{2}} = \begin{pmatrix} (-)^{2m-1}\sqrt{\frac{l+m+\frac{1}{2}}{2l+1}}\,Y_l^{m-\frac{1}{2}} \\ \sqrt{\frac{l-m+\frac{1}{2}}{2l+1}}\,Y_l^{m+\frac{1}{2}} \end{pmatrix} \tag{1.6.4}$$

for $l = j - \frac{1}{2}$ and

$$\Omega^m_{jj+\frac{1}{2}} = \begin{pmatrix} -\sqrt{\frac{l-m+\frac{1}{2}}{2l+1}}\,Y_l^{m-\frac{1}{2}} \\ \sqrt{\frac{l+m+\frac{1}{2}}{2l+1}}\,Y_l^{m+\frac{1}{2}} \end{pmatrix} \tag{1.6.5}$$

for $l = j + \frac{1}{2}$. Since m is always half-integer, the signs in the first lines have definite values.

These spherical harmonics satisfy the following eigenvalue equations

$$\boldsymbol{j}^2\,\Omega^m_{jl} = j(j+1)\,\Omega^m_{jl} \tag{1.6.6}$$

$$= (\boldsymbol{L}+\boldsymbol{s})^2\Omega^m_{jl} = (\boldsymbol{L}^2 + 2\boldsymbol{L}\boldsymbol{s} + \boldsymbol{s}^2)\Omega^m_{jl}. \tag{1.6.7}$$

Since the eigenvalues of $\boldsymbol{L}^2$ are $l(l+1)$ and of $\boldsymbol{s}^2$ is $3/4$, we find

$$\boldsymbol{L}\boldsymbol{\sigma}\,\Omega^m_{jl} = \left[j(j+1) - l(l+1) - \tfrac{3}{4}\right]\Omega^m_{jl}$$

$$\stackrel{\text{def}}{=} -(1+\kappa)\Omega^m_{jl}. \tag{1.6.8}$$

For $l = j - \frac{1}{2}$ we have $\kappa = -(j+\frac{1}{2})$ and for $l = j + \frac{1}{2}$, $\kappa = j + \frac{1}{2}$. In addition we must determine the effect of the operator

$$\frac{\boldsymbol{\sigma}\cdot\boldsymbol{x}}{r} = \sigma_1 \sin\vartheta\cos\varphi + \sigma_2\sin\vartheta\sin\varphi + \sigma_3\cos\vartheta. \tag{1.6.9}$$

In the standard representation of the Pauli matrices we have

$$\frac{\boldsymbol{\sigma}\cdot\boldsymbol{x}}{r}\Omega^m_{jj-\frac{1}{2}} = \begin{pmatrix} \cos\vartheta & \sin\vartheta\cos\varphi - i\sin\vartheta\sin\varphi \\ \sin\vartheta\cos\varphi + i\sin\vartheta\sin\varphi & -\cos\vartheta \end{pmatrix}\Omega^m_{jj-\frac{1}{2}}$$

$$= \begin{pmatrix} \cos\vartheta & e^{-i\varphi}\sin\vartheta \\ e^{i\varphi}\sin\vartheta & -\cos\vartheta \end{pmatrix}\begin{pmatrix} \sqrt{\frac{l+m+\frac{1}{2}}{2l+1}}\,Y_l^{m-\frac{1}{2}} \\ \sqrt{\frac{l-m+\frac{1}{2}}{2l+1}}\,Y_l^{m+\frac{1}{2}} \end{pmatrix}. \tag{1.6.10}$$

Using the expressions for the spherical harmonics, we obtain for the first component

$$\sqrt{\frac{l+m+\frac{1}{2}}{2l+1}}\cos\vartheta(-)^{m-\frac{1}{2}}\sqrt{\frac{(2l+1)(l-m+\frac{1}{2})!}{4\pi(l+m-\frac{1}{2})!}}\,P_l^{m-\frac{1}{2}}(\cos\vartheta)e^{i(m-\frac{1}{2})\varphi}$$

$$+\sqrt{\frac{l-m+\frac{1}{2}}{2l+1}}\sin\vartheta(-)^{m+\frac{1}{2}}\sqrt{\frac{(2l+1)(l-m-\frac{1}{2})!}{4\pi(l+m+\frac{1}{2})!}}P_l^{m+\frac{1}{2}}(\cos\vartheta)e^{i(m-\frac{1}{2})\varphi}$$

$$=(-)^{m-\frac{1}{2}}\sqrt{\frac{(2l+3)(l+1-m+\frac{1}{2})!}{4\pi(l+1+m-\frac{1}{2})!}}e^{i(m-\frac{1}{2})\varphi}P_{l+1}^{m-\frac{1}{2}}\sqrt{\frac{l+1-m+\frac{1}{2}}{2l+3}}. \tag{1.6.11}$$

This is just the negative first component of $\Omega^m_{jj+\frac{1}{2}}$. A similar calculation can be carried out for the second component (Problem 1.15). We finally get

$$\frac{\boldsymbol{\sigma}\cdot x}{r}\Omega^m_{jj-\frac{1}{2}}=-\Omega^m_{jj+\frac{1}{2}}. \tag{1.6.12}$$

Multiplying this equation by $\boldsymbol{\sigma}\cdot\boldsymbol{x}/r$ we shall obtain

$$\frac{\boldsymbol{\sigma}\cdot\boldsymbol{x}}{r}\Omega^m_{jj+\frac{1}{2}}=-\Omega^m_{jj-\frac{1}{2}}. \tag{1.6.13}$$

We are now in the position to determine the eigenvectors of the Dirac Hamiltonian

$$H=\begin{pmatrix} m & \boldsymbol{\sigma}\cdot\boldsymbol{p} \\ \boldsymbol{\sigma}\cdot\boldsymbol{p} & -m\end{pmatrix}+eV(r),\quad \boldsymbol{p}=-i\frac{\partial}{\partial\boldsymbol{x}}, \tag{1.6.14}$$

with an attractive Coulomb potential

$$V(r)=-\frac{Ze}{4\pi r},\quad eV=-\frac{Z\alpha}{r}.$$

We have chosen $\hbar=1=c$, α is the fine-structure constant

$$\alpha=\frac{e^2}{4\pi}, \tag{1.6.15}$$

which will describe the fine-structure of the hydrogen spectrum (see (1.6.41)). Instead of the eigenvalue equation

$$H\psi=E\psi \tag{1.6.16}$$

we first consider

$$H^2\psi=E^2\psi,\quad \text{where} \tag{1.6.17}$$

$$H^2=m^2+\boldsymbol{p}^2+2eV\begin{pmatrix} m & \boldsymbol{\sigma}\cdot\boldsymbol{p} \\ \boldsymbol{\sigma}\cdot\boldsymbol{p} & -m\end{pmatrix}-ie\begin{pmatrix} 0 & \boldsymbol{\sigma}\cdot\boldsymbol{\partial}V \\ \boldsymbol{\sigma}\cdot\boldsymbol{\partial}V & 0\end{pmatrix}+e^2V^2. \tag{1.6.18}$$

The first matrix in (1.6.18) is $H-eV$ again (1.6.14) and in the second we use

$$e\frac{\partial V}{\partial\boldsymbol{x}}=\frac{Z\alpha}{r^2}\frac{\boldsymbol{x}}{r}.$$

This leads to

$$H^2\psi = \left[\boldsymbol{p}^2 + m^2 - 2\frac{Z\alpha}{r}E - \frac{(Z\alpha)^2}{r^2} - i\frac{Z\alpha}{r^2}\begin{pmatrix} 0 & \frac{\boldsymbol{\sigma}\cdot\boldsymbol{x}}{r} \\ \frac{\boldsymbol{\sigma}\cdot\boldsymbol{x}}{r} & 0 \end{pmatrix}\right]\psi. \tag{1.6.19}$$

For the solution of (1.6.19) we make the following ansatz

$$\psi(r,\vartheta,\varphi) = \begin{pmatrix} f(r)\Omega_{jl}^m \\ g(r)\Omega_{jl'}^m \end{pmatrix}, \tag{1.6.20}$$

where

$$l = j - \tfrac{1}{2}, \quad l' = j + \tfrac{1}{2}. \tag{1.6.21}$$

A second possibility is to interchange the values of l and l' in (1.6.21), hence there remains a twofold degeneracy of the states. Using (1.6.12-13), we arrive at

$$\left[-\frac{1}{r}\frac{d^2}{dr^2}r + m^2 - \frac{2Z\alpha E}{r} + \frac{1}{r^2}M\right]\begin{pmatrix} f \\ g \end{pmatrix} = E^2\begin{pmatrix} f \\ g \end{pmatrix}. \tag{1.6.22}$$

Here M is the following 2×2 matrix

$$M = \begin{pmatrix} (j-\frac{1}{2})(j+\frac{1}{2}) - (Z\alpha)^2 & iZ\alpha \\ iZ\alpha & (j+\frac{1}{2})(j+\frac{3}{2}) - (Z\alpha)^2 \end{pmatrix}. \tag{1.6.23}$$

We are looking for square-integrable solutions of (1.6.22), that means

$$\int_0^\infty dr\, r^2 |f(r)|^2 < \infty, \quad \int_0^\infty dr\, r^2 |g(r)|^2 < \infty. \tag{1.6.24}$$

This problem can be reduced to the radial Schrödinger equation for the hydrogen atom

$$\left[-\frac{1}{r}\frac{d^2}{dr^2}r + \frac{\lambda(\lambda+1)}{r^2} - \frac{2k}{r}\right]\varphi = \varepsilon\,\varphi. \tag{1.6.25}$$

As is well-known, the eigenvalues of this problem are given by

$$\varepsilon_n = -\frac{k^2}{n^2} \tag{1.6.26}$$

with corresponding eigenfunctions

$$\varphi_{\lambda n}(r) = e^{-\frac{k}{n}r} r^\lambda L_{\lambda n}(r). \tag{1.6.27}$$

Here $n = 1, 2, \ldots$ is the principal quantum number, $\lambda = 0, 1, \ldots$ the angular momentum quantum number and $L_{\lambda n}(r)$ are polynomials of degree $n - \lambda - 1$, the so-called generalized Laguerre polynomials. For the sake of completeness we give a derivation of these results in Appendix A by means of raising and lowering operators, which leads in a simple way to all properties of the eigenfunctions (1.6.27) needed below. It is important to note that for non-integer λ and n, but $n - \lambda - 1 \geq 0$ integer, $\varphi_{\lambda n}$ (1.6.27) is also a square-integrable solution and one gets all square-integrable solutions in this way.

To reduce (1.6.22) to (1.6.25), we have to identify

$$k = Z\alpha E, \quad E > 0 \tag{1.6.28}$$

$$\varepsilon = E^2 - m^2 = -\frac{k^2}{n'} = -(Z\alpha)^2 \frac{E^2}{n'^2}. \tag{1.6.29}$$

λ or rather $\lambda(\lambda+1)$ is determined by diagonalizing the 2×2 matrix M (1.6.23): The eigenvalues of M are equal to

$$\Omega_\pm = (j+\tfrac{1}{2})^2 - (Z\alpha)^2 \pm \sqrt{(j+\tfrac{1}{2})^2 - (Z\alpha)^2}. \tag{1.6.31}$$

Introducing

$$\gamma \stackrel{\text{def}}{=} \sqrt{(j+\tfrac{1}{2})^2 - (Z\alpha)^2}, \tag{1.6.32}$$

we have

$$\Omega_\pm = \gamma^2 \pm \gamma = \lambda^2 + \lambda.$$

This leads to two λ-values

$$\lambda_\pm = \begin{cases} \gamma \\ \gamma - 1 \end{cases} \stackrel{\text{def}}{=} j \pm \tfrac{1}{2} - \delta_j, \tag{1.6.33}$$

with

$$\delta_j = j + \tfrac{1}{2} - \gamma. \tag{1.6.34}$$

There are two Schrödinger problems involved with different angular momentum values, the latter are shifted from the integer values $l = j \pm \frac{1}{2}$ by a small amount δ_j (1.6.34). To get a square-integrable solution, the principal quantum number must be shifted accordingly

$$n' = n - \delta_j = n - j - \tfrac{1}{2} + \gamma. \tag{1.6.35}$$

Here n is the non-relativistic principal quantum number which is an integer $= 1, 2, \ldots$. The energy eigenvalues for the Dirac equation now follow from (1.6.29)

$$E^2\left(1 + \frac{(Z\alpha)^2}{n'^2}\right) = m^2$$

$$E_{nj} = \frac{m}{\sqrt{1 + \frac{(Z\alpha)^2}{n'^2}}}. \tag{1.6.36}$$

This is the fine-structure formula of Sommerfeld.

For a discussion of the fine-structure of the hydrogen-like spectra, we expand (1.6.36) for $Z\alpha \ll 1$:

$$E_{nj} = m\left[1 - \frac{Z^2\alpha^2}{2n'^2} + \frac{3}{8}\frac{(Z\alpha)^4}{n'^4} - \ldots\right], \tag{1.6.37}$$

where

$$n' = n - j - \tfrac{1}{2} + \gamma, \quad n \geq j + \tfrac{1}{2}. \tag{1.6.38}$$

We must also expand γ (1.6.32)

$$\gamma = (j+\frac{1}{2})\sqrt{1-\frac{Z^2\alpha^2}{(j+\frac{1}{2})^2}} = j+\frac{1}{2}-\frac{Z^2\alpha^2}{2(j+\frac{1}{2})}+\dots, \tag{1.6.39}$$

which leads to

$$n' = n - \frac{Z^2\alpha^2}{2(j+\frac{1}{2})} + O(Z\alpha)^4. \tag{1.6.40}$$

Substituting this into (1.6.37) and expanding again we arrive at

$$E_{nj} = m\left[1-\frac{Z^2\alpha^2}{2n^2}\left(1+\frac{Z^2\alpha^2}{n(j+\frac{1}{2})}-\frac{3}{4}\frac{Z^2\alpha^2}{n^2}\right)+O(Z\alpha)^6\right]. \tag{1.6.41}$$

The leading term is the rest energy of the electron, the next term $\sim (Z\alpha)^2$ agrees with the Balmer formula of the non-relativistic Schrödinger equation. The terms of $O(Z\alpha)^4$ give the fine-structure. They depend on j, hence, the angular momentum degeneracy of the non-relativistic theory is partially lifted. There remains the degeneracy $l = j \pm \frac{1}{2}$ mentioned above that corresponds to different parity, except in the boundary case $j+\frac{1}{2} = n$ where only $l = j - \frac{1}{2} = n-1$ is possible. There is of course the usual $2j+1$-fold degeneracy of the unbroken spherical symmetry (labelled by the magnetic quantum number $m = -j, \dots +j$) and an additional twofold degeneracy due to the spin of the electron. The latter is lifted by spin-dependent interaction with the nuclear spin (hyperfine-structure). The next terms $O(Z\alpha)^6$ in (1.6.41) are of the same order as radiative corrections coming from quantum electrodynamics.

We now turn to the determination of the eigenfunctions. The diagonalization

$$M = W\begin{pmatrix} \Omega_+ & 0 \\ 0 & \Omega_- \end{pmatrix} W^{-1} \tag{1.6.42}$$

of the matrix M (1.6.23) is achieved by

$$W = \begin{pmatrix} 1 & 1 \\ \dfrac{j+\frac{1}{2}+\gamma}{iZ\alpha} & \dfrac{j+\frac{1}{2}-\gamma}{iZ\alpha} \end{pmatrix}. \tag{1.6.43}$$

The columns of this matrix are eigenvectors of M

$$M\begin{pmatrix} f \\ g \end{pmatrix} = \Omega\begin{pmatrix} f \\ g \end{pmatrix}, \tag{1.6.44}$$

but not yet normalized. Since

$$W^{-1}\begin{pmatrix} f \\ g \end{pmatrix} = \begin{pmatrix} \varphi_+ \\ \varphi_- \end{pmatrix}, \tag{1.6.45}$$

where

$$\varphi_\pm = \varphi_{\lambda_\pm n'}(r) \tag{1.6.46}$$

are Schrödinger eigenfunctions (1.6.27), it follows that

$$\begin{pmatrix} f \\ g \end{pmatrix} = W \begin{pmatrix} \varphi_+ \\ \varphi_- \end{pmatrix}. \tag{1.6.47}$$

This implies

$$f(r) = \varphi_+ + \varphi_- = r^{\gamma-1}(rL_{\lambda_+ n'} + L_{\lambda_- n'}) \exp\left[-\sqrt{m^2 - E^2}r\right] \tag{1.6.48}$$

$$g(r) = \frac{1}{iZ\alpha}\left[(j + \tfrac{1}{2} + \gamma)\varphi_+ + (j + \tfrac{1}{2} - \gamma)\varphi_-\right]. \tag{1.6.49}$$

Here $\lambda_+ = \lambda_- + 1 = \gamma$ and the polynomials in (1.6.48) have both degree $n' - \lambda_- - 1 = n - j - \frac{1}{2} \geq 1$, where n is the non-relativistic principal quantum number.

These results are not yet the desired components of the Dirac eigenvectors, because the relative normalization between φ_+ and φ_- is not determined by the quadratic eigenvalue equation (1.6.17), but only by the original one (1.6.16). We must therefore substitute our solution into

$$H\psi = \begin{pmatrix} m + eV & \boldsymbol{\sigma} \cdot \boldsymbol{p} \\ \boldsymbol{\sigma} \cdot \boldsymbol{p} & -m + eV \end{pmatrix} \begin{pmatrix} f\Omega^m_{j,j-\frac{1}{2}} \\ g\Omega^m_{j,j+\frac{1}{2}} \end{pmatrix} = E\psi. \tag{1.6.50}$$

We need

$$\begin{aligned} \boldsymbol{\sigma} \cdot \boldsymbol{p}\, g\Omega^m_{jl} &= \frac{\boldsymbol{\sigma} \cdot \boldsymbol{x}}{r}\Big(\frac{\boldsymbol{\sigma} \cdot \boldsymbol{x}}{r}\boldsymbol{\sigma} \cdot \boldsymbol{p}\Big) g\Omega^m_{jl} \\ &= \frac{\boldsymbol{\sigma} \cdot \boldsymbol{x}}{r}\Big(\frac{\boldsymbol{x} \cdot \boldsymbol{p}}{r} + i\boldsymbol{\sigma}\big[\frac{\boldsymbol{x}}{r} \wedge \boldsymbol{p}\big]\Big) g\Omega^m_{jl}. \end{aligned} \tag{1.6.51}$$

The second operator contains the angular momentum $\boldsymbol{L}$, so that by means of (1.6.8) we get

$$= \left[-i\frac{\partial g}{\partial r} - i(1 + \kappa)\frac{g}{r}\right]\frac{\boldsymbol{\sigma} \cdot \boldsymbol{x}}{r}\Omega^m_{jl} = -\left[\ldots\right]\Omega^m_{jl'}. \tag{1.6.52}$$

In particular, we have

$$\boldsymbol{\sigma} \cdot \boldsymbol{p}\, g\Omega^m_{j,j+\frac{1}{2}} = i\left[\frac{dg}{dr} + (j + \tfrac{3}{2})\frac{g}{r}\right]\Omega^m_{j,j-\frac{1}{2}}. \tag{1.6.53}$$

We now obtain for the first component in (1.6.50)

$$\Big(E - m + \frac{Z\alpha}{r}\Big)f = i\Big[g' + (j + \tfrac{3}{2})\frac{g}{r}\Big]. \tag{1.6.54}$$

As we know from Appendix A, the radial Schrödinger eigenfunctions

$$\varphi_\pm = \frac{1}{r}u_{\lambda_\pm, n'} \tag{1.6.55}$$

satisfy the following relations

$$a_\lambda u_{\lambda n'} \stackrel{\text{def}}{=} \Big(\frac{d}{dr} + \frac{\lambda}{r} - \frac{k}{\lambda}\Big)u_{\lambda n'} = C_- u_{\lambda-1, n'} \tag{1.6.56}$$

$$a_\lambda^+ u_{\lambda-1,n'} \stackrel{\text{def}}{=} \left(-\frac{d}{dr} + \frac{\lambda}{r} - \frac{k}{\lambda}\right) u_{\lambda-1,n'} = C_+ u_{\lambda,n'}, \tag{1.6.57}$$

where a_λ and a_λ^+ are lowering and raising operators. In our case $\lambda_+ = \gamma$, $\lambda_- = \gamma - 1$ we find

$$\varphi_+' = C_- \varphi_- - \frac{\gamma+1}{r}\varphi_+ + \frac{k}{\gamma}\varphi_+ \tag{1.6.58}$$

$$\varphi_-' = -C_+ \varphi_+ + \frac{\gamma-1}{r}\varphi_- - \frac{k}{\gamma}\varphi_-. \tag{1.6.59}$$

Inserting this into (1.6.54), the equation is identically fulfilled if the constants $C_\pm$ have the following values

$$C_+ = \frac{Z\alpha}{j+\frac{1}{2}-\gamma}\left(E\frac{j+\frac{1}{2}}{\gamma} + m\right) \tag{1.6.60}$$

$$C_- = \frac{Z\alpha}{j+\frac{1}{2}+\gamma}\left(E\frac{j+\frac{1}{2}}{\gamma} - m\right). \tag{1.6.61}$$

Substituting

$$\varphi_{\lambda_+ n'} = e^{-\frac{k}{n'}r} r^\gamma L_{\lambda_+ n'} \tag{1.6.62}$$

$$\varphi_{\lambda_- n'} = e^{-\frac{k}{n'}r} r^{\gamma-1} L_{\lambda_- n'} \tag{1.6.63}$$

into (1.6.59), we obtain the following relation

$$L_{\lambda_+ n'} = \frac{1}{C_+ r}\left\{k\left(\frac{1}{n'} - \frac{1}{\gamma}\right)L_{\lambda_- n'} - L'_{\lambda_- n'}\right\}. \tag{1.6.64}$$

$L_{\lambda_- n'}$ can be determined by the recursion relation (A.23) of the Schrödinger equation. This relation gets started with

$$L_{n'-1,n'} = C \tag{1.6.65}$$

where the constant C is determined by the overall normalization.

As an example we calculate the ground state of an hydrogen-like atom. This is a $1S_{\frac{1}{2}}$ state in the non-relativistic terminology, that means $n = 1$, $j = \frac{1}{2}$. We have

$$\gamma = \sqrt{1-(Z\alpha)^2} = \lambda_+ \tag{1.6.66}$$

and $n' = \gamma$. Hence, $L_{\lambda_+ n'}$ vanishes according to (1.6.64) because $L_{\lambda_- n'} = C$. With

$$E^2 = \frac{m^2}{1+\frac{Z^2\alpha^2}{\gamma^2}} = m^2(1 - Z^2\alpha^2) \tag{1.6.67}$$

we finally get

$$f(r) = \varphi_-(r) = C r^{\gamma-1} e^{-mZ\alpha r} \tag{1.6.68}$$

$$g(r) = C\frac{1-\gamma}{iZ\alpha} r^{\gamma-1} e^{-mZ\alpha r} = \frac{C}{i}\frac{Z\alpha}{1+\gamma} r^{\gamma-1} e^{-mZ\alpha r}. \tag{1.6.69}$$

Due to the small nominator in (1.6.69), $g(r)$ is the small component. Since $\gamma - 1 < 0$, the radial functions have a weak singularity at $r = 0$, in contrast to the Schrödinger wave function. This shows that the Coulomb potential is more singular for the Dirac equation. In fact, for $Z\alpha > 1$ all results break down (1.6.67). The reason is that the Dirac operator is no longer self-adjoint. The Schrödinger operator, on the other hand, is self-adjoint for all Z.

1.7 Problems

1.1. How does a Dirac spinor transform under rotation about 2π?

1.2. Compute the product of a Lorentz boost in 1-direction times a boost in the 2-direction in the spinor representation $SL(2,\mathbb{C})$. Write the result in the form $R\Lambda(\chi)$, where R is a rotation (the so-called Thomas precession) and $\Lambda(\chi)$ a Lorentz boost.

1.3. Determine the solution of Weyl's equation

$$(\partial_0 + \boldsymbol{\sigma} \cdot \boldsymbol{\partial})\psi = 0 \tag{1.7.1}$$

in the form of a plane wave

$$\psi(x) = e^{-ipx} u(p), \quad px = p^0 x^0 - \boldsymbol{p} \cdot \boldsymbol{x}.$$

a) What is the connection between energy $E = p^0$ and momentum $\boldsymbol{p}$ of the particle and its helicity $\boldsymbol{\sigma} \cdot \boldsymbol{p}/|\boldsymbol{p}|$ as a function of E?

b) Calculate $u(p)$ and verify that the spatially reflected solution does not satisfy (1.7.1).

c) Determine the general solution of (1.7.1).

1.4. Consider a wave packet solution of Weyl's equation:

$$\psi(x) = \int d^3p\, \hat{\psi}(\boldsymbol{p}) e^{-i(Et - \boldsymbol{p}\cdot\boldsymbol{x})}, \quad \hat{\psi} = \begin{pmatrix} \hat{\psi}_1 \\ \hat{\psi}_2 \end{pmatrix}. \tag{1.7.2}$$

The position of the wave packet is

$$\boldsymbol{X} = \int \psi^+(x) \boldsymbol{x} \psi(x)\, d^3x.$$

Show:

$$\frac{\partial}{\partial t}\boldsymbol{X} = \int d^3p\, \hat{\psi}(\boldsymbol{p})^+ \frac{\partial E}{\partial \boldsymbol{p}} \hat{\psi}(\boldsymbol{p}).$$

How does the wave packet move in the cases $E > 0$ and $E < 0$? Are both cases physically exceptable?

1.5. Define the γ-algebra by

$$\gamma^\mu\gamma^\nu + \gamma^\nu\gamma^\mu = 2g^{\mu\nu} \tag{1.7.3}$$

$$\gamma^{\mu+} = g^{\mu\mu}\gamma^\mu. \tag{1.7.4}$$

Prove: a)

$$\gamma_\mu\gamma^\mu = 4, \quad \gamma^{\mu+} = \gamma^0\gamma^\mu\gamma^0.$$

b) Define: $\not{p} \stackrel{\text{def}}{=} \gamma^\mu p_\mu$. Show:

$$\gamma_\mu \not{p}\gamma^\mu = -2\not{p}, \quad \gamma_\mu \not{p}\not{q}\gamma^\mu = 4p\cdot q \tag{1.7.5}$$

$$\gamma_\mu \not{k}\not{p}\not{q}\gamma^\mu = -2\not{q}\not{p}\not{k}, \tag{1.7.6}$$

where always summation over double indices is assumed.

c)

$$\not{p}\not{q} = p\cdot q - ip_\mu q_\nu \sigma^{\mu\nu} \tag{1.7.7}$$

$$\text{with} \quad \sigma^{\mu\nu} = \frac{i}{2}(\gamma^\mu\gamma^\nu - \gamma^\nu\gamma^\mu).$$

d) $\gamma^5 \stackrel{\text{def}}{=} \gamma^0\gamma^1\gamma^2\gamma^3$ satisfies:

$$(\gamma^5)^2 = 1, \quad \gamma^{5+} = \gamma^5 \tag{1.7.8}$$

$$\gamma^\mu\gamma^5 + \gamma^5\gamma^\mu = 0, \quad \left[\gamma^5, \sigma^{\mu\nu}\right] = 0. \tag{1.7.9}$$

1.6. Show that

$$\sigma^{kl} = \varepsilon^{klm}\begin{pmatrix}\sigma_m & 0\\ 0 & \sigma_m\end{pmatrix} \tag{1.7.10}$$

holds in the spinor and standard representations.

1.7. Consider the γ-algebra as a multiplicative group and determine its equivalence classes of conjugated elements.

1.8. Prove:

$$\operatorname{Tr}\gamma^{\mu_1}\ldots\gamma^{\mu_n} = 0, \tag{1.7.11}$$

if n is odd and

$$\operatorname{Tr}\gamma^{\mu_1}\ldots\gamma^{\mu_n} = \operatorname{Tr}\gamma^{\mu_n}\ldots\gamma^{\mu_2}\gamma^{\mu_1} \tag{1.7.12}$$

if n is even. Calculate $\operatorname{Tr}\gamma^\mu\gamma^\nu$ and $\operatorname{Tr}\gamma^{\mu_1}\gamma^{\mu_2}\gamma^{\mu_3}\gamma^{\mu_4}$. What is the rule to determine the signs of the terms? Make a conjecture for the trace of $2n$ γ-matrices. How many terms appear?

1.9. Calculate: $\operatorname{Tr}\gamma^5$, $\operatorname{Tr}\gamma^5\gamma^\mu$, $\operatorname{Tr}\gamma^5\gamma^\mu\gamma^\nu$, $\operatorname{Tr}\gamma^5\gamma^{\mu_1}\gamma^{\mu_2}\gamma^{\mu_3}$ and $\operatorname{Tr}\gamma^5\gamma^{\mu_1}\gamma^{\mu_2}\gamma^{\mu_3}\gamma^{\mu_4}$.

1.10. a) Derive the adjoint Dirac equation

$$(-i\partial_\mu - eA_\mu)\overline{\psi}\gamma^\mu = m\overline{\psi} \tag{1.7.13}$$

from the Dirac equation.

b) Derive (1.7.13) from the variational principle by variation of ψ.

c) Prove current conservation by means of (1.7.13).

1.11. Majorana-representation: Construct a representation of the γ-algebra, such that the Dirac equation assumes the following form

$$\frac{\partial\psi}{\partial t} = (\tilde{\alpha}_0 m + \tilde{\alpha}_1\partial_1 + \tilde{\alpha}_2\partial_2 + \tilde{\alpha}_3\partial_3)\psi, \tag{1.7.14}$$

where $\tilde{\alpha}_0$, $\tilde{\alpha}_1$, $\tilde{\alpha}_2$, $\tilde{\alpha}_3$ are real matrices. Determine the $\tilde{\alpha}$ and the corresponding $\tilde{\gamma}$-matrices in the spinor representation. Show that the equation with electromagnetic potentials remain complex.

1.12. Show that the Dirac equation with $m = 0$ is invariant under the global gauge transformation

$$\psi \longrightarrow e^{i(\Lambda+\lambda\gamma^5)}\psi. \tag{1.7.15}$$

Make the equation also invariant with respect to local gauge transformations by introducing two additional vector fields A_μ and Z_μ. What is the gauge-invariant Lagrangian of this theory? Determine the Noether currents of the global gauge invariance.

1.13. Show that the Dirac current can be decomposed into a convective current and a spin current

$$j^\mu = \frac{ie}{2m}\left[\overline{\psi}\partial^\mu\psi - (\partial^\mu\overline{\psi})\psi\right] - \frac{e^2}{m}A^\mu\overline{\psi}\psi$$

$$+\frac{e}{2m}\partial_\nu\overline{\psi}\sigma^{\mu\nu}\psi, \tag{1.7.16}$$

which are separately conserved and gauge invariant.

1.14. a) Derive the quadratic Dirac equation

$$\left[\left(i\partial_\mu - eA_\mu\right)\left(i\partial^\mu - eA^\mu\right) - \frac{e}{2}\sigma^{\mu\nu}F_{\mu\nu}\right]\psi = m^2\psi \tag{1.7.17}$$

from the Dirac equation and identify the magnetic moment of the electron.

b) Show that (1.7.17) follows from the Lagrangian

$$L = \left[(-i\partial_\mu - eA_\mu)\overline{\psi}\right](i\partial^\mu - eA^\mu)\psi$$

$$+\overline{\psi}\left(\frac{e}{2}\sigma^{\mu\nu}F_{\mu\nu} - m^2\right)\psi. \tag{1.7.18}$$

c) Show that the Noether current for (1.7.18), which expresses the global gauge invariance of L, is just the convective current of Problem 1.12.

1.15. Verify the second component of Eq.(1.6.12).

1.16. Calculate the fine structure splitting between the $2S_{1/2}$ and $2P_{3/2}$ states in hydrogen in Hertz.

1.17. a) Set up the Klein-Gordon equation for a charged spin 0 particle in an electromagnetic potential $A^{\mu}(x)$ by means of gauge invariance.

b) What is the corresponding Lagrangian variational principle? Determine the conserved Noether current.

c) Discuss the non-relativistic limit by means of the Ansatz

$$\psi = \varphi \exp -\frac{i}{\hbar} mc^2 t. \tag{1.7.19}$$

Show that the leading order in $1/c$ is given by the Schrödinger equation.

1.18. Calculate the bound state energies for the Klein-Gordon equation with an attractive Coulomb potential

$$eV = -\frac{Z\alpha}{4\pi r}. \tag{1.7.20}$$

Compare the result with the eigenvalues for the Dirac equation.

2. Field Quantization

The preceding chapter, entitled "Relativistic Quantum Mechanics" did not really deal with quantum mechanics. The theory so far has no consistent interpretation because it suffers from two defects: (i) The hamiltonian operators are unbounded from below and, (ii) there is no many-particle formulation of the classical Dirac theory. These defects will now be removed. We concentrate first on the second problem (ii). In the course of its treatment, problem (i) will be solved, too.

There is a powerful method, called second quantization in Fock space, which transforms a one-particle theory into a many-particle theory. This method is first described in great generality and then applied to quantization of the Dirac field in an external classical electromagnetic field. If the external field is time-dependent it is impossible, in general, to implement the time evolution of the quantized electron-positron field in a fixed (Fock) Hilbert space. The more one thinks about this situation, the more one is led to the conclusion that one should not insist on a detailed description of the system in time. From the physical point of view, this is not so surprising because in contrast to non-relativistic quantum mechanics, the time behavior of a relativistic system with creation and annihilation of particles is unobservable. Essentially only scattering experiments are possible, therefore we retreat to scattering theory. One learns modesty in field theory.

The scattering operator **S** in Fock space can be explicitly constructed from the one-particle S-matrix S (Sect. 2.3). Is is uniquely determined up to a phase. The phase is fixed by the requirement of causality of **S** (Sect. 2.7). This causal phase is connected with vacuum polarization (Sect. 2.9), leading to a non-perturbative understanding of this phenomenon in the external field case which is free from any divergences. For the first time we realize the crucial fact that causality makes QED finite. The non-perturbative treatment of the external field problem is an essential tool for strong field quantum electrodynamics. Finally we must quantize the radiation field (Sect. 2.11). This is a delicate subject because we want to maintain explicit Lorentz covariance. In the last section we discuss the interesting Casimir effect.

2.1 Second Quantization in Fock Space

Let $\mathcal{H}$ be the one-particle Hilbert space, for example $L^2(\mathbb{R}^3)$ in case of a spinless particle, and $\varphi_1(\boldsymbol{x}) \in \mathcal{H}$. A two-particle state is then an element of the tensor product space

$$\varphi_2(\boldsymbol{x}_1, \boldsymbol{x}_2) \in \mathcal{H}_2 = \mathcal{H} \otimes \mathcal{H} \tag{2.1.1}$$

and similarly for a n-particle state

$$\varphi_n \in \mathcal{H}^{\otimes n}. \tag{2.1.2}$$

However, identical particles obey Bose or Fermi statistics, that means the state must be symmetric

$$S_n^+ \varphi_n = \frac{1}{n!} \sum_\pi \varphi_n(\boldsymbol{x}_{\pi 1}, \ldots, \boldsymbol{x}_{\pi n}) \tag{2.1.3}$$

or antisymmetric

$$S_n^- \varphi_n = \frac{1}{n!} \sum_\pi (-)^\pi \varphi_n(\boldsymbol{x}_{\pi 1}, \ldots, \boldsymbol{x}_{\pi n}). \tag{2.1.4}$$

The sums run over all $n!$ permutations π of the n particles, $(-)^\pi$ is equal to $+1$ if π is an even permutation of $1, 2, \ldots, n$ and -1 if π is odd. The operators $S_n^\pm$ (2.1.3, 4) are projection operators in the n-particle space (2.1.2):

$$(S_n^\pm)^2 = S_n^\pm$$

$$(\psi_n, S_n^\pm \varphi_n) = (S_n^\pm \psi_n, \varphi_n).$$

The true physical n-particle spaces are

$$\mathcal{H}_n^\pm = S_n^\pm \mathcal{H}^{\otimes n}, \tag{2.1.5}$$

where the symmetric tensor product (+) corresponds to bosons and the antisymmetric tensor product (−) to fermions.

In order to describe all multi-particle states simultaneously, one introduces the Fock space

$$\mathcal{F}^\pm = \oplus_{n=0}^\infty \mathcal{H}_n^\pm, \tag{2.1.6}$$

where

$$\mathcal{H}_0 = \{\alpha \Omega\}, \; \alpha \in \mathbb{C} \tag{2.1.7}$$

is the one-dimensional space consisting of the vacuum Ω. An element of $\mathcal{F}$ is an infinite sequence of states

$$\Phi = (\varphi_0, \varphi_1, \varphi_2, \ldots) \in \mathcal{F} \tag{2.1.8}$$

with

$$\varphi_0 = \alpha \Omega, \; \varphi_1 \in \mathcal{H}_1, \ldots, \varphi_n \in \mathcal{H}_n^\pm, \ldots .$$

Since the Fock space (2.1.6) is a direct sum, the scalar product is given by

$$(\Phi, \Psi) = \sum_{n=0}^{\infty} (\varphi_n, \psi_n)_n, \tag{2.1.9}$$

where $(,)_n$ denotes the scalar product in $\mathcal{H}_n$. $\mathcal{F}$ is the space of all multi-particle states (2.1.8) with finite norm

$$\sum_n \|\varphi\|_n^2 \stackrel{\text{def}}{=} \|\Phi\|^2 < \infty. \tag{2.1.10}$$

which is a Hilbert space.

Now we consider operators in Fock space which are always printed in bold face letters. We begin with the particle number operator $\mathbf{N}$ defined as follows: Let $\Phi = (\varphi_0, \varphi_1, \ldots) \in \mathcal{F}$, then

$$(\mathbf{N}\Phi)_n = n\varphi_n. \tag{2.1.11}$$

This is an unbounded operator with domain

$$D(\mathbf{N}) = \left\{\Phi \in \mathcal{F} \mid \sum_n n^2 \|\varphi_n\|_n^2 = \|\mathbf{N}\Phi\|^2 < \infty\right\}. \tag{2.1.12}$$

With this domain we have a positive selfadjoint operator in $\mathcal{F}$. All operators $\mathbf{A}$ commuting with $\mathbf{N}$

$$[\mathbf{A}, \mathbf{N}] = 0 \tag{2.1.13}$$

do not change the particle numbers of the states.

Now we introduce operators which do change particle number. The so-called emission (or creation) operators $a^*(f)$ create a particle with "wave function" $f \in \mathcal{H}_1$

$$a^*(f)\Omega = f$$

$$(a^*(f)\Phi)_n = \sqrt{n} S_n^{\pm}(f \otimes \varphi_{n-1}), \quad n = 1, 2, \ldots \,. \tag{2.1.14}$$

For the moment, the asterisk $*$ is only a notation, later on it will get the meaning of the adjoint. Due to the square root of n in (2.1.14), the emission operators are unbounded for bosons, but for fermions they are bounded. To see this, let us write down the antisymmetrization of the product

$$S_n^- f(x_1)\varphi_{n-1}(x_2, \ldots, x_n) = \frac{1}{n}\Big[f(x_1)\varphi_{n-1}(x_2, \ldots, x_n)$$

$$- \sum_{j=2}^{n} f(x_j)\varphi_{n-1}(x_2, \ldots, \underset{j-1}{x_1}, \ldots, x_n)\Big]. \tag{2.1.15}$$

The subscript $j-1$ means that x_1 is permuted to the place $j-1$. In (2.1.15) only the $n-1$ transpositions of x_1 with one coordinate $x_2, x_3, \ldots, x_n$ have to be carried out, because φ_{n-1} is already antisymmetric. It is easily seen that the norm squared of (2.1.15) is

$$\|S_n^- f \otimes \varphi_{n-1}\|_n^2 = \frac{1}{n^2}[n\|f\|_1^2 \|\varphi_{n-1}\|_{n-1}^2 - \text{pos.}], \tag{2.1.16}$$

where "pos." stands for a positive number. Then

$$\|S_n^- f \otimes \varphi_{n-1}\|_n^2 \leq \frac{1}{n}\|f\|_1^2 \|\varphi_{n-1}\|_{n-1}^2$$

and

$$\|a^*(f)\Phi\|^2 \leq \sum_n n\frac{1}{n}\|f\|_1^2\|\varphi_{n-1}\|_{n-1}^2 = \|f\|_1^2\|\Phi\|^2. \tag{2.1.17}$$

The emission operator $a^*(f)$ is therefore a bounded operator defined on the whole of $\mathcal{F}^-$.

Together with the emission operators, one considers the so-called field "operators" $a^*(\boldsymbol{x})$ and $\hat{a}^*(\boldsymbol{k})$ in momentum space. These are not operators in $\mathcal{F}$ but operator-valued functionals which express the linear mapping $f \to a^*(f)$ as follows

$$a^*(f) = \int d^3x\, a^*(\boldsymbol{x})f(\boldsymbol{x}) = \int d^3k\, \hat{a}^*(\boldsymbol{k})\hat{f}(\boldsymbol{k}). \tag{2.1.18}$$

The hat always denotes Fourier transformation. Let f_j be a complete orthonormal basis in $\mathcal{H}_1$ and the complex conjugate f_j^*, too, assuming for a moment $\mathcal{H}_1 = L^2$. Then (2.1.18) can be written as a formal complex scalar product

$$a^*(f_j) = (f_j^*, a^*(\boldsymbol{x})).$$

Since this expression has the form of a Fourier coefficient, one can write the field operators as formal sums

$$a^*(\boldsymbol{x}) = \sum_j a^*(f_j)f_j(\boldsymbol{x})^* = \sum_j a^*(f_j^*)f_j(\boldsymbol{x}) \tag{2.1.19}$$

$$\hat{a}^*(\boldsymbol{k}) = \sum_j a^*(\hat{f}_j)\hat{f}_j(\boldsymbol{k})^* = \sum_j a^*(\hat{f}_j^*)\hat{f}_j(\boldsymbol{k}). \tag{2.1.20}$$

These expressions are independent of the basis f_j. They are useful to define a formal expression in the field operators as an operator in Fock space. Using the definition (2.1.14), we find

$$\begin{aligned}(\hat{a}^*(\boldsymbol{k})\Phi)_n(\boldsymbol{k}_1,\ldots,\boldsymbol{k}_n) &= \sum_j \hat{f}_j(\boldsymbol{k}) \\ &\times \sqrt{n}S_n^{\mp}\left(\hat{f}_j^*(\boldsymbol{k}_1)\varphi_{n-1}(\boldsymbol{k}_2,\ldots,\boldsymbol{k}_n)\right) \\ &= \sqrt{n}S_n^{\mp}\left(\delta(\boldsymbol{k}-\boldsymbol{k}_1)\varphi_{n-1}(\boldsymbol{k}_2,\ldots,\boldsymbol{k}_n)\right),\end{aligned} \tag{2.1.20a}$$

where the completeness relation for the $\hat{f}_j$ has been taken into account. Due to the δ-distribution, the result is not a vector in Fock space, hence, $\hat{a}^*(\boldsymbol{k})$

is not an operator. However, it is a quadratic form: if all φ_n and ψ_n are continuous functions, then the scalar product $(\Psi, \hat{a}^*(\boldsymbol{k})\Phi)$ is well-defined.

Next we consider absorption (annihilation) operators. Let us first assume $\mathcal{H}_1 = L^2(\mathbb{R}^3)$. Then the operator

$$(a(f)\Phi)_n(\boldsymbol{x}_1, \ldots, \boldsymbol{x}_n) = \sqrt{n+1} \int d^3x\, f(\boldsymbol{x})^* \varphi_{n+1}(\boldsymbol{x}, \boldsymbol{x}_1, \ldots, \boldsymbol{x}_n) \qquad n = 0, 1, 2, \ldots, \tag{2.1.21}$$

absorbs a particle with wave function f. On the vacuum $\Omega = (1, 0, 0, \ldots)$ one defines

$$a(f)\Omega = 0 \quad , \quad \forall f \in \mathcal{H}_1. \tag{2.1.22}$$

Let us determine the adjoint of $a(f)$:

$$(\Psi, a(f)\Phi) = \sum_n \sqrt{n+1} \int d^3x_1 \ldots d^3x_n d^3x f(\boldsymbol{x})^*$$

$$\times \Psi_n^*(\boldsymbol{x}_1, \ldots, \boldsymbol{x}_n)\varphi_{n+1}(\boldsymbol{x}, \boldsymbol{x}_1, \ldots, \boldsymbol{x}_n) =$$

$$\sum_n \sqrt{n+1} \int d^3x\, d^3x_1 \ldots d^3x_n [S_{n+1}^{\pm} f(\boldsymbol{x})\psi_n(\boldsymbol{x}_1, \ldots, \boldsymbol{x}_n]^* \varphi_{n+1}(\boldsymbol{x}, \boldsymbol{x}_1, \ldots \boldsymbol{x}_n)$$

$$= (a^*(f)\Psi, \Phi). \tag{2.1.23}$$

This implies

$$a(f) = a^*(f)^+, \tag{2.1.24}$$

which can be taken as the abstract definition of the absorption operators. Note that $a(f)$ is antilinear in f. According to (2.1.24) we write

$$a^+(f) = a(f)^+ \tag{2.1.25}$$

instead of $a^*(f)$ from now on. The corresponding field operators are defined as in (2.1.18, 20)

$$a(f) = \int d^3x\, f(\boldsymbol{x})^* a(\boldsymbol{x}) \tag{2.1.26}$$

$$a(\boldsymbol{x}) = \sum_j a(f_j) f_j(\boldsymbol{x}). \tag{2.1.27}$$

Inserting the definition (2.1.21), we find by means of the completeness relation

$$\big(a(\boldsymbol{x})\Phi\big)_n(\boldsymbol{x}_1, \ldots, \boldsymbol{x}_n) = \sum_j f_j(\boldsymbol{x})$$

$$\times \int d^3x'\, f_j^*(\boldsymbol{x}')\varphi_{n+1}(\boldsymbol{x}', \boldsymbol{x}_1, \ldots, \boldsymbol{x}_n)$$

$$= \sqrt{n+1}\varphi_{n+1}(\boldsymbol{x}, \boldsymbol{x}_1, \ldots, \boldsymbol{x}_n). \tag{2.1.27a}$$

If

$$\sum_n n\|\varphi_n\|^2 < \infty,$$

the result is a vector in Fock space, so that $a(\boldsymbol{x})$ is an operator in contrast to (2.1.20a).

Now we shall express arbitrary operators in Fock space in terms of emission and absorption operators. We claim that the particle number operator is just

$$\mathsf{N} = \int d^3x\, a^+(\boldsymbol{x})a(\boldsymbol{x}). \tag{2.1.28}$$

In fact, by means of (2.1.19) and (2.1.27) one has

$$\mathsf{N} \stackrel{\text{def}}{=} \sum_{jk} a^+(f_j)a(f_k)(f_j,\, f_k) = \sum_j a^+(f_j)a(f_j)$$

and, since

$$(a^+(f_j)a(f_j)\Phi)_n = \sqrt{n}S_n^{\pm} f_j(\boldsymbol{x}_1) \int d^3x\, \sqrt{n} f_j^*(\boldsymbol{x})\varphi_n(\boldsymbol{x}, \boldsymbol{x}_2, \dots, \boldsymbol{x}_n), \tag{2.1.29}$$

one finds

$$(\mathsf{N}\Phi)_n = nS_n^{\pm}\varphi_n(\boldsymbol{x}_1, \dots, \boldsymbol{x}_n\,).$$

Similarly, we claim that any one-particle operator $A(\boldsymbol{x})$ is lifted to Fock space by the following formula

$$\mathsf{A} = \int d^3x\, a^+(\boldsymbol{x})A(\boldsymbol{x})a(\boldsymbol{x}) \tag{2.1.30}$$

$$\stackrel{\text{def}}{=} \sum_{jk} (f_j,\, A(\boldsymbol{x})f_k)_1 a^+(f_j)a(f_k). \tag{2.1.31}$$

As before (2.1.29), we have

$$\begin{aligned}
(\mathsf{A}\Phi)_n &= n\sum_{jk} S_n^{\pm} f_j(\boldsymbol{x}_1) \int d^3x'\, f_k(\boldsymbol{x}')^*\varphi_n(\boldsymbol{x}', \boldsymbol{x}_2, \dots, \boldsymbol{x}_n) \\
&\quad \times \int d^3x\, f_j(\boldsymbol{x})^* A(\boldsymbol{x}) f_k(\boldsymbol{x}) \\
&= n\sum_k S_n^{\pm} A(\boldsymbol{x}_1) f_k(\boldsymbol{x}_1) \int d^3x'\, f_k(\boldsymbol{x}')^*\varphi_n(\boldsymbol{x}', \boldsymbol{x}_2, \dots, \boldsymbol{x}_n) \\
&= nS_n^{\pm} A(\boldsymbol{x}_1)\varphi_n(\boldsymbol{x}_1, \dots, \boldsymbol{x}_n\,) \\
&= \sum_{m=1}^{n} A(\boldsymbol{x}_m)\varphi_n(\boldsymbol{x}_1, \dots, \boldsymbol{x}_n\,),
\end{aligned} \tag{2.1.32}$$

which is the correct extension of $A(\boldsymbol{x})$ to a multi-particle state. Suppose the operator $A(\boldsymbol{x})$ to be bounded on $\mathcal{H}_1$, then

$$\|(\mathbf{A}\Phi)_n\|_n \leq n\|A\|_1\|\varphi_n\|_n, \tag{2.1.33}$$

and

$$\|\mathbf{A}\Phi\|^2 \leq \|A\|_1^2 \sum_n n^2\|\varphi_n\|_n^2 = \|A\|_1^2\|\mathbf{N}\Phi\|^2. \tag{2.1.34}$$

Hence, the domain of $\mathbf{A}$ contains the domain of the particle number operator

$$D(\mathbf{A}) \supseteq D(\mathbf{N}), \tag{2.1.35}$$

which is a dense set.

Next we shall consider two-particle operators, as for example a two-body potential $V(\boldsymbol{x}, \boldsymbol{x}')$. By the same reasoning as above, one obtains

$$\mathbf{V} = \frac{1}{2}\int d^3x\, d^3x'\, a^+(\boldsymbol{x}')a^+(\boldsymbol{x})V(\boldsymbol{x}, \boldsymbol{x}')a(\boldsymbol{x})a(\boldsymbol{x}')$$

$$= \frac{1}{2}\sum_{jkj'k'} (f_j f_k, V f_{j'} f_{k'})a^+(f_j)a^+(f_k)a(f_{k'})a(f_{j'}), \tag{2.1.36}$$

operating on the n-particle sector as follows

$$(\mathbf{V}\Phi)_n = \sum_{j<k} V(\boldsymbol{x}_j, \boldsymbol{x}_k)\varphi_n(\boldsymbol{x}_1, \ldots, \boldsymbol{x}_n). \tag{2.1.37}$$

The expressions (2.1.31) and (2.1.36) have a formal similarity with ordinary quantum mechanical matrix elements. They are in so-called normal ordered form which means that all absorption operators are standing to the right and all emission operators to the left. Such normal ordering, also called Wick ordering, will appear in the following over and over again. We shall prove below (see Corollary 1.3) that any operator in Fock space can be expressed in terms of emission and absorption operators.

The most important properties of emission and absorption operators are their commutation relations in case of Bose particles or anticommutation relations in case of fermions. Both cases can be treated together. Using the concrete definitions (2.1.14, 21), one obtains

$$([a(f), a^+(g)]_\mp\Phi)_n = \sqrt{n+1}\int d^3x\, f(\boldsymbol{x})^*\sqrt{n+1}\, S_{n+1}^\pm\, g(\boldsymbol{x})\varphi_n(\boldsymbol{x}_1, \ldots, \boldsymbol{x}_n)$$

$$\mp\sqrt{n}S_n^\pm g(\boldsymbol{x}_1)\int d^3x\, \sqrt{n}f(\boldsymbol{x})^*\varphi_n(\boldsymbol{x}, \boldsymbol{x}_2, \ldots, \boldsymbol{x}_n). \tag{2.1.38}$$

Let us compute the first term T_1 on the right of (2.1.38). Since

$$S_{n+1}^\pm\, g(\boldsymbol{x})\varphi_n(\boldsymbol{x}_1, \ldots, \boldsymbol{x}_n) = \frac{1}{n+1}\Big[g(\boldsymbol{x})\varphi_n(\boldsymbol{x}_1, \ldots, \boldsymbol{x}_n)$$

$$\pm\sum_{j=1}^{n} g(\boldsymbol{x}_j)\varphi_n(\boldsymbol{x}_1, \ldots, \underset{j}{\boldsymbol{x}}, \ldots, \boldsymbol{x}_n)\Big],$$

the first term becomes

$$T_1 = (f,\, g)\varphi_n \pm \Big[g(\boldsymbol{x}_1) \int d^3x\, f(\boldsymbol{x})^* \varphi_n(\boldsymbol{x}, \boldsymbol{x}_2, \ldots, \boldsymbol{x}_n)$$

$$\pm \sum_{j=2}^{n} g(\boldsymbol{x}_j) \int d^3x\, f(\boldsymbol{x})^* \varphi_n(\boldsymbol{x}, \boldsymbol{x}_2 \ldots, \underset{j}{\boldsymbol{x}_1}, \ldots, \boldsymbol{x}_n) \Big]. \tag{2.1.39}$$

The square bracket in (2.1.39) is just the second term in (2.1.38), hence

$$[a(f),\, a^+(g)]_\mp \Phi = (f,\, g)_1 \Phi, \tag{2.1.40}$$

where Φ must be in the domain of the left side. In the same way one computes

$$[a(f),\, a(g)]_\mp = 0 = [a(f)^+,\, a(g)^+]_\mp. \tag{2.1.41}$$

Equation (2.1.40) is equivalent to the following commutation of the field operators

$$[\hat{a}(\boldsymbol{k}),\, \hat{a}^+(\boldsymbol{k}')] = \delta(\boldsymbol{k} - \boldsymbol{k}'). \tag{2.1.41a}$$

In the case of anticommutators, relation (2.1.41) implies

$$a^+(f) a^+(f) = 0,$$

that means, it is impossible to create two particles in the same state. This is the Pauli principle for fermions. Accordingly, anticommutators lead to Fermi statistics, Bose statistics requires commutators. The fundamental importance of the commutation relations (2.1.40,41) is contained in the following

Theorem 1.1. Every irreducible representation of (2.1.40,41) with vacuum Ω (2.1.22) is unitary equivalent to the Fock representation constructed above.

Note that the vacuum is not assumed to be unique. Irreducible means that there is no subspace in $\mathcal{F}$ which is invariant under all $a(f)$, $a^+(f)$. That is to say, there is no projection operator P with

$$[P,\, a(f)]_- = 0 = [P,\, a(f)^+]_-, \tag{2.1.42}$$

for all $f \in \mathcal{H}_1$. To prove this theorem, it is convenient to show first the irreducibility of the Fock representation:

Theorem 1.2. Let A be a closed operator in $\mathcal{F}^\pm$ which is defined on Ω and satisfies

$$[A,\, a(f)]_- = 0 = [A,\, a(f)^+]_-\,,\ \forall f \in \mathcal{H}_1. \tag{2.1.43}$$

Then it follows

$$A = \alpha \mathbb{1}\,,\ \alpha \in \mathbb{C}. \tag{2.1.44}$$

An operator which is a multiple of $\mathbb{1}$ (2.1.44) is called a C-number.

Proof. We first construct a basis in $\mathcal{F}$. If Ω is the vacuum, the one-particle sector is spanned by

$$\frac{1}{\sqrt{1!}}a(f)^+\Omega\,,\ f\in\mathcal{H}_1. \tag{2.1.45}$$

The two particle states are given by

$$\frac{1}{\sqrt{2!}}a(f_1)^+a(f_2)^+\Omega=(0,0,S_2f_1(\boldsymbol{x}_1)f_2(\boldsymbol{x}_2),0,\ldots), \tag{2.1.46}$$

and so on. Accordingly, the vectors

$$\mathcal{F}_0=\left\{\frac{1}{\sqrt{n!}}\prod_{j=1}^{n}a(f_j)^+\Omega\ \mid\ \forall n,f_j\right\} \tag{2.1.47}$$

form a basis in $\mathcal{F}$, if the f_j are a basis is $\mathcal{H}_1$. Next we note that a vector $\Phi\in\mathcal{F}$ satisfying

$$a(f)\Phi=0\,,\quad \text{for all}\quad f\in\mathcal{H}_1 \tag{2.1.48}$$

must be the vacuum, because

$$\int d^3x\,f(\boldsymbol{x})^*\varphi_{n+1}(\boldsymbol{x},\boldsymbol{x}_1,\ldots,\boldsymbol{x}_n)=0\,,\,n=0,1,\ldots$$

implies

$$\varphi_n=0\quad\text{for}\quad n=1,2,\ldots,\qquad\text{that means}\quad \Phi=c\Omega.$$

We now consider

$$Aa(f)\Omega=0=a(f)A\Omega\,,\quad \forall f\in\mathcal{H}_1$$

which implies $A\Omega=\alpha\Omega$ (2.1.48). Similarly, we find on the one-particle sector

$$Aa(f)^+\Omega=a(f)^+A\Omega=\alpha a(f)^+\Omega,$$

and in the same way, $A=\alpha\mathbb{1}$ on the dense set $\mathcal{F}_0$(2.1.47). Since A is closed, the same is true on the whole of $\mathcal{F}$. This completes the proof of Theorem 1.2. □

Let $\mathcal{A}$ be the (weakly closed) algebra generated by $a(f),a(f)^+$ (von Neumann algebra). The commutant $\mathcal{A}'$ of $\mathcal{A}$ is the algebra of all bounded operators commuting with all elements of $\mathcal{A}$. According to Theorem 1.2, this commutant is trivial

$$\mathcal{A}'=\alpha\mathbb{1}. \tag{2.1.49}$$

Then the double commutant $\mathcal{A}''$ is the algebra of all bounded operators on $\mathcal{F}$. Since $\mathcal{A}''=\mathcal{A}$ (see any book on von Neumann algebras), we have proven

Corollary 1.3. Any (bounded) operator on Fock space can be expressed in terms of absorption and emission operators a,a^+.

Proof of Theorem 1.1. This proof is of general interest because we construct the Fock space $\mathcal{F}$ from the commutation relations (2.1.40,41) and the vacuum Ω. Ω is defined by

$$a(f)\Omega = 0\,, \quad \forall f \in \mathcal{H}_1. \tag{2.1.50}$$

As in the proof of Theorem 1.2 we consider

$$\frac{1}{\sqrt{1!}}a(f)^{+}\Omega \quad \text{for all} \quad f \in \mathcal{H}_1. \tag{2.1.51}$$

Since

$$\begin{aligned}(a^{+}(f)\Omega,\, a^{+}(g)\Omega) &= (\Omega,\, a(f)a^{+}(g)\Omega)\\ &= (f,\, g) \pm (\Omega,\, a^{+}(g)a(f)\Omega) = (f,\, g),\end{aligned}$$

the space of vectors (2.1.51) is unitarily equivalent to the one-particle sector $(0, f, 0, \ldots)$. Proceeding in the same way with

$$\frac{1}{\sqrt{2!}}a^{+}(f_1)a^{+}(f_2)\Omega, \tag{2.1.52}$$

one gets by repeated application of the commutation relations

$$\begin{aligned}\frac{1}{2!}(a^{+}(f_1)a^{+}(f_2)\Omega,\, a^{+}(g_1)a^{+}(g_2)\Omega) &= \frac{1}{2!}(\Omega,\, a(f_2)a(f_1)a^{+}(g_1)a^{+}(g_2)\Omega)\\ = \frac{1}{2!}[(f_1,\, g_1)(f_2,\, g_2) \pm (f_1,\, g_2)(f_2,\, g_1)] &= (S_2^{\pm}(f_1 \otimes f_2),\, S_2^{\pm}(g_1 \otimes g_2)).\end{aligned}$$

This shows the unitary equivalence to the two-particle sector. It is easy to see by induction that the vectors

$$(a^{+}(f)\Phi)_n \quad \text{correspond to} \quad \sqrt{n}S_n^{\pm}(f \otimes \varphi_{n-1}). \tag{2.1.53}$$

This equivalence with the Fock representation proves the theorem. □

Now we shall introduce a dynamical time evolution in Fock space. We start from a one-particle dynamics given by a Schrödinger equation

$$i\frac{df}{dt} = Hf\,, \quad (\hbar = 1). \tag{2.1.54}$$

This equation is solved by a unitary group in the one-particle space $\mathcal{H}_1$

$$f(t) = e^{-iHt}f(0) \stackrel{\text{def}}{=} Uf. \tag{2.1.55}$$

We take the function $f(-t)$ as test function for the field operators

$$a^{+}(f(-t))\Phi = a^{+}(U^{-1}f)\Phi. \tag{2.1.56}$$

Since

$$(a^{+}(U^{-1}f)\Phi)_n = \sqrt{n}S_n(U^{-1}f \otimes \varphi_{n-1}), \tag{2.1.57}$$

it is appropriate to define an operator $\mathbf{U}$ on Fock space as follows:

$$\mathbf{U}\Omega = \Omega$$

$$(\mathbf{U}\Phi)_n = (\otimes_{j=1}^n U)\varphi_n\,, \quad n = 1, 2, \ldots\,. \tag{2.1.58}$$

If φ_n is a product state

$$\varphi_n = S_n(f_1 \otimes \ldots \otimes f_n), \tag{2.1.59}$$

then

$$(\otimes_{j=1}^n U)\varphi_n = S_n(Uf_1 \otimes \ldots \otimes Uf_n). \tag{2.1.60}$$

In the same way, one defines $\mathbf{U}^{-1}$, both operators are unitary on $\mathcal{F}$. A glance at (2.1.57) shows that

$$a^+(U^{-1}f) = \mathbf{U}^{-1}a^+(f)\mathbf{U}. \tag{2.1.61}$$

One sometimes calls $\mathbf{U}$ the second quantization of U. Taking the adjoint of (2.1.61), we have

$$a(U^{-1}f) = \mathbf{U}^{-1}a(f)\mathbf{U} \quad , \quad \forall f \in \mathcal{H}_1. \tag{2.1.62}$$

Note that this construction is possible for any unitary U.

The time evolution (2.1.56) can now be written as follows

$$a^+(f(-t)) = \mathbf{U}_t^{-1}a^+(f(0))\mathbf{U}_t \stackrel{\text{def}}{=} a_t^+(f(0)). \tag{2.1.63}$$

Differentiating with respect to t, using the product rule in (2.1.60), we obtain

$$i\frac{d}{dt}\mathbf{U}_t = \mathbf{H}\mathbf{U}_t = \mathbf{U}_t\mathbf{H}, \tag{2.1.64}$$

where

$$(\mathbf{H}\Phi)_n = \sum_{j=1}^n \mathbb{1} \otimes \ldots \otimes \underset{j}{H} \otimes \ldots \mathbb{1}\varphi_n$$
$$= \sum_{j=1}^n H(x_j)\varphi_n(x_1, \ldots, x_n). \tag{2.1.65}$$

In agreement with (2.1.31, 32), this is the second-quantized Hamiltonian

$$\mathbf{H} = \sum_{jj'}(f_j, Hf_{j'})a^+(f_j)a(f_{j'}). \tag{2.1.66}$$

Differentiating (2.1.63), one finds the Heisenberg equations of motion

$$i\frac{d}{dt}a_t^+(f) = -\mathbf{H}\mathbf{U}_t^{-1}a^+(f)\mathbf{U}_t + \mathbf{U}_t^{-1}a^+(f)\mathbf{U}_t\mathbf{H} = [a_t^+(f), \mathbf{H}]. \tag{2.1.67}$$

Without subscript, the square brackets always mean the commutator. In view of (2.1.67), the field operators $a_t(f)$, $a_t(\boldsymbol{x})$, *etc.* are called Heisenberg field operators. In contrast to non-relativistic quantum mechanics, in field theory the Heisenberg picture is more natural than the Schrödinger picture:

The states, i.e. the Fock space, remain unchanged, the observables change with time. The solution of (2.1.64) is the unitary transformation

$$\mathbf{U}_t = e^{-i\mathbf{H}t}, \tag{2.1.68}$$

and (2.1.63) then reads as follows

$$a_t^+(f) = e^{-i\mathbf{H}t} a^+(f) e^{i\mathbf{H}t}. \tag{2.1.69}$$

The connection with experiments is given by expectation values in Fock space, for example

$$\langle \mathbf{A} \rangle_\Phi \stackrel{\text{def}}{=} (\Phi, \mathbf{A}\Phi). \tag{2.1.70}$$

To get a Poincaré invariant quantization, we must modify the linear dependence on the test functions a little bit. Let us introduce the emission operator

$$\tilde{a}^+(f(-t)) = \int a^+(\boldsymbol{k}) \hat{f}(-t, \boldsymbol{k}) \frac{d^3k}{\sqrt{2E}} = \int a^+(\boldsymbol{k}) e^{iEt} \hat{f}(\boldsymbol{k}) \frac{d^3k}{\sqrt{2E}}$$

$$= (2\pi)^{-3/2} \int \frac{d^3k}{\sqrt{2E}} \int d^3x \, a^+(\boldsymbol{k}) e^{iEt - i\boldsymbol{k}\boldsymbol{x}} f(\boldsymbol{x}). \tag{2.1.71}$$

This leads to the following emission operator in x-space

$$\varphi^{(+)}(t, \boldsymbol{x}) = (2\pi)^{-3/2} \int \frac{d^3k}{\sqrt{2E}} a^+(\boldsymbol{k}) e^{iEt - i\boldsymbol{k}\boldsymbol{x}}. \tag{2.1.72}$$

Its adjoint gives the corresponding absorptive part

$$\varphi^{(-)}(x) = (2\pi)^{-3/2} \int \frac{d^3k}{\sqrt{2E}} a(\boldsymbol{k}) e^{-iEx_0 + i\boldsymbol{k}\boldsymbol{x}} \tag{2.1.73}$$

$$= \varphi^{(+)}(x)^+.$$

It now follows from (2.1.41a) that the commutator

$$[\varphi^{(-)}(x), \varphi^{(+)}(y)] = \frac{1}{(2\pi)^3} \int \frac{d^3k}{2E} e^{-iE(x_0 - y_0) + i\boldsymbol{k}(\boldsymbol{x}-\boldsymbol{y})}$$

$$= \frac{1}{(2\pi)^3} \int d^4k \, \delta(k^2 - m^2) \Theta(k^0) e^{-ik(x-y)} \tag{2.1.74}$$

is Poincaré invariant, if we choose the relativistic energy-momentum relation

$$E = k^0 = +\sqrt{\boldsymbol{k}^2 + m^2}. \tag{2.1.75}$$

$\varphi^{(+)}$, $\varphi^{(-)}$ and the total scalar field

$$\varphi(x) = \varphi^{(+)}(x) + \varphi^{(-)}(x) \tag{2.1.76}$$

are then solutions of the Klein-Gordon equation

$$(\square + m^2)\varphi(x) = 0. \tag{2.1.77}$$

Since $\varphi^+ = \varphi$, we have constructed the hermitean quantized scalar field. It describes neutral particles with spin 0.

The charged scalar field is obtained by using a second set of emission and absorption operators $b^+(\boldsymbol{k}), b(\boldsymbol{k})$ to describe the antiparticles. Instead of (2.1.76) we then get the following field operator

$$\tilde{\varphi}(x) = (2\pi)^{-3/2} \int \frac{d^3k}{\sqrt{2E}} \Big[a(\boldsymbol{k}) e^{-ikx} + b^+(\boldsymbol{k}) e^{ikx} \Big]. \tag{2.1.78}$$

The b, b^+ satisfy the same commutation rules as before (2.1.41a), and all b's commute with all a's. The commutation rule for $\tilde{\varphi}$ is then given by

$$[\tilde{\varphi}(x), \tilde{\varphi}(y)^+] = \frac{1}{(2\pi)^3} \int d^4k \, \delta(k^2 - m^2) \mathrm{sgn}\,(k^0) e^{-ik(x-y)}. \tag{2.1.79}$$

The invariant distributions (2.1.74, 79) are studied in detail in Sect. 2.3.

2.2 Quantization of the Dirac Field

The general method of second quantization, described in the last section, is now applied to the Dirac field. In this case, the one-particle Hilbert space is

$$\mathcal{H}_1 = (L^2(\mathbb{R}^3))^4 \ni f = \begin{pmatrix} f_1 \\ f_2 \\ f_3 \\ f_4 \end{pmatrix}, \tag{2.2.1}$$

with scalar product (see (1.4.8))

$$(f, g) = \int d^3x \, f(\boldsymbol{x})^+ g(\boldsymbol{x}). \tag{2.2.2}$$

Using the adjoint spinors, we can also write

$$(f^+, g^+) = (g, f). \tag{2.2.3}$$

The absorption operators are

$$a(f) = \int d^3x \, f(\boldsymbol{x})^+ a(\boldsymbol{x}), \tag{2.2.4}$$

where the spinor field $a(\boldsymbol{x})$ in (2.2.4) is a 4×1 column matrix. Similarly, the emission operators are given by

$$a(f)^+ = \int d^3x \, a(\boldsymbol{x})^+ f(\boldsymbol{x}). \tag{2.2.5}$$

They depend linearly on f, whereas $a(f)$ in (2.2.4) is antilinear in f.

Now we shall consider a one-particle dynamics in $\mathcal{H}_1$, given by the Hamiltonian (1.5.24)

$$H = \boldsymbol{\alpha} \cdot \boldsymbol{p} + \beta m + e(V(\boldsymbol{x}) - \boldsymbol{\alpha} \cdot \boldsymbol{A}(\boldsymbol{x})) \tag{2.2.6}$$

$$\hbar = c = 1.$$

Here we quantize the Dirac field in a time-independent external electromagnetic field. This external field problem is of considerable practical interest and, of course, the free field is trivially included. The spectrum of the Hamiltonian (2.2.6) consists of a positive and negative part. Let $P_\pm$ be the projection operators on the corresponding spectral subspaces $\mathcal{H}_\pm = P_\pm \mathcal{H}_1$

$$P_+ + P_- = \mathbb{1} \quad , \quad P_+ P_- = 0.$$

$$\mathcal{H}_1 = P_+ \mathcal{H}_1 + P_- \mathcal{H}_1. \tag{2.2.7}$$

We split the spinor field according to this decomposition

$$a(f) = b(P_+ f) + d(P_- f) \quad , \quad a(f)^+ = b(P_+ f)^+ + d(P_- f)^+, \tag{2.2.8}$$

setting

$$b(P_- f) = 0 = d(P_+ f).$$

In agreement with (2.1.41), the Fermi operators fulfill the following anticommutation relations

$$\{b(f), b(g)^+\} = (P_+ f, P_+ g)$$

$$\{d(f), d(g)^+\} = (P_- f, P_- g), \quad \text{otherwise } 0. \tag{2.2.9}$$

The curly brackets always mean anticommutators from now on.

Let f_j be a basis in $\mathcal{H}_+$ and g_j a basis in $\mathcal{H}_-$, both sets of basis vectors belonging to the domain of H in (2.2.6). Since the subspaces $\mathcal{H}_\pm$ are invariant subspaces with respect to H, the corresponding second quantized Hamiltonian is

$$\tilde{\mathsf{H}} = \sum_{jk} [(f_j, H f_k) b(f_j)^+ b(f_k) + (g_j, H g_k) d(g_j)^+ d(g_k)], \tag{2.2.10}$$

in accordance with (2.1.66). This Fock space operator coincides with H on $\mathcal{H}_1$ and, hence, is also unbounded from below. The second term in (2.2.10) is responsible for this. This leads us to the cure of the defect: We interchange the role of d^+ and d which produces a minus sign, since $dd^+ = -d^+ d$. This has no influence on the algebraic structure, however, the vacuum defined by

$$b(f)\Omega = 0, \; d(f)\Omega = 0, \quad \forall f \in \mathcal{H}_1 \tag{2.2.11}$$

is completely changed. Instead of (2.2.8), we therefore consider

$$\psi(f)^+ = b(f_+) + d(f_-^+)^+, \quad \text{with} \quad f_\pm = P_\pm f. \tag{2.2.12}$$

The adjoint in the test function of d is necessary in order to have a antilinear dependence on f. To simplify notation, we omit this adjoint in the following, assuming the new convention that $d(f_-)^+$ is antilinear in f and $d(f_-)$ linear. The quantized Dirac fields then are

$$\psi(f) = b(f_+) + d(f_-)^+$$
$$\psi(f)^+ = b(f_+)^+ + d(f_-). \tag{2.2.13}$$

The anticommutation relations follow from (2.2.9):

$$\{\psi(f),\, \psi(g)^+\} = \{b(f_+),\, b(g_+)^+\} + \{d(f_-)^+,\, d(g_-)\}$$
$$= (f_+,\, g_+) + (g_-^+,\, f_-^+) = (f,\, g), \tag{2.2.14}$$

where (2.2.3) has been used. As before, one introduces the corresponding field operators (distributions)

$$\psi(f) = \int d^3x\, f(\boldsymbol{x})^+ \psi(\boldsymbol{x}) \tag{2.2.15}$$

$$\psi(f)^+ = \int d^3x\, \psi^+(\boldsymbol{x}) f(\boldsymbol{x}). \tag{2.2.16}$$

To get a positive Hamiltonian for the spin $\frac{1}{2}$ Dirac field, it is essential to quantize it with anticommutators. This is part of the so-called spin-statistics theorem.

Next we consider the time evolution of the Dirac field. In view of (2.1.69), one defines

$$\psi_t^+(f) \stackrel{\text{def}}{=} \psi(e^{iHt} f)^+ = e^{i\mathbf{H}t} \psi(f)^+ e^{-i\mathbf{H}t}. \tag{2.2.17}$$

The Hamiltonian $\mathbf{H}$ has the same form as (2.2.10), but with d and d^+ interchanged

$$\mathbf{H} = \sum_{jk} [(f_j,\, Hf_k) b(f_j)^+ b(f_k) + (g_j,\, Hg_k) d(g_j) d(g_k)^+]$$

$$= \sum_{jk} [(f_j,\, Hf_k) b(f_j)^+ b(f_k) - (g_j,\, Hg_k) d(g_k)^+ d(g_j) + \text{C-number}]. \tag{2.2.18}$$

The unimportant C-number will be dropped, because it gives no contribution in (2.2.17). Strictly speaking, the first line in (2.2.18) is only a formal expression not a well-defined operator, because the C-number is infinite. The well-defined Hamiltonian $\mathbf{H}$ is the normally ordered expression without the C-number. Since

$$\mathbf{H}|_{P_+\mathcal{H}_1} = H|_{P_+\mathcal{H}_1}\,,\ \mathbf{H}|_{P_-\mathcal{H}_1} = -H|_{P_-\mathcal{H}_1} \geq 0, \tag{2.2.19}$$

this operator is bounded from below. It is normally ordered

$$\mathbf{H} = \int d^3x\, :\psi(\boldsymbol{x})^+ H \psi(\boldsymbol{x}):, \tag{2.2.20}$$

which is always denoted by double dots. The Heisenberg equations of motion follow from (2.2.17)

$$i\frac{d}{dt}\psi_t^+(f) = [\psi_t^+(f), \mathbf{H}]. \tag{2.2.21}$$

The combination of emission with absorption operators in the Dirac field (2.2.13) has far-reaching consequences which we shall investigate later on (pair production and annihilation, vacuum polarization). To get a particle interpretation, we study the expression for the charge which classically was of the following form (see (1.5.22))

$$Q = e\int d^3x\, \psi(t,\boldsymbol{x})^+\psi(t,\boldsymbol{x}),$$

and after quantization, applying (2.1.27),

$$= e\sum_{jk}\int d^3x\,[b(f_j)^+f_j(\boldsymbol{x})^+ + d(g_j)g_j(\boldsymbol{x})^+][b(f_k)f_k(\boldsymbol{x}) + d(g_k)^+g_k(\boldsymbol{x})]$$

$$= e\sum_j[b(f_j)^+b(f_j) + d(g_j)d(g_j)^+]$$

$$= e\sum_j[b(f_j)^+b(f_j) - d(g_j)^+d(g_j)] + \text{C-number}.$$

The C-number will be omitted, such that the vacuum expectation value vanishes

$$(\Omega,\, \mathbf{Q}\Omega) = 0. \tag{2.2.22}$$

The resulting charge operator in Fock space is again normally ordered

$$\mathbf{Q} = e\sum_j[b(f_j)^+b(f_j) - d(g_j)^+d(g_j)]$$

$$= e\int d^3x\, :\psi(x)^+\psi(x):\,. \tag{2.2.23}$$

The d operators appear with a minus sign, they therefore describe antiparticles. To be definite, we refer to the b's as electron and to the d's as positron emission and absorption operators.

Now we shall consider the special case of the free Dirac field. According to (1.4.37, 38), the one-particle wave function is

$$f(\boldsymbol{x}) = (2\pi)^{-3/2}\int d^3p\,[f_{s+}(\boldsymbol{p})u_s(\boldsymbol{p})e^{i\boldsymbol{p}\cdot\boldsymbol{x}} + f_{s-}(\boldsymbol{p})v_s(\boldsymbol{p})e^{-i\boldsymbol{p}\cdot\boldsymbol{x}}], \tag{2.2.24}$$

where

$$f_{s+}(\boldsymbol{p}) = u_s^+(\boldsymbol{p})\hat{f}(\boldsymbol{p})$$

$$f_{s-}(\boldsymbol{p}) = v_s^+(\boldsymbol{p})\hat{f}(-\boldsymbol{p}), \tag{2.2.25}$$

and the hat means the ordinary Fourier transform. Let f_j, g_j be a complete orthonormal system in $\mathcal{H}_1 = \mathcal{H}_+ \oplus \mathcal{H}_-$, then the Dirac field is given by (2.1.27)

$$\psi(\boldsymbol{x}) = \sum_j [\psi(f_j)f_j(\boldsymbol{x}) + \psi(g_j)g_j(\boldsymbol{x})]$$

$$= \sum_j [b(f_j)f_j(\boldsymbol{x}) + d(g_j)^+ g_j(\boldsymbol{x})]. \tag{2.2.26}$$

Using (2.2.25), we obtain

$$\sum_j d(g_j)^+ g_{js}(\boldsymbol{p}) = v_s^+(\boldsymbol{p}) \sum_j d(g_j)^+ \hat{g}_j(-\boldsymbol{p})$$
$$\stackrel{\text{def}}{=} v_s^+(\boldsymbol{p})\hat{d}(-\boldsymbol{p})^+ \stackrel{\text{def}}{=} d_s(\boldsymbol{p})^+ \tag{2.2.27}$$

and similarly

$$b_s(\boldsymbol{p}) = \sum_j b(f_j)f_{js}(\boldsymbol{p}). \tag{2.2.28}$$

Here we have introduced the field operator-distributions in momentum space, satisfying the following anticommutation relations

$$\{b_s(\boldsymbol{p}),\, b_{s'}(\boldsymbol{p}')^+\} = \sum_{jk} (f_j,\, f_k) f_{js}(\boldsymbol{p}) f_{ks'}(\boldsymbol{p}')^*$$

$$= \delta^3(\boldsymbol{p} - \boldsymbol{p}')\delta_{ss'} \tag{2.2.29}$$

$$\{d_s(\boldsymbol{p})^+,\, d_{s'}(\boldsymbol{p}')\} = \delta^3(\boldsymbol{p} - \boldsymbol{p}')\delta_{ss'}, \tag{2.2.30}$$

and zero otherwise.

The time evolution of the free Dirac field follows from (2.2.7)

$$\psi(t, \boldsymbol{x}) \stackrel{\text{def}}{=} \sum_j \Big[\psi(e^{iH_0 t} f_j) f_j(\boldsymbol{x}) + \psi(e^{iH_0 t} g_j) g_j(\boldsymbol{x})\Big]$$

$$= \sum_j \Big[b(P_+ e^{iH_0 t} f_j) f_j(\boldsymbol{x}) + d(P_- e^{iH_0 t} g_j)^+ g_j(\boldsymbol{x})\Big]. \tag{2.2.31}$$

We compute the first term

$$\psi_t^{(-)}(f_j) \stackrel{\text{def}}{=} b(P_+ e^{iH_0 t} f_j) = \int d^3p\, (P_+ e^{iH_0 t} f_j)_s^*(\boldsymbol{p}) b_s(\boldsymbol{p})$$

$$= \int d^3p (e^{iEt} u_s^+(\boldsymbol{p}) \hat{f}_j(\boldsymbol{p}))^* b_s(\boldsymbol{p}) = \int d^3p\, e^{-iEt} \hat{f}_j(\boldsymbol{p})^+ u_s(\boldsymbol{p}) b_s(\boldsymbol{p}).$$

Using

$$f_j(\boldsymbol{x}) = (2\pi)^{-3/2} \int d^3p' \hat{f}_j(\boldsymbol{p}') e^{i\boldsymbol{p}'\cdot\boldsymbol{x}},$$

we arrive at

$$\psi^{(-)}(x) = (2\pi)^{-3/2} \int d^3p\, e^{-i(Et-\boldsymbol{p}\cdot\boldsymbol{x})} u_s(\boldsymbol{p}) b_s(\boldsymbol{p})$$

$$= (2\pi)^{-3/2} \int d^3p\, e^{-ipx} u_s(\boldsymbol{p}) b_s(\boldsymbol{p}). \tag{2.2.32}$$

The second term in (2.2.31) is treated in the same way:

$$\psi_t^{(+)}(g_j) \stackrel{\text{def}}{=} d(P_- e^{iH_0 t} g_j)^+ = \int d^3p\, (P_- e^{iH_0 t} g_j)_s^* d_s(\boldsymbol{p})^+$$

$$= \int d^3p\, (e^{-iEt} v_s^+(\boldsymbol{p}) \hat{g}_j(-\boldsymbol{p}))^* d_s(\boldsymbol{p})^+ = \int d^3p\, e^{iEt} \hat{g}_j(-\boldsymbol{p})^+ v_s(\boldsymbol{p}) d_s(\boldsymbol{p})^+.$$

With

$$g_j(\boldsymbol{x}) = (2\pi)^{-3/2} \int d^3p' \hat{g}_j(-\boldsymbol{p}') e^{-i\boldsymbol{p}'\cdot\boldsymbol{x}},$$

one gets

$$\psi^{(+)}(x) = (2\pi)^{-3/2} \int d^3p\, e^{i(Et-\boldsymbol{p}\cdot\boldsymbol{x})} v_s(\boldsymbol{p}) d_s(\boldsymbol{p})^+$$

$$= (2\pi)^{-3/2} \int d^3p\, e^{ipx} v_s(\boldsymbol{p}) d_s(\boldsymbol{p})^+. \tag{2.2.33}$$

Totally, the time-dependent Dirac field is

$$\psi(x) = (2\pi)^{-3/2} \int d^3p\, [b_s(\boldsymbol{p}) u_s(\boldsymbol{p}) e^{-ipx} + d_s(\boldsymbol{p})^+ v_s(\boldsymbol{p}) e^{ipx}]. \tag{2.2.34}$$

The result has the same form as a classical solution (1.4.35) of the Dirac equation. The first term $\psi^{(-)}$ contains the electron absorption operator, whereas the second term $\psi^{(+)}$ describes positron creation. This splitting of the fields will always be necessary in the further development of the theory. The adjoint Dirac field is obtained from (2.2.34)

$$\psi^+(x) = (2\pi)^{-3/2} \int d^3p\, [b_s^+(\boldsymbol{p}) u_s(\boldsymbol{p})^+ e^{ipx} + d_s(\boldsymbol{p}) v_s(\boldsymbol{p})^+ e^{-ipx}].$$

Multiplying by γ^0, we get the Dirac adjoint

$$\overline{\psi}(x) = \psi^+(x)\gamma^0 = \overline{\psi}^{(+)} + \overline{\psi}^{(-)}$$

$$\overline{\psi}^{(+)} = (2\pi)^{-3/2} \int d^3p\, b_s^+(\boldsymbol{p}) \overline{u}_s(\boldsymbol{p}) e^{ipx}$$

$$\overline{\psi}^{(-)}(x) = (2\pi)^{-3/2} \int d^3p\, d_s(\boldsymbol{p}) \overline{v}_s(\boldsymbol{p}) e^{-ipx}. \tag{2.2.35}$$

It is now easy to compute the anticommutation relations for the free Dirac field for arbitrary times. With the aid of (2.2.29, 30), one finds

$$\{\psi_a^{(-)}(x), \overline{\psi}_b^{(+)}(y)\} = (2\pi)^{-3} \int d^3p\, u_{sa}(\boldsymbol{p}) \overline{u}_{sb}(\boldsymbol{p}) e^{-ip(x-y)}. \tag{2.2.36}$$

The integral here is understood as a distributional Fourier transform. In the result (2.2.36), the covariant positive spectral projection operator (1.4.43) appears

$$\{\psi^{(-)}(x), \overline{\psi}^{(+)}(y)\} = (2\pi)^{-3} \int \frac{d^3p}{2E} (\not{p} + m) e^{-ip(x-y)} \stackrel{\text{def}}{=} \frac{1}{i} S^{(+)}(x-y). \tag{2.2.37}$$

In the same way, one obtains the other non-vanishing anticommutator

$$\{\psi^{(+)}(x), \overline{\psi}^{(-)}(y)\} \stackrel{\text{def}}{=} \frac{1}{i} S^{(-)}(x-y) = (2\pi)^{-3} \int \frac{d^3p}{2E} (\not{p} - m) e^{ip(x-y)}. \tag{2.2.38}$$

This gives the anticommutation relation for the total Dirac field

$$\{\psi(x), \overline{\psi}(y)\} = \frac{1}{i} S(x-y), \tag{2.2.39}$$

with

$$S(x) = S^{(-)}(x) + S^{(+)}(x). \tag{2.2.40}$$

The Lorentz invariant distributions $S^{(+)}$, $S^{(-)}$, S, appearing here, will be further studied in the following section.

Next we shall consider charge conjugation in Fock space. We remember that the charge-conjugate spinor (1.5.52)

$$\psi_C = C\gamma^0\psi^* = U_C\psi \tag{2.2.41}$$

has been introduced in the one-particle theory, where U_C is an anti-unitary operator satisfying

$$U_C H_0 U_C^{-1} = -H_0, \tag{2.2.42}$$

and, hence,

$$U_C P_\pm U_C^{-1} = P_\mp. \tag{2.2.43}$$

The corresponding second-quantized operator $\mathbf{U}_C$ is defined by

$$\mathbf{U}_C \psi(f) \mathbf{U}_C^{-1} = \psi(f_C)^+, \tag{2.2.44}$$

and is assumed to be unitary, so that

$$\mathbf{U}_C \psi(f)^+ \mathbf{U}_C^{-1} = \psi(f_C) \quad , \quad \forall f \in \mathcal{H}_1. \tag{2.2.45}$$

Proposition 2.1. $\mathbf{U}_C$ is uniquely determined by (2.2.44, 45) up to a phase.

Proof. If $\tilde{\mathbf{U}}_C$ is another operator in $\mathcal{F}$, satisfying (2.2.44), then

$$\mathbf{U}_C \psi(f) \mathbf{U}_C^{-1} = \tilde{\mathbf{U}}_C \psi(f) \tilde{\mathbf{U}}_C^{-1},$$

$$\tilde{\mathbf{U}}_C^{-1} \mathbf{U}_C \psi(f) = \psi(f) \tilde{\mathbf{U}}_C^{-1} \mathbf{U}_C \quad , \quad \forall f \in \mathcal{H}_1,$$

and the same is true for all $\psi^+(f)$. It follows from the irreducibility of the Fock representation (Theorem 1.2 (2.1.43, 44)) that

$$\tilde{\mathbf{U}}_C^{-1}\mathbf{U}_C = \alpha \mathbb{1} \quad \text{i.e.} \quad \mathbf{U}_C = \alpha \tilde{\mathbf{U}}_C. \tag{2.2.46}$$

Since both operators are assumed as unitary, it follows $|\alpha| = 1$. □

The adjoint has been chosen in (2.2.44), in order to get the same transformation law for the field operator as in the one-particle theory (see (2.2.53)). Let us go on to further restrict the phase factor α in (2.2.46). If we require invariance of the vacuum

$$\mathbf{U}_C \Omega = \Omega = \tilde{\mathbf{U}}_C \Omega, \tag{2.2.47}$$

then it follows $\alpha = 1$, and the unitary charge conjugation operator in Fock space is uniquely determined.

In order to show the existence and to obtain the physical interpretation of charge conjugation $\mathbf{U}_C$, we substitute

$$\psi(f) = b(P_+ f) + d(P_- f)^+ \tag{2.2.48}$$

into (2.2.44). If $f = f_- \in P_- \mathcal{H}_1$, then

$$P_- U_C f_- = U_C P_+ f_- = 0$$

and we get from (2.2.44)

$$\mathbf{U}_C d(f_-)^+ \mathbf{U}_C^{-1} = b(U_C f_-)^+ \quad \text{or}$$

$$\mathbf{U}_C d(f_-)^+ = b(U_C f_-)^+ \mathbf{U}_C. \tag{2.2.49}$$

By (2.2.47), this shows that positron states are mapped onto electron states

$$\mathbf{U}_C d(f_-)^+ \Omega = b(U_C f_-)^+ \Omega, \tag{2.2.50}$$

and vice versa.

We now want to formulate charge conjugation for the distributional field operators. We write (2.2.44) in the form

$$\int d^3x\, f_a^*(\boldsymbol{x})(\mathbf{U}_C \psi(\boldsymbol{x}) \mathbf{U}_C^{-1})_a = \int d^3x\, \psi^+(\boldsymbol{x}) C\gamma^0 f^*(\boldsymbol{x}), \tag{2.2.51}$$

where (2.2.41) has been used. Since $C\gamma^0 = -\gamma^0 C$, this becomes

$$= -\int d^3x\, \overline{\psi}(\boldsymbol{x}) C f(\boldsymbol{x})^* = -\int d^3x\, (C^T \overline{\psi}(\boldsymbol{x})^T)_a f_a(\boldsymbol{x})^*$$

$$= \int d^3x\, (C\overline{\psi}(\boldsymbol{x})^T)_a f_a(\boldsymbol{x})^*, \tag{2.2.52}$$

because $C^T = -C$. We conclude that

$$\mathbf{U}_C \psi(\boldsymbol{x}) \mathbf{U}_C^{-1} = C\overline{\psi}(\boldsymbol{x})^T. \tag{2.2.53}$$

This is the desired relation, which may also be written as

$$C^{-1}\psi(\boldsymbol{x}) = \mathbf{U}_C^{-1}\overline{\psi}(\boldsymbol{x})^T\mathbf{U}_C = \mathbf{U}_C\overline{\psi}(\boldsymbol{x})^T\mathbf{U}_C^{-1} \quad \text{or}$$

$$\mathbf{U}_C\overline{\psi}(\boldsymbol{x})\mathbf{U}_C^{-1} = \psi(\boldsymbol{x})^T C. \tag{2.2.54}$$

The reason why we have discussed charge conjugation only for the free field and not in the presence of an external potential is the following: The potential would be changed under charge conjugation ($e \to -e$), which causes a change of the Fock representation. Then, $\mathbf{U}_C$ maps between different Fock spaces. This complicates the discussion without gaining much. Nevertheless, we shall be able to discuss the full content of charge conjugation in the S-matrix by means of (2.2.44, 45). The other discrete symmetries (parity and time-reversal) are discussed in Sect. 4.4.

We close this section by a discussion of Poincaré covariance of the free Dirac field. As we have found in Sect. 1.4 (1.4.63), there is a unitary representation of the Poincaré group $\mathcal{P}_+^\uparrow$ in $\mathcal{H}_1$:

$$(U(a,\Lambda)f)(x) = S(\Lambda)f(\Lambda^{-1}(x-a)).$$

It can be lifted into Fock space $\mathcal{F}$ as usual

$$\psi(U(a,\Lambda)f) \stackrel{\text{def}}{=} \mathbf{U}(a,\Lambda)\psi(f)\mathbf{U}(a,\Lambda)^{-1} \quad , \quad \forall f \in \mathcal{H}_1$$

$$= \int d^3x\, (S(\Lambda)f(\Lambda^{-1}(x-a)))^+\psi(x). \tag{2.2.55}$$

As in (1.4.59), this may be represented in invariant form

$$= \int d\sigma^\mu(x)\overline{(S(\Lambda)f(\Lambda^{-1}(x-a)))}\,\gamma_\mu\psi(x)$$

$$= \int d\sigma^\mu(x)\overline{f}(\Lambda^{-1}(x-a))S(\Lambda)^{-1}\gamma_\mu\psi(x). \tag{2.2.56}$$

Inserting here SS^{-1} in front of $\psi(x)$ and using the transformed coordinates $x' = \Lambda^{-1}(x-a)$, we find with the aid of (1.2.34)

$$= \int d\sigma^\mu(x)\overline{f}(x')\Lambda_\mu{}^\nu\gamma_\nu S(\Lambda)^{-1}\psi(\Lambda x'+a)$$

$$= \int d\sigma^\nu(x')\overline{f}(x')\gamma_\nu S(\Lambda^{-1})\psi(\Lambda x'+a) = \int d^3x\, f(x)^+S(\Lambda^{-1})\psi(\Lambda x+a). \tag{2.2.57}$$

This is equivalent to the following transformation law for the field operator

$$\mathbf{U}(a,\Lambda)\psi(x)\mathbf{U}(a,\Lambda)^{-1} = S(\Lambda^{-1})\psi(\Lambda x+a), \tag{2.2.58}$$

in formal agreement with the one-particle theory (1.2.32).

2.3 Discussion of the Commutation Functions

The various commutation functions introduced in the last section will now be discussed in some detail. In particular, the support properties of these functions in x-space will be very important later on.

The function $S^{(+)}(x)$ was defined in (2.2.37) as the following distributional Fourier transform

$$S^{(+)}(x) \stackrel{\text{def}}{=} \frac{i}{(2\pi)^3} \int \frac{d^3p}{2E} (\not{p} + m) e^{-ipx}$$

$$= \frac{i}{(2\pi)^3} (i\not{\partial}_x + m) \int \frac{d^3p}{2E} e^{-ipx} \stackrel{\text{def}}{=} (i\not{\partial}_x + m) D^{(+)}(x). \tag{2.3.1}$$

We have expressed $S^{(+)}(x)$ by the simpler scalar distribution $D^{(+)}(x)$. In order to show that $D^{(+)}(x)$ is indeed a Lorentz scalar, we rewrite the 3-dimensional Fourier integral in (2.3.1) in 4-dimensional form. For this purpose, the p_0-integration must be introduced trivially by means of the following δ-distribution

$$\delta(p_0^2 - E^2) = \delta(p^2 - m^2) = \frac{\delta(p_0 - E)}{2E} + \frac{\delta(p_0 + E)}{2E}. \tag{2.3.2}$$

Since we have $p_0 = E$ in (2.3.1), the second term in (2.3.2) has to be cut off by a Θ-function:

$$D^{(+)}(x) = \frac{i}{(2\pi)^3} \int d^4p\, \delta(p^2 - m^2) \Theta(p^0) e^{-ipx}. \tag{2.3.3}$$

This is obviously Lorentz invariant. The result shows that the 3-dimensional integration $\int d^3p/2E$ in (2.3.1) is equivalent to the Lorentz invariant integration over one shell of the mass hyperboloid $p^2 = m^2$.

The function $S^{(-)}(x)$ defined in (2.2.38) is treated in the same way

$$S^{(-)}(x) = \frac{i}{(2\pi)^3} \int \frac{d^3p}{2E} (\not{p} - m) e^{ipx}$$

$$= -\frac{i}{(2\pi)^3} (i\not{\partial}_x + m) \int \frac{d^3p}{2E} e^{ipx} \stackrel{\text{def}}{=} (i\not{\partial}_x + m) D^{(-)}(x). \tag{2.3.4}$$

The new invariant distribution $D^{(-)}(x)$ may be written as

$$D^{(-)}(x) = -\frac{i}{(2\pi)^3} \int d^4p\, \delta(p^2 - m^2)\, \Theta(p^0) e^{ipx}$$

$$= -\frac{i}{(2\pi)^3} \int d^4p\, \delta(p^2 - m^2)\, \Theta(-p^0) e^{-ipx}. \tag{2.3.5}$$

The total anti-commutator (2.2.39) of the Dirac field is then

$$S(x) = S^{(-)}(x) + S^{(+)}(x) \stackrel{\text{def}}{=} (i\not{\partial} + m)D(x). \tag{2.3.6}$$

The function

$$D(x) = D^{(-)}(x) + D^{(+)}(x) = \frac{i}{(2\pi)^3}\int d^4p\,\delta(p^2 - m^2)\,\text{sgn}\,p^0 e^{-ipx} \tag{2.3.7}$$

is the Jordan-Pauli function.

We now proceed to investigate the Jordan-Pauli function (2.3.7) in x-space. For this purpose, it is convenient to utilize the original 3-dimensional Fourier representations

$$D(x) = \frac{i}{(2\pi)^3}\int \frac{d^3p}{2E}\,e^{i\boldsymbol{p}\cdot\boldsymbol{x}}(e^{-iEx_0} - e^{iEx_0})$$

$$= (2\pi)^{-3}\int \frac{d^3p}{2E}\,2\sin Ex_0\,e^{i\boldsymbol{p}\cdot\boldsymbol{x}} \quad , \quad E = +\sqrt{\boldsymbol{p}^2 + m^2}. \tag{2.3.8}$$

The distributional Fourier transform in (2.3.8) must be understood as the following weak (distributional) limit ($p = |\boldsymbol{p}|$)

$$= (2\pi)^{-2}\text{w} - \lim_{P\to\infty}\int_0^P dp\frac{\boldsymbol{p}^2}{\sqrt{\boldsymbol{p}^2+m^2}}\sin\sqrt{\boldsymbol{p}^2+m^2}x_0\int_{-1}^{+1} d\cos\theta\,e^{ipr\cos\theta}$$

$$= \frac{2}{(2\pi)^2 r}\,\text{w} - \lim_{P\to\infty}\int_0^P dp\,\frac{p}{\sqrt{\boldsymbol{p}^2+m^2}}\sin\sqrt{\boldsymbol{p}^2+m^2}x_0\sin pr$$

$$= -\frac{1}{2\pi^2 r}\frac{d}{dr}\int_0^\infty \frac{dp}{E}\sin Ex_0\cos pr$$

$$= -\frac{1}{8\pi^2 r}\frac{d}{dr}\int_{-\infty}^\infty \frac{dp}{E}\left[\sin(Ex_0 + pr) + \sin(Ex_0 - pr)\right]. \tag{2.3.9}$$

The two integrals in (2.3.9) are abbreviated as I_+, I_-. Using the substitutions

$$p = m\sinh z \quad , \quad E = \sqrt{\boldsymbol{p}^2 + m^2} = m\cosh z, \tag{2.3.10}$$

we obtain

$$I_\pm = \int_{-\infty}^\infty dz\,\sin(x_0 m\cosh z \pm rm\sinh z). \tag{2.3.11}$$

In order to evaluate this integral, we must distinguish three cases: (i) For $x_0 > r$, we write

$$x_0 = a\cosh\tau \quad , \quad r = a\sinh\tau \quad , \quad a = \sqrt{x_0^2 - r^2}, \tag{2.3.12}$$

and get

$$I_\pm = \int\limits_{-\infty}^{\infty} dz \, \sin[ma\cosh(z \pm \tau)] = \pi J_0(ma), \tag{2.3.13}$$

where J_0 is the Bessel function of order 0. (ii) For $x_0 < -r$, we write

$$x_0 = -a\cosh\tau \quad , \quad r = a\sinh\tau, \tag{2.3.14}$$

and obtain in the same way

$$I_\pm = -\pi J_0(ma). \tag{2.3.15}$$

(iii) In the remaining case $-r < x_0 < r$, we use the substitutions

$$x_0 = b\sinh\tau \quad , \quad r = b\cosh\tau \quad , \quad b = \sqrt{r^2 - x_0^2}, \tag{2.3.16}$$

then

$$I_\pm = \int\limits_{-\infty}^{\infty} dz \, \sin[mb\sinh(\tau \pm z)] = 0. \tag{2.3.17}$$

Summing up, we find the following expression for the Jordan-Pauli function

$$\begin{aligned} D(x) &= -\frac{1}{4\pi r}\frac{d}{dr}\mathrm{sgn}x_0 \Theta(x^2) J_0\left(m\sqrt{x_0^2 - r^2}\right) \\ &= \frac{1}{2\pi}\mathrm{sgn}x_0 \frac{d}{d(x^2)}\Theta(x^2) J_0\left(m\sqrt{x^2}\right) \\ &= \frac{1}{2\pi}\mathrm{sgn}x_0 \left[\delta(x^2) - \Theta(x^2)\frac{m}{2\sqrt{x^2}} J_1\left(m\sqrt{x^2}\right)\right], \end{aligned} \tag{2.3.18}$$

where $J_0(0) = 1$ has been used and J_1 is the Bessel function of order 1. We emphasize the fact that $D(x)$ vanishes for space-like x:

$$\mathrm{supp}\, D = \{x \in \mathbb{M} \mid x^2 \geq 0\}. \tag{2.3.19}$$

Since time-like vectors (2.3.19) are causally connected with $x = 0$, one says that $D(x)$ has a causal support. The same is then true for $S(x)$ (2.3.6). The sign-function in (2.3.18) can be decomposed as follows

$$\mathrm{sgn}\,(x_0) = \Theta(x_0) - \Theta(-x_0).$$

Accordingly, the Jordan-Pauli function has the *causal* decomposition into retarded and advanced functions

$$D(x) = D^{\mathrm{ret}}(x) - D^{\mathrm{av}}(x) \quad , \quad \text{with} \tag{2.3.20}$$

$$D^{\mathrm{ret}}(x) = \Theta(x^0)D(x) \quad , \quad D^{\mathrm{av}}(x) = \Theta(-x^0)D(x). \tag{2.3.21}$$

These are well-defined tempered distributions because by means of (2.3.8) they can be expressed as four-dimensional distributional Fourier transforms:

$$D^{\text{ret}}(x) = \frac{\Theta(x^0)}{(2\pi)^3}\int \frac{d^3p}{2E}\, 2\sin Ex^0\, e^{i\mathbf{p}\cdot\mathbf{x}}$$

$$= (2\pi)^{-4}\int d^4p \frac{e^{-ipx}}{m^2 - p^2 - ip^0 0} \tag{2.3.22}$$

$$D^{\text{av}}(x) = (2\pi)^{-4}\int d^4p \frac{e^{-ipx}}{m^2 - p^2 + ip^0 0}. \tag{2.3.23}$$

Note that the 4-dimensional distributional Fourier transformation has by definition a factor $(2\pi)^{-2}$. Therefore, the Fourier transforms of (2.3.22, 23) are given by

$$\hat{D}^{\text{ret}}(p) = (2\pi)^{-2}\frac{1}{m^2 - p^2 - ip^0 0} \tag{2.3.24}$$

$$\hat{D}^{\text{av}}(p) = (2\pi)^{-2}\frac{1}{m^2 - p^2 + ip^0 0}. \tag{2.3.25}$$

Similarly, one introduces retarded and advanced distributions for the spinor field

$$S^{\text{ret}}(x) \stackrel{\text{def}}{=} (i\partial\!\!\!/ + m)D^{\text{ret}}(x) = -(2\pi)^{-4}\int d^4p \frac{p\!\!\!/ + m}{p^2 - m^2 + ip^0 0}e^{-ipx} \tag{2.3.26}$$

$$S^{\text{av}}(x) \stackrel{\text{def}}{=} (i\partial\!\!\!/ + m)D^{\text{av}}(x) = -(2\pi)^{-4}\int d^4p \frac{p\!\!\!/ + m}{p^2 - m^2 - ip^0 0}e^{-ipx}. \tag{2.3.27}$$

In p-space, we have

$$\hat{S}^{\text{ret}}(p) = -(2\pi)^{-2}\frac{p\!\!\!/ + m}{p^2 - m^2 + ip^0 0} \tag{2.3.28}$$

$$\hat{S}^{\text{av}}(p) = -(2\pi)^{-2}\frac{p\!\!\!/ + m}{p^2 - m^2 - ip^0 0}.$$

We now proceed to investigate the function $D^{(+)}(x)$ (2.3.3). Starting from the 3-dimensional Fourier integral

$$D^{(+)}(x) = \frac{i}{(2\pi)^3}\int \frac{d^3p}{2E}\, e^{i\mathbf{p}\cdot\mathbf{x}}(\cos Ex_0 - i\sin Ex_0) \tag{2.3.29}$$

$$\stackrel{\text{def}}{=} D_1(x) + \frac{1}{2}D(x), \tag{2.3.30}$$

where D_1 is the term with $\cos Ex_0$ in (2.3.29), we observe that only this term remains to be evaluated, because the sin-term gives the Jordan-Pauli function (2.3.8). By the same method as above (2.3.9), we arrive at

$$D_1(x) = -\frac{i}{16\pi^2 r}\frac{d}{dr}\int_{-\infty}^{\infty}\frac{dp}{E}\left[\cos(Ex_0 + pr) + \cos(Ex_0 - pr)\right]. \tag{2.3.31}$$

To compute the two integrals

$$I_{\pm} = \int_{-\infty}^{\infty} dz \cos(mx_0 \cosh z \pm mr \sinh z),$$

in (2.3.31), we use the same substitutions as in the three cases above. The results in the cases (i) $x_0 > r$ and (ii) $x_0 < -r$ agree

$$I_{\pm} = \int_{-\infty}^{\infty} dz \cos(ma \cosh z) = -\pi N_0(ma). \tag{2.3.32}$$

Here $N_0(z)$ is the Neumann function of order 0. In the case (iii) $-r < x_0 < r$, we do not get 0 as in (2.3.17), but

$$I_{\pm} = \int_{-\infty}^{\infty} dz \cos(mb \sinh z) = 2K_0(mb),$$

where $K_0(z)$ is the modified Bessel function. This yields

$$D_1(x) = \frac{i}{4\pi} \frac{d}{d(x^2)} \left[-\Theta(x^2) N_0\left(m\sqrt{x^2}\right) + \Theta(-x^2) \frac{2}{\pi} K_0\left(m\sqrt{-x^2}\right) \right]. \tag{2.3.33}$$

To evaluate the distributional derivative, we need the asymptotic behaviour of the Bessel functions at 0

$$N_0\left(m\sqrt{x^2}\right) \longrightarrow \frac{2}{\pi} \left(\log\left(\frac{m}{2}\sqrt{x^2}\right) + \gamma \right)$$

$$K_0\left(m\sqrt{-x^2}\right) \longrightarrow -\left(\log\left(\frac{m}{2}\sqrt{-x^2}\right) + \gamma \right), \quad x^2 \to 0, \text{or } m \to 0 \tag{2.3.34}$$

where γ is Euler's constant. Using these asymptotic expressions in (2.3.33), we realize that there is no jump at $x^2 = 0$ in the square bracket. Consequently, the Θ-functions in (2.3.33) must not be differentiated

$$D_1(x) \longrightarrow -\frac{i}{2\pi^2} \frac{d}{dx^2} \left(\log \frac{m}{2} \sqrt{|x^2|} - \gamma \right) = -\frac{i}{4\pi^2} \mathrm{P} \frac{1}{x^2}. \tag{2.3.35}$$

The differentiation of Bessel functions of order 0 gives Bessel functions of order 1. This leads us to

$$D_1(x) = \frac{i}{4\pi} \mathrm{P} \left[\Theta(x^2) \frac{m}{2\sqrt{x^2}} N_1\left(m\sqrt{x^2}\right) + \Theta(-x^2) \frac{m}{\pi\sqrt{-x^2}} K_1\left(m\sqrt{-x^2}\right) \right],$$

and

$$D^{(+)}(x) = \frac{1}{2} D(x) + D_1(x)$$

$$= \frac{1}{4\pi} \left\{ \operatorname{sgn} x^0 \left[\delta(x^2) - \Theta(x^2) \frac{m}{2\sqrt{x^2}} J_1\left(m\sqrt{x^2}\right) \right] + i \mathrm{P} \left[\Theta(x^2) \frac{m}{2\sqrt{x^2}} N_1\left(m\sqrt{x^2}\right) \right. \right.$$

$$+\frac{m}{\pi\sqrt{-x^2}}\Theta(-x^2)K_1\Big(m\sqrt{-x^2}\Big)\Big]\Big\}. \tag{2.3.36}$$

For the function $D^{(-)}$, we readily get

$$D^{(-)}(x) = \frac{1}{2}D(x) - D_1(x)$$

$$= \frac{1}{4\pi}\Big\{\operatorname{sgn} x^0\Big[\delta(x^2) - \Theta(x^2)\frac{m}{2\sqrt{x^2}}J_1\Big(m\sqrt{x^2}\Big)\Big] - i\,\mathrm{P}\Big[\Theta(x^2)\frac{m}{2\sqrt{x^2}}N_1\Big(m\sqrt{x^2}\Big)$$

$$+\frac{m}{\pi\sqrt{-x^2}}\Theta(-x^2)K_1\Big(m\sqrt{-x^2}\Big)\Big]\Big\} \tag{2.3.37}$$

$$= -D^{(+)}(-x) = D^{(+)}(x)^*. \tag{2.3.38}$$

These distributions do not vanish for space-like x ($x^2 < 0$). Therefore, they are not causal. For later reference, we also write down the Fourier transforms, which are given by (2.3.3, 5):

$$\hat{D}^{(+)}(p) = \frac{i}{2\pi}\Theta(p^0)\delta(p^2 - m^2) \tag{2.3.39}$$

$$\hat{D}^{(-)}(p) = -\frac{i}{2\pi}\Theta(-p^0)\delta(p^2 - m^2). \tag{2.3.40}$$

For completeness, we already mention the so-called Feynman propagators here:

$$D^F(x) \stackrel{\text{def}}{=} D^{\text{ret}}(x) - D^{(-)}(x) = D^{\text{av}}(x) + D^{(+)}(x) \tag{2.3.41}$$

$$S^F(x) \stackrel{\text{def}}{=} S^{(-)}(x) - S^{\text{ret}}(x) = -S^{(+)}(x) - S^{\text{av}}(x) \tag{2.3.42}$$

$$= -(i\partial\!\!\!/ + m)D^F(x). \tag{2.3.43}$$

The different signs in these two definitions agree with the widely adopted convention, which is convenient in the applications. The expressions in p-space are the following:

$$D^F(x) = (2\pi)^{-4}\int d^4p\frac{e^{-ipx}}{m^2 - p^2 - i0} \tag{2.3.44}$$

$$\hat{D}^F(p) = (2\pi)^{-2}\frac{1}{m^2 - p^2 - i0} \tag{2.3.45}$$

$$S^F(x) = (2\pi)^{-4}\int d^4p\frac{p\!\!\!/ + m}{p^2 - m^2 + i0}e^{-ipx} \tag{2.3.46}$$

$$\hat{S}^F(p) = (2\pi)^{-2}\frac{p\!\!\!/ + m}{p^2 - m^2 + i0}. \tag{2.3.47}$$

The Feynman propagators are often called "causal propagators". In view of our above discussion, we find this terminology misleading because D^F and S^F do not have a causal support.

The commutation functions for mass zero particles are of particular interest. Taking the limit $m \to 0$ in (2.3.18, 36, 38, 41) using (2.3.35) we obtain

$$D_0(x) = \frac{1}{2\pi} \operatorname{sgn} x^0 \delta(x^2) \tag{2.3.48}$$

$$D_0^{\text{ret}}(x) = \frac{1}{2\pi}\Theta(x^0)\delta(x^2), \quad D_0^{\text{av}}(x) = \frac{1}{2\pi}\Theta(-x^0)\delta(x^2) \tag{2.3.49}$$

$$\begin{aligned} D_0^{(\pm)}(x) &= \frac{1}{4\pi}\operatorname{sgn} x^0 \delta(x^2) \mp \mathrm{P}\frac{i}{4\pi^2 x^2} \\ &= \mp \frac{i}{4\pi^2} \frac{1}{(x^0 \mp i0)^2 - \boldsymbol{x}^2} \end{aligned} \tag{2.3.50}$$

$$D_0^F(x) = \frac{1}{4\pi}\delta(x^2) + \mathrm{P}\frac{i}{4\pi^2 x^2} = \frac{i}{4\pi^2}\frac{1}{x^2 + i0}. \tag{2.3.51}$$

For later use we also determine the transformation properties with respect to the matrix C (1.5.45) of charge conjugation:

$$\begin{aligned} C^{-1}S(x)C &= (-i\gamma^{\mu T}\partial_\mu + m)D(x) \\ &= -\Big(i\gamma^{\mu T}\frac{\partial}{\partial(-x^\mu)} + m\Big)D(-x) = -S^T(-x), \end{aligned} \tag{2.3.52}$$

$$\begin{aligned} \gamma^5 C S(x) C^{-1}\gamma^5 &= \gamma^5(-i\gamma^{\mu T}\partial_\mu + m)D(x)\gamma^5 \\ &= (i\gamma^{\mu T}\partial_\mu + m)D(x) = (i\gamma^{0*}\partial_0 - i\gamma^{k*}\partial_k + m)D(x) \\ &= \Big(i\gamma^0\frac{\partial}{\partial(-x^0)} + i\gamma^k\partial_k + m\Big)^* D(x) = -S(-x^0, \boldsymbol{x})^*, \end{aligned} \tag{2.3.53}$$

because $D(x)$ (2.3.18) changes sign under time-reversal.

2.4 The Scattering Operator (S-Matrix) in Fock Space

In Sect. 2.2 we have quantized the Dirac field in a *time-independent* electromagnetic potential $A^\mu(\boldsymbol{x})$. The Fock vacuum was defined by the relations

$$\begin{aligned} b(f)\Omega &= \psi(P_+ f)\Omega = 0 \\ d(f)\Omega &= \psi(P_- f)\Omega = 0 \quad , \quad \forall f \in \mathcal{H}_1. \end{aligned} \tag{2.4.1}$$

Here, $P_\pm$ are the projection operators on the positive and negative spectral subspaces of the one-particle Hamiltonian H (2.6), respectively. Consequently, for different potentials $A^\mu(\boldsymbol{x})$, we have different vacua (2.4.1) and, hence, different Fock representations. We now turn to *time-dependent* external fields. In this case, we must expect that the Fock representation will change continuously with time. To avoid this somewhat odd construction, we retreat to scattering theory. As discussed in the introduction to this part, this is not a severe restriction from the physical point of view, because the detailed time behavior of a relativistic system with creation and annihilation of particles is not observable. Later on, we shall also realize some mathematical facts which support this point of view.

We start from a one-particle dynamics, defined by a time-dependent Hamiltonian

$$H(t) = H_0 + V(t), \tag{2.4.2}$$

where

$$V(t) = e(V(t, \boldsymbol{x}) - \boldsymbol{\alpha} \cdot \boldsymbol{A}(t, \boldsymbol{x})). \tag{2.4.3}$$

The potentials are assumed to vanish for $t \longrightarrow \mp\infty$ in such a way that the wave operators

$$W_{\substack{\text{in} \\ \text{out}}} = s - \lim_{t \to \mp\infty} U(t, 0)^+ e^{-iH_0 t} \tag{2.4.4}$$

exist, together with a unitary S-matrix

$$S = W_{\text{out}}^+ W_{\text{in}}. \tag{2.4.5}$$

We refer to Sect. 0.3 of the preliminary Chap. 0 for a brief discussion of this matter. Since by assumption, we have the free dynamics for $t \longrightarrow \pm\infty$, we base second quantization on the Fock representation of the free Dirac field

$$\psi(f) = b(P_+^0 f) + d(P_-^0 f)^+. \tag{2.4.6}$$

Here, $P_\pm^0$ denote the projection operators corresponding to the free Dirac Hamiltonian H_0. We drop the zero in the following.

The second quantized S-matrix in Fock space is now defined by

$$\psi(S^+ f) = \mathbf{S}^{-1} \psi(f) \mathbf{S}, \tag{2.4.7}$$

$$\psi(S^+ f)^+ = \mathbf{S}^{-1} \psi(f)^+ \mathbf{S} \quad , \quad \forall f \in \mathcal{H}_1, \tag{2.4.8}$$

if it exists. We have taken the adjoint S^+ in the test functions because $\psi(f)$ is antilinear in f. As in proposition 2.1, the definitions (2.4.7, 8) imply that $\mathbf{S}$ is unitary and uniquely determined up to a phase $e^{i\lambda}$. However, this phase $\lambda[A]$ is physical because it depends on the external potential $A^\mu(x)$, which the experimentalist can easily vary. This will lead us to the interesting discussion of vacuum polarization in Sects. 2.8–10.

We now turn to the explicit construction of $\mathbf{S}$. Since $\mathbf{S}$ is essentially unique, we make an Ansatz and determine the unknown quantities in such a way that (2.4.7, 8) are satisfied. To carry this out, we have to develop some algebra. Let A be a bounded operator in $\mathcal{H}_1$ and f_j a basis in $P_+\mathcal{H}_1$, g_k is a basis in $P_-\mathcal{H}_1$. Then we define the operator

$$Ab^+b \stackrel{\text{def}}{=} \sum_{jk} (f_j, Af_k)_1 b(f_j)^+ b(f_k). \tag{2.4.9}$$

One easily verifies that this definition is independent of the choice of the basis. The operator (2.4.9) is at least defined for all vectors in the domain $D(\mathbf{N})$ (2.1.12) of the particle number operator. Similarly, let

$$Ab^+d^+ \stackrel{\text{def}}{=} \sum_{jk} (f_j, Ag_k)_1 b(f_j)^+ d(g_k)^+$$

$$Ad^+d \stackrel{\text{def}}{=} \sum_{jk} (g_k, Ag_j)_1 d(g_j)^+ d(g_k). \tag{2.4.10}$$

Every test function f_j must appear once linear and once anti-linear in these definitions. Next we compute the exponential

$$e^{Ab^+d^+} = \sum_{n=0}^{\infty} \frac{1}{n!} (Ab^+d^+)^n$$

$$= \sum_{n=0}^{\infty} \frac{1}{n!} \sum_{\substack{j_1,\dots j_n \\ k_1,\dots k_n}} (f_{j_1}, Ag_{k_1}) \dots (f_{j_n}, Ag_{k_n}) b_{j_1}^+ d_{k_1}^+ \dots b_{j_n}^+ d_{k_n}^+ \tag{2.4.11}$$

$$= \dots\dots b_{j_1}^+ b_{j_2}^+ \dots b_{j_n}^+ d_{k_n}^+ d_{k_{n-1}}^+ \dots d_{k_1}^+ . \tag{2.4.12}$$

Note that in reordering the Fermi operators in (2.4.11), one carries out an even number of transpositions which leads to the same sign in (2.4.12).

We now proceed to compute the commutator of $b(f)$ with the exponential (2.4.12). By successive commutation, we shall obtain

$$b(f)e^{Ab^+d^+} = \sum_n \frac{1}{n!} \sum \dots b(f) b_{j_1}^+ \dots b_{j_n}^+ d_{k_n}^+ \dots d_{k_1}^+$$

$$= e^{Ab^+d^+} b(f) + \sum_n \frac{1}{n!} \sum (f_{j_1}, Ag_{k_1}) \dots (f_{j_n}, Ag_{k_n})$$

$$\times \sum_{m=1}^{n} (f, f_{j_m})(-)^{m-1} b_{j_1}^+ \dots \not{b}_{j_m}^+ \dots b_{j_n}^+ d_{k_n}^+ \dots d_{k_m}^+ \dots d_{k_1}^+ . \tag{2.4.13}$$

The slash in (2.4.13) means that the operator with index j_m is missing. The minus sign in (2.4.13) $(-)^{m-1}$ is compensated by anti-commuting $d_{k_m}^+$ to the right. There remains to calculate the partial sum

$$\sum_{j_m k_m} (f_{j_m}, Ag_{k_m})(f, f_{j_m}) d_{k_m}^+ = \sum_{k_m} (P_+ f, Ag_{k_m}) d_{k_m}^+$$

$$= d^+(A^+ P_+ f). \tag{2.4.14}$$

There are n such terms in (2.4.13) which changes the $1/n!$ into $1/(n-1)!$. Then, the second sum in (2.4.13) is again an exponential series, hence, we end up with the following commutation rule

$$b(f)e^{Ab^+d^+} = e^{Ab^+d^+}(b(f) + d(A^+P_+f)^+). \tag{2.4.15}$$

In the same way, one finds

$$d(f)e^{Ab^+d^+} = e^{Ab^+d^+}(d(f) - b(AP_-f)^+). \tag{2.4.16}$$

Next we want to take the hermitean conjugate of these relations. It follows easily from the definition (2.4.10) that

$$(Ab^+d^+)^+ = A^+db \tag{2.4.17}$$

and

$$(e^{Ab^+d^+})^+ = e^{A^+db}. \tag{2.4.18}$$

We take the adjoint of (2.4.15), interchanging A with A^+,

$$e^{Adb}b(f)^+ = \big(b(f)^+ + d(AP_+f)\big)e^{Adb}. \tag{2.4.19}$$

Since the d operator commutes with the exponential, we may also write

$$b(f)^+e^{Adb} = e^{Adb}\big(b(f)^+ - d(AP_+f)\big). \tag{2.4.20}$$

Similarly, we get from (2.4.16)

$$e^{Adb}d(f)^+ = \big(d(f)^+ - b(A^+P_-f)\big)e^{Adb} \tag{2.4.21}$$

$$d(f)^+e^{Adb} = e^{Adb}\big(d(f)^+ + b(A^+P_-f)\big). \tag{2.4.22}$$

Exponentials with mixed emission and absorption operators, like

$$e^{Bb^+b} = \sum_n \sum_{\substack{j_1\cdots j_n\\ k_1\cdots k_n}} (f_{j_1}, Bf_{k_1})\dots(f_{j_n}, Bf_{k_n})\cdot b_{j_1}^+b_{k_1}\dots b_{j_n}^+b_{k_n} \tag{2.4.23}$$

must be normally ordered:

$$:e^{Bb^+b}: \stackrel{\text{def}}{=} \dots b_{j_1}^+\dots b_{j_n}^+b_{k_1}\dots b_{k_n}(-)^{1+2+\cdots(n-1)}$$

$$= \dots b_{j_1}^+\dots b_{j_n}^+b_{k_n}\dots b_{k_1}. \tag{2.4.24}$$

The commutator with $b(f)$ can be computed as above

$$b(f):e^{Bb^+b}: = :e^{Bb^+b}:b(f)$$

$$+\dots\sum_{m=1}^{n}(f, f_{j_m})(-)^{m-1}b_{j_1}^+\dots \not{b}_{j_m}^+\dots b_{j_n}^+b_{k_n}\dots b_{k_m}\dots b_{k_1}. \tag{2.4.25}$$

The operator b_{k_m} is again anti-commuted to the right which compensates the minus sign. The partial sum to be calculated in (2.4.25) is

$$\sum_{j_mk_m}(f_{j_m}, Bf_{k_m})(f, f_{j_m})b_{k_m} = \sum_{k_m}(P_+f, Bf_{k_m})b_{k_m}$$

$$= b(B^+ P_+ f),$$

which yields the following commutation rule:

$$b(f) : e^{Bb^+b} :=: e^{Bb^+b} : (b(f) + b(B^+ P_+ f))$$
$$=: e^{Bb^+b} : b((1 + B^+)P_+ f). \tag{2.4.26}$$

Substituting $1 + B = A$, $B = A - 1$, this may be written as

$$b(f) : e^{(A-1)b^+b} :=: e^{(A-1)b^+b} : b(A^+ P_+ f). \tag{2.4.27}$$

In the same way, one proves

$$d(f) : e^{(1-A)dd^+} :=: e^{(1-A)dd^+} : d(AP_- f). \tag{2.4.28}$$

For computing the hermitean adjoint of these relations, we use

$$: e^{Ab^+b} :^+ =: e^{A^+b^+b} : .$$

Interchanging A^+ with A, we obtain

$$: e^{(A-1)b^+b} : b(f)^+ = b(AP_+ f)^+ : e^{(A-1)b^+b} : \quad , \tag{2.4.29}$$

by means of the formula (2.4.27). Substituting

$$P_+ AP_+ f = f' \quad \text{or} \quad f = (A_{++})^{-1} f',$$

we find

$$b(f')^+ : e^{(A-1)b^+b} :=: e^{(A-1)b^+b} : b(A_{++}^{-1} f')^+, \tag{2.4.30}$$

where $A_{++} \stackrel{\text{def}}{=} P_+ AP_+$. The commutation rule

$$d(g)^+ : e^{(1-A)dd^+} :=: e^{(1-A)dd^+} : d(A_{--}^{+-1} g)^+ \tag{2.4.31}$$

is derived in the same way.

An important warning is in place here: Normal ordering is not a linear operation! A counterexample is

$$: (b^+b + bb^+) : \neq : b^+b : + : bb^+ : \tag{2.4.32}$$

for any Fermi operator b, because the left side is equal to $: 1 :$ whereas the right side is 0. That means, every normal ordered expression has its own definition as a certain sum of normal products.

We are now ready to construct the S-matrix **S** in Fock space, satisfying (2.4.7, 8). Considering $f \in P_+ \mathcal{H}_1$, we decompose the test function

$$S^+ f = P_+ S^+ f + P_- S^+ f = S_{++}^+ f + S_{-+}^+,$$

where

$$S^+_{++} = P_+ S^+ P_+ \quad , \quad S^+_{-+} \stackrel{\text{def}}{=} P_- S^+ P_+ = (S^+)_{-+}. \tag{2.4.33}$$

Then, it follows from (2.4.7) that

$$b(f)\mathbf{S} = \mathbf{S}[b(S^+_{++}f) + d(S^+_{-+}f)^+]. \tag{2.4.34}$$

Similarly, we get for $g \in P_-\mathcal{H}_1$

$$d(g)^+\mathbf{S} = \mathbf{S}[b(S^+_{+-}g) + d(S^+_{--}g)^+], \tag{2.4.35}$$

and the second defining Eq. (2.4.8) yields

$$b(f)^+\mathbf{S} = \mathbf{S}[b(S^+_{++}f)^+ + d(S^+_{-+}f)] \tag{2.4.36}$$

$$d(g)\mathbf{S} = \mathbf{S}[b(S^+_{+-}g)^+ + d(S^+_{--}g)]. \tag{2.4.37}$$

We make the following Ansatz

$$\mathbf{S} = Ce^{A_1 b^+ d^+} : e^{(A_2-1)b^+b} : \quad : e^{(1-A_3)dd^+} : e^{A_4 db} \tag{2.4.38}$$

and try to determine $A_1, \ldots, A_4$ in such a way that the Eqs. (2.4.34–37) are satisfied. C is a normalization factor. The Ansatz is suggested by physical reasoning: The first factor describes electron-positron pair creation, the second one electron scattering, the third one positron scattering and the last one pair annihilation. The expression (2.4.38) is normally ordered, which is most convenient for applications. First, we substitute (2.4.38) into (2.4.35), because here the positron emission operator $d(g)^+$ can be commuted furthest to the right

$$d(g)^+\mathbf{S} = \ldots d(g)^+ : e^{(1-A_3)dd^+} : \ldots$$
$$= \ldots : e^{(1-A_3)dd^+} : d(A^{+-1}_{3--}g)^+ e^{A_4 db},$$

by means of the commutation rule (2.4.31). Using (2.4.22), we commute with the last factor

$$= \mathbf{S}[d(A^{+-1}_{3--}g)^+ + b(A_4^+ A^{+-1}_{3--}g)], \tag{2.4.39}$$

and this must be equal to

$$= \mathbf{S}[d(S^+_{--}g)^+ + b(S^+_{+-}g)],$$

according to (2.4.35). This enables us to determine

$$A^+_{3--} = (S^+_{--})^{-1} = (S^{-1}_{--})^+.$$

We drop the subscript $--$ at A_3 because, according to its definition in (2.4.38), this operator anyhow maps from $P_-\mathcal{H}_1$ into itself,

$$A_3 = S^{-1}_{--}. \tag{2.4.40}$$

Identifying the test functions of b in (2.4.39), we arrive at

$$A_4^+ S^+_{--} = S^+_{+-} \quad \text{or} \quad A_4^+ = S^+_{+-}(S^+_{--})^{-1},$$

$$A_4 = S_{--}^{-1} S_{-+}. \tag{2.4.41}$$

The inverse operators have always to be formed on the subspaces where the operators are defined.

Next we consider (2.4.36). By means of the formulae (2.4.30) and (2.4.20) we shall obtain

$$\begin{aligned} b(f)^+ \mathbf{S} &= \dots b(f)^+ : e^{(A_2-1)b^+b} : \dots \\ &= \dots : e^{(A_2-1)b^+b} : b(A_{2++}^{-1} f)^+ : e^{(1-A_3)dd^+} : e^{A_4 db} \\ &= \mathbf{S}[b(A_{2++}^{-1} f)^+ - d(A_{4-+} A_{2++}^{-1} f)], \end{aligned} \tag{2.4.42}$$

which must be equal to

$$= \mathbf{S}[b(S_{++}^+ f)^+ + d(S_{-+}^+ f)].$$

This leads to

$$A_2 = (S_{++}^+)^{-1} \quad \text{and} \tag{2.4.43}$$

$$A_4 = -S_{-+}^+ A_{2++} = -S_{-+}^+ S_{++}^{+-1}. \tag{2.4.44}$$

For consistency, the result (2.4.44) must agree with (2.4.41). This is in fact true in virtue of the unitarity of S. Writing

$$S = \begin{pmatrix} S_{++} & S_{+-} \\ S_{-+} & S_{--} \end{pmatrix} \quad , \quad S^+ = \begin{pmatrix} S_{++}^+ & S_{+-}^+ \\ S_{-+}^+ & S_{--}^+ \end{pmatrix}, \tag{2.4.45}$$

the unitarity relation $SS^+ = \mathbf{1}$ may be represented in the form

$$S_{++} S_{++}^+ + S_{+-} S_{-+}^+ = P_+ \tag{2.4.46}$$

$$S_{++} S_{+-}^+ + S_{+-} S_{--}^+ = 0 \tag{2.4.47}$$

$$S_{-+} S_{++}^+ + S_{--} S_{-+}^+ = 0 \tag{2.4.48}$$

$$S_{-+} S_{+-}^+ + S_{--} S_{--}^+ = P_-. \tag{2.4.49}$$

In virtue of (2.4.48), we obviously have

$$S_{--}^{-1} S_{-+} + S_{-+}^+ (S_{++}^+)^{-1} = 0, \tag{2.4.50}$$

which shows that (2.4.44) is correct.

We now turn to (2.4.34). Using the commutation rules (2.4.15, 27, 28) and (2.4.22), we shall obtain

$$\begin{aligned} b(f)\mathbf{S} &= C e^{A_1 b^+ d^+} [b(f) + d(A_1^+ P_+ f)^+] : e^{(A_2-1)b^+b} : \dots \\ &= \dots : e^{(A_2-1)b^+b} : [b(A_2^+ P_+ f) + d(A_1^+ P_+ f^+] : e^{(1-A_3)dd^+} : \dots \\ &= \dots : e^{(1-A_3)dd^+} : [b(A_2^+ P_+ f) + d((A_{3--}^+)^{-1} A_{1-+}^+ f)^+] e^{A_4 db} \\ &= \mathbf{S}[b(A_{2++}^+ f) + d((A_{3--}^+)^{-1} A_{1-+}^+ f)^+ + b(A_{4+-}^+ (A_{3--}^+)^{-1} A_{1-+}^+ f)] \\ &= \mathbf{S}[b(S_{++}^+ f) + d(S_{-+}^+ f)^+]. \end{aligned} \tag{2.4.51}$$

We compare the test functions of d^+, which gives

$$S^+_{-+} = (A^+_{3--})^{-1} A^+_{1-+} \quad \text{or} \quad A^+_1 = A^+_{3--} S^+_{-+}$$

$$A_1 = S_{+-}(S_{--})^{-1}. \tag{2.4.52}$$

We have thus succeeded to determine all unknown operators $A_1, \ldots, A_4$. It is straight-forward to check that all remaining relations are fulfilled in virtue of the unitarity relations (2.4.46–49).

There remains to determine the normalization factor C in (2.4.38). Since $\mathbf{S}$ is unitary in $\mathcal{F}$, one has

$$1 = (\mathbf{S}\Omega, \mathbf{S}\Omega) = |C|^2 (e^{A_1 b^+ d^+}\Omega, e^{A_1 b^+ d^+}\Omega). \tag{2.4.53}$$

It is highly non-trivial that the scalar product (2.4.53) is actually finite. We will see that this requirement leads to a severe restriction on S, which was an arbitrary unitary operator up to now. To compute (2.4.53), it is convenient to use the polar decomposition of A_1

$$A_1 = U|A_1|, \tag{2.4.54}$$

where U is unitary and

$$A_1^+ A_1 = |A_1| U^+ U |A_1| = |A_1|^2. \tag{2.4.55}$$

Let us assume that the positive selfadjoint operator $|A_1|$ has a discrete spectrum

$$|A_1| = \sum_j \mu_j (\varphi_j, \cdot) \varphi_j, \tag{2.4.56}$$

where φ_j is a complete orthonormal system in $\mathcal{H}_1$ and $\mu_j \geq 0$. We will shortly see that this requirement is necessary for a finite normalization constant (2.4.53). Then (2.4.54) may be written as

$$A_1 = \sum_j \mu_j (\varphi_j, \cdot) U \varphi_j = \sum_j \mu_j (\varphi_j, \cdot) \psi_j. \tag{2.4.57}$$

Since A_1 is a operator from $\mathcal{H}_-$ into $\mathcal{H}_+$, φ_j is a basis in $\mathcal{H}_-$ and ψ_j a basis in $\mathcal{H}_+$. The exponent in (2.4.53) greatly simplifies

$$A_1 b^+ d^+ = \sum_{jk} (\psi_j, A_1 \varphi_k) b(\psi_j)^+ d(\varphi_k)^+$$

$$= \sum_j \mu_j b(\psi_j)^+ d(\varphi_j)^+. \tag{2.4.58}$$

This operator has diagonal form, therefore, the exponential simply becomes

$$e^{A_1 b^+ d^+} = \prod_j e^{\mu_j b_j^+ d_j^+} = \prod_j (1 + \mu_j b_j^+ d_j^+), \tag{2.4.59}$$

because the higher powers of the exponential series vanish (Pauli principle). Using this in (2.4.53), only the quadratic terms are different from 0

$$(e^{A_1 b^+ d^+}\Omega,\, e^{A_1 b^+ d^+}\Omega) = \prod_j ((1+\mu_j b_j^+ d_j^+)\Omega,\, (1+\mu_j b_j^+ d_j^+)\Omega)$$

$$= \prod_j (1+\mu_j^2) = \det(1+|A_1|^2) = \det(1+A_1^+ A_1) = \frac{1}{|C|^2}. \qquad (2.4.60)$$

This result can also be proven without the spectral decomposition (2.4.57) by using some more abstract definition of the determinant (*M. Reed, B. Simon, Methods of Modern Mathematical Physics, Vol. 4 (1978), p. 323*). The infinite determinant exists if and only if $A_1^+ A_1$ is a trace-class operator, or A_1 is a Hilbert-Schmidt operator. Since $(S_{--})^{-1}$ must be bounded, this is equivalent to the condition that $S_{+-} = P_+ S P_-$ is a Hilbert-Schmidt operator. This leads us to the basic

Theorem 4.1. The S-matrix $\mathbf{S}$ in Fock space exists, if and only if $P_+ S P_-$ is a Hilbert-Schmidt operator. It is then given by

$$\mathbf{S} = C e^{S_{+-} S_{--}^{-1} b^+ d^+} : e^{(S_{++}^{+-1} - 1) b^+ b} :: e^{(1 - S_{--}^{-1}) d d^+} : e^{S_{--}^{-1} S_{-+} d b}. \qquad (2.4.61)$$

This theorem can also be applied to the question of the existence of a unitary time evolution $\mathbf{U}$ in Fock space, defined by

$$\psi(U(t)f) = \mathbf{U}(t)\psi(f)\mathbf{U}(t)^{-1}$$

$$\psi(U(t)f)^+ = \mathbf{U}(t)\psi(f)^+\mathbf{U}(t)^{-1}, \qquad (2.4.62)$$

where $U(t)$ is the one-particle time evolution. This $\mathbf{U}(t)$ does not exist in general, because the necessary condition for it is that $P_+ U(t) P_-$ is a Hilbert-Schmidt operator, which is generally not true (cf. *G. Nenciu, G. Scharf, Helv. Phys. Acta 51 (1978) 412*). For the S-matrix, on the other hand, the Hilbert-Schmidt condition can be satisfied for a large class of external fields, as will be discussed in the following section. This is the mathematical reason for the retreat to scattering theory. In the general discussion of second quantization in Sect. 2.1, we found that any unitary one-particle operator could be second-quantized. Why is this impossible in case of the time-evolution? The reason for this is the mixture of absorption and emission operators (2.2.2), which we were forced to introduce for the Dirac field. If the Hilbert-Schmidt condition is not fulfilled, then the vector

$$e^{A_1 b^+ d^+}\Omega \notin \mathcal{F} \qquad (2.4.63)$$

is not in Fock space. That means, the external field produces too many pairs. Such external fields are unphysical, they would be modified by pair creation.

Let us rewrite the factor $|C|^2$ in different forms. Since

$$1 + A_1^+ A_1 = 1 + (S_{--}^+)^{-1} S_{-+}^+ S_{+-} (S_{--})^{-1}$$
$$= (S_{--}^+)^{-1} [S_{--}^+ S_{--} + S_{-+}^+ S_{+-}] (S_{--})^{-1} = (S_{--} S_{--}^+)^{-1},$$

we find

$$|C|^2 = \det(S_{--} S_{--}^+) = \det(1 - S_{-+} S_{+-}^+). \tag{2.4.64}$$

Using the general identity

$$\det(1 - AA^+) = \det(1 - A^+ A), \tag{2.4.65}$$

we may also write

$$|C|^2 = \det(S_{--} S_{--}^+) = \det(1 - S_{-+} S_{+-}^+) = \det(1 - S_{-+}(S_{-+})^+)$$
$$= \det(1 - S_{+-}^+ S_{-+}) = \det(S_{++}^+ S_{++}) = \det(S_{++} S_{++}^+)$$
$$= \det(1 - S_{+-} S_{-+}^+). \tag{2.4.66}$$

The behaviour under charge conjugation follows from $U_C P_+ U_C^{-1} = P_-$ (2.2.43):

$$|C|^2(-e) = \det(S_{--} S_{--}^+)(-e) = \det(U_C S_{++} S_{++}^+(e) U_C^{-1})$$
$$= \det(S_{++} S_{++}^+(e)) = |C|^2(e). \tag{2.4.67}$$

The fact that $|C|^2$ only contains even powers of e, is called Furry's theorem.

The Hilbert-Schmidt operator S_{+-} has the following canonical form (*M. Reed B. Simon, Methods of Modern Mathematical Physics, Vol. 1, (1978) p. 203*):

$$S_{+-} = \sum_n \lambda_n (\varphi_n^-, \cdot) \varphi_n^+, \tag{2.4.68}$$

where φ_n^-, φ_n^+ are complete orthonormal systems in $\mathcal{H}_-$, $\mathcal{H}_+$, respectively. The singular values λ_n may be chosen real and nonnegative, by absorbing a phase in φ_n^+, say. Unitarity of S implies $\lambda_n \leq 1$, such that $0 \leq \lambda_n \leq 1$ and the Hilbert-Schmidt property means

$$\sum_n \lambda_n^2 < \infty. \tag{2.4.69}$$

We want to derive similar normal forms for the other parts of S (2.4.45). For the adjoint of (2.4.68) we obtain

$$S_{-+}^+ = \sum_n \lambda_n (\varphi_n^+, \cdot) \varphi_n^-. \tag{2.4.70}$$

By unitarity, we find

$$S_{--}^+ S_{--} = P_- - S_{-+}^+ S_{+-} = P_- - \sum_n \lambda_n^2 (\varphi_n^-, \cdot) \varphi_n^-$$

$$= \sum_n (1 - \lambda_n^2)(\varphi_n^-, \cdot)\varphi_n^-,$$

and, therefore,

$$S_{--} = \sum_n \sqrt{1 - \lambda_n^2}(\varphi_n^-, \cdot)\psi_n^-, \tag{2.4.71}$$

where ψ_n^- is another basis in $\mathcal{H}_-$ which is completely fixed by (2.4.71), if there is no degeneracy in the spectrum of S_{--}. This we assume for the sake of simplicity from now on.

Next we construct S_{++}. Starting with the unitarity relation

$$S_{++}S_{++}^+ = P_+ - S_{+-}S_{-+}^+,$$

we get

$$S_{++}S_{++}^+ = P_+ - \sum_n \lambda_n^2(\varphi_n^+, \cdot)\varphi_n^+ = \sum_n (1 - \lambda_n^2)(\varphi_n^+, \cdot)\varphi_n^+, \tag{2.4.72}$$

which leads to

$$S_{++} = \sum_n \sqrt{1 - \lambda_n^2}(\psi_n^+, \cdot)\varphi_n^+. \tag{2.4.73}$$

Here again, ψ_n^+ is another basis in $\mathcal{H}_+$ which is fixed by (2.4.73). Finally, S_{+-} is obtained from

$$S_{+-}^+ S_{-+} = P_+ - S_{++}^+ S_{++} = P_+ - \sum_n (1 - \lambda_n^2)(\psi_n^+, \cdot)\psi_n^+$$

$$= \sum_n \lambda_n^2(\psi_n^+, \cdot)\psi_n^+,$$

giving

$$S_{-+} = \sum_n \lambda_n(\psi_n^+, \cdot)\tilde{\psi}_n^-. \tag{2.4.74}$$

Here, $\tilde{\psi}_n^-$ is a third basis in $\mathcal{H}_-$.

All four pieces of S are now determined, but the remaining unitarity relations give additional constraints. From

$$S_{-+}S_{+-}^+ = P_- - S_{--}S_{--}^+ = P_- - \sum_n (1 - \lambda_n^2)(\psi_n^-, \cdot)\psi_n^-$$

$$= \sum_n \lambda_n^2(\psi_n^-, \cdot)\psi_n^-,$$

we get

$$S_{-+} = \sum_n \lambda_n(\tilde{\psi}_n^+, \cdot)\psi_n^-. \tag{2.4.75}$$

If there is no degeneracy, it follows from (2.4.74, 75)

$$\tilde{\psi}_n^+ = e^{-i\alpha_n}\psi_n^+$$

$$\tilde{\psi}_n^- = e^{i\alpha_n}\psi_n^-, \tag{2.4.76}$$

with α_n real, so that

$$S_{-+} = \sum_n \lambda_n e^{i\alpha_n}(\psi_n^+, \cdot)\psi_n^- .$$

Using this in

$$S_{++}^+ S_{+-} = -S_{+-}^+ S_{--},$$

we find $e^{i\alpha_n} = -1$ for all n, hence

$$S_{-+} = -\sum_n \lambda_n(\psi_n^+, \cdot)\psi_n^- . \tag{2.4.77}$$

All remaining unitarity relations are then identically satisfied.

Comparing (2.4.77) with (2.4.68), we notice a lack of symmetry in the expressions. This can be removed by redefining $\psi_n^- \longrightarrow -\psi_n^-$. Then we have the following normal form of the one-particle scattering matrix (2.4.45)

$$S_{+-} = \sum_n \lambda_n(\varphi_n^-, \cdot)\varphi_n^+$$

$$S_{-+} = \sum_n \lambda_n(\psi_n^+, \cdot)\psi_n^-$$

$$S_{++} = \sum_n \sqrt{1-\lambda_n^2}(\psi_n^+, \cdot)\varphi_n^+$$

$$S_{--} = -\sum_n \sqrt{1-\lambda_n^2}(\varphi_n^-, \cdot)\psi_n^- . \tag{2.4.78}$$

This normal form is a consequence of unitarity and the Hilbert-Schmidt property alone. It contains two orthonormal systems φ_n^-, ψ_n^- and φ_n^+, ψ_n^+ in $\mathcal{H}_-$ and $\mathcal{H}_+$, respectively, and corresponding singular values λ_n.

Finally, we want to discuss two general properties of the S-operator **S** in Fock space. The first one is charge conservation. The charge operator **Q** was introduced in (2.2.23):

$$\mathbf{Q} = e\sum_n [b(\varphi_n^+)^+ b(\varphi_n^+) - d(\psi_n^-)^+ d(\psi_n^-)], \tag{2.4.79}$$

where we have used basis vectors from the normal form of S (2.4.78). By means of the formulae (2.4.34, 36), we shall obtain

$$\begin{aligned} b(\varphi_n^+)^+ b(\varphi_n^+)\mathbf{S} &= \mathbf{S}[b(S_{++}^+\varphi_n^+)^+ + d(S_{-+}^+\varphi_n^+)][b(S_{++}^+\varphi_n^+) + d(S_{-+}^+\varphi_n^+)^+] \\ &= \mathbf{S}\Big[\sqrt{1-\lambda_n^2}\, b_n^+ + \lambda_n d_n\Big]\Big[\sqrt{1-\lambda_n^2}\, b_n + \lambda_n d_n^+\Big]. \end{aligned} \tag{2.4.80}$$

Here, the normal form (2.4.78) has been applied, and we have written

$$b_n = b(\psi_n^+) \quad , \quad d_n = d(\varphi_n^-)$$

for shortness. Similarly, we find with the aid of (2.4.37, 35)

$$d(\psi_n^-)^+ d(\psi_n^-)\mathbf{S} = \mathbf{S}\Big[\lambda_n b_n - \sqrt{1-\lambda_n^2}\, d_n^+\Big]\Big[\lambda_n b_n^+ - \sqrt{1-\lambda_n^2}\, d_n\Big]. \quad (2.4.81)$$

The two results obtained give

$$[b(\varphi_n^+)^+ b(\varphi_n^+) - d(\psi_n^-)^+ d(\psi_n^-)]\mathbf{S} = \mathbf{S}[b_n^+ b_n - d_n^+ d_n].$$

Summing over n, we have proven charge conservation of $\mathbf{S}$

$$\mathbf{QS} = \mathbf{SQ}. \quad (2.4.82)$$

The last subject of this long section is gauge invariance. As we have found at the end of Chap. 1 (1.5.60), the one-particle S-matrix is gauge invariant

$$S[A'] = S[A],$$

provided the gauge function Λ in the gauge transformation

$$A_\mu(x) \longrightarrow A'_\mu(x) = A_\mu(x) - \partial_\mu \Lambda(x)$$

vanishes for $t \to \pm\infty$. The same is then true for the Fock space scattering operator $\mathbf{S}$,

$$\mathbf{S}[A'] = \mathbf{S}[A] \quad (2.4.83)$$

because it is given by S according to Theorem 4.1.

2.5 Perturbation Theory

In the first part of this section, we will prove the Hilbert-Schmidt property of Theorem 4.1. This is a rather technical matter. Nevertheless, it is worthwhile to discuss it also from the physical point of view, because we shall obtain an answer to the question which external fields are physically meaningful.

The perturbation series for the one-particle scattering matrix S was discussed in Sect. 0.3 (0.3.18). Let the time-dependent Hamiltonian be $H(t) = H_0 + H_1(t)$, then

$$S = \sum_{n=0}^{\infty} (-i)^n \int_{-\infty}^{+\infty} dt_1 \int_{-\infty}^{t_1} dt_2 \cdots \int_{-\infty}^{t_{n-1}} dt_n \tilde{H}_1(t_1)\ldots\tilde{H}_1(t_n) \stackrel{\text{def}}{=} \sum_{n=0}^{\infty} S^{(n)}, \quad (2.5.1)$$

where

$$\tilde{H}_1(t) = e^{iH_0 t} H_1(t) e^{-iH_0 t} \quad (2.5.2)$$

is the interaction operator in the interaction picture. We have

$$H_1(t) = e(V - \boldsymbol{\alpha}\cdot\boldsymbol{A})(t, \boldsymbol{x}), \quad (2.5.3)$$

and $S^{(n)}$ (2.5.1) is the n-th order in e. H_0 is simplest in momentum space, where it is given by the following matrix multiplication operator

$$H_0(\boldsymbol{p}) = \boldsymbol{\alpha} \cdot \boldsymbol{p} + \beta m. \tag{2.5.4}$$

The exponentials in (2.5.2) are also matrix multiplications in p-space, but $H_1(t)$ is a convolution

$$(H_1(t)f)(\boldsymbol{p}) = (2\pi)^{-3/2} \int d^3p' H_1(t, \boldsymbol{p} - \boldsymbol{p}')f(\boldsymbol{p}'). \tag{2.5.5}$$

The n-th order term $S^{(n)}$ may be written in p-space as follows

$$S^{(n)}(\boldsymbol{p}, \boldsymbol{q}) = \frac{(-i)^n}{(2\pi)^{3n/2}} \int_{-\infty}^{+\infty} dt_1 \int_{-\infty}^{t_1} dt_2 \cdots \int_{-\infty}^{t_{n-1}} dt_n \int d^3p_1 \cdots \int d^3p_{n-1}$$

$$e^{it_1 H_0(\boldsymbol{p})} H_1(t_1, \boldsymbol{p} - \boldsymbol{p}_1) e^{-i(t_1 - t_2)H_0(\boldsymbol{p}_1)} H_1(t_2, \boldsymbol{p}_1 - \boldsymbol{p}_2)$$
$$\ldots e^{-i(t_{n-1} - t_n)H_0(\boldsymbol{p}_{n-1})} H_1(t_n, \boldsymbol{p}_{n-1} - \boldsymbol{q}) e^{-it_n H_0(\boldsymbol{q})}. \tag{2.5.6}$$

The projection operators are also simple in p-space (see (1.4.40))

$$P_+(\boldsymbol{p})H_0(\boldsymbol{p}) = \frac{1}{2E}(E + \boldsymbol{\alpha} \cdot \boldsymbol{p} + \beta m)(\boldsymbol{\alpha} \cdot \boldsymbol{p} + \beta m)$$
$$= \frac{1}{2E}[E(\boldsymbol{\alpha} \cdot \boldsymbol{p} + \beta m) + E^2] = E(\boldsymbol{p})P_+(\boldsymbol{p}), \tag{2.5.7}$$

and

$$P_-(\boldsymbol{p})H_0(\boldsymbol{p}) = -E(\boldsymbol{p})P_-(\boldsymbol{p}), \tag{2.5.8}$$
$$P_\pm(\boldsymbol{p})e^{itH_0(\boldsymbol{p})} = e^{\pm itE(\boldsymbol{p})}P_\pm(\boldsymbol{p}). \tag{2.5.9}$$

This enables us to write $P_+ S^{(n)} P_-$ in the form

$$S_{+-}^{(n)}(\boldsymbol{p}, \boldsymbol{q}) = \frac{(-i)^n}{(2\pi)^{3n/2}} \int_{-\infty}^{+\infty} dt_1 \int_{-\infty}^{t_1} dt_2 \cdots \int_{-\infty}^{t_{n-1}} dt_n \, e^{it_1 E(\boldsymbol{p}) + it_n E(\boldsymbol{q})}$$

$$\int d^3p_1 \cdots \int d^3p_{n-1} \, P_+(\boldsymbol{p}) H_1(t_1, \boldsymbol{p} - \boldsymbol{p}_1) e^{-i(t_1 - t_2)H_0(\boldsymbol{p}_1)}$$

$$\ldots e^{-i(t_{n-1} - t_n)H_0(\boldsymbol{p}_{n-1})} H_1(t_n, \boldsymbol{p}_{n-1} - \boldsymbol{q}) P_-(\boldsymbol{q}). \tag{2.5.10}$$

In order to estimate this integral operator, we introduce center of mass and relative coordinates in time

$$s_1 = \frac{1}{n}(t_1 + t_2 + \ldots + t_n) \quad , \quad -\infty < s_1 < \infty$$

$$s_j = t_j - t_{j-1} \quad , \quad j = 2, 3, \ldots n \quad , \quad -\infty < s_j \leq 0. \tag{2.5.11}$$

The inverse transformation is

$$t_1 = s_1 - \frac{1}{n}\sum_{j=2}^{n}(n-j+1)s_j \quad , \quad t_n = s_1 + \frac{1}{n}\sum_{j=2}^{n}(j-1)s_j$$

$$t_k = s_1 + \frac{1}{n}\sum_{j=2}^{k}(j-1)s_j - \frac{1}{n}\sum_{j=k+1}^{n}(n-j+1)s_j,$$

$$2 \le k \le n-1. \tag{2.5.12}$$

Since the Jacobian is

$$\det\left(\frac{\partial s_j}{\partial t_k}\right) = 1,$$

we may write (2.5.10) as follows

$$S_{+-}^{(n)}(\boldsymbol{p},\boldsymbol{q}) = \int_{-\infty}^{+\infty} ds_1 \int_{-\infty}^{0} ds_2 \ldots \int_{-\infty}^{0} ds_n\, e^{is_1(E(\boldsymbol{p})+E(\boldsymbol{q}))}$$

$$\times \exp -\frac{i}{n}\sum_{j=2}^{n} s_j[(n-j+1)E(\boldsymbol{p}) - (j-1)E(\boldsymbol{q})]$$

$$\times T_n(s_1, \ldots s_n; \boldsymbol{p}, \boldsymbol{q}). \tag{2.5.13}$$

We shall partially integrate this expression twice with respect to s_1. We shall integrate the first exponential and differentiate T_n. The boundary terms are zero, because H_1 vanishes for $s_1 = \pm\infty$. Since $t_{k+1} - t_k = s_{k+1}$, the variable s_1 appears in T_n in the potentials H_1 only. Hence, only the potentials must be differentiated. Returning to the original variables $t_1, \ldots, t_n$, using

$$\frac{\partial}{\partial s_1} = \sum_{k=1}^{n} \frac{\partial t_k}{\partial s_1}\frac{\partial}{\partial t_k} = \sum_{k=1}^{n}\frac{\partial}{\partial t_k},$$

we end up with

$$S_{+-}^{(n)}(\boldsymbol{p},\boldsymbol{q}) = \int_{-\infty}^{+\infty} dt_1 \int_{-\infty}^{t_1} dt_2 \cdots \int_{-\infty}^{t_{n-1}} dt_n \frac{e^{it_1E(\boldsymbol{p})+it_nE(\boldsymbol{q})}}{(E(\boldsymbol{p})+E(\boldsymbol{q}))^2}$$

$$\times T_n^{(2)}(t_1, \ldots t_n; \boldsymbol{p}, \boldsymbol{q}), \tag{2.5.14}$$

where

$$T_n^{(2)} = -\frac{(-i)^n}{(2\pi)^{3n/2}} \sum_{j_1,\cdots j_n} \int d^3p_1 \cdots \int d^3p_{n-1}\, P_+(\boldsymbol{p}) H_1^{j_1}(t_1, \boldsymbol{p}-\boldsymbol{p}_1)$$

$$\times e^{-i(t_1-t_2)H_0(\boldsymbol{p}_1)} \ldots e^{-i(t_{n-1}-t_n)H_0(\boldsymbol{p}_{n-1})} H_1^{j_n}(t_n, \boldsymbol{p}_{n-1}-\boldsymbol{q})P_-(\boldsymbol{q}). \tag{2.5.15}$$

The subscripts $j_k = 0, 1, 2$ denote the totally two derivatives

$$\sum_{k=1}^{n} j_k = 2. \tag{2.5.16}$$

There are n^2 terms in (2.5.15).

We first estimate the matrix kernel of (2.5.14) in the $\mathbb{C}^4$ norm, which we denote by $|\cdot|$. Since the projection operators obey $|P_\pm| \leq 1$, we obviously have

$$\left| \frac{e^{it_1 E(\boldsymbol{p})+it_n E(\boldsymbol{q})}}{(E(\boldsymbol{p})+E(\boldsymbol{q}))^2} T_n^{(2)} \right| \leq \frac{1}{E(\boldsymbol{p})^2} |T_n^{(2)}|$$

$$\leq \frac{1}{(2\pi)^{3n/2} E(\boldsymbol{p})^2} \sum_{j_1,\cdots j_n} \int d^3 p_1 \cdots \int d^3 p_{n-1} \, |H_1^{j_1}(t_1, \boldsymbol{p}-\boldsymbol{p}_1)|$$

$$\times |H_1^{j_2}(t_2, \boldsymbol{p}_1-\boldsymbol{p}_2)| \ldots |H_1^{j_n}(t_n, \boldsymbol{p}_{n-1}-\boldsymbol{q})|. \tag{2.5.17}$$

With the new integration variables $\boldsymbol{k}_i = \boldsymbol{p}_i - \boldsymbol{q}$, this may be written as

$$= \frac{1}{(2\pi)^{3n/2} E(\boldsymbol{p})^2} \sum \int d^3 k_1 \ldots d^3 k_{n-1} |H_1^{j_1}(t_1, \boldsymbol{p}-\boldsymbol{q}-\boldsymbol{k}_1)|$$

$$\times |H_1^{j_2}(t_2, \boldsymbol{k}_1-\boldsymbol{k}_2)| \ldots |H_1^{j_n}(t_n, \boldsymbol{k}_{n-1})|$$

$$= \frac{1}{(2\pi)^{3n/2} E(\boldsymbol{p})^2} \sum \Big(|H_1^{j_1}(t_1)| * |H_1^{j_2}(t_2)| * \ldots * |H_1^{j_n}(t_n)| \Big)(\boldsymbol{p}-\boldsymbol{q}). \tag{2.5.18}$$

The asterisks mean convolutions. Abbreviating the integral operator (2.5.14) without the t-integrals by $S_n^{(2)}$, we are now able to estimate its Hilbert-Schmidt norm

$$\|S_n^{(2)}\|_{HS}^2 \stackrel{\text{def}}{=} \int d^3 p \int d^3 q |S_n^{(2)}(\boldsymbol{p}, \boldsymbol{q})|^2$$

$$\leq \frac{1}{(2\pi)^{3n}} \int \frac{d^3 p}{E(\boldsymbol{p})^4} \int d^3 q' |\sum \big(\ldots\big)(\boldsymbol{q}')|^2, \tag{2.5.19}$$

where the dots represent the convolution product of (2.5.18). The integral

$$\int \frac{d^3 p}{E(\boldsymbol{p})^4} = 4\pi \int_0^\infty dp \frac{p^2}{(p^2+m^2)^2} = \frac{\pi^2}{m}$$

is finite and this explains why we have carried out two integrations by part.

The second factor in (2.5.19) is a L^2 norm

$$\| \sum |H_1^{j_1}| * |H_1^{j_2}| * \ldots * |H_1^{j_n}| \, \|_2$$

$$\leq \sum \| \, |H_1^{j_1}| * \ldots * |H_1^{j_n}| \, \|_2. \tag{2.5.20}$$

The convolutions are estimated by Young's inequality

$$\|f * g\|_2 \leq \|f\|_1 \|g\|_2 \tag{2.5.21}$$

in terms of L^2 and L^1 norms:

$$\leq \sum \|H_1^{j1}\|_1 \|H_1^{j2}\|_1 \ldots \cdot \|H_1^{jn}\|_2. \tag{2.5.22}$$

Let us now assume that

$$H_1 = e(V - \boldsymbol{\alpha} \cdot \boldsymbol{A})(t, \boldsymbol{p}) \tag{2.5.23}$$

is two times differentiable in t, and

$$\|H_1^j\|_p \leq F(t) \quad \text{for} \quad p = 1, 2 \tag{2.5.24}$$

and time derivatives $j = 0, 1, 2$, and $F(t)$ is supposed to decrease to 0 at infinity such that

$$\int_{-\infty}^{+\infty} F(t)\, dt = b < \infty. \tag{2.5.25}$$

Then we have

$$\|S_n^{(2)}(t_1, \ldots t_n)\|_{HS}^2 \leq \ (2\pi)^{-3n} \frac{\pi^2}{m} (n^2 F(t_1) \cdot \ldots \cdot F(t_n))^2. \tag{2.5.26}$$

We finally obtain for the Hilbert-Schmidt norm of (2.5.14)

$$\|S_{+-}^{(n)}\|_{HS} \leq \int_{-\infty}^{+\infty} dt_1 \int_{-\infty}^{t_1} dt_2 \cdots \int_{-\infty}^{t_{n-1}} dt_n \|S_n^{(2)}(t_1, \ldots t_n)\|_{HS}$$

$$\leq (2\pi)^{-3n/2} \frac{\pi}{\sqrt{m}} n^2 \int_{-\infty}^{+\infty} dt_1 \int_{-\infty}^{t_1} dt_2 \cdots \int_{-\infty}^{t_{n-1}} dt_n F(t_1) F(t_2) \ldots \cdot F(t_n)$$

$$= (2\pi)^{-3n/2} \frac{\pi}{\sqrt{m}} n^2 \frac{b^n}{n!}. \tag{2.5.27}$$

In virtue of the factorial, the sum over n converges for any value of b (2.5.25)

$$\|S_{+-}\|_{HS} \leq \sum_{n=0}^{\infty} \|S_{+-}^{(n)}\|_{HS} < \infty.$$

We have therefore proven the following basic

Theorem 5.1. Let $A^\mu(t, \boldsymbol{p})$ be two times differentiable in t and

$$\left\| \frac{\partial^j}{\partial t^j} A^\mu(t, \boldsymbol{p}) \right\|_q \leq F(t) \tag{2.5.28}$$

for $j = 0, 1, 2$ and $q = 1, 2$, with

$$\int_{-\infty}^{+\infty} F(t)dt < \infty. \tag{2.5.29}$$

Then $S_{+-}[A]$ is a Hilbert-Schmidt operator and the perturbation series for S_{+-} converges in the Hilbert-Schmidt norm.

The proof of the theorem goes also through with potentials which are only piecewise two times differentiable. The first time derivative may have finitely many discontinuities, but $A^\mu(t)$ must be continuous. For potentials which are discontinuous in time, Theorem 5.1 is not true. Such potentials are unphysical. Since (2.5.28) implies that the interaction $H_1(t)$ is a bounded operator in $\mathcal{H}_1$, it follows from (2.5.29) that the perturbation series (2.5.1) converges in operator norm (see (0.3.17)).

For later applications, we transform the t-integrals in (2.5.6), using the retarded Green's function of the free Dirac equation

$$S_R(t, \boldsymbol{p}) \stackrel{\text{def}}{=} \Theta(t) e^{-itH_0(\boldsymbol{p})}. \tag{2.5.30}$$

Substituting this into (2.5.6), all t-integrations may be extended from $-\infty$ up to $+\infty$. Then the t-integrals can be carried out by means of the distributional Fourier transform

$$\hat{S}_R(p_0, \boldsymbol{p}) = \frac{1}{\sqrt{2\pi}} \lim_{\varepsilon \downarrow 0} \int_0^\infty e^{ip_0 t - \varepsilon t} e^{-itH_0(\boldsymbol{p})} dt. \tag{2.5.31}$$

Inserting $P_+ + P_- = 1$ between the exponentials and taking (2.5.9) into account, this may be written as

$$= \frac{1}{\sqrt{2\pi}} \lim_{\varepsilon \downarrow 0} \left[P_+(\boldsymbol{p}) \int_0^\infty e^{(ip_0 - \varepsilon - iE(\boldsymbol{p}))t} dt + P_-(\boldsymbol{p}) \int_0^\infty e^{(ip_0 - \varepsilon + iE(\boldsymbol{p}))t} dt \right]$$

$$= \frac{i}{\sqrt{2\pi}} \left[\frac{P_+(\boldsymbol{p})}{p_0 - E(\boldsymbol{p}) + i0} + \frac{P_-(\boldsymbol{p})}{p_0 + E(\boldsymbol{p}) + i0} \right]$$

$$= \frac{i}{\sqrt{2\pi} 2E} \left[\frac{E\gamma^0 - \boldsymbol{\gamma} \cdot \boldsymbol{p} + m}{p^0 - E + i0} + \frac{E\gamma^0 + \boldsymbol{\gamma} \cdot \boldsymbol{p} - m}{p^0 + E + i0} \right] \gamma^0. \tag{2.5.32}$$

This can be written in covariant form by introducing

$$p^0 = E \quad , \quad p^2 = p^{02} - \boldsymbol{p}^2 \quad , \quad \not{p} = \gamma^0 p^0 - \boldsymbol{\gamma} \cdot \boldsymbol{p} :$$

$$\hat{S}_R(p) = \frac{i}{\sqrt{2\pi}} \frac{\not{p} + m}{p^2 - m^2 + ip^0 0} \gamma^0 = \frac{i}{\sqrt{2\pi}} S^{\text{ret}}(p) \gamma^0. \tag{2.5.33}$$

Here the retarded function (2.3.28) appears. We have written it without the hat because the right powers of 2π for the Fourier transform are lacking. The γ^0 in (2.5.33) combines with the potential to give the covariant expression

$$\gamma^0 H_1 = e\gamma^0(V - \boldsymbol{\alpha}\cdot\boldsymbol{A}) = e(\gamma^0 V - \boldsymbol{\gamma}\cdot\boldsymbol{A}) = e\not{A}. \tag{2.5.34}$$

In order to compute the projected parts of $S^{(n)}$, we use (2.5.9) for t_1 and t_n and set

$$p^0 \stackrel{\text{def}}{=} \pm E(\boldsymbol{p}) \quad , \quad q^0 \stackrel{\text{def}}{=} \pm E(\boldsymbol{q}). \tag{2.5.35}$$

We substitute

$$S_R(t,\boldsymbol{p}_1) = \frac{1}{\sqrt{2\pi}}\int dp_1^0 \hat{S}_R(p_1^0,\boldsymbol{p}_1)e^{-ip_1^0 t}\gamma^0 = \frac{i}{2\pi}\int dp_1^0 S^{\text{ret}}(p_1)e^{-ip_1^0 t}\gamma^0, \tag{2.5.36}$$

$$\not{A}(t_1,\boldsymbol{p}-\boldsymbol{p}_1) = \frac{1}{\sqrt{2\pi}}\int dk_1 \hat{\not{A}}(k_1,\boldsymbol{p}-\boldsymbol{p}_1)e^{-ik_1 t} \tag{2.5.37}$$

into (2.5.10) and compute the t_1 integral:

$$\int_{-\infty}^{+\infty} dt_1 e^{it_1 p^0} e^{-ik_1 t_1} e^{-ip_1^0(t_1-t_2)} = 2\pi\delta(p^0 - p_1^0 - k_1)e^{ip_1^0 t_2}. \tag{2.5.38}$$

This leads to $k_1 = p^0 - p_1^0$ in (2.5.37). Then, the t_2 integral is carried out and so on. We finally end up with

$$S_{\pm\pm}^{(n)}(\boldsymbol{p},\boldsymbol{q}) = -i(2\pi)^{-2n+1}P_\pm(\boldsymbol{p})\gamma^0 \int d^4p_1 \cdots d^4p_{n-1}\, e^n$$

$$\times \hat{\not{A}}(p-p_1)S^{\text{ret}}(p_1)\hat{\not{A}}(p_1-p_2)\dots S^{\text{ret}}(p_{n-1})\hat{\not{A}}(p_{n-1}-q)P_\pm(\boldsymbol{q}). \tag{2.5.39}$$

This result is the starting point for applications in the following sections.

2.6 Electron Scattering

As a first application, we consider scattering of electrons in the external field. The incoming electron with wave function Φ is described by the Fock state

$$b(\Phi)^+\Omega \stackrel{\text{def}}{=} b_1^+\Omega, \tag{2.6.1}$$

and the outgoing electron by

$$b(\Psi)^+\Omega \stackrel{\text{def}}{=} b_2^+\Omega. \tag{2.6.2}$$

We must compute the S-matrix element

$$\mathbf{S}_{fi} = (b_2^+\Omega,\, \mathbf{S}b_1^+\Omega) = (\Omega,\, b_2\mathbf{S}b_1^+\Omega), \tag{2.6.3}$$

the subscript fi stands for final, initial. We substitute the expression (2.4.38) for $\mathbf{S}$ in here and use the commutation rules (2.4.15) (2.4.27) to move b_2 in (2.6.3) to the right. The first step is

$$b_2 e^{A_1 b^+ d^+} = e^{A_1 b^+ d^+}(b_2 + d(A_1^+ P_+ \Psi)^+). \tag{2.6.4}$$

The d^+ which commutes with the exponential, can be taken to the left factor Ω of the scalar product (2.6.3), where it gives zero. Then we are left with

$$\mathbf{S}_{fi} = \ldots : e^{(A_2 - 1)b^+ b} : b(A_2^+ P_+ \Psi) \ldots$$

$$= (\Omega, \mathbf{S} b(A_2^+ P_+ \Psi) b(\Phi)^+ \Omega) = (A_2^+ P_+ \Psi, \Phi)(\Omega, \mathbf{S}\Omega) = C(\Psi, A_2 \Phi), \tag{2.6.5}$$

where A_2 is given by (2.4.43)

$$A_2 = S_{++}^{+-1} = (S_{++}^{-1})^+. \tag{2.6.6}$$

The inverse in (2.6.6) can be computed in perturbation theory by means of the following general formula

$$S^{-1} = \Big(1 + \sum_{n=1}^{\infty} S_n\Big)^{-1}$$

$$= 1 + \sum_n \sum_{n_1 + \ldots + n_j = n} (-)^j S_{n_1} S_{n_2} \ldots S_{n_j}. \tag{2.6.7}$$

We write the n-th order term as

$$(S^{-1})_n = -(S_n - S_{n-1}S_1) + S_{n-2}(S_2 - S_1 S_1)$$

$$+ S_{n-3}(S_3 - S_2 S_1 - S_1 S_2 + S_1 S_1 S_1) + \ldots \tag{2.6.8}$$

Let us consider the first bracket in this, inserting the perturbative expressions given by (2.5.39):

$$\Big((S_{++})_n - (S_{++})_{n-1}(S_{++})_1\Big)(\boldsymbol{p}, \boldsymbol{q}) = -\frac{ie^n}{(2\pi)^{2n-1}} P_+(\boldsymbol{p})\gamma^0$$

$$\times \int d^4p_1 \cdots d^4p_{n-1}\, \not{A}(p - p_1) S^{\rm ret}(p_1) \ldots S^{\rm ret}(p_{n-1}) \not{A}(p_{n-1} - q) P_+(\boldsymbol{q})$$

$$+ \frac{e^n}{(2\pi)^{2n-3+1}} P_+(\boldsymbol{p})\gamma^0 \int d^4p_1 \cdots d^4p_{n-1}\, \not{A}(p - p_1) S^{\rm ret}(p_1) \ldots$$

$$\ldots S^{\rm ret}(p_{n-2}) \not{A}(p_{n-2} - p_{n-1}) P_+(p_{n-1}) \gamma^0 \delta^1(p_{n-1}^0 - E_{p_{n-1}}) \not{A}(p_{n-1} - q) P_+(\boldsymbol{q}) \tag{2.6.9}$$

$$= -\frac{ie^n}{(2\pi)^{2n-1}} P_+(\boldsymbol{p})\gamma^0 \int d^4p_1 \cdots d^4p_{n-1}\, \not{A}(p - p_1) S^{\rm ret}(p_1) \ldots$$

$$\ldots \not{A}(p_{n-2} - p_{n-1}) [S^{\rm ret}(p_{n-1}) + 2\pi i P_+(p_{n-1}) \gamma^0 \delta^1(p_{n-1}^0 - E_{p_{n-1}})]$$

$$\times \not{A}(p_{n-1} - q) P_+(\boldsymbol{q}). \tag{2.6.10}$$

The δ-distribution in (2.6.9) is due to the fact that the product of two integral operators in $\mathcal{H}_1$ is the 3-dimensional integral of the kernels. We concentrate on the square bracket in (2.6.10) and substitute $S^{\rm ret}$ from (2.5.32,33):

$$[\dots] = \left[\frac{P_+(\boldsymbol{p})}{p^0 - E_p + i0} + \frac{P_-(\boldsymbol{p})}{p^0 + E_p + i0} + 2\pi i P_+(\boldsymbol{p})\delta^1(p^0 - E_p)\right]\gamma^0$$

$$= \left[\frac{P_+(\boldsymbol{p})}{p^0 - E_p - i0} + \frac{P_-(\boldsymbol{p})}{p^0 + E_p + i0}\right]\gamma^0$$

$$= \frac{1}{2E}\left[\frac{E\gamma^0 - \boldsymbol{\gamma}\cdot\boldsymbol{p} + m}{p^0 - E - i0} + \frac{E\gamma^0 + \boldsymbol{\gamma}\cdot\boldsymbol{p} - m}{p^0 + E + i0}\right]$$

$$= \frac{\not p + m}{p^2 - m^2 - i0} \stackrel{\text{def}}{=} S^{AF}(p). \tag{2.6.11}$$

The resulting distribution differs from the Feynman propagator (2.3.47) in the sign of $i0$, apart from the powers of 2π. Let us call (2.6.11) the anti-Feynman propagator.

It is now evident that by the above mechanism, all S^{ret} in S^{-1}_{++} are transformed into S^{AF}

$$(S^{-1}_{++})(\boldsymbol{p}, \boldsymbol{q}) = i\sum_n e^n (2\pi)^{1-2n} P_+(\boldsymbol{p})\gamma^0 \int d^4p_1 \cdots d^4p_{n-1}$$

$$\not A(p - p_1)S^{AF}(p_1)\dots S^{AF}(p_{n-1})\not A(p_{n-1} - q)P_+(\boldsymbol{q}). \tag{2.6.12}$$

In view of (2.6.6), we must calculate the adjoint of this. Remembering

$$\not A^+ = \gamma^0 \not A \gamma^0,$$

and introducing the Feynman propagator by

$$\gamma^0 S^{AC}(p)^+ = \frac{\not p + m}{p^2 - m^2 + i0}\gamma^0 = S^F(p)\gamma^0, \tag{2.6.13}$$

we end up with

$$(S^{-1+}_{++})(\boldsymbol{p}, \boldsymbol{q}) = -2\pi i \sum_n (2\pi)^{-2n} P_+(\boldsymbol{p})\gamma^0 \int d^4p_1 \cdots d^4p_{n-1}$$

$$\times\, e\not A(p - p_1)S^F(p_1)\dots S^F(p_{n-1})e\not A(p_{n-1} - q)P_+(\boldsymbol{q}). \tag{2.6.14}$$

We now return to the S-matrix element for electron scattering (2.6.5). The initial state Φ is of the following form

$$\mathcal{H}_+ \ni \Phi(\boldsymbol{x}) = (2\pi)^{-3/2}\int d^3p\, \Phi_s(\boldsymbol{p})u_s(\boldsymbol{p})e^{ip\cdot x}, \tag{2.6.15}$$

$$\|\Phi\|^2 = \sum_{s=\pm1} \|\Phi_s\|^2 = 1,$$

and similarly for the final state Ψ. The matrix element (2.6.5) will be calculated in p-space

$$C(\Psi,\, A_2\Phi) = \int d^3p \int d^3q\, \Psi_s^*(\boldsymbol{p})S_{s\sigma}(\boldsymbol{p}, \boldsymbol{q})\Phi_\sigma(\boldsymbol{q}), \tag{2.6.16}$$

with

$$S_{s\sigma}(\boldsymbol{p},\boldsymbol{q}) \stackrel{\text{def}}{=} u_s^+(\boldsymbol{p})A_2(\boldsymbol{p},\boldsymbol{q})u_\sigma(\boldsymbol{q}). \tag{2.6.17}$$

The distributional S-matrix element (2.6.17) contains the scattering information which is independent of the initial and final states. As we will see below, for a static (time-independent) potential, it has the following structure

$$S_{s\sigma}(\boldsymbol{p},\boldsymbol{q}) = M_{s\sigma}(\boldsymbol{p},\boldsymbol{q})\delta(E_p - E_q), \tag{2.6.18}$$

the δ-distribution is a consequence of energy conservation in the scattering process. The transition probability is given by

$$p_{fi} = |\mathsf{S}_{fi}|^2. \tag{2.6.19}$$

Regarding (2.6.18), there arises the problem of squaring a δ- distribution. To avoid this nasty operation, we continue to calculate with wave packets

$$p_{fi} = \int d^3p\, d^3q\, d^3p'\, d^3q' \Psi^*_{s_1}(\boldsymbol{p})S_{s_1\sigma_1}(\boldsymbol{p},\boldsymbol{q})\Phi_{\sigma_1}(\boldsymbol{q}) \\ \times \Psi_{s_2}(\boldsymbol{p}')S_{s_2\sigma_2}(\boldsymbol{p}',\boldsymbol{q}')^*\Phi_{\sigma_2}(\boldsymbol{q}')^*. \tag{2.6.20}$$

The use of wave packets is not only desirable for mathematical reasons, it also gives us interesting insight into the physics of the scattering process.

We sum (2.6.19) over a complete system of one-electron final states by means of the completeness relation

$$\sum_f \psi^f_{s_1}(\boldsymbol{p})^*\psi^f_{s_2}(\boldsymbol{p}') = \delta_{s_1s_2}\delta^3(\boldsymbol{p}-\boldsymbol{p}'). \tag{2.6.21}$$

For the S-matrix element (2.6.18), this leads to the expression

$$\sum_f p_{fi} = \int d^3p\, d^3q\, d^3q' M_{s_1\sigma_1}(\boldsymbol{p},\boldsymbol{q})\delta(E_p - E_q) \\ \times M_{s_1\sigma_2}(\boldsymbol{p},\boldsymbol{q}')^*\delta(E_p - E_{q'})\Phi_{\sigma_1}(\boldsymbol{q})\Phi_{\sigma_2}(\boldsymbol{q}')^*. \tag{2.6.22}$$

The incoming electron state is assumed to be

$$\Phi_{s_1}(\boldsymbol{q}) = \delta_{s_1s_i}\varphi(\boldsymbol{q}), \tag{2.6.23}$$

where s_i is the initial spin and $\varphi(\boldsymbol{q})$ is supposed to be sharply peaked at the initial momentum $\boldsymbol{p}_i$. The width of $\varphi(\boldsymbol{q})$ must be small compared with the scale over which M is varying. Then the integral (2.6.22) can be separated

$$\sum_f p_{fi} = \sum_{s_1}\int d^3p\, |M_{s_1s_i}(\boldsymbol{p},\boldsymbol{p}_i)|^2 \\ \times \int d^3q\, d^3q'\delta(E_p - E_q)\delta(E_p - E_{q'})\varphi(\boldsymbol{q})\varphi(\boldsymbol{q}')^*. \tag{2.6.24}$$

The second integral depends only on the initial state, we abbreviate it by $F(|\boldsymbol{p}|)$. Writing

$$\sum_{s_1} |M_{s_1 s_i}(\boldsymbol{p}, \boldsymbol{p}_i)|^2 \stackrel{\text{def}}{=} f(\boldsymbol{p}) = f(|\boldsymbol{p}|, \vartheta, \varphi), \tag{2.6.25}$$

we perform the p-integral in spherical coordinates

$$\int d^3p\, f(\boldsymbol{p}) F(|\boldsymbol{p}|) = \int d^2\Omega\, f(|\boldsymbol{p}_i|, \vartheta, \varphi)$$

$$\times \int_0^\infty d|\boldsymbol{p}| \boldsymbol{p}^2 \delta(E_p - E_q)\delta(E_p - E_{q'})\varphi(\boldsymbol{q})\varphi(\boldsymbol{q}')^*. \tag{2.6.26}$$

Using $|\boldsymbol{p}|d|\boldsymbol{p}| = E_p dE_p$, the $|\boldsymbol{p}|$-integral can be carried out, which leads to $E_p = E_q \approx E_i$ in the slowly varying quantities

$$\sum_f p_{fi} = \sum_{s_1} \int d^2\Omega\, |M_{s_1 s_i}(|\boldsymbol{p}_i|, \vartheta, \varphi, \boldsymbol{p}_i)|^2 E_i |\boldsymbol{p}_i|$$

$$\times \int d^3q d^3q' \varphi(\boldsymbol{q})\varphi(\boldsymbol{q}')^* \delta(E_q - E_{q'}). \tag{2.6.27}$$

By means of the Fourier representation of the δ-distribution

$$\delta(E_q - E_{q'}) = \frac{1}{2\pi} \int e^{-i(E_q - E_{q'})t} dt, \tag{2.6.28}$$

(2.6.27) can be rewritten as

$$\sum_f p_{fi} = (2\pi)^2 E_i |\boldsymbol{p}_i| \sum_{s_1} \int d^2\Omega |M_{s_1 s_i}|^2 \int |\varphi(t, \boldsymbol{0})|^2 dt, \tag{2.6.29}$$

where

$$\varphi(t, \boldsymbol{x}) = (2\pi)^{-3/2} \int d^3q\, e^{-i(E_q t - \boldsymbol{q}\cdot\boldsymbol{x})} \varphi(\boldsymbol{q}) \tag{2.6.30}$$

is a free wave packet in x-space. If this wave packet is very sharp in momentum space, then the spreading in x-space in the course of time can be neglected. The wave packet is then shifted with velocity $\boldsymbol{v}$ without changing its shape

$$\varphi(t, \boldsymbol{x}) = \varphi_0(\boldsymbol{x} + \boldsymbol{v}t). \tag{2.6.31}$$

Scattering on a localized potential is only possible if there is a beam of electrons coming in, not just one. We therefore average (2.6.29) with respect to $\boldsymbol{x}$ in the wave packet (2.6.31) over a cylinder of radius R perpendicular to the beam direction:

$$\sum_f p_{fi}(R) = \frac{1}{\pi R^2} \int_{|\boldsymbol{x}_\perp| \le R} d^2x_\perp \int dt |\varphi_0(\boldsymbol{x} + \boldsymbol{v}t|^2 \ldots . \tag{2.6.32}$$

The cross section is defined by

$$\sigma = \lim_{R\to\infty} \pi R^2 \sum_f p_{fi}(R), \tag{2.6.33}$$

which obviously has the dimension of an area. In the limit $R \to \infty$, the integrals in (2.6.32) can be easily performed

$$\int d^2x_\perp \int dt |\varphi_0(\boldsymbol{x}_\perp + \boldsymbol{v}t)|^2 = \frac{1}{|\boldsymbol{v}|} \int d^3x \, |\varphi_0|^2 = \frac{1}{|\boldsymbol{v}|}, \tag{2.6.34}$$

which yields

$$\sigma = (2\pi)^2 \frac{E_i |\boldsymbol{p}_i|}{v} \sum_{s_1} \int d^2\Omega |M_{s_1 s_i}(|\boldsymbol{p}_i|, \vartheta, \varphi, \boldsymbol{p}_i)|^2. \tag{2.6.35}$$

The velocity of the incoming electrons is given by

$$\boldsymbol{v} = \frac{\boldsymbol{p}_i}{E_i}, \tag{2.6.36}$$

and we write $E_i = E = E_f$. Without the integrations over the angles, the quantity (2.6.35) is the differential cross section

$$\frac{d\sigma}{d\Omega} = (2\pi)^2 E^2 |M_{s_f s_i}(\boldsymbol{p}_f, \boldsymbol{p}_i)|^2. \tag{2.6.37}$$

We now turn to the computation of the differential cross section (2.6.37) in lowest order of perturbation theory. The normalization factor $C = 1 + O(e^2 A^2)$ in (2.6.16) can be put $= 1$. According to (2.6.17), we have in first order

$$\begin{aligned} S^{(1)}_{s\sigma}(\boldsymbol{p}, \boldsymbol{q}) &= -\frac{ie}{2\pi} u_s^+(\boldsymbol{p}) \gamma^0 A\!\!\!/(p-q) u_\sigma(\boldsymbol{q}) \\ &= -\frac{ie}{2\pi} \overline{u}_s(\boldsymbol{p}) \gamma_\mu u_\sigma(\boldsymbol{q}) A^\mu(p-q), \end{aligned} \tag{2.6.38}$$

where $p^0 = E_p$, $q^0 = E_q$. We specialize to a static potential

$$\begin{aligned} A^\mu(p) &= (2\pi)^{-2} \int d^4x \, e^{i(p^0 t - \boldsymbol{p}\cdot\boldsymbol{x})} A^\mu(\boldsymbol{x}) \\ &= (2\pi)^{-1} \delta(p^0) \int d^3x \; e^{-i\boldsymbol{p}\cdot\boldsymbol{x}} A^\mu(\boldsymbol{x}) \\ &= (2\pi)^{\frac{1}{2}} \hat{A}^\mu(\boldsymbol{p}) \delta(p^0). \end{aligned} \tag{2.6.39}$$

Strictly speaking, static potentials are not allowed in our treatment of the external field problem (cf. Theorem 5.1). But they can always be included if they are regarded as a limit of time-dependent potentials. One must only be careful in taking this limit sufficiently late in the calculation. Here, in (2.6.38), we can safely use (2.6.39)

$$S^{(1)}_{s\sigma}(\boldsymbol{p},\boldsymbol{q}) = -\frac{ie}{\sqrt{2\pi}}\overline{u}_s(\boldsymbol{p})\gamma_\mu u_\sigma(\boldsymbol{q})\hat{A}^\mu(\boldsymbol{p}-\boldsymbol{q})\delta(E_p - E_q). \tag{2.6.40}$$

Without the δ-distribution, this is just $M_{s\sigma}$, introduced in (2.6.18). Hence, by means of (2.6.37), we get the differential cross section

$$\frac{d\sigma}{d\Omega} = 2\pi E^2 \overline{u}_{s_f}(\boldsymbol{p}_f)\gamma_\mu u_{s_i}(\boldsymbol{p}_i)\overline{u}_{s_i}(\boldsymbol{p}_i)\gamma_\nu u_{s_f}(\boldsymbol{p}_f)$$
$$\times e^2 \hat{A}^\mu(\boldsymbol{p}_i - \boldsymbol{p}_f)\hat{A}^\nu(\boldsymbol{p}_f - \boldsymbol{p}_i). \tag{2.6.41}$$

For simplicity, let us assume that polarizations are not observed. Then we must average over the initial spin s_i and sum over the final spin s_f

$$\frac{d\sigma}{d\Omega} = 2\pi E^2 \frac{1}{2}\sum_{s_i}\sum_{s_f}\cdots \quad .$$

Using (1.4.43), the spin sums can be expressed as a trace

$$\frac{d\sigma}{d\Omega} = \frac{2\pi}{8}\mathrm{tr}\,[\gamma_\mu(\not{p}_i + m)\gamma_\nu(\not{p}_f + m)]e^2\hat{A}^\mu(\boldsymbol{p}_i - \boldsymbol{p}_f)\hat{A}^\nu(\boldsymbol{p}_f - \boldsymbol{p}_i). \tag{2.6.42}$$

The trace in (2.6.42) is computed by means of the formulae

$$\mathrm{tr}\,\gamma_\mu\gamma_\nu = 4g_{\mu\nu}$$

$$\mathrm{tr}\,\gamma_\mu\gamma_\alpha\gamma_\nu\gamma_\beta = 4(g_{\mu\alpha}g_{\nu\beta} + g_{\alpha\nu}g_{\beta\mu} - g_{\mu\nu}g_{\alpha\beta}). \tag{2.6.43}$$

All products of an odd number of γ-matrices have trace zero. We finally end up with

$$\frac{d\sigma}{d\Omega} = \frac{2\pi}{2}[p_i^\mu p_f^\nu + p_i^\nu p_f^\mu - g^{\mu\nu}(p_i p_f - m^2)]$$
$$\times e^2\hat{A}_\mu(\boldsymbol{p}_i - \boldsymbol{p}_f)\hat{A}_\nu(\boldsymbol{p}_f - \boldsymbol{p}_i). \tag{2.6.44}$$

For comparison with well-known results of quantum mechanics, we consider the special case of the Coulomb potential

$$A^0(\boldsymbol{x}) = -\frac{Ze}{|\boldsymbol{x}|}, \tag{2.6.45}$$

$$\hat{A}^0(\boldsymbol{p}) = -\sqrt{\frac{2}{\pi}}\frac{Ze}{\boldsymbol{p}^2}. \tag{2.6.46}$$

Then, the differential cross section (2.6.44) becomes

$$\frac{d\sigma}{d\Omega} = 2Z^2e^4(E^2 + m^2 + \boldsymbol{p}_i\cdot\boldsymbol{p}_f)\frac{1}{(\boldsymbol{p}_i - \boldsymbol{p}_f)^4}. \tag{2.6.47}$$

Let $\vartheta = \angle(\boldsymbol{p}_i, \boldsymbol{p}_f)$ be the scattering angle, then, in virtue of $|\boldsymbol{p}_i| = |\boldsymbol{p}_f| \stackrel{\text{def}}{=} |\boldsymbol{p}|$, we obviously have

$$(\boldsymbol{p}_i - \boldsymbol{p}_f)^2 = \boldsymbol{p}_i{}^2 + \boldsymbol{p}_f{}^2 - 2\boldsymbol{p}_i \cdot \boldsymbol{p}_f = 2|\boldsymbol{p}|^2(1 - \cos\vartheta)$$

$$= 4|\boldsymbol{p}|^2 \sin^2\frac{\vartheta}{2} \quad , \quad \text{and}$$

$$\frac{d\sigma}{d\Omega} = 4Z^2e^4m^2 \frac{1 + \frac{\boldsymbol{p}^2}{m^2}\cos^2\frac{\vartheta}{2}}{16\boldsymbol{p}^4\sin^4\frac{\vartheta}{2}}. \tag{2.6.48}$$

In the non-relativistic limit $\boldsymbol{p}^2 \ll m^2$, introducing the non-relativistic energy

$$E_0 = \frac{\boldsymbol{p}^2}{2m} = \frac{m}{2}\boldsymbol{v}^2, \tag{2.6.49}$$

the result becomes

$$\frac{d\sigma}{d\Omega} \approx \frac{Z^2e^4}{16E_0^2\sin^4\frac{\vartheta}{2}} = \frac{Z^2e^4}{4m^2v^4\sin^4\frac{\vartheta}{2}}. \tag{2.6.50}$$

This is Rutherford's scattering formula.

2.7 Pair Production

The creation of pairs of electrons and positrons by time-dependent electromagnetic fields is the first effect which is typical for relativistic quantum field theory. Abbreviating the emission operators of the pair by

$$b_1^+ = b(\Phi)^+ \quad , \quad d_1^+ = d(\Psi)^+, \tag{2.7.1}$$

we have to compute the following S-matrix element

$$\mathsf{S}_{fi} = (b_1^+ d_1^+ \Omega, \mathsf{S}\Omega) = C(b_1^+ d_1^+ \Omega, e^{A_1 b^+ d^+}\Omega)$$

$$= C(\Omega, d_1 b_1 e^{A_1 b^+ d^+}\Omega). \tag{2.7.2}$$

The commutation rules (2.4.15, 16) enable us to reduce the right-hand side to the form

$$= C(\Omega, d_1 e^{A_1 b^+ d^+} d(A_1^+ P_+ \Phi)^+ \Omega)$$

$$= C(\Omega, e^{A_1 b^+ d^+}(d_1 - b^+) d(A_1^+ P_+ \Phi)^+ \Omega). \tag{2.7.3}$$

We shift the exponential and the b^+ to the left factor of the scalar product and arrive at

$$\mathsf{S}_{fi} = C(\Omega, d(\Psi) d(A_1^+ P_+ \Phi)^+ \Omega) = C(\Phi, A_1 \Psi), \tag{2.7.4}$$

where A_1 is given by (2.4.52)

$$A_1 = S_{+-}(S_{--})^{-1} = -(S_{++}^{-1})^+(S_{-+})^+. \tag{2.7.5}$$

Using the spectral decomposition (2.4.78), the first expression in (2.7.5) leads to the following result

$$\mathsf{S}_{fi} = -\left(\prod_n \nu_n\right)\sum_n \frac{\lambda_n}{\nu_n}(\psi_n^-, \Psi)(\Phi, \varphi_n^+), \tag{2.7.6}$$

with

$$\nu_n = \sqrt{1-\lambda_n^2}. \tag{2.7.7}$$

The transition probability is now given by

$$|\mathsf{S}_{fi}|^2 = \left(\prod_n \nu_n\right)^2 \sum_n \frac{\lambda_n}{\nu_n}(\Phi, \varphi_n^+)(\psi_n^-, \Psi)$$

$$\times \sum_m \frac{\lambda_m}{\nu_m}(\varphi_m^+, \Phi)(\Psi, \psi_m^-). \tag{2.7.8}$$

We want to compute the energy spectrum of the created positrons. Then, we have to sum over a complete set of electron states $\Phi \in \mathcal{H}_+$

$$P_{\text{pos}}(\Psi) = \left(\prod_n \nu_n\right)^2 \sum_n \frac{\lambda_n^2}{\nu_n^2}|(\psi_n^-, \Psi)|^2$$

$$= \left(\prod_n \nu_n\right)^2 \sum_n \frac{\lambda_n^2}{\nu_n^2}\left|\int d^3p\, \psi_n^-(\mathbf{p})^* \Psi(\mathbf{p})\right|^2. \tag{2.7.9}$$

This is the production rate of positrons in the state Ψ. The total positron production rate is obtained by summing over a complete set of positron states $\Psi \in \mathcal{H}_-$

$$P_{\text{tot}} = \left(\prod_n \nu_n\right)^2 \sum_n \frac{\lambda_n^2}{\nu_n^2}\int d^3p\, |\psi_n^-(\mathbf{p})|^2. \tag{2.7.10}$$

In order to get the positron spectrum, we express the momentum integral in terms of energy and angles

$$\int d|\mathbf{p}|\mathbf{p}^2 d\Omega = \int dE E|\mathbf{p}|d\Omega.$$

Then we arrive at the following formula for the positron spectrum

$$\frac{dP}{dE} = \left(\prod_n \nu_n\right)^2 \sum_n \frac{\lambda_n^2}{\nu_n^2} E|\mathbf{p}| \int d\Omega |\psi_n^-(\mathbf{p})|^2. \tag{2.7.11}$$

The quantities entering in (2.7.11) follow from the spectral decomposition (2.4.78) of the one-particle S-matrix

$$S_{-+} = \sum_n \lambda_n(\psi_n^+, \cdot)\psi_n^-. \tag{2.7.12}$$

The exact formula (2.7.11) is useful for the calculation of pair production in strong electromagnetic fields, as they are for example produced in heavy ion collisions.

We now return to weak fields and compute (2.7.4, 5) in perturbation theory. We take the adjoint of (2.5.39), using

$$S^{\text{ret}}(p)^+ = \frac{\gamma^0 \not{p} \gamma^0 + m}{p^2 - m^2 - ip^0 0} = \gamma^0 S^{\text{av}}(p)\gamma^0, \tag{2.7.13}$$

and obtain

$$(S_{-+})_n^+(\boldsymbol{p},\boldsymbol{q}) = i(2\pi)^{1-2n} P_+(\boldsymbol{p})\gamma^0 \int d^4p_1 \cdots d^4p_{n-1}$$

$$\not{A}(p-p_1)S^{\text{av}}(p_1)\ldots S^{\text{av}}(p_{n-1})\not{A}(p_{n-1}-q)P_-(\boldsymbol{q}). \tag{2.7.14}$$

The integral kernel for (2.7.5) is given to n-th order by

$$(S_{++}^{+-1}S_{+-}^+)_n(\boldsymbol{p},\boldsymbol{q}) = \int d^3q' \, [(S_{++}^{+-1})_0(\boldsymbol{p},\boldsymbol{q}')(S_{+-}^+)_n(\boldsymbol{q}',\boldsymbol{q})$$

$$+(S_{++}^{+-1})_1(\boldsymbol{p},\boldsymbol{q}')(S_{+-}^+)_{n-1}(\boldsymbol{q}',\boldsymbol{q}) + \ldots + (S_{++}^{+-1})_{n-1}(\boldsymbol{p},\boldsymbol{q}')(S_{+-}^+)_1(\boldsymbol{q}',\boldsymbol{q})]. \tag{2.7.15}$$

Here we insert (2.6.14) and (2.7.14). It is not hard to see by means of the formula

$$S^{\text{av}}(p) - 2\pi i P_+(p)\gamma^0\delta^1(p^0 - E_p) = S^F(p) \tag{2.7.16}$$

that all advanced propagators in (2.7.14) are changed into Feynman propagators. This leads to the following expression for (2.7.5)

$$A_1(\boldsymbol{p},\boldsymbol{q}) = -2\pi i \sum_n (2\pi)^{-2n} P_+(\boldsymbol{p})\gamma^0 \int d^4p_1 \cdots d^4p_{n-1}$$

$$e\not{A}(p-p_1)S^F(p_1)\ldots S^F(p_{n-1})e\not{A}(p_{n-1}-q)P_-(\boldsymbol{q}). \tag{2.7.17}$$

We now turn to the calculation of the S-matrix element (2.7.4). Assuming the wave functions of the produced pair to be

$$\mathcal{H}_+ \ni \Phi(\boldsymbol{x}) = (2\pi)^{-3/2} \int d^3p \, \Phi_s(\boldsymbol{p})u_s(\boldsymbol{p})e^{i\boldsymbol{p}\cdot\boldsymbol{x}} \tag{2.7.18}$$

$$\mathcal{H}_- \ni \Psi(\boldsymbol{x}) = (2\pi)^{-3/2} \int d^3q \, \Psi_\sigma(\boldsymbol{q})v_\sigma(\boldsymbol{q})e^{-i\boldsymbol{q}\cdot\boldsymbol{x}}, \tag{2.7.19}$$

we find

$$\mathsf{S}_{fi} = \int d^3p \, d^3q \, \Phi_s^*(\boldsymbol{p})S_{s\sigma}(\boldsymbol{p},\boldsymbol{q})\Psi_\sigma(-\boldsymbol{q}), \tag{2.7.20}$$

where

$$S_{s\sigma}(\boldsymbol{p},\boldsymbol{q}) = C\, u_s^+(\boldsymbol{p})A_1(\boldsymbol{p},\boldsymbol{q})v_\sigma(-\boldsymbol{q}). \tag{2.7.21}$$

The transition probability becomes

$$p_{fi} = |\mathbf{S}_{fi}|^2 = \int d^3p\, d^3q d^3p' d^3q' \Phi^*_{s_1}(\boldsymbol{p}) S_{s_1\sigma_1}(\boldsymbol{p},\boldsymbol{q}) \Psi_{\sigma_1}(-\boldsymbol{q})$$

$$\times \Phi_{\sigma_2}(\boldsymbol{p}') S_{s_2\sigma_2}(\boldsymbol{p}',\boldsymbol{q}')^* \Psi_{\sigma_2}(-\boldsymbol{q}')^*. \tag{2.7.22}$$

The total transition probability is obtained by summing over the final states f

$$P = \sum_{s\sigma} \int d^3p\, d^3q |S_{s\sigma}(\boldsymbol{p},-\boldsymbol{q})|^2$$

$$= C^2 \sum_{s\sigma} \int d^3p\, d^3q |u_s^+(\boldsymbol{p}) A_1(\boldsymbol{p},-\boldsymbol{q}) v_\sigma(\boldsymbol{q})|^2, \tag{2.7.23}$$

where the completeness relation (2.6.21) has been used.

In lowest order ($n = 1$), we have

$$A_1(\boldsymbol{p},-\boldsymbol{q}) = -i(2\pi)^{-1} P_+(\boldsymbol{p}) \gamma^0 e\not{A}(p+q) P_-(-\boldsymbol{q}), \tag{2.7.24}$$

according to (2.7.17), which yields

$$P = (2\pi)^{-2} \sum_{s\sigma} \int d^3p\, d^3q\, \overline{u}_s(\boldsymbol{p}) \gamma^\mu v_\sigma(\boldsymbol{q}) [\overline{u}_s(\boldsymbol{p}) \gamma^\nu v_\sigma(\boldsymbol{q})]^+$$

$$\times eA_\mu(p+q) eA_\nu(-p-q). \tag{2.7.25}$$

As in (2.6.42), the spin sum can be expressed as a trace

$$\sum_{s\sigma} \overline{u}_s(\boldsymbol{p}) \gamma^\mu v_\sigma(\boldsymbol{q}) v_\sigma(\boldsymbol{q})^+ \gamma^0 \gamma^\nu u_s(\boldsymbol{p})$$

$$= \operatorname{tr} \frac{1}{2E_p}(\not{p}+m)\gamma^\mu \frac{1}{2E_q}(\not{q}-m)\gamma^\nu. \tag{2.7.26}$$

Then we arrive at

$$P = (2\pi)^{-2} \int \frac{d^3p}{2E_p} \frac{d^3q}{2E_q} \operatorname{tr}\left[(\not{p}+m)\gamma^\mu(\not{q}-m)\gamma^\nu\right]$$

$$\times eA_\mu(p+q) eA_\nu(-p-q). \tag{2.7.27}$$

Using $p+q=k$ as a new integration variable, the result can be written as

$$P = (2\pi)^{-2} \int d^4k\, T^{\mu\nu}(k) eA_\mu(k) eA_\nu(-k), \tag{2.7.28}$$

where

$$T^{\mu\nu}(k) = \int \frac{d^3p}{2E_p} \frac{d^3q}{2E_q} \delta(p+q-k) \operatorname{tr}\left[(\not{p}+m)\gamma^\mu(\not{q}-m)\gamma^\nu\right]. \tag{2.7.29}$$

We now proceed to investigate the tensor (2.7.29), which will also be important in other connection. The 3-dimensional integrations in (2.7.29) may be written as integrals over Minkowski space with the Lorentz-invariant measure

$$\int \frac{d^3p}{2E_p} \ldots = \int d^4p\, \delta(p^2 - m^2)\Theta(p^0) \ldots, \tag{2.7.30}$$

in virtue of (2.3.2). Then, one integration in (2.7.29) can be carried out by means of the δ-distribution in (2.7.29)

$$T^{\mu\nu}(k) = \int d^4p\, \delta(p^2 - m^2)\Theta(p^0)\delta((k-p)^2 - m^2)\Theta(k^0 - p^0)$$
$$\times \operatorname{tr}\left[(\not{p} + m)\gamma^\mu(\not{k} - \not{p} - m)\gamma^\nu\right]. \tag{2.7.31}$$

The trace will be computed utilizing the formulae (2.6.43)

$$\operatorname{tr}\left(p_\alpha(k_\beta - p_\beta)\gamma^\alpha\gamma^\mu\gamma^\beta\gamma^\nu - m^2\gamma^\mu\gamma^\nu\right)$$
$$= 4[p^\mu k^\nu + p^\nu k^\mu - 2p^\mu p^\nu - p\cdot k g^{\mu\nu}]. \tag{2.7.32}$$

It is now evident that (2.7.31) is a second rank Lorentz tensor. Furthermore, it follows from (2.7.32) that

$$k_\mu \operatorname{tr}[\ldots] = (k^2 - 2kp)p^\nu = 0,$$

because $k^2 - 2kp = (k-p)^2 - m^2 = 0$, in virtue of the second δ-distribution in (2.7.31). This yields

$$k_\mu T^{\mu\nu}(k) = 0. \tag{2.7.33}$$

This property of $T^{\mu\nu}$ expresses gauge invariance. In fact, a gauge transformation

$$A_\mu \to A_\mu + ik_\mu \Lambda \tag{2.7.34}$$

does not alter the probability P (2.7.28) of pair creation.

The second rank tensor $T^{\mu\nu}(k)$, depending on the four vector k only, has the following general form

$$T^{\mu\nu}(k) = A(k^2)g^{\mu\nu} + B(k^2)k^\mu k^\nu, \tag{2.7.35}$$

where the functions A and B must be Lorentz invariant and, therefore, are functions of k^2. This often used theorem is not so simple to prove (see *K. Hepp, Helvetica Phys. Acta 36, 355 (1966)*). Without using it, the final result (2.7.36) can be obtained by calculating in a special Lorentz system as below and transforming to a general system by a Lorentz transformation. From (2.7.33), we conclude that

$$0 = A(k^2)k^\nu + B(k^2)k^2k^\nu,$$

hence, $A = -k^2B$, i.e.

$$T^{\mu\nu}(k) = (k^\mu k^\nu - k^2 g^{\mu\nu})B(k^2). \tag{2.7.36}$$

In order to determine the unknown $B(k^2)$, we calculate

$$\sum_{\mu} T^{\mu}{}_{\mu} = -3k^2 B(k^2)$$

$$= 4\int d^4p\,\delta(p^2-m^2)\Theta(p^0)\delta(k^2-2kp)\Theta(k^0-p^0)(2kp-2p^2-4pk), \quad (2.7.37)$$

which gives

$$B(k^2) = \frac{4}{3}\left(1+\frac{2m^2}{k^2}\right)\int d^4p\,\delta(p^2-m^2)\Theta(p^0)\delta(k^2-2kp)\Theta(k^0-p^0), \quad (2.7.38)$$

taking the two δ-distributions into account in the last factor in (2.7.37). Since p and q are time-like, $p^2 = m^2 = q^2$, so is $k = p + q$. There exists, therefore, a reference system such that $k = (k_0, \mathbf{0})$. In this Lorentz frame, (2.7.38) assumes the following simple form

$$B(k_0^2) = \frac{4}{3}\left(1+\frac{2m^2}{k_0^2}\right)\int \frac{d^3p}{2E}\delta(k_0^2 - 2k_0E)\Theta(k_0 - E). \quad (2.7.39)$$

It follows that

$$E = \sqrt{\boldsymbol{p}^2 + m^2} = \frac{k_0}{2} \quad , \quad |\boldsymbol{p}| = \sqrt{\frac{k_0^2}{4} - m^2},$$

and

$$B(k_0^2) = \frac{2}{3}\left(1+\frac{2m^2}{k_0^2}\right)4\pi\Theta(k_0^2 - 4m^2)\int_0^\infty d|\boldsymbol{p}|\frac{\boldsymbol{p}^2}{E}\delta\Big(2k_0\Big(\frac{k_0}{2} - E\Big)\Big)\Theta(k_0 - E)$$

$$= \frac{8\pi}{3}\left(1+\frac{2m^2}{k_0^2}\right)\Theta(k_0^2 - 4m^2)\int_m^\infty dE|\boldsymbol{p}|\frac{\delta(E-\frac{k_0}{2})}{2k_0}\Theta\Big(\frac{k_0}{2}\Big)$$

$$= \frac{2\pi}{3}\left(1+\frac{2m^2}{k_0^2}\right)\sqrt{1-\frac{4m^2}{k_0^2}}\Theta(k_0^2 - 4m^2)\Theta(k_0). \quad (2.7.40)$$

The result in an arbitrary Lorentz frame then reads

$$B(k^2) = \frac{2\pi}{3}\left(1+\frac{2m^2}{k^2}\right)\sqrt{1-\frac{4m^2}{k^2}}\Theta(k^2 - 4m^2)\Theta(k_0). \quad (2.7.41)$$

Finally, we substitute (2.7.41) into (2.7.36) and end up with the following expression for the total probability of pair production (2.7.28)

$$P = (2\pi)^{-2}\frac{2\pi}{3}\int d^4k\left(1+\frac{2m^2}{k^2}\right)\sqrt{1-\frac{4m^2}{k^2}}\Theta(k^2-4m^2)\Theta(k^0)$$

$$\times (k^\mu k^\nu - k^2 g^{\mu\nu})eA_\mu(k)eA_\nu(-k). \quad (2.7.42)$$

Because of the first Θ-function, the integral in (2.7.42) extends over momenta $k^2 > 4m^2$, only. The physical reason for this lies in the fact that the energy which the external field must supply to create a pair, must be bigger than $2m$. Since the potential is real in x-space, we have $A_\nu(-k) = A_\nu(k)^*$. Consequently, Eq. (2.7.42) can be written in the symmetrical form

$$P = \frac{e^2}{(2\pi)^2}\frac{\pi}{3}\int d^4k\left(1+\frac{2m^2}{k^2}\right)\sqrt{1-\frac{4m^2}{k^2}}\,\Theta(k^2-4m^2)$$

$$\times\,(k^\mu k^\nu - k^2 g^{\mu\nu})A_\mu(k)A_\nu(k)^*. \tag{2.7.43}$$

2.8 The Causal Phase of the S-Matrix

The unitary scattering operator $\mathbf{S}$ in Fock space has been defined in Sect. 2.4 (2.4.7, 8) up to a phase $\exp i\lambda$, only. Since $\mathbf{S}$ describes the scattering and particle production processes in an external electromagnetic potential A, the phase $\lambda[A]$ must depend on A. However, this dependence is not yet specified by the theory developed so far. Some additional input is necessary, in order to specify it. This very important new input is causality.

We first remind the reader of causality in the one-particle theory (cf. Sect. 0.3). Let us consider a potential

$$A^\mu(x) = A_1^\mu(x) + A_2^\mu(x), \tag{2.8.1}$$

which is the sum of two parts with disjoint supports in time

$$\operatorname{supp} A_1 \subset (-\infty, r] \quad , \quad \operatorname{supp} A_2 \subset [r, +\infty). \tag{2.8.2}$$

That is to say, A_1 vanishes for times $t > r$ and A_2 for $t < r$. Then the one-particle S-matrix factorizes

$$S[A] = W_{\rm out}^+ W_{\rm in} = U_0(0,\infty)U(\infty,-\infty)U_0(-\infty,0) \tag{2.8.3}$$

$$= U_0(0,\infty)U(\infty,r)U_0(r,0)U_0(0,r)U(r,-\infty)U_0(-\infty,0)$$

$$= S_2S_1 \quad , \quad S_j \stackrel{\rm def}{=} S[A_j], \tag{2.8.4}$$

where U and U_0 are the unitary propagators with and without external field, respectively. The arguments $\pm\infty$ in (2.8.4) stand for the corresponding strong limits $t \to \pm\infty$. A similar factorization should hold for the S-operator $\mathbf{S}$ in Fock space.

From the defining relation (2.4.7), applied to S_1, S_2 and S, we conclude that

$$\psi(S^+f) = \psi(S_1^+S_2^+f) = \mathbf{S}_1^{-1}\psi(S_2^+f)\mathbf{S}_1$$

$$= \mathbf{S}_1^{-1}\mathbf{S}_2^{-1}\psi(f)\mathbf{S}_2\mathbf{S}_1 = \mathbf{S}^{-1}\psi(f)\mathbf{S}, \tag{2.8.5}$$

for all $f \in \mathcal{H}_1$, and similarly for ψ^+. It follows from the irreducibility of the Fock representation (see proof of Propos. 2.1 (2.2.46)) that

$$\mathbf{S} = e^{i\psi_{12}}\mathbf{S}_2\mathbf{S}_1, \tag{2.8.6}$$

where ψ_{12} is an undetermined phase. To fix this phase, it is only necessary to specify the vacuum expectation value of (2.8.6). The requirement $\psi_{12} = 0$ leads us to the following causality condition

$$(\Omega, \mathbf{S}\Omega) = (\Omega, \mathbf{S}_2\mathbf{S}_1\Omega). \tag{2.8.7}$$

We call (2.8.7) global causality condition for the Fock space S-operator in contrast to a differential condition, to be derived below. It follows from the foregoing reasoning that (2.8.7) can be fulfilled and that it implies the operator condition

$$\mathbf{S}[A_1 + A_2] = \mathbf{S}[A_2]\,\mathbf{S}[A_1] \tag{2.8.8}$$

for any two potentials with disjoint supports in time (2.8.2).

We now turn to the derivation of the differential condition. Let

$$\operatorname{supp} f \subset (-\infty, r] \quad , \quad \operatorname{supp} g \subset [r, +\infty) \tag{2.8.9}$$

be two auxiliary potentials, then

$$\mathbf{S}[A_1 + \varepsilon_1 f + A_2 + \varepsilon_2 g] = \mathbf{S}[A_2 + \varepsilon_2 g]\mathbf{S}[A_1 + \varepsilon_1 f], \tag{2.8.10}$$

according to (2.8.8). We conclude that

$$\begin{aligned} &\mathbf{S}^+[A + \varepsilon_2 g]\,\mathbf{S}[A + \varepsilon_1 f + \varepsilon_2 g] \\ &= \mathbf{S}^+[A_1]\,\mathbf{S}^+[A_2 + \varepsilon_2 g]\,\mathbf{S}[A_2 + \varepsilon_2 g]\,\mathbf{S}[A_1 + \varepsilon_1 f] \\ &= \mathbf{S}^+[A_1]\,\mathbf{S}[A_1 + \varepsilon_1 f] \end{aligned} \tag{2.8.11}$$

is independent of ε_2. Hence, differentiating with respect to ε_1 and ε_2 at 0, we find

$$\frac{\partial}{\partial \varepsilon_2}\bigg|_0 \mathbf{S}^+[A + \varepsilon_2 g]\frac{\partial}{\partial \varepsilon_1}\bigg|_0 \mathbf{S}[A + \varepsilon_1 f + \varepsilon_2 g] = 0. \tag{2.8.12}$$

This can be written in terms of functional derivatives as follows

$$\int d^4x \int d^4y\, g_\nu(y) f_\mu(x) \frac{\delta}{\delta A_\nu(y)}\mathbf{S}^+[A]\frac{\delta}{\delta A_\mu(x)}\mathbf{S}[A] = 0, \tag{2.8.13}$$

or

$$\frac{\delta}{\delta A_\nu(y)}\mathbf{S}^+[A]\frac{\delta}{\delta A_\mu(x)}\mathbf{S}[A] = 0 \quad \text{for} \quad x^0 < r < y^0, \tag{2.8.14}$$

$$\text{if} \quad A(t, \boldsymbol{x}) = 0 \tag{2.8.15}$$

in a neighbourhood of $t = r$. If condition (2.8.15) is dropped, one gets Bogoliubov's causality condition (*N.N. Bogoliubov, D.V. Shirkov, Introduction to the Theory of Quantized Fields, New York, N.Y. 1959.*)

$$\frac{\delta}{\delta A_\nu(y)}\left(\Omega, \mathbf{S}^+\frac{\delta \mathbf{S}}{\delta A_\mu(x)}\Omega\right) = 0 \quad \text{for} \quad x^0 < y^0, \tag{2.8.16}$$

and *arbitrary* potential $A(x)$. This local causality condition is stronger than (2.8.14) plus (2.8.15) or (2.8.8). Consequently, it is at the moment not clear whether it can be satisfied by choosing the phase $\varphi[A]$ properly. We are going to show that this is in fact possible.

Let us write

$$\mathbf{S} = e^{i\varphi}\tilde{\mathbf{S}}, \tag{2.8.17}$$

where $\tilde{\mathbf{S}}$ is the unitary S-operator, determined by Theorem 4.1 (2.4.61), without phase. In view of (2.8.16), we must compute

$$\frac{\delta}{\delta A_\nu(y)}\left(\mathbf{S}\Omega,\ \frac{\delta \mathbf{S}}{\delta A_\mu(x)}\Omega\right) = \frac{\delta}{\delta A_\nu(y)}\Big[e^{-i\varphi}\frac{\delta e^{i\varphi}}{\delta A_\mu(x)}(\tilde{\mathbf{S}}\Omega,\ \tilde{\mathbf{S}}\Omega)$$

$$+(\tilde{\mathbf{S}}\Omega,\ \frac{\delta \tilde{\mathbf{S}}}{\delta A_\mu(x)}\Omega)\Big] = i\frac{\delta^2\varphi}{\delta A_\nu(y)\delta A_\mu(x)} + \frac{\delta}{\delta A_\nu(y)}\left(\tilde{\mathbf{S}}\Omega,\ \frac{\delta \tilde{\mathbf{S}}}{\delta A_\mu(x)}\Omega\right). \tag{2.8.18}$$

We claim that the last term in (2.8.18) is purely imaginary. In fact, differentiating

$$(\tilde{\mathbf{S}}\Omega,\ \tilde{\mathbf{S}}\Omega) = 1,$$

we obtain

$$\left(\frac{\delta \tilde{\mathbf{S}}}{\delta A_\mu(x)}\Omega,\ \tilde{\mathbf{S}}\Omega\right) + \left(\tilde{\mathbf{S}}\Omega,\ \frac{\delta \tilde{\mathbf{S}}}{\delta A_\mu(x)}\Omega\right) = 0, \tag{2.8.19}$$

and consequently

$$\left(\tilde{\mathbf{S}}\Omega,\ \frac{\delta \tilde{\mathbf{S}}}{\delta A_\mu(x)}\Omega\right) = -\left(\frac{\delta \tilde{\mathbf{S}}}{\delta A_\mu(x)}\Omega,\ \tilde{\mathbf{S}}\Omega\right) = -\left(\tilde{\mathbf{S}}\Omega,\ \frac{\delta \tilde{\mathbf{S}}}{\delta A_\mu(x)}\Omega\right)^*. \tag{2.8.20}$$

This is imaginary and the same is obviously true for $\mathbf{S}$. Hence, the real part of the causality condition (2.8.16) is automatically satisfied. The imaginary part can be fulfilled by appropriate choice of φ in (2.8.18).

We now proceed to construct the causal phase φ in lowest order perturbation theory. Starting from

$$\tilde{\mathbf{S}}\Omega = Ce^{A_1 b^+ d^+}\Omega = C\Big(\Omega + \sum_{mn}(S_{+-}S_{--}^{-1})_{mn} b_m^+ d_n^+ \Omega + \dots\Big), \tag{2.8.21}$$

we observe that C and S_{--}^{-1} can be put equal to 1 in lowest order. The derivative becomes

$$\frac{\delta}{\delta A_\mu(x)}\tilde{\mathbf{S}}\Omega = \frac{\delta C}{\delta A_\mu(x)}\Big[\Omega + \sum_{mn}(S_{+-})_{mn} b_m^+ d_n^+ \Omega + \dots\Big]$$

$$+C\sum_{mn}\left(\frac{\delta S_{+-}}{\delta A_\mu(x)}\right)_{mn} b_m^+ d_n^+ \Omega + \dots, \tag{2.8.22}$$

where the second term in the square bracket can be dropped because the factor in front is already $O(A)$. This yields

$$\left(\tilde{\mathbf{S}}\Omega, \frac{\delta \tilde{\mathbf{S}}}{\delta A_\mu(x)}\Omega\right) = C\frac{\delta C}{\delta A_\mu(x)} + C^2 \sum_{mn} (S_{+-})^+_{nm} \left(\frac{\delta S_{+-}}{\delta A_\mu(x)}\right)_{mn} + \dots$$

$$= C\frac{\delta C}{\delta A_\mu(x)} + \mathrm{Tr}\left(S^+_{-+}\frac{\delta S_{+-}}{\delta A_\mu(x)}\right) + \dots . \tag{2.8.23}$$

The first term on the right-hand side is real. Since the left-hand side is purely imaginary, the first term must cancel against the real part of the second one

$$(\tilde{\mathbf{S}}\Omega, \frac{\delta \tilde{\mathbf{S}}}{\delta A_\mu(x)}\Omega) = i \,\mathrm{Im}\,\mathrm{Tr}\left(S^+_{-+}\frac{\delta S_{+-}}{\delta A_\mu(x)}\right) + \dots . \tag{2.8.24}$$

It follows from the local causality condition (2.8.16), in virtue of (2.8.18) that

$$F(x,y) \stackrel{\mathrm{def}}{=} \frac{\delta^2 \varphi}{\delta A_\nu(y)\delta A_\mu(x)} + \mathrm{Im}\frac{\delta}{\delta A_\nu(y)}\mathrm{Tr}\left((S_{+-})^+\frac{\delta S_{+-}}{\delta A_\mu(x)}\right)$$

$$= 0 \quad \text{for} \quad x^0 < y^0. \tag{2.8.25}$$

We call a function $F(x,y)$ which vanishes for $x^0 < y^0$ a causal function.

We next compute the second term in (2.8.25). In lowest order of perturbation theory, we have

$$S^{(1)}_{+-}(\boldsymbol{p},\boldsymbol{q}) = -i(2\pi)^{-1}P_+(\boldsymbol{p})\gamma^0 e\not{A}(p+q)P_-(-\boldsymbol{q})$$

$$= -i(2\pi)^{-3}P_+(\boldsymbol{p})\gamma^0\gamma^\nu \int d^4x\, eA_\nu(x)e^{i(p+q)x}P_-(-\boldsymbol{q}), \tag{2.8.26}$$

by means of (2.7.24). The functional derivative becomes

$$\frac{\delta S^{(1)}_{+-}}{\delta A_\nu(x)} = -ie(2\pi)^{-3}P_+(\boldsymbol{p})\gamma^0\gamma^\nu e^{i(p+q)x}P_-(-\boldsymbol{q}). \tag{2.8.27}$$

Using the adjoint

$$S^{(1)}_{+-}(\boldsymbol{p},\boldsymbol{q})^+ = ie(2\pi)^{-3}P_-(-\boldsymbol{q})\int d^4y\, A_\nu(y)e^{-i(p+q)y}\gamma^0\gamma^\nu P_+(\boldsymbol{p}), \tag{2.8.28}$$

and the formula

$$\mathrm{Tr}\,K = \int d^3q\, K(\boldsymbol{q},\boldsymbol{q}) \tag{2.8.29}$$

for the trace of an integral operator, we get for $x \neq y$

$$\mathrm{Tr}\frac{\delta}{\delta A_\nu(y)}(S_{+-})^+\frac{\delta S_{+-}}{\delta A_\mu(x)} = e^2(2\pi)^{-6}\int d^3q \int d^3p$$

$$\times\, e^{i(p+q)(x-y)}\mathrm{tr}\, P_-(-\boldsymbol{q})\gamma^0\gamma^\nu P_+(\boldsymbol{p})\gamma^0\gamma^\mu P_-(-\boldsymbol{q}). \tag{2.8.30}$$

Since the trace is unchanged under cyclic permutation, substituting the projection operators by their covariant expressions, this may be written in covariant form as

$$= e^2(2\pi)^{-6}\int d^4p\,d^4q\,e^{i(p+q)(x-y)}\frac{\delta(q^0-E_q)}{2E_q}\frac{\delta(p^0-E_p)}{2E_p}$$

$$\times\,\mathrm{tr}\,(\not{q}-m)\gamma^\nu(\not{p}+m)\gamma^\mu. \tag{2.8.31}$$

Introducing the new variable of integration $k=p+q$, this may be represented in the form

$$\mathrm{Tr}\,\frac{\delta}{\delta A_\nu(y)}(S_{+-})^+\frac{\delta S_{+-}}{\delta A_\mu(x)} = e^2(2\pi)^{-6}\int d^4k\,e^{ik(x-y)}T^{\mu\nu}(k), \tag{2.8.32}$$

where $T_{\mu\nu}(k)$ is just the tensor (2.7.31) for pair creation:

$$T^{\mu\nu}(k) = \int d^4p d^4q\,\delta(p+q-k)\delta(q^2-m^2)\Theta(q^0)$$

$$\times\,\delta(p^2-m^2)\theta(p^0)\mathrm{tr}\,\ldots$$

$$= \int d^4p\,\delta((k-p)^2-m^2)\theta(k^0-p^0)\delta(p^2-m^2)\theta(p^0)\mathrm{tr}\,\ldots\quad.$$

The result (2.7.41) can immediately be taken over

$$T^{\mu\nu}(k) = (k^\mu k^\nu - k^2 g^{\mu\nu})\frac{2\pi}{3}\Big(1+\frac{2m^2}{k^2}\Big)\sqrt{1-\frac{4m^2}{k^2}}\,\Theta(k^2-4m^2)\Theta(k^0). \tag{2.8.33}$$

We now return to (2.8.25). The causal function $F(x,y)$ assumes the following form

$$F(x,y) = \frac{\delta^2\varphi}{\delta A_\nu(y)\delta A_\mu(x)} + e^2(2\pi)^{-6}\mathrm{Im}\int d^4k\,e^{ik(x-y)}T^{\mu\nu}(k)$$

$$= \ldots + e^2(2\pi)^{-6}\int\limits_{k^0\geq 0} d^4k\,\sin k(x-y)T^{\mu\nu}(k), \tag{2.8.34}$$

because $T^{\mu\nu}(k)$ (2.8.33) is real. To write the last term as a complex Fourier transform, we must continue $T^{\mu\nu}(k)$ antisymmetrically to $k^0<0$

$$F(x,y) = \frac{\delta^2\varphi}{\delta A_\nu(y)\delta A_\mu(x)} + e^2(2\pi)^{-6}\frac{i}{2}\int d^4k\,e^{-ik(x-y)}P^{\mu\nu}(k), \tag{2.8.35}$$

with

$$P^{\mu\nu}(k) \stackrel{\mathrm{def}}{=} (k^\mu k^\nu - k^2 g^{\mu\nu})\frac{2\pi}{3}\Big(1+\frac{2m^2}{k^2}\Big)\sqrt{1-\frac{4m^2}{k^2}}\,\Theta(k^2-4m^2)\mathrm{sgn}\,k^0. \tag{2.8.36}$$

The Fourier transform of a causal function, vanishing for $x^0-y^0=t<0$, satisfies a dispersion relation. We do not suppose that the reader is familiar with this technique and therefore make a short digression to present the results used in the following. Let

$$F(t) = (2\pi)^{-1} \int d\omega \, e^{-i\omega t} R(\omega) = 0 \quad \text{for} \quad t < 0. \tag{2.8.37}$$

In general, this is a distributional Fourier transform. The inverse Fourier transform

$$R(\omega) = \int_0^\infty F(t) e^{i\omega t} dt \tag{2.8.38}$$

is an analytic function of ω, regular in the upper half plane $\operatorname{Im}\omega \geq 0$, since the integral in (2.8.38) runs only over positive t. By Cauchy's theorem, we have

$$R(z) = \frac{1}{2\pi i} \int_\Gamma \frac{R(\zeta)}{\zeta - z} d\zeta \quad , \quad \operatorname{Im} z > 0, \tag{2.8.39}$$

where Γ is a closed contour, consisting of a part of the real axis and a large semi-circle with z in the interior. Assuming

$$|R(\zeta)| \leq \frac{\text{const}}{|\zeta|} \quad \text{for} \quad |\zeta| \to \infty, \tag{2.8.40}$$

the integral over the large semi-circle vanishes in the limit (2.8.40) and the integral over the real axis converges

$$R(z) = \frac{1}{2\pi i} \int_{-\infty}^{+\infty} \frac{R(\zeta)}{\zeta - z} d\zeta. \tag{2.8.41}$$

Writing $z = \omega + i\varepsilon$, and taking the limit $\varepsilon \to 0$ (in the sense of distributions), we shall obtain

$$R(\omega) = \frac{1}{2\pi i} \int_{-\infty}^{+\infty} \frac{R(\zeta)}{\zeta - \omega - i0} d\zeta. \tag{2.8.42}$$

In this context, we remember the distributional identity

$$\frac{1}{x \pm i0} = \text{P}\frac{1}{x} \mp i\pi\delta(x), \tag{2.8.43}$$

where P means that the principal value integral must be taken, if (2.8.43) is applied to a test function. It enables us to transform (2.8.42) into

$$R(\omega) = \frac{1}{2\pi i} \text{P} \int_{-\infty}^{+\infty} \frac{R(\zeta)}{\zeta - \omega} d\zeta + \frac{1}{2} R(\omega)$$

$$R(\omega) = \frac{1}{\pi i} \text{P} \int_{-\infty}^{+\infty} \frac{R(\zeta)}{\zeta - \omega} d\zeta. \tag{2.8.44}$$

Separating real and imaginary parts, we arrive at the following dispersion relations

$$\operatorname{Re} R(\omega) = \frac{1}{\pi} \,\mathrm{P} \int\limits_{-\infty}^{+\infty} \frac{\operatorname{Im} R(\zeta)}{\zeta - \omega} d\zeta \tag{2.8.45}$$

$$\operatorname{Im} R(\omega) = -\frac{1}{\pi} \,\mathrm{P} \int\limits_{-\infty}^{+\infty} \frac{\operatorname{Re} R(\zeta)}{\zeta - \omega} d\zeta. \tag{2.8.46}$$

If instead of (2.8.40), the function $R(\zeta)$ is bounded by

$$|R(\zeta)| \leq \text{const } |\zeta|^n \quad \text{for} \quad |\zeta| \to \infty, \tag{2.8.47}$$

then, instead of $R(z)$, we consider

$$\frac{R(z)}{(z - \omega_0 + i\varepsilon)^{n+1}} \quad \text{with} \quad \varepsilon > 0,\, \omega_0 \quad \text{real.} \tag{2.8.48}$$

The pole of (2.8.48) in the lower half plan does not disturb the foregoing arguments. Now we obtain

$$\frac{R(\omega)}{(\omega - \omega_0 + i\varepsilon)^{n+1}} = \frac{1}{\pi i} \,\mathrm{P} \int\limits_{-\infty}^{+\infty} \frac{R(\zeta)}{(\zeta - \omega_0 + i\varepsilon)^{n+1}} \frac{d\zeta}{\zeta - \omega}. \tag{2.8.49}$$

Taking the limit $\varepsilon \to 0$ in the sense of distributions and using the n times differentiated Eq. (2.8.43)

$$\frac{1}{(x + i0)^{n+1}} = \mathrm{P} \frac{1}{x^{n+1}} - (-)^n \frac{i\pi}{n!} \delta^{(n)}(x), \tag{2.8.50}$$

the δ-term in (2.8.49) becomes

$$\int\limits_{-\infty}^{+\infty} \delta^{(n)}(\zeta - \omega_0) \frac{R(\zeta)}{\zeta - \omega} d\zeta = (-)^n \frac{d^n}{d\zeta^n} \frac{R(\zeta)}{\zeta - \omega}\bigg|_{\zeta = \omega_0}.$$

By means of the Leibnitz product rule, we get

$$= n! \sum_{\nu=0}^{n} (-)^\nu \frac{R^{(\nu)}(\omega_0)}{\nu!} \frac{1}{(\omega_0 - \omega)^{n-\nu+1}}. \tag{2.8.51}$$

Multiplying (2.8.49) by $(\omega - \omega_0)^{n+1}$, we arrive at the dispersion relations with subtractions

$$R(\omega) = \frac{(\omega - \omega_0)^{n+1}}{\pi i} \,\mathrm{P} \int\limits_{-\infty}^{+\infty} \frac{R(\zeta)}{(\zeta - \omega_0)^{n+1}} \frac{d\zeta}{\zeta - \omega}$$

$$+\sum_{\nu=0}^{n}\frac{R^{(\nu)}(\omega_0)}{\nu!}(\omega-\omega_0)^\nu, \tag{2.8.52}$$

or, taking the imaginary part,

$$\operatorname{Im}R(\omega)=-\frac{(\omega-\omega_0)^{n+1}}{\pi}\,\mathrm{P}\int_{-\infty}^{+\infty}\frac{\operatorname{Re}R(\zeta)}{(\zeta-\omega_0)^{n+1}}\frac{d\zeta}{\zeta-\omega}$$

$$+\sum_{\nu=0}^{n}\frac{\operatorname{Im}R^{(\nu)}(\omega_0)}{\nu!}(\omega-\omega_0)^\nu. \tag{2.8.53}$$

The sum consists of the first $n+1$ terms of the Taylor series of the function on the left-hand side.

We now return to the causal function (2.8.35). Since $P^{\mu\nu}(k)$, appearing there in the Fourier integral, is real (2.8.36)

$$P^{\mu\nu}(k)\stackrel{\mathrm{def}}{=}\left(\frac{k^\mu k^\nu}{k^2}-g^{\mu\nu}\right)P(k), \tag{2.8.54}$$

it alone cannot be the Fourier transform of a causal function. The lacking imaginary part must be supplied by the first term containing the phase $\varphi[A]$,

$$\frac{\delta^2\varphi}{\delta A_\nu(y)\delta A_\mu(x)}=e^2(2\pi)^{-6}\frac{i}{2}\int d^4k\,e^{-ik(x-y)}\left(\frac{k^\mu k^\nu}{k^2}-g^{\mu\nu}\right)iQ(k),$$

$$\text{for}\quad x^0<y^0. \tag{2.8.55}$$

where $iQ(k)$ is determined by $P(k)$ by a dispersion relation. We calculate again in the special Lorentz system, with $k=(\omega,\mathbf{0})$. Since

$$|P(\omega)|\le \mathrm{const}\,|\omega|^2, \tag{2.8.56}$$

we have $n=2$ in (2.8.53) and take $\omega_0=0$

$$Q(\omega)=-\frac{\omega^3}{\pi}\,\mathrm{P}\int_{-\infty}^{+\infty}\frac{P(\zeta)}{\zeta^3}\frac{d\zeta}{\zeta-\omega}. \tag{2.8.57}$$

According to (2.8.36), $P(\zeta)$ is odd because of $\operatorname{sgn}k_0$, hence

$$Q(\omega)=-\frac{2\omega^4}{\pi}\,\mathrm{P}\int_0^\infty d\zeta\frac{P(\zeta)}{\zeta^3(\zeta^2-\omega^2)}$$

$$=-\frac{\omega^4}{\pi}\frac{2\pi}{3}\,\mathrm{P}\int_{4m^2}^\infty ds\frac{s+2m^2}{s^2(s-\omega^2)}\sqrt{1-\frac{4m^2}{s}}, \tag{2.8.58}$$

where the new variable of integration $s = \zeta^2$ has been introduced. The subtraction terms in (2.8.53) would give an additional polynomial $C_0 + C_1\omega + C_2\omega^2$ in (2.8.58). This would lead to local terms $\sim \delta^{(\nu)}(x^0 - y^0)$ in (2.8.55). Since such terms only contribute for $x^0 = y^0$, they are not determined by the local causality condition. The non-perturbative construction of the S-matrix discussed in the next section implies that these local terms must be 0. For this reason, we omit them.

It is very interesting to investigate (2.8.57) in t-space ($t = x^0 - y^0$). To compute the inverse distributional Fourier transform, denoted by F^{-1}, we need

$$F^{-1}\Big[-\omega^3\,\mathrm{P}\frac{1}{\omega' - \omega}\Big](t) = (i\partial_t)^3 F^{-1}\Big[\mathrm{P}\frac{1}{\omega - \omega'}\Big]$$

$$= (i\partial_t)^3 e^{-i\omega' t} F^{-1}\Big[\mathrm{P}\frac{1}{\omega}\Big] = -i\sqrt{\frac{\pi}{2}}(i\partial_t)^3 \mathrm{sgn}\, t\, e^{-i\omega' t}. \tag{2.8.59}$$

For the inverse Fourier transform of (2.8.57) we now get

$$\check{Q}(t) = -\frac{i}{\sqrt{2\pi}}(i\partial_t)^3 \mathrm{sgn}\, t\, \mathrm{P}\int_{-\infty}^{+\infty} d\omega' \frac{P(\omega')}{\omega'^3} e^{-i\omega' t}.$$

In carrying out the differentiation, we obtain local terms $\sim \delta^{(\nu)}(t)$ as above in (2.8.55) from the differentiation of $\mathrm{sgn}\, t$ which are omitted again. Then we arrive at

$$\check{Q}(t) = -\frac{i}{\sqrt{2\pi}} \mathrm{sgn}\, t \int_{-\infty}^{+\infty} d\omega' P(\omega') e^{-i\omega' t} = -i\, \mathrm{sgn}\, t\, \check{P}(t). \tag{2.8.60}$$

The causal function $F(x, y)$ (2.8.35) is obtained by adding

$$\check{P}(t) + i\check{Q}(t) = \check{P}(t)(1 + \mathrm{sgn}\, t)$$

$$= \check{P}(t) 2\Theta(t). \tag{2.8.61}$$

That means, the acausal part $t < 0$ is simply cut off by the Θ-function. This is a first example of distribution splitting. In the construction of the S-matrix of full QED, this operation will play an important role. The doubling in (2.8.61) for $t > 0$ is due to the symmetric continuation of (2.8.55). The simplest way to derive (2.8.60) is directly in x-space from (2.8.35). But the explicit computation of Q is best done in p-space as above.

The result (2.8.58) in an arbitrary Lorentz system reads

$$Q(k) = -\frac{2}{3} k^4\, \mathrm{P} \int_{4m^2}^{\infty} ds \frac{s + 2m^2}{s^2(s - k^2)} \sqrt{1 - \frac{4m^2}{s}}, \tag{2.8.62}$$

to be used in (2.8.55). Since Q is even, $Q(-k) = Q(k)$, only $\cos k(x - y)$ survives in the Fourier integral. Consequently, the result for the second derivative of φ (2.8.55) is real and symmetric in x, y, as it should be. It therefore holds for all x and y, not only for $x^0 < y^0$. Then, the causal phase is obtained by two integrations

$$\varphi[A] = \frac{1}{2}\int d^4x d^4y \frac{\delta^2\varphi}{\delta A_\nu(y)\delta A_\mu(x)} A_\nu(y)A_\mu(x) + O(A^4)$$

$$= -\frac{e^2}{4}(2\pi)^{-2}\int d^4k A_\mu(k)^* \left(\frac{k^\mu k^\nu}{k^2} - g^{\mu\nu}\right) Q(k)A_\nu(k), \tag{2.8.63}$$

The phase $\varphi[A]$ must be even in A, which is again Furry's theorem (see below (3.11.24) in Chap. 3).

The S-operator in Fock space $\mathbf{S}[A]$ is now completely determined

$$\mathbf{S} = Ce^{i\varphi}e^{A_1 b^+ d^+} : e^{(A_2-1)b^+ b} :: e^{(1-A_3)dd^+} : e^{A_4 db}. \tag{2.8.64}$$

We want to combine the two C-number factors in front. They appear in the vacuum expectation value

$$(\Omega, \mathbf{S}\Omega) = Ce^{i\varphi}(\Omega, e^{A_1 b^+ d^+}\Omega) = Ce^{i\varphi}. \tag{2.8.65}$$

The absolute square

$$|(\Omega, \mathbf{S}\Omega)|^2 = C^2 = 1 - P \tag{2.8.66}$$

must be equal to 1 minus the total probability P of pair creation (2.7.43), because the external field can change the vacuum state only into pair states. Hence,

$$C^2 = 1 - \frac{e^2}{(2\pi)^2}\frac{\pi}{3}\int d^4k\,(k^2 + 2m^2)\sqrt{1 - \frac{4m^2}{k^2}}\Theta(k^2 - 4m^2)$$

$$\times\left(\frac{k^\mu k^\nu}{k^2} - g^{\mu\nu}\right)A_\mu(k)A_\nu(k)^* + O(A^4). \tag{2.8.67}$$

We write this as an exponential

$$C = \exp\left[-e^2(2\pi)^{-2}\frac{\pi}{6}\int d^4k \ldots + O(A^4)\right], \tag{2.8.68}$$

in order to combine it with the phase $e^{i\varphi}$. In fact, introducing

$$\Pi(k) = -\frac{2}{3}k^4\int\limits_{4m^2}^{\infty} ds \frac{s + 2m^2}{s^2(s - k^2 - i0)}\sqrt{1 - \frac{4m^2}{s}} \tag{2.8.69}$$

$$= Q(k) - i\pi\frac{2}{3}\Theta(k^2 - 4m^2)(k^2 + 2m^2)\sqrt{1 - \frac{4m^2}{k^2}} \tag{2.8.70}$$

$$= Q(k) - iP(k)\operatorname{sgn} k^0, \tag{2.8.71}$$

we obviously have

$$Ce^{i\varphi} = \exp\left[-i\frac{e^2}{4}(2\pi)^{-2}\int d^4k\, A_\mu(k)^*\left(\frac{k^\mu k^\nu}{k^2} - g^{\mu\nu}\right)\Pi(k)A_\nu(k) + O(A^4)\right]. \tag{2.8.72}$$

2.9 Non-Perturbative Construction of the Causal Phase

The construction of the causal phase $\varphi[A]$ in the last section was perturbative. Although perturbation theory is under full control in the external field problem, this approach is not entirely satisfactory. We have first differentiated the global causality condition (2.8.8) twice to get the local condition (2.8.16). From the local condition we obtained the second derivative of $\varphi[A]$ (2.8.55), and then we had to integrate two times to find $\varphi[A]$. Clearly, the most natural approach would be to use the global causality condition (2.8.8). This is precisely what we shall do in this section. We follow essentially the work of *G. Scharf and W.F. Wreszinski, Nuovo Cimento, 93 A, 1 (1986).* This section is somewhat technical, the reader may skip it for the first reading.

The idea of the construction is quite simple: For technical reasons the external potential is assumed to be different from 0 in the finite time interval $[0,T]$ only. We divide this interval in N intervals of length $\Delta t = T/N$. On each of the small intervals $\Delta_j, j = 1, \dots N$, the potential is approximated by an auxiliary potential A^j_μ which interpolates linearly in time at every point in space. The intervals Δ_j are then translated such that intervals of length $2\cdot\Delta_1 t$ appear between them. In those time intervals the potential A^j_μ is linearly switched off, say during the time interval $[t_j, t_j+\Delta_1 t]$, and A^{j+1}_μ is linearly switched on during $[t_j+\Delta_1 t, t_j+2\cdot\Delta_1 t]$. The resulting potentials $\tilde{A}^j_\mu$

$$\tilde{A}^j_\mu(t,\boldsymbol{x}) = \begin{cases} A_\mu(t_j-\Delta t) + \frac{A_\mu(t_j)-A_\mu(t_j-\Delta t)}{\Delta t}(t-t_j+\Delta t) \\ \qquad\qquad \text{for} \quad t_j-\Delta t \le t \le t_j, \\ \frac{A_\mu(t_j-\Delta t)}{\Delta_1 t}(t-t_j+\Delta t+\Delta_1 t) \text{ for } t_j-\Delta t-\Delta_1 t \le t \le t_j-\Delta t, \\ \frac{A_\mu(t_j)}{\Delta_1 t}(t_j+\Delta_1 t - t) \quad \text{for} \quad t_j \le t \le t_j+\Delta_1 t, \end{cases}$$

$$j = 1, \dots, N, \tag{2.9.1}$$

are perfectly good scattering potentials on $[t_j-\Delta t-\Delta_1 t \le t \le t_j+\Delta_1 t$, with scattering operators S^j and causal phases $\varphi[\tilde{A}^j_\mu]$. The switching off in between allows successive application of the global causality condition (2.8.8)

$$(\Omega,\, \mathbf{S}[A_1+A_2]\Omega) = (\Omega,\, \mathbf{S}[A_1]\,\mathbf{S}[A_2]\Omega). \tag{2.9.2}$$

Finally, we perform the limit $\Delta_1 t \to 0$ and then $\Delta t \to 0$.

Let us consider two potentials $A^{(j)}$, $j=1,2$, with disjoint compact supports in time and corresponding S-operators on Fock space

$$\mathbf{S}_1 = C_1 e^{i\varphi_1} e^{A_1 b^+ d^+} : e^{A_2 b^+ b} :: e^{A_3 d d^+} : e^{A_4 d b}, \tag{2.9.3}$$

$$\mathbf{S}_2 = C_2 e^{i\varphi_2} e^{B_1 b^+ b^+} : e^{B_2 b^+ b} :: e^{B_3 dd^+} : e^{B_4 db}. \tag{2.9.4}$$

The operator $\mathbf{S}$

$$\mathbf{S} = C\, e^{i\varphi} e \ldots \tag{2.9.5}$$

corresponds to the total potential

$$A(x) = A^{(1)}(x) + A^{(2)}(x). \tag{2.9.6}$$

Inserting all this into (2.9.2), we get

$$(\Omega,\, \mathbf{S}\Omega) = Ce^{i\varphi} = (\Omega,\, \mathbf{S}_2\mathbf{S}_1\Omega)$$

$$= C_1 C_2 e^{i(\varphi_1+\varphi_2)} (e^{B_4^+ b^+ d^+}\Omega,\, e^{A_1 b^+ d^+}\Omega). \tag{2.9.7}$$

In order to compute the scalar product in (2.9.7), we first expand the exponentials

$$(\,,\,) = 1 + \sum_{n=1}^{\infty} \sum_{\substack{p_1<\cdots<p_n\\ q_1<\cdots<q_n}} [\det(B_4^+)_{p_j q_k}]^* \det(A_1)_{p_j q_k}. \tag{2.9.8}$$

Here we can drop the adjoint and the complex conjugate operations and use the multiplication theorem for determinants

$$(\,,\,) = 1 + \sum_n \sum_{\substack{p_1<\cdots<p_n\\ q_1<\cdots<q_n}} \det\left(\sum_{m=1}^{n} (A_1)_{p_j q_m} (B_4)_{q_m p_k}\right). \tag{2.9.9}$$

By multilinearity of the determinant, we obtain

$$= 1 + \sum_n \sum_{\substack{p_1<\cdots<p_n\\ q_1<\cdots<q_n}} \sum_{m_1\cdots m_n=1}^{n} \det((A_1)_{p_j q_{m_j}} (B_4)_{q_{m_k} p_k})$$

$$= 1 + \sum_n \sum_{\substack{p_1<\cdots<p_n\\ q_1<\cdots<q_n}} \det((A_1)_{p_j q_j} (B_4)_{q_k p_k}), \tag{2.9.10}$$

where the added diagonal terms in q vanish. Again by multilinearity we have

$$(\,,\,) = 1 + \sum_n \sum_{p_1<\cdots<p_n} \det\left(\sum_q (A_1)_{p_j q} (B_4)_{q p_k}\right)$$

$$= 1 + \sum_n \sum_{p_1<\cdots<p_n} \det(A_1 B_4)_{p_j p_k} = 1 + \sum_{n=1}^{\infty} \frac{1}{n!} \sum_{p_1\cdots p_n=1}^{\infty} \det(A_1 B_4)_{p_j p_k}$$

$$= \det(1 + A_1 B_4) = \det(1 + B_4 A_1). \tag{2.9.11}$$

We now return to the causality condition (2.9.7)

$$e^{i\varphi} = e^{i(\varphi_1+\varphi_2)} \frac{C_1 C_2}{C} \det(1 + B_4 A_1). \tag{2.9.12}$$

After construction, the last two factors must combine to form a complex number λ with $|\lambda| = 1$. It is instructive to verify this explicitly: Let us denote the one-particle S-operators for $A^{(1)}, A^{(2)}$ by S and T, then, using the expression (2.4.64), we arrive at

$$\lambda = \left[\frac{\det(S_{--}S^+_{--}) \det(T_{--}T^+_{--})}{\det(TS)_{--}(TS)^+_{--}} \right]^{1/2} \det(1 + T^{-1}_{--}T_{-+}S_{+-}S^{-1}_{--}). \tag{2.9.13}$$

We take the logarithm, using the well-known identity $\mathrm{Tr} \log A = \log \det A$,

$$\log \lambda = \frac{1}{2} \mathrm{Tr} \log(S_{--}S^+_{--}) + \frac{1}{2} \mathrm{Tr} \log(T_{--}T^+_{--}) -$$

$$-\frac{1}{2} \mathrm{Tr} \log(TS)_{--}(TS)^+_{--} + \mathrm{Tr} \log(1 + T^{-1}_{--}T_{-+}S_{+-}S^{-1}_{--}) \tag{2.9.14}$$

and calculate its real part. There remains to compute

$$\mathrm{Re}\, \mathrm{Tr} \log(1 + T^{-1}_{--}T_{-+}S_{+-}S^{-1}_{--})$$

$$= \frac{1}{2} \mathrm{Tr} \log(1 + S^{-1+}_{--}S^+_{-+}T^+_{+-}T^{-1+}_{--})(1 + T^{-1}_{--}T_{-+}S_{+-}S^{-1}_{--}), \tag{2.9.15}$$

where the polar decomposition (2.4.54) was used. Finally we get

$$\frac{1}{2} \mathrm{Tr} \log S^{-1+}_{--}(S^+_{--}T^+_{--} + S^+_{-+}T^+_{+-})T^{-1+}_{--}T^{-1}_{--}(T_{--}S_{--} + T_{-+}S_{+-})S^{-1}_{--}$$

$$= \frac{1}{2} \mathrm{Tr} \log S^{-1+}_{--}(TS)^+_{--}(T_{--}T^+_{--})^{-1}(TS)_{--}S^{-1}_{--}. \tag{2.9.16}$$

This can be rewritten as

$$\frac{1}{2} \log \det[(S_{--}S^+_{--})^{-1}(TS)^+_{--}(TS)_{--}(T_{--}T^+_{--})^{-1}]$$

$$= \frac{1}{2} \log \det(TS)_{--}(TS)^+_{--} - \frac{1}{2} \log \det(S_{--}S^+_{--}) - \frac{1}{2} \log \det(T_{--}T^+_{--}). \tag{2.9.17}$$

Substituting this into (2.9.14), we conclude that $\mathrm{Re} \log \lambda = 0$, or $|\lambda| = 1$.

The only content of (2.9.12) is its argument, that means, the causality condition precisely determines the phase

$$\varphi = \varphi_1 + \varphi_2 + \arg \det(1 + B_4 A_1)$$

$$= \varphi_1 + \varphi_2 + \mathrm{Im} \log \det(1 + T^{-1}_{--}T_{-+}S_{+-}S^{-1}_{--}). \tag{2.9.18}$$

Similarly as in (2.9.16), this can be expressed entirely by $--$ operators

$$\varphi = \varphi_1 + \varphi_2 + \mathrm{Im} \log \det[T^{-1}_{--}(TS)_{--}S^{-1}_{--}]. \tag{2.9.19}$$

In contrast to (2.9.17), however, we cannot write here the last term as

$$-\mathrm{Im}\,\log\det T_{--} + \mathrm{Im}\,\log\det(TS)_{--} - \mathrm{Im}\,\log\det S_{--}, \tag{2.9.20}$$

because $S_{--} - 1$ is not trace class and the same is true for the other operators in (2.9.20).

The result (2.9.18) can be successively applied to a potential divided into N pieces as (2.9.1),

$$\varphi = \sum_{j=1}^{N} \Big\{ \varphi[\tilde{A}_j^\mu] + \mathrm{Im}\,\log\det[1 + (S^j_{--})^{-1} S^j_{-+} (S(j-1))_{+-} (S(j-1))^{-1}_{--}] \Big\}, \tag{2.9.21}$$

where

$$S(j-1) = \begin{cases} \prod_{l=j-1}^{1} S^l, & j \geq 2, \\ 0 \quad \text{for} \quad j = 1. \end{cases} \tag{2.9.22}$$

We transform the last term in (2.9.21) as in (2.9.15)

$$\begin{aligned} 1 &+ (S^j_{--})^{-1} S^j_{-+} S(j-1)_{+-} S(j-1)^{-1}_{--} \\ &= (S^j_{--})^{-1} [S^j_{--} S(j-1)_{--} + S^j_{-+} S(j-1)_{+-}] S(j-1)^{-1}_{--} \\ &= (S^j_{--})^{-1} S(j)_{--} S(j-1)^{-1}_{--}. \end{aligned} \tag{2.9.23}$$

This enables us to carry out the sum in (2.9.21) by induction:

$$\log\det(S^{j+1}_{--})^{-1} S(j+1)_{--} S(j)^{-1}_{--} + \log\det S(j)_{--} \prod_{l=1}^{j} (S^l_{--})^{-1}$$

$$= \log\det(S^{j+1}_{--})^{-1} S(j+1)_{--} \prod_{l=1}^{j} (S^l_{--})^{-1} = \log\det S(j+1)_{--} \prod_{l=1}^{j+1} (S^l_{--})^{-1}. \tag{2.9.24}$$

This gives the following result

$$\varphi = \sum_{j=1}^{N} \varphi[\tilde{A}_j^\mu] + \mathrm{Im}\,\log\det S(N)_{--} \prod_{j=1}^{N} (S^j_{--})^{-1}. \tag{2.9.25}$$

In the reference mentioned above, the various transformations are rigorously justified.

Equation (2.9.25) is only useful if we have some mean to determine the phases of the elementary potentials $\tilde{A}_j$. For this reason the limits $\Delta t, \Delta_1 t \to 0$ are performed. Then it is sufficient to insert for $\varphi[\tilde{A}_j^\mu]$ the result in second-order perturbation theory, derived in the last section (2.8.63)

$$\varphi_2[A] = \frac{e^2}{4} (2\pi)^{-2} \int d^4k\, Q(k) |A^\mu(k)|^2, \tag{2.9.26}$$

where the Lorentz gauge $k_\mu A^\mu(k) = 0$ has been assumed for simplicity. Then we have the following expression for the phase

$$\varphi[A] = \lim_{\Delta t\to 0}\lim_{\Delta_1 t\to 0}\left\{\sum_{j=1}^{N}\varphi_2[\tilde{A}_j^\mu] + \operatorname{Im}\log\det S(N)_{--}\prod_{j=1}^{N}(S^j_{--})^{-1}\right\}. \tag{2.9.27}$$

In the limit $\Delta_1 t \to 0$, each $\varphi_2(\tilde{A}_j^\mu)$ diverges because the potential becomes discontinuous in time. There must, therefore, take place compensation with the last term, say Φ_N, in (2.9.27). In fact, writing

$$\Phi_N = \operatorname{Im}\log\det[1+T(N)], \tag{2.9.28}$$

the determinant also diverges in the limit $\Delta_1 t \to 0$, but the regularized determinant

$$\det_2[1+T(N)] \stackrel{\text{def}}{=} \det\{[1+T(N)]\exp[-\operatorname{Tr}T(N)]\} \tag{2.9.29}$$

has a limit. We therefore split off the trace

$$\Phi_N = \operatorname{Im}\log\det_2[1+T(N)] + \operatorname{Im}\operatorname{Tr}T(N), \tag{2.9.30}$$

and there remains to discuss the compensation between this trace and φ_2.

The compensation in question must occur in second-order perturbation theory

$$T(N)_2 = \sum_{m>n} S^m_{-+}S^n_{+-}, \tag{2.9.31}$$

where

$$S^m_{-+}(\boldsymbol{q},\boldsymbol{p}) = -i(2\pi)^{-1}P_-(\boldsymbol{q})\gamma^0 e\tilde{A\!\!\!/}^m(q-p)P_+(\boldsymbol{p}). \tag{2.9.32}$$

We compute the trace

$$\operatorname{Tr}S^m_{-+}S^n_{+-} = -(2\pi)^{-2}e^2\operatorname{tr}\int d^3q\,d^3p$$

$$P_-(\boldsymbol{q})\gamma^0\gamma^\nu\tilde{A}^m_\nu(q-p)P_+(\boldsymbol{p})\gamma^0\gamma^\mu\tilde{A}^n_\mu(p-q). \tag{2.9.33}$$

Changing the integration variable $\boldsymbol{q}\to-\boldsymbol{q}$ and going over to covariant expressions, we shall obtain

$$\operatorname{Tr}S^m_{-+}S^n_{+-} = -(2\pi)^{-2}e^2\int\frac{d^3q}{2E_q}\frac{d^3p}{2E_p}$$

$$\times\operatorname{tr}(q\!\!\!/-m)\gamma^\nu(p\!\!\!/+m)\gamma^\mu\tilde{A}^m_\nu(-p-q)\tilde{A}^n_\mu(p+q)$$

$$= -(2\pi)^{-2}e^2\int d^4q\,d^4p\,\delta(q^2-m^2)\Theta(q^0)\delta(p^2-m^2)\Theta(p^0)\operatorname{tr}\ldots$$

Using $k = p+q$ as a new integration variable, we write this in the form

$$\operatorname{Tr} S^m_{-+} S^n_{+-} = -e^2(2\pi)^{-2} \int d^4k\, T^{\mu\nu}(k) \tilde{A}^m_\nu(k)^* \tilde{A}^n_\mu(k), \tag{2.9.34}$$

where the tensor $T^{\mu\nu}(k)$ is well-known to us from pair production (2.7.31) and (2.8.33)

$$T^{\mu\nu}(k) = (k^\mu k^\nu - k^2 g^{\mu\nu}) \frac{2\pi}{3}\Big(1 + \frac{2m^2}{k^2}\Big)\sqrt{1 - \frac{4m^2}{k^2}}\,\Theta(k^2 - 4m^2)\Theta(k^0). \tag{2.9.35}$$

In view of (2.9.27), we have to calculate the imaginary part

$$\operatorname{Im}\operatorname{Tr} S^m_{-+} S^n_{+-} = -e^2(2\pi)^{-2} \int d^4k\, T^{\mu\nu}(k) \operatorname{Im} \tilde{A}^m_\nu(k)^* \tilde{A}^n_\mu(k). \tag{2.9.36}$$

The factor $\Theta(k^0)$ in (2.9.35) restricts the integration to $k^0 > 0$. But we want to integrate over whole k-space. Since

$$\tilde{A}(-k) = \tilde{A}(k)^*, \tag{2.9.37}$$

the imaginary part in (2.9.36) is odd in k^0. We therefore continue (2.9.35) antisymmetrically to negative k^0, as in (2.8.36),

$$P^{\mu\nu}(k) = \Big(\frac{k^\mu k^\nu}{k^2} - g^{\mu\nu}\Big) P(k) \tag{2.9.38}$$

$$P(k) = \operatorname{sgn} k^0 \Theta(k^2 - 4m^2) \frac{2\pi}{3}(k^2 + 2m^2)\sqrt{1 - \frac{4m^2}{k^2}}, \tag{2.9.39}$$

which is identical with (2.8.54). As above, we use the Lorentz gauge

$$k^\mu \tilde{A}^n_\mu(k) = 0, \tag{2.9.40}$$

$$\operatorname{Im}\operatorname{Tr} S^m_{-+} S^n_{+-} = \frac{e^2}{2}(2\pi)^{-2} \int d^4k\, P(k) \operatorname{Im} \tilde{A}^m_\mu(k)^* \tilde{A}^{n\mu}(k). \tag{2.9.41}$$

The imaginary part can be taken out of the integral. Since P (2.9.39) is odd, this may be written as

$$= \frac{e^2}{2i}(2\pi)^{-2} \int d^4k\, P(k) \tilde{A}^m_\mu(k)^* \tilde{A}^{n\mu}(k),$$

taking (2.9.37) into account. In x-space, we obtain

$$\operatorname{Im}\operatorname{Tr} S^m_{-+} S^n_{+-} = \frac{e^2}{2i}(2\pi)^{-4} \int d^4x d^4y\, \tilde{A}^m_\mu(x) \check{P}(x-y) \tilde{A}^{n\mu}(y). \tag{2.9.42}$$

For $m > n$, we have $x^0 - y^0 = t > 0$ and according to (2.8.60)

$$\check{P}(x-y) = i\check{Q}(x-y), \quad \text{hence,}$$

$$\operatorname{Im}\operatorname{Tr} S^m_{-+} S^n_{+-} = \frac{e^2}{2}(2\pi)^{-4} \int d^4x d^4y\, \tilde{A}^m_\mu(x) \check{Q}(x-y) \tilde{A}^{n\mu}(y) =$$

$$= \frac{e^2}{2}(2\pi)^{-2}\int d^4k\, Q(k)\tilde{A}_\mu^m(k)^*\tilde{A}^{n\mu}(k)$$

$$= \frac{e^2}{2}(2\pi)^{-2}\int d^4k\, Q(k)\mathrm{Re}\,[\tilde{A}_\mu^m(k)^*\tilde{A}^{n\mu}(k)], \tag{2.9.43}$$

because $Q(k)$ is even in k. Now we see the compensation of divergences in (2.9.27):

$$\sum_j \varphi_2[\tilde{A}_j^\mu] + \mathrm{Im}\,\mathrm{Tr}\,T(N)_2 = \frac{e^2}{4}(2\pi)^{-2}\int d^4k\, Q(k)$$

$$\times\Big[\sum_j |\tilde{A}_j^\mu(k)|^2 + 2\sum_{m>n}\mathrm{Re}\,\tilde{A}_\mu^m(k)^*\tilde{A}^{n\mu}(k)\Big]$$

$$= \frac{e^2}{4}(2\pi)^{-2}\int d^4k\, Q(k)\Big|\sum_j \tilde{A}_j^\mu(k)\Big|^2. \tag{2.9.44}$$

This converges for $\Delta_1 t \to 0$ to the second-order phase of the continuous potential

$$A(N)^\mu = \sum_{j=1}^N A_j^\mu. \tag{2.9.45}$$

The limit would not exist, however, if $Q(k)$ (2.8.58) would contain subtraction terms

$$Q(k) + C_{1\mu}k^\mu + C_2 k^2. \tag{2.9.46}$$

The contribution of a constant term C_0 in (2.9.46) vanishes in the limit. Consequently, there remains no freedom in the definition of $\varphi[A]$. It is an important observation that this non-perturbative approach determines the S-matrix uniquely, in contrast to perturbation theory.

Using (2.9.30) in (2.9.27), we arrive at the following expression for the causal phase

$$\varphi[A] = \lim_{\Delta t\to 0}\{\varphi_2[A(N)] + \mathrm{Im}\,\log\det_2(1+T(N))$$

$$+ \lim_{\Delta_1 t\to 0}\mathrm{Im}\,\mathrm{Tr}\,(T(N) - T(N)_2)\}. \tag{2.9.47}$$

For a discussion of the remaining limits, we refer to the paper cited at the beginning of this section.

2.10 Vacuum Polarization

The definition of a current density is always a problem in QED. The usual definition

$$j^\mu(x) = e : \overline{\psi}(x)\gamma^\mu\psi(x) : \tag{2.10.1}$$

with normal ordering with respect to free field operators b, d cannot be exactly true (see problem 2.16). The best definition is obtained by means of the well-defined scattering operator $\mathbf{S}[A]$:

$$j^\mu(x) \stackrel{\text{def}}{=} i\mathbf{S}^+ \frac{\delta \mathbf{S}[A]}{\delta A_\mu(x)}. \tag{2.10.2}$$

This is in fact a hermitean operator

$$j^\mu(x) = \frac{i}{2}\left(\mathbf{S}^+ \frac{\delta \mathbf{S}}{\delta A_\mu(x)} - \frac{\delta \mathbf{S}^+}{\delta A_\mu(x)}\mathbf{S}\right), \tag{2.10.3}$$

because

$$\frac{\delta}{\delta A_\mu(x)}(\mathbf{S}^+\mathbf{S}) = 0. \tag{2.10.4}$$

The physical interpretation as the current density follows from the fact that it is a conserved quantity as a consequence of gauge invariance: The gauge invariance of the S-operator (2.4.82)

$$\mathbf{S}[A_\mu - \varepsilon\partial_\mu \Lambda] = \mathbf{S}[A_\mu] \tag{2.10.5}$$

implies by differentiation with respect to ε at $\varepsilon = 0$

$$-\int d^4x \frac{\delta}{\delta A_\mu(x)}\mathbf{S}[A]\partial_\mu \Lambda = 0 = \int d^4x\, \Lambda(x)\partial_\mu \frac{\delta}{\delta A_\mu(x)}\mathbf{S}[A], \tag{2.10.6}$$

supposed $\Lambda(x)$ vanishes at infinity. Since Λ is arbitrary, we may conclude

$$\partial_\mu \frac{\delta \mathbf{S}}{\delta A_\mu(x)} = 0 \quad \text{and} \tag{2.10.7}$$

$$\partial_\mu j^\mu(x) = 0. \tag{2.10.8}$$

It is our aim now to calculate the vacuum expectation value of this current density

$$(\Omega, j^\mu(x)\Omega) = i\Big(\mathbf{S}\Omega, \frac{\delta \mathbf{S}}{\delta A_\mu(x)}\Omega\Big). \tag{2.10.9}$$

Inserting (2.8.17)

$$\mathbf{S} = \tilde{\mathbf{S}}e^{i\varphi}, \quad \frac{\delta \mathbf{S}}{\delta A_\mu(x)} = i\frac{\delta\varphi}{\delta A_\mu(x)}\mathbf{S} + e^{i\varphi}\frac{\delta\tilde{\mathbf{S}}}{A_\mu(x)}, \tag{2.10.10}$$

we find

$$(\Omega, j^\mu(x)\Omega) = i\left[i\frac{\delta\varphi}{\delta A_\mu(x)} + \left(\tilde{\mathbf{S}}\Omega, \frac{\delta\tilde{\mathbf{S}}}{\delta A_\mu(x)}\Omega\right)\right]. \tag{2.10.11}$$

Here the last term was calculated in Sect. 2.8 (2.8.24), we have therefore in lowest order of perturbation theory

$$(\Omega, j^\mu(x)\Omega) = -\frac{\delta\varphi}{\delta A_\mu(x)} - \operatorname{Im}\operatorname{Tr}\left(S^+_{-+}\frac{\delta S_{+-}}{\delta A_\mu(x)}\right). \tag{2.10.12}$$

By means of formulae (2.8.26, 27), we shall obtain

$$\operatorname{Tr}\left(S^+_{-+}\frac{\delta S_{+-}}{\delta A_\mu(x)}\right)^{(1)} = e^2(2\pi)^{-4}\int d^3q d^3p\, e^{i(p+q)x}$$

$$\times \operatorname{tr} P_-(-\boldsymbol{q})\gamma^0\gamma^\nu P_+(\boldsymbol{p})\gamma^0\gamma^\mu P_-(-\boldsymbol{q})A_\nu(-p-q)$$

$$= e^2(2\pi)^{-4}\int d^4q d^4p\, e^{i(p+q)x}\frac{\delta(q^0-E_q)}{2E_q}\frac{\delta(p^0-E_p)}{2E_p}\cdot$$

$$\times \operatorname{tr}(\not{q}-m)\gamma^\nu(\not{p}+m)\gamma^\mu A_\nu(-p-q)$$

$$= e^2(2\pi)^{-4}\int d^4k\, e^{ikx}T^{\mu\nu}(k)A_\nu(-k), \tag{2.10.13}$$

where (2.8.31, 32) have been taken into account. The subscript $^{(1)}$ refers to first order perturbation theory.

Since $T^{\mu\nu}$ (2.8.33) is real with support in $k^0 > 0$, the imaginary part in (2.10.12) becomes

$$\operatorname{Im}\operatorname{Tr}\left(S^+_{-+}\frac{\delta S_{+-}}{\delta A_\mu(x)}\right)^{(1)} = e^2(2\pi)^{-4}\int\limits_{k^0>0} d^4k\, T^{\mu\nu}(k)\operatorname{Im}(e^{ikx}A_\nu(-k)). \tag{2.10.14}$$

The last imaginary part in (2.10.14) is odd in k, therefore, we must continue $T^{\mu\nu}$ antisymmetrically to $k^0 < 0$ as in (2.8.36), which leads to $P^{\mu\nu}$ (2.8.36, 54)

$$= e^2(2\pi)^{-4}\frac{i}{2}\int d^4k\left(\frac{k^\mu k^\nu}{k^2} - g^{\mu\nu}\right)P(k)e^{-ikx}A_\nu(k). \tag{2.10.15}$$

The first term in (2.10.12)

$$\frac{\delta\varphi}{\delta A_\mu(x)} = \int\frac{\delta^2\varphi}{\delta A_\nu(y)\delta A_\mu(x)}A_\nu(y)d^4y + O(A^3)$$

is computed by means of (2.8.55)

$$= e^2(2\pi)^{-6}\frac{i}{2}\int d^4y\int d^4k\, e^{-ik(x-y)}A_\nu(y)\left(\frac{k^\mu k^\nu}{k^2} - g^{\mu\nu}\right)iQ(k)$$

$$= e^2(2\pi)^{-4}\frac{i}{2}\int d^4k\, e^{-ikx}A_\nu(k)\left(\frac{k^\mu k^\nu}{k^2} - g^{\mu\nu}\right)iQ(k). \tag{2.10.16}$$

Substituting all this into (2.10.12), we arrive at the desired vacuum expectation value

$$(\Omega,\, j^\mu(x)\Omega) = -e^2(2\pi)^{-4}\frac{i}{2}\int d^4k\, e^{-ikx} A_\nu(k)$$

$$\times \Big(\frac{k^\mu k^\nu}{k^2} - g^{\mu\nu}\Big)(P+iQ) \tag{2.10.17}$$

$$\stackrel{\text{def}}{=} \frac{e^2}{2}(2\pi)^{-4}\int d^4k\, e^{-ikx}\Big(\frac{k^\mu k^\nu}{k^2} - g^{\mu\nu}\Big)\tilde{\Pi}(k)A_\nu(k). \tag{2.10.18}$$

The function

$$\tilde{\Pi}(k) = -\frac{2}{3}k^4 \int\limits_{4m^2}^{\infty} ds \frac{s+2m^2}{s^2(s-k^2-ik^0 0)}\sqrt{1-\frac{4m^2}{s}} \tag{2.10.19}$$

differs from $\Pi(k)$ (2.8.69) because the imaginary part is odd in k. The result in momentum space is

$$(\Omega,\, \hat{j}^\mu(k)\Omega) = \frac{e^2}{2}(2\pi)^{-2}\Big(\frac{k^\mu k^\nu}{k^2} - g^{\mu\nu}\Big)\tilde{\Pi}(k)A_\nu(k)$$

$$\stackrel{\text{def}}{=} \hat{j}^\mu_{\text{pol}}(k), \tag{2.10.20}$$

this is the current density of vacuum polarization. Since the real part of (2.10.20) is even and the imaginary part is odd, $j^\mu_{\text{pol}}(x)$ becomes real, as it should.

The current density (2.10.20) is the source of the vacuum polarization potential [1]

$$\Box A^\mu_{\text{pol}}(x) = j^\mu_{\text{pol}}(x), \tag{2.10.21}$$

or after Fourier transformation

$$-k^2 A^\mu_{\text{pol}}(k) = j^\mu_{\text{pol}}(k)$$

$$A^\mu_{\text{pol}}(k) = -\frac{e^2}{8\pi^2}\Big(\frac{k^\mu k^\nu}{k^2} - g^{\mu\nu}\Big)\frac{\tilde{\Pi}(k)}{k^2}A_\nu(k). \tag{2.10.22}$$

For a scalar potential we have

$$A^0_{\text{pol}}(k) = -\frac{\alpha}{2\pi}\frac{\mathbf{k}^2}{k^2}\frac{\tilde{\Pi}(k)}{k^2}A^0(k), \tag{2.10.23}$$

where

$$\alpha = \frac{e^2}{4\pi} \approx \frac{1}{137} \tag{2.10.24}$$

is the fine-structure constant. We want to evaluate this for the important case of a Coulomb potential

---

[1] Remember that we use Heaviside–Lorentz rational units.

$$A^0(x) = \frac{Ze}{4\pi|\boldsymbol{x}|}\exp(-\varepsilon t^2). \tag{2.10.25}$$

The Gaussian factor has been introduced to switch the potential on and off in time. We take the limit $\varepsilon \to 0$ at the end of the calculation. The four-dimensional Fourier transform is

$$\hat{A}^0(k) = \frac{Ze}{4\pi\sqrt{\pi\varepsilon}}\frac{1}{|\boldsymbol{k}|^2}\exp(-k_0^2/4\varepsilon), \tag{2.10.26}$$

and the corresponding vacuum polarization potential becomes

$$A^0_{\rm pol}(k) = -\frac{Ze\alpha}{8\pi^2\sqrt{\pi\varepsilon}}\frac{\tilde{\Pi}(k)}{k^4}\exp(-k_0^2/4\varepsilon). \tag{2.10.27}$$

In order to compute the potential in x-space, we notice that for $\varepsilon \to 0$ the Gaussian factor in (2.10.27) approaches a δ-distribution $\sim \delta(k_0)$. Only the real part of $\tilde{\Pi}$ contributes which coincides with that of Π (2.8.69)

$$A^0_{\rm pol}(\boldsymbol{x}) = -\frac{Ze\alpha}{(2\pi)^4}\int d^3k \frac{\operatorname{Re}\Pi(-\boldsymbol{k}^2)}{\boldsymbol{k}^4}e^{i\boldsymbol{k}\cdot\boldsymbol{x}}. \tag{2.10.28}$$

Since $\operatorname{Re}\Pi = Q$ according to (2.8.71), we see that for a static potential the vacuum polarization is determined by the causal phase alone. Integration over the angles leads to the expression

$$A^0_{\rm pol}(\boldsymbol{x}) = -\frac{2Ze\alpha}{(2\pi)^3 r}\int\limits_0^\infty d|\boldsymbol{k}|\frac{\sin|\boldsymbol{k}|r}{|\boldsymbol{k}|^3}\operatorname{Re}\Pi(-\boldsymbol{k}^2) \tag{2.10.29}$$

$$= -\frac{Ze\alpha}{(2\pi)^3 r}\operatorname{Re}\int\limits_{-\infty}^{+\infty} dy\, \Pi(-y^2)\frac{\sin yr}{y^3}. \tag{2.10.30}$$

The remaining integral can be calculated by shifting the contour of integration. It follows from (2.8.69),

$$\Pi(-y^2) = -\frac{2}{3}y^4\int\limits_{4m^2}^\infty ds \frac{s+2m^2}{s^2(s+y^2-i0)}\sqrt{1-\frac{4m^2}{s}} \tag{2.10.31},$$

that Π has branch points at $y = \pm 2im$. We take the cuts along the positive and negative imaginary axis to $\pm i\infty$. Substituting $\sin yr$ in (2.10.30) by exponentials and closing the contour of integration in the upper and lower half plane, respectively, we shall obtain

$$A^0_{\rm pol}(\boldsymbol{x}) = -\frac{Ze\alpha}{(2\pi)^3 r}\operatorname{Re} i\left[\int\limits_{2m}^\infty dz\, \Pi(z^2 - iz0)\frac{e^{-zr}}{z^3} + \int\limits_{-\infty}^{-2m} dz\, \Pi(z^2 - iz0)\frac{e^{zr}}{z^3}\right] \tag{2.10.32}$$

$$= -\frac{2Ze\alpha}{(2\pi)^3 r}\operatorname{Re} i \int_{2m}^{\infty} dz\, \Pi(z^2 - i0)\frac{e^{-zr}}{z^3} \tag{2.10.33}$$

$$= \frac{Ze\alpha}{(2\pi)^3 r}\int_{4m^2}^{\infty} ds \frac{\operatorname{Im}\Pi(s-i0)}{s^2}\exp(-r\sqrt{s}). \tag{2.10.34}$$

The $i0$ is important to specify, on which side of the cut the integrals must be evaluated. Using (2.8.70)

$$\operatorname{Im}\Pi(s-i0) = \frac{2}{3}\pi\,\Theta(s-4m^2)(s+2m^2)\sqrt{1-\frac{4m^2}{s}}, \tag{2.10.35}$$

we finally get

$$A^0_{\mathrm{pol}}(\boldsymbol{x}) = \frac{1}{3}\frac{Ze\alpha}{(2\pi)^2 r}\int_{4m^2}^{\infty} ds\left(1+\frac{2m^2}{s}\right)\frac{\sqrt{s-4m^2}}{s^{3/2}}\exp(-r\sqrt{s}), \tag{2.10.36}$$

or, with the new variable of integration $s = 4m^2\zeta^2$,

$$A^0_{\mathrm{pol}}(\boldsymbol{x}) = \frac{1}{6\pi}\frac{Ze\alpha}{r}\int_1^{\infty} d\zeta\,(1+\frac{1}{2\zeta^2})\frac{\sqrt{\zeta^2-1}}{\zeta^2}e^{-2mr\zeta}. \tag{2.10.37}$$

This polarization potential modifies the original Coulomb potential. The total potential generated by a classical point charge Ze is therefore given by

$$A^0_{\mathrm{tot}}(r) = \frac{Ze}{4\pi r}\left[1+\frac{2\alpha}{3\pi}\int_1^{\infty} d\zeta\, e^{-2mr\zeta}\left(1+\frac{1}{2\zeta^2}\right)\frac{\sqrt{\zeta^2-1}}{\zeta^2}\right]. \tag{2.10.38}$$

Since the polarization potential vanishes exponentially for large r, one observes the original potential at distances r large compared with the Compton wave length $\hbar/mc$. However, for small distances, it seems as if the effective charge at $r=0$ were bigger. The simple classical picture is that the central charge is partially screened by opposite polarization charges. The influence of this screening becomes smaller if one approaches the central charge. The vacuum around the classical source acts like a polarizable medium which explains the term vacuum polarization. In a way this is a resurrection of Maxwell's aether. Note that the total vacuum polarization charge vanishes

$$\int \varrho_{\mathrm{pol}}(\boldsymbol{x})\, d^3x = (2\pi)^{3/2}\hat{\varrho}_{\mathrm{pol}}(0) = 0. \tag{2.10.39}$$

This follows from (2.10.19) since

$$\varrho_{\mathrm{pol}}(\boldsymbol{x}) = -\triangle A^0_{\mathrm{pol}}(\boldsymbol{x}) = \frac{Ze\alpha}{(2\pi)^4}\int d^3k\,\frac{\operatorname{Re}\Pi(-\boldsymbol{k}^2)}{\boldsymbol{k}^2}\, e^{i\boldsymbol{k}\cdot\boldsymbol{x}} \tag{2.10.40}$$

and

$$\hat{\varrho}_{\text{pol}}(\boldsymbol{k}) = \frac{Ze\alpha}{(2\pi)^{5/2}} \frac{\operatorname{Re} \Pi(-\boldsymbol{k}^2)}{\boldsymbol{k}^2}, \tag{2.10.41}$$

(see also (3.6.29)).

2.11 Quantization of the Radiation Field

One finds various essentially different methods of quantization of the radiation field in the literature. This very fact shows already that there are some difficulties with the subject. They are due to the vanishing mass of the photon and its vector character. The different possible procedures depend on the choice, which part of the classical Maxwell theory one would like to retain in the quantum theory.

In view of later extensions to non-abelian gauge theories, it is appropriate to take the vector potential $A^\mu(x)$ as the fundamental field. Our starting point then are the covariant field equations for this vector field, discussed at the end of Sect. 1.2, namely the Lorentz gauge condition (1.2.40)

$$\partial_\mu A^\mu(x) = 0 \tag{2.11.1}$$

and the wave equation

$$\partial_\nu \partial^\nu A^\mu(x) = \square A^\mu(x) = 0. \tag{2.11.2}$$

Since the four components of A are not coupled in the wave equation, various other gauge conditions are possible, for example

$$A^0 = 0 \quad \text{(temporal gauge)} \tag{2.11.3}$$

$$A^3 = 0 \quad \text{(axial gauge)}$$

$$\boldsymbol{\partial} \cdot \boldsymbol{A} = 0 \quad \text{(Coulomb gauge).} \tag{2.11.4}$$

For the free radiation field, the conditions (2.11.3) and (2.11.4) can be simultaneously satisfied. This gauge is called radiation gauge. Then the electric and magnetic fields

$$\boldsymbol{E} = -\frac{\partial \boldsymbol{A}}{\partial t} \quad , \quad \boldsymbol{B} = \boldsymbol{\partial} \wedge \boldsymbol{A} \tag{2.11.5}$$

obey Maxwell's equations. Condition (2.11.4), written in momentum space

$$\boldsymbol{k} \cdot \hat{\boldsymbol{A}}(t, \boldsymbol{k}) = 0,$$

expresses the fact that the radiation field is transversal. There are no longitudinal and scalar photons ($A^0 = 0$) in this gauge.

An important property of the radiation gauge is its essential uniqueness. In fact, it follows from (2.11.5) that

$$\boldsymbol{\partial} \wedge \boldsymbol{B} = \boldsymbol{\partial} \wedge (\boldsymbol{\partial} \wedge \boldsymbol{A}) = \operatorname{grad} \operatorname{div} \boldsymbol{A} - \Delta \boldsymbol{A} = - \Delta \boldsymbol{A}, \tag{2.11.6}$$

where (2.11.4) has been used. Let us assume that $\boldsymbol{B}$ goes to zero faster than $|\boldsymbol{x}|^{-1-\varepsilon}$ $(\varepsilon > 0)$ and $\boldsymbol{A}$ tends to zero arbitrarily slowly for $|\boldsymbol{x}| \to \infty$, t fixed. Then we get the following unique solution of the Laplace equation (2.11.6) for the vector potential

$$\boldsymbol{A}(t, \boldsymbol{x}) = \frac{1}{4\pi} \int \frac{(\boldsymbol{\partial} \wedge \boldsymbol{B})(t, \boldsymbol{y})}{|\boldsymbol{x} - \boldsymbol{y}|} d^3 y. \tag{2.11.7}$$

The existence of the integral can be easily shown by partial integration.

In contrast to the radiation gauge, the Lorentz condition (2.11.1) does not fix the gauge uniquely. In fact, if

$$\tilde{A}^{\mu}(x) = A^{\mu}(x) + \partial^{\mu} \Lambda(x)$$

is another potential with $\partial_{\mu} \tilde{A}^{\mu} = 0$, then

$$\partial_{\mu} \partial^{\mu} \Lambda = \Box \Lambda = 0. \tag{2.11.8}$$

But the homogeneous wave equation has non-trivial solutions vanishing for $|\boldsymbol{x}| \to \infty$ in contrast to the Laplace equation. The merit of the Lorentz gauge is of course its covariance.

Every gauge condition above has the following property: If it is satisfied in an arbitrarily small time interval, it is automatically satisfied for all t. For example, if $A^0(t = 0) = 0$ and $(\partial_t A^0)(t = 0) = 0$ for all $\boldsymbol{x}$, then A^0 vanishes identically for all t as solution of the wave equation. Since the wave equation is second order in time, it is not enough to specify the gauge at one instant of time, but an arbitrarily small finite time interval is sufficient for this. In any case, the gauge condition is a boundary condition. It is unimportant for the dynamical time evolution. We therefore omit it in the process of quantization.

We quantize $A^{\mu}(x)$ as four independent real scalar fields. Let

$$A^{\mu}(t, \boldsymbol{x}) = (2\pi)^{-3/2} \int \frac{d^3 k}{\sqrt{2\omega}} \left(a^{\mu}(\boldsymbol{k}) e^{-i(\omega t - \boldsymbol{k} \cdot \boldsymbol{x})} + a^{\mu}(\boldsymbol{k})^* e^{i(\omega t - \boldsymbol{k} \cdot \boldsymbol{x})} \right), \tag{2.11.9}$$

be a real solution of the wave equation. With

$$\omega(\boldsymbol{k}) = |\boldsymbol{k}| \stackrel{\text{def}}{=} k^0 \quad , \quad c = 1 \tag{2.11.10}$$

the exponent can be written in covariant form $k \cdot x$. The reason for the factor $\sqrt{2\omega}$ will soon become clear. After quantization, $A^{\mu}(x)$ becomes an operator-valued distribution. We therefore probe (2.11.9) with a real test function $f(\boldsymbol{x})$ and write

$$A^{\mu}(t, f) = \int d^3 x \, A^{\mu}(t, \boldsymbol{x}) f(\boldsymbol{x}) \tag{2.11.11}$$

in $\boldsymbol{k}$-space

$$\hat{f}(\boldsymbol{k}) = (2\pi)^{-3/2} \int d^3x\, f(\boldsymbol{x}) e^{-i\boldsymbol{k}\cdot\boldsymbol{x}}, \quad \hat{f}(-\boldsymbol{k}) = \hat{f}(\boldsymbol{k})^* \tag{2.11.12}$$

$$A^\mu(t, f) = \int \frac{d^3k}{\sqrt{2\omega}} \Big(a^\mu(\boldsymbol{k}) \hat{f}(\boldsymbol{k})^* e^{-i\omega t} + a^\mu(\boldsymbol{k})^* \hat{f}(\boldsymbol{k}) e^{i\omega t} \Big). \tag{2.11.13}$$

Now let us consider

$$\tilde{a}^\mu(\hat{f}) = \int \frac{d^3k}{\sqrt{2\omega}} \hat{f}(\boldsymbol{k})^* a^\mu(\boldsymbol{k}) \tag{2.11.14}$$

as operators in photon Fock space, $\mu = 0, 1, 2, 3$, obeying the commutation relations

$$[\tilde{a}^\mu(\hat{f}),\, \tilde{a}^\nu(\hat{g})^+] = \delta^\mu_\nu \int \frac{d^3k}{2\omega} \hat{f}(\boldsymbol{k})^* \hat{g}(\boldsymbol{k}), \tag{2.11.15}$$

and 0 otherwise. This relation reads in distributional form as follows

$$[a^\mu(\boldsymbol{k}),\, a^\nu(\boldsymbol{k}')^+] = \begin{cases} \delta(\boldsymbol{k} - \boldsymbol{k}') & \text{for} \quad \mu = \nu, \\ 0 & \text{for} \quad \mu \neq \nu. \end{cases} \tag{2.11.16}$$

Then, as we know from Sect. 2.1, $a^{\mu +}$ are emission operators and a^μ absorption operators for photons.

There is, however, a serious difficulty with Lorentz covariance in this approach: If we retain the classical expression (2.11.9) in the form

$$A^\mu(x) = (2\pi)^{-3/2} \int \frac{d^3k}{\sqrt{2\omega}} \Big(a^\mu(\boldsymbol{k}) e^{-ikx} + a^\mu(\boldsymbol{k})^+ e^{ikx} \Big), \tag{2.11.17}$$

we obtain the following commutator

$$[A^\mu(x),\, A^\nu(y)] = (2\pi)^{-3} \int \frac{d^3k}{\sqrt{2\omega}} \int \frac{d^3k'}{\sqrt{2\omega'}}$$

$$\times \Big\{ [a^\mu(\boldsymbol{k}),\, a^\nu(\boldsymbol{k}')^+]\, e^{-ikx + ik'y} + [a^\mu(\boldsymbol{k})^+,\, a^\nu(\boldsymbol{k}')]\, e^{ikx - ik'y} \Big\}$$

$$= \delta^\mu_\nu\, (2\pi)^{-3} \int \frac{d^3k}{2\omega} \Big(e^{-ik(x-y)} - e^{ik(x-y)} \Big)$$

$$= \delta^\mu_\nu\, (2\pi)^{-3} \int d^4k\, \delta(k^2) e^{-ik(x-y)} \operatorname{sgn}(k^0)$$

$$= \delta^\mu_\nu\, \frac{1}{i} D_0(x - y). \tag{2.11.18}$$

The Lorentz invariant Jordan-Pauli function (2.3.7) for mass 0 appears here. This was the reason for the factor $\sqrt{2\omega}$ in (2.11.9). However, the right-hand side is not a second rank Lorentz tensor as the left-hand side. We should have $g^{\mu\nu}$ instead of δ^μ_ν. The simplest way to remedy this defect is to change the definition of A^0 into

$$A^0(x) = (2\pi)^{-3/2} \int \frac{d^3k}{\sqrt{2\omega}} \left(a^0(\boldsymbol{k}) e^{-ikx} - a^0(\boldsymbol{k})^+ e^{ikx} \right). \tag{2.11.19}$$

This makes A^0 a skew-adjoint operator instead of self-adjoint. As we shall discuss below, the physical Hilbert space of QED will be defined in such a way that all expectation values of A^0 (and of any quantity derived from it) vanish. Then the non-self-adjointness of A^0 (2.11.19) causes no problems.

We remind the reader of the quantization of the Dirac field, where we also had to modify the classical expression, namely by interchanging absorption and emission operators (2.2.2). This was necessary there, in order to get the energy bounded from below. Here we must change the classical expression for reasons of covariance. In any case, in constructing a quantum theory, one must always be prepared to depart from the classical theory. There is no logical route to go over from classical to quantum physics. The correctness of a quantization procedure can only be proven by working out its consequences. The same method as used for the Dirac field could also be applied here, namely the interchange of a^0 and a^{0+} in (2.11.16). However, the commutation rule $[a^0, a^{0+}] = -1$ leads to a space with indefinite metric (Gupta-Bleuler method). This is too high a price which we are not willing to pay for a plus sign in (2.11.19).

With the new definition (2.11.19), the commutation relations for the radiation field are

$$[A^\mu(x), A^\nu(y)] = g^{\mu\nu} i D_0(x-y). \tag{2.11.20}$$

We need also the commutators of the absorption and emission parts alone. Let

$$A^\mu_-(x) = (2\pi)^{-3/2} \int \frac{d^3k}{\sqrt{2\omega}} a^\mu(\boldsymbol{k}) e^{-ikx} \tag{2.11.21}$$

$$A^\mu_+(x) = (2\pi)^{-3/2} \int \frac{d^3k}{\sqrt{2\omega}} a^\mu(\boldsymbol{k})^+ e^{ikx} \cdot \begin{cases} -1 & \text{for } \mu = 0 \\ 1 & \text{for } \mu = 1,2,3, \end{cases} \tag{2.11.22}$$

then the only non-vanishing commutators are

$$[A^\mu_-(x), A^\nu_+(y)] = g^{\mu\nu} i D_0^{(+)}(x-y) \tag{2.11.23}$$

$$[A^\mu_+(x), A^\nu_-(y)] = g^{\mu\nu} i D_0^{(-)}(x-y). \tag{2.11.24}$$

We will briefly discuss the time evolution of the free radiation field. After construction $A^\mu(x)$ is a solution of the wave equation, therefore one may define

$$i\frac{\partial}{\partial t} A^n(t,\boldsymbol{x}) = (2\pi)^{-3/2} \int \frac{d^3k}{\sqrt{2\omega}} \left[\omega a^n(\boldsymbol{k}) e^{-i\omega t + i\boldsymbol{k}\cdot\boldsymbol{x}} \right.$$
$$\left. - \omega a^n(\boldsymbol{k})^+ e^{i\omega t - i\boldsymbol{k}\cdot\boldsymbol{x}} \right] \stackrel{\text{def}}{=} [A^n, \mathsf{H}_0] \tag{2.11.25}$$

and

$$i\frac{\partial}{\partial t} A^0(t,\boldsymbol{x}) = (2\pi)^{-3/2} \int \frac{d^3k}{\sqrt{2\omega}} \left[\omega a^0(\boldsymbol{k}) e^{-i\omega t + i\boldsymbol{k}\cdot\boldsymbol{x}} \right.$$

$$+\omega a^0(\boldsymbol{k})^+ e^{i\omega t - i\boldsymbol{k}\cdot\boldsymbol{x}}] \stackrel{\text{def}}{=} [A^0(t,\boldsymbol{x}),\, \mathbf{H}_0]. \tag{2.11.26}$$

The operator $\mathbf{H}_0$ is uniquely determined by these two equations up to an additive constant. It is easy to verify that the positive definite operator

$$\mathbf{H}_0 = \int d^3k\, \omega(\boldsymbol{k}) \sum_{\mu=0}^{3} a^\mu(\boldsymbol{k})^+ a^\mu(\boldsymbol{k}) \tag{2.11.27}$$

satisfies (2.11.25) (2.11.26). As far as positive definiteness of the energy is concerned, our procedure of quantization of the radiation field is satisfactory.

Hitherto we have not imposed any gauge condition. Now we shall investigate the Lorentz condition (2.11.1). As an operator equation, it cannot hold on the whole photon Fock space $\mathcal{F}$, because

$$\partial_\mu A^\mu(x)\Omega = \partial_\mu A^\mu_+(x)\Omega \neq 0.$$

However, the vacuum expectation value of (2.11.1) vanishes

$$(\Omega,\, \partial_\mu A^\mu \Omega) = (\Omega,\, \partial_\mu A^\mu_+ \Omega) = (\mp \partial_\mu A^\mu_- \Omega,\, \Omega) = 0,$$

where the upper minus sign corresponds to $\mu = 0$. The same is true for a large class of states. The expression

$$\partial_\mu A^\mu(x) = (2\pi)^{-3/2} \int \frac{d^3k}{\sqrt{2\omega}} \Big[-i\Big(\omega a^0(\boldsymbol{k}) + k_j a^j(\boldsymbol{k})\Big)e^{-ikx}$$
$$+i\Big(-\omega a^0(\boldsymbol{k})^+ + k_j a^j(\boldsymbol{k})^+\Big)e^{ikx}\Big] \tag{2.11.28}$$

may be written in a more transparent form by introducing the absorption and emission operators for longitudinal photons

$$a_\parallel(\boldsymbol{k}) \stackrel{\text{def}}{=} \frac{k_j}{\omega} a^j(\boldsymbol{k}), \tag{2.11.29}$$

$$\partial_\mu A^\mu(x) = (2\pi)^{-3/2} \int d^3k \sqrt{\frac{\omega}{2}} \Big[-i\Big(a^0(\boldsymbol{k}) + a_\parallel(\boldsymbol{k})\Big)e^{-ikx}$$
$$+i\Big(-a^0(\boldsymbol{k})^+ + a_\parallel(\boldsymbol{k})^+\Big)e^{ikx}\Big]. \tag{2.11.30}$$

If Φ, Φ' are states without scalar and longitudinal photons, that means,

$$a^0\Phi = 0 \quad , \quad a_\parallel \Phi = 0, \tag{2.11.31}$$

then we obviously have

$$(\Phi,\, \partial_\mu A^\mu(x)\Phi') = 0.$$

The subspace of states Φ (2.11.31) without scalar and longitudinal photons is called the physical subspace $\mathcal{F}_{\text{phys}}$ of the radiation field. Only such states are taken as incoming and outgoing photon states in scattering theory. In fact, as will be shown in Sect. 3.11 (3.11.25), the S-matrix of full QED is only unitary if it is restricted to $\mathcal{F}_{\text{phys}}$. As mentioned above, all expectation values of A^0 indeed vanish on $\mathcal{F}_{\text{phys}}$. Therefore, we must not worry about the non-hermitean character of A^0. The best way to specify asymptotic transverse photon states is by means of polarization vectors. Let

$$\varepsilon_\mu(\boldsymbol{k}) = (0, \boldsymbol{\varepsilon}) \quad , \quad \boldsymbol{k} \cdot \boldsymbol{\varepsilon}(\boldsymbol{k}) = 0 \qquad \boldsymbol{\varepsilon}^2 = 1 \tag{2.11.32}$$

be a transverse polarization vector. Then a one-photon state is given by

$$\varphi = \int d^3k\, f(\boldsymbol{k})\varepsilon_\mu(\boldsymbol{k})a_\mu(\boldsymbol{k})^+\Omega. \tag{2.11.33}$$

We have written the two summation indices μ downstairs because the sum in (2.11.33) is not a Minkowski scalar product, since $a_\mu(\boldsymbol{k})$ is not a vector field.

We now turn to Poincaré covariance of the radiation field. For this purpose it is convenient to consider the field operators as distributions over real test functions $\varphi_\mu(x) \in S(\mathbb{R}^4)$ in Minkowski space. The latter are classical four-vector fields. We introduce the operator

$$A(\varphi) = \int d^4x\, \varphi_\mu(x)A^\mu(x) \tag{2.11.34}$$

in Fock space. As in (2.2.62), the Poincaré transformation of the test functions

$$\varphi'_\mu(x') = \Lambda_\mu{}^\nu\varphi_\nu(x) = \Lambda_\mu{}^\nu\varphi_\nu(\Lambda^{-1}(x'-a)) \tag{2.11.35}$$

can be lifted into Fock space by the definition

$$\begin{aligned} \mathbf{U}(a,\Lambda)A(\varphi)\mathbf{U}(a,\Lambda)^{-1} &\stackrel{\text{def}}{=} A(\varphi') \\ &= \int d^4x'\, \Lambda_\mu{}^\nu\varphi_\nu(\Lambda^{-1}(x'-a))A^\mu(x') \\ &= \int d^4x\, \varphi_\nu(x)\left(\Lambda^{-1}\right)^\nu{}_\mu A^\mu(\Lambda x + a). \end{aligned} \tag{2.11.36}$$

In the last equation, (0.2.7) was used. This leads to the following transformation law

$$\mathbf{U}(a,\Lambda)A^\nu(x)\mathbf{U}(a,\Lambda)^{-1} = \left(\Lambda^{-1}\right)^\nu{}_\mu A^\mu(\Lambda x + a). \tag{2.11.37}$$

Although we have used the suggestive letter $\mathbf{U}$ in these equations, the representation $\mathbf{U}(a,\Lambda)$, defined by (2.11.37), is not unitary. This is a consequence of the non-selfadjointness of $A^0(x)$. However, it is possible to define another conjugation K in Fock space such that (2.11.34) becomes self-conjugate

$$A(\varphi)^K = A(\varphi) \tag{2.11.38}$$

and $\mathbf{U}(a,\Lambda)$ is pseudo-unitary in the following sense

$$\mathbf{U}(a,\Lambda)^K = \mathbf{U}(a,\Lambda)^{-1}. \tag{2.11.39}$$

In order to achieve this, we introduce the bounded operator

$$\eta = (-1)^{\mathsf{N}_0}, \tag{2.11.40}$$

where N_0 is the particle number operator for scalar photons. We obviously have

$$\eta^+ = \eta \quad , \quad \eta^2 = 1, \tag{2.11.41}$$

and η anticommutes with the emission and absorption operators for scalar photons

$$a^0(\varphi)\eta = -\eta a^0(\varphi) \quad , \quad a^0(\varphi)^+\eta = -\eta a^0(\varphi)^+, \tag{2.11.42}$$

because these operators change the number of scalar photons by one. It commutes with all other emission and absorption operators. The conjugation K is now defined by

$$A^K = \eta A^+ \eta \tag{2.11.43}$$

for any (densely defined) operator in Fock space. It has all desired properties

$$(A+B)^K = A^K + B^K \quad , \quad (AB)^K = B^K A^K$$

$$A^{KK} = A \quad , \quad (\lambda A)^K = \lambda^* A^K, \tag{2.11.44}$$

for λ complex. Furthermore, if A can be inverted then

$$(A^{-1})^K = (A^K)^{-1}.$$

Since the skew-adjointness of A^0 is compensated by anticommuting with η, $A(\varphi)$ is indeed self-conjugate (2.11.38). It follows from (2.11.42) and (2.11.19) that

$$A^\mu(x) = (2\pi)^{-3/2} \int \frac{d^3k}{\sqrt{2\omega}} \left(a^\mu(\boldsymbol{k}) e^{-ikx} + a^\mu(\boldsymbol{k})^K e^{ikx} \right) \tag{2.11.45}$$

from which the self-conjugacy is evident. The pseudo-unitarity (2.11.39) then follows by taking the conjugate of (2.11.36). From the physical point of view, the conjugation K is as good as the adjoint, because the matrix elements are the same on the physical subspace $\mathcal{F}_{\mathrm{phys}}$: For $\Phi, \Psi \in \mathcal{F}_{\mathrm{phys}}$, we have

$$(\Phi,\, A^K\Psi) = (\Phi,\, \eta A^+ \eta \Psi) = (\eta\Phi,\, A^+\Psi) = (\Phi,\, A^+\Psi), \tag{2.11.46}$$

because the number of scalar photons is zero in $\mathcal{F}_{\mathrm{phys}}$. Accordingly, a self-conjugate operator has real expectation values on $\mathcal{F}_{\mathrm{phys}}$. We are now fully reconciled with the lack of self-adjointness in $A^\mu(x)$.

However the pseudo-unitarity (2.11.37) cannot be the whole story. The physical content of relativistic invariance is the fact that two observers in uniform motion relative to each other observe the same physics. Accordingly, proper Poincaré transformations must give rise to a *unitary* mapping between physical states in $\mathcal{F}_{\mathrm{phys}}$. We are now going to show how this comes about.

In the discussion so far, it was not necessary to use a concrete realization of the photon Fock space $\mathcal{F}$. To introduce such a representation, we consider a general time-dependent one-photon state

$$\Phi = \int d^3k \hat{f}_\mu(\boldsymbol{k}) e^{i\omega t} a_\mu(\boldsymbol{k})^+ \Omega \tag{2.11.47}$$

and represent it by the four-vector potential f

$$f^\mu(t, \boldsymbol{x}) = (2\pi)^{-3/2} \int \frac{d^3k}{\sqrt{2\omega}} \hat{f}^\mu(\boldsymbol{k}) e^{i(\omega t - \boldsymbol{k}\cdot\boldsymbol{x})}. \tag{2.11.48}$$

This is a complex solution of the wave equation with "positive frequencies", according to the usual terminology. The denominator $\sqrt{2\omega}$ has again been introduced for reasons of covariance as in (2.11.9). It is convenient to define the Fock space scalar product by means of time derivatives in such a way that the factors 2ω drop out:

$$(f, g) \stackrel{\text{def}}{=} \sum_{\mu=0}^{3} i \int d^3x \, [f_\mu^* \partial_t g_\mu - (\partial_t f_\mu^*) g_\mu]. \tag{2.11.49}$$

In fact, from (2.11.48) we then obtain

$$(f, g) = \sum_{\mu=0}^{3} \int d^3k \, \hat{f}_\mu(\boldsymbol{k})^* \hat{g}_\mu(\boldsymbol{k}). \tag{2.11.50}$$

This is the usual positive definite L^2 scalar product, in agreement with (Φ, Ψ) computed from (2.11.47) by means of the commutation relations (2.11.16). Furthermore, since (2.11.50) is constant in time, the time evolution is unitary. But the sum over μ in (2.11.49) is not a Minkowski product. We therefore expect troubles with Lorentz invariance. Nevertheless, as in (1.4.59), we rewrite (2.11.49) in covariant form as a surface integral

$$(f, g) = \sum_{\mu=0}^{3} i \int_{t=\text{const}} d\sigma^\nu(x) \, [f_\mu^* \partial_\nu g_\mu - (\partial_\nu f_\mu^*) g_\mu], \tag{2.11.51}$$

which can be taken over an arbitrary smooth space-like surface Q. This is a consequence of Gauss' theorem

$$\left(\int_{t=\text{const}} - \int_Q \right) = i \int d^4x \, \partial^\nu \Big[f_\mu^* \partial_\nu g_\mu - (\partial_\nu f_\mu^*) g_\mu \Big] = 0, \tag{2.11.52}$$

because f_μ^* and g_μ are solutions of the wave equation. The generalization of the construction to many-photon states is straightforward.

The vectors $\Phi \in \mathcal{F}_{\text{phys}}$, defined by (2.11.31), obviously obey the radiation gauge, for example,

$$\Phi_0 = 0 \quad , \quad \partial_j \Phi_j(x) = 0, \tag{2.11.53}$$

in the case of a one-photon state. Although this condition depends on the frame of reference, the whole space $\mathcal{F}_{\text{phys}}$ is the same for all observers, as we will shortly see. The photon states which satisfy the Lorentz condition (2.11.1) form a subspace $\mathcal{F}_L$ in $\mathcal{F}$ (L stands for Lorentz). Any one-photon state Ψ in $\mathcal{F}_L$ is obtained from some physical state $\Phi \in \mathcal{F}_{\text{phys}}$ by a gauge transformation

$$\Psi_\mu(x) = \Phi_\mu(x) + \partial_\mu \Lambda(x). \tag{2.11.54}$$

Φ is uniquely determined by Ψ, because the radiation gauge (2.11.7) is unique. This means, geometrically speaking, that there is a *fibration* in $\mathcal{F}_L$: $\mathcal{F}_{\text{phys}}$ is the base space, a general $\Psi \in \mathcal{F}_L$ is connected with a unique $\Phi \in \mathcal{F}_{\text{phys}}$ by a fibre which consists of gauge equivalent states. The projection on $\mathcal{F}_{\text{phys}}$ along the fibres, that means by gauge transformations (2.11.54), is different from the orthogonal projection in the Hilbert space sense. The latter is not used here.

Let us now consider two one-photon states Ψ, $\Phi \in \mathcal{F}_{\text{phys}}$. Since the zero components Φ_0, Ψ_0 vanish due to the radiation gauge, the scalar product (2.11.51) can be written as

$$(\Phi, \Psi) = -(\Phi_\mu, \Psi^\mu), \tag{2.11.55}$$

where the usual summation convention is applied if the greek indices are written up and down. Under a Poincaré transformation (2.11.35) the states are rotated out of $\mathcal{F}_{\text{phys}}$, in general, because the transformed states Φ'_μ, Ψ'^μ have non-vanishing zero components. Nevertheless, (2.11.55) remains invariant

$$(\Phi, \Psi) = -(\Phi'_\mu, \Psi'^\mu), \tag{2.11.56}$$

because this is now a Minkowski product which renders the bilinear form (2.11.51) invariant. Furthermore, the transformed states are in $\mathcal{F}_L$ because they satisfy the Lorentz gauge condition

$$0 = \partial^\mu \Phi_\mu = \partial'^\mu \Phi'_\mu, \tag{2.11.57}$$

and the same for Ψ'. Next we transform back into $\mathcal{F}_{\text{phys}}$ by projecting along the fibres, that means by gauge transformations (2.11.54)

$$\Phi'_\mu = \tilde{\Phi}_\mu + \partial_\mu \chi \quad , \quad \Psi'_\mu = \tilde{\Psi}_\mu + \partial_\mu \Lambda. \tag{2.11.58}$$

The new states are in the radiation gauge

$$\tilde{\Phi}_0 = 0 \quad , \quad \partial_j \tilde{\Phi}_j = 0 \tag{2.11.59}$$

and the same for $\tilde{\Psi}$. Let us now consider

$$-(\Phi'_\mu, \Psi'^\mu) = -(\tilde{\Phi}_\mu, \tilde{\Psi}^\mu) - (\partial_\mu \chi, \tilde{\Psi}^\mu) - (\tilde{\Phi}_\mu, \partial^\mu \Lambda) - (\partial_\mu \chi, \partial^\mu \Lambda). \tag{2.11.60}$$

Rewriting the scalar products in the non-covariant form (2.11.49), we see that the second and third term vanish by (3-dimensional) partial integration. The gauge functions χ, Λ satisfy the wave equation (2.11.8) because the gauge transformations are within the Lorentz class. Then the last term in (2.11.60) can be transformed as follows

$$\begin{aligned}(\partial_\mu\chi,\, \partial^\mu\Lambda) &= (\partial_0\chi,\, \partial_0\Lambda) - (\partial_j\chi,\, \partial_j\Lambda)\\ &= (\partial_0\chi,\, \partial_0\Lambda) + \tfrac{1}{2}(\triangle\chi,\, \Lambda) + \tfrac{1}{2}(\chi,\, \triangle\Lambda)\\ &= (\partial_0\chi,\, \partial_0\Lambda) + \tfrac{1}{2}(\partial_0^2\chi,\, \Lambda) + \tfrac{1}{2}(\chi,\, \partial_0^2\Lambda)\\ &= \tfrac{1}{2}\partial_0^2(\chi,\, \Lambda) = 0,\end{aligned} \tag{2.11.61}$$

because the scalar product between two solutions of the wave equation is constant in time (2.11.50). Hence we arrive at

$$(\Phi,\, \Psi) = -(\tilde{\Phi}_\mu,\, \tilde{\Psi}^\mu) = (\tilde{\Phi},\, \tilde{\Psi}), \tag{2.11.62}$$

which is the desired unitary mapping in $\mathcal{F}_{\rm phys}$. The whole process is schematically illustrated in Fig. 1. It shows how the pseudo-unitary Lorentz transformation $\mathbf{U}(\Lambda)$ cooperates with gauge transformations to give a unitary transformation $\tilde{\mathbf{U}}(\Lambda)$ in $\mathcal{F}_{\rm phys}$. This establishes the important fact that $\mathcal{F}_{\rm phys}$ is independent of the reference frame.

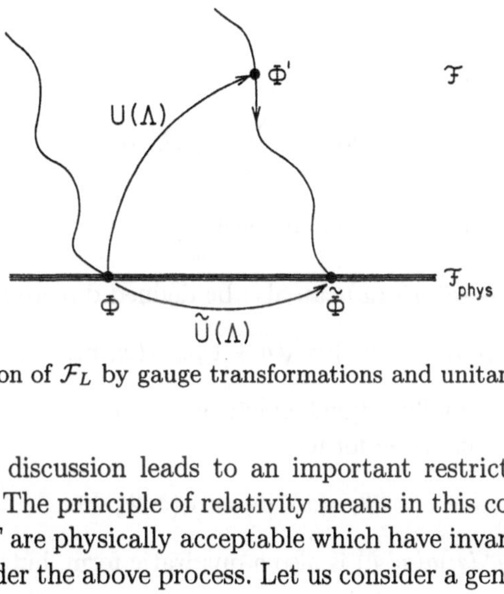

Fig. 1. Fibration of $\mathcal{F}_L$ by gauge transformations and unitarity in $\mathcal{F}_{\rm phys}$

The preceding discussion leads to an important restriction on the observables in QED. The principle of relativity means in this context that only such observables T are physically acceptable which have invariant matrix elements $(\Phi,\, T\Psi)$ under the above process. Let us consider a general one-photon observable

$$T = \int d^4x_1\, d^4x_2\ :\ A_\mu(x_1)\, t^{\mu\nu}(x_1, x_2)\, A_\nu(x_2)\ :, \tag{2.11.63}$$

which is written in normally ordered form. Using (2.11.17) and (2.11.47), we find

$$(\Phi, T\Psi) = \int d^4x_1\, d^4x_2\, f_\mu(x_1)^* t^{\mu\nu}(x_1, x_2)\, g_\nu(x_2), \tag{2.11.63}$$

where f, g are given by (2.11.48). This remains invariant under Poincaré transformations if $t^{\mu\nu}$ is a tensor in Minkowski space. But in addition, it must be invariant under gauge transformations (2.11.58)

$$(\Phi, T\Psi) = (\Phi', T'\Psi') = (\tilde{\Phi}, T'\tilde{\Psi}). \tag{2.11.64}$$

By partial integration, this is guaranteed by the following conditions

$$\frac{\partial}{\partial x_1^\mu} t^{\mu\nu}(x_1, x_2) = 0 = \frac{\partial}{\partial x_2^\nu} t^{\mu\nu}(x_1, x_2), \tag{2.11.65}$$

which express the gauge invariance of T. Only Poincaré and gauge invariant quantities have a direct physical meaning.

2.12 Problems

2.1. Prove the commutation relations

$$[a^+(f), a^+(g)]_\mp = 0 = [a(f), a(g)]_\mp. \tag{2.12.1}$$

What commutation relations do the field operators $a(\boldsymbol{x})$, $a^+(\boldsymbol{y})$ and $\hat{a}(\boldsymbol{k})$, $\hat{a}^+(\boldsymbol{k}')$ obey?

2.2. Show that the definition

$$a(\boldsymbol{x}) = \sum_j a(f_j) f_j(\boldsymbol{x}) \tag{2.12.2}$$

is independent of the choice of the basis f_j.

2.3. The field operator $a(\boldsymbol{x})$ can also be defined directly:

$$(a(\boldsymbol{x})\Phi)_n(\boldsymbol{x}_1, \ldots, \boldsymbol{x}_n) = \sqrt{n+1}\, \varphi_{n+1}(\boldsymbol{x}, \boldsymbol{x}_1, \ldots, \boldsymbol{x}_n). \tag{2.12.3}$$

Show that the formally adjoint operator is not an operator in Fock space. However, it is a quadratic form

$$(\Psi, a(\boldsymbol{x})^+\Phi) = (a(\boldsymbol{x})\Psi, \Phi). \tag{2.12.4}$$

Show that $(\Psi, a^+(\boldsymbol{y})a(\boldsymbol{x})\Phi)$ is also a quadratic form, but $(\Psi, a(\boldsymbol{x})a^+(\boldsymbol{y})\Phi)$ is not.

2.4. Show that the emission operator $a^+(f)$ for fermions

$$(a^+(f)\Phi)_n = \sqrt{n}\, S_n^-(f \otimes \varphi_{n-1}) \tag{2.12.5}$$

is a bounded operator in Fock space.

2.5. Let

$$H(\boldsymbol{x}_1, \boldsymbol{x}_2) = H(\boldsymbol{x}_1) + H(\boldsymbol{x}_2) + V(\boldsymbol{x}_1, \boldsymbol{x}_2) \tag{2.12.6}$$

be a 2-particle Hamiltonian. What is the corresponding second-quantized operator in Fock space? How does it operate on the n-particle sector $\mathcal{H}_n$?

2.6. Occupation number representation: Let f_k, $k = 1, 2, \ldots$ be a complete orthonormal system in the 1-particle space $\mathcal{H}_1$. We define a basis in $\mathcal{H}_n$ by

$$C\,|n_1, n_2, \ldots\rangle \stackrel{\text{def}}{=} S_n^{\pm} f_{k_1} \otimes \ldots \otimes f_{kn} \tag{2.12.7}$$

$$k_1 \leq k_2 \leq \ldots \leq k_n, \quad \sum_{j=1}^{\infty} n_j = n, \tag{2.12.8}$$

where n_j is the number of particles in the state f_j and C a normalization constant.

a) Calculate C such that $|n_1, n_2, \ldots\rangle$ is normalized to 1.

b) Show that without the constraint (2.12.8) one has a non-countable set of vectors. How is this possible inspite of the separability of the Fock space?

2.7. How do the emission and absorption operators operate in the occupation number representation a) for bosons, b) for fermions? Show that the operators for fermions are bounded.

2.8. Use the Green's function

$$\frac{1}{i} S^{(+)}(x - y)\gamma^0 = (\not\partial - im)\gamma^0 D^{(+)}(x - y) \tag{2.12.9}$$

to express a time-dependent solution of the Dirac equation in the positive spectral subspace $P_+\mathcal{H}_1$. Consider a solution that has compact support in $\mathbb{R}^3$ at $t = 0$. Show that the support for $t > 0$ is not compact. Is this a contradiction to causality?

2.9. Show that the definition of the Fock space operator

$$Ab^+b \stackrel{\text{def}}{=} \sum_{jk} (f_j, Af_k) b(f_j)^+ b(f_k) \tag{2.12.10}$$

is independent of the choice of the basis f_j.

2.10. Prove:

$$\begin{aligned} d(f)e^{Ab^+d^+} &= e^{Ab^+d^+}(d(f) - b(A_{+-}f)^+) \\ d(f) : e^{(1-A)dd^+} : &=: e^{(1-A)dd^+} : d(A_{--}f) \\ d(f)^+ : e^{(1-A)dd^+} : &=: e^{(1-A)dd^+} : d(A_{--}^{+-1}f)^+, \end{aligned} \tag{2.12.11}$$

where $A_{+-} = P_+ A P_-$, etc.

2.11. Verify that the S-matrix (2.4.61) satisfies also the equation

$$d(g)\mathbf{S} = \mathbf{S}[b(S^+_{+-}g)^+ + d(S^+_{--}g)]. \tag{2.12.12}$$

2.12. Express the S-matrix element for positron scattering in an external potential by the 1-particle S-matrix.

2.13. Calculate the differential cross section for positron scattering in an external potential in lowest order perturbation theory. Compare the result with electron scattering. What do you expect in higher orders comparing the two processes?

2.14. Relate the S-matrix element of problem 2.12 to electron scattering by means of charge conjugation.

2.15. Verify by explicit calculation that the real part of the causality condition (2.8.16) is automatically satisfied. To compute the functional derivative of the normalization factor

$$|C|^2 = \det(1 - S^+_{-+}S_{+-}), \tag{2.12.13}$$

use

$$\frac{\delta |C|^2}{\delta A_\mu(x)} = -2\mathrm{Re}\ \mathrm{Tr}\, S^+_{-+}\frac{\delta S_{+-}}{\delta A_\mu(x)}. \tag{2.12.14}$$

Prove this equation.

2.16. The electromagnetic current density was defined by

$$j^\mu(x) = i\mathbf{S}^+\frac{\delta \mathbf{S}}{\delta A_\mu(x)}. \tag{2.12.15}$$

Show that in lowest order perturbation theory you get the usual expression

$$j^\mu(x) = e : \overline{\psi}(x)\gamma^\mu\psi(x) : + O(e^2). \tag{2.12.16}$$

3. Causal Perturbation Theory

In this chapter we deal with full QED which means that both the electron-positron field and the radiation field are quantized. In the preceding chapter, in particular in Sect. 2.4, we learned the important lesson that it is very difficult to give a precise meaning to interacting field operators directly. We have to retreat to scattering theory, instead, and we shall now construct the scattering matrix of full QED. The S-matrix maps the asymptotically incoming, free fields on the outgoing ones and, hence, it should be possible to express it by the well-defined free fields. We already emphasized the importance of causality for this purpose (Sects. 2.8–10). Here we shall see causality in its full strength: The S-matrix is essentially completely determined by causality and translation invariance, supposed the coupling, i.e. the first order of the perturbation series, is given.

The program of constructing the S-matrix by means of causality goes back to E.C.G. Stückelberg (see the historical introduction). He introduced a "macroscopic" causality condition and made use of unitarity in addition. Later on N.N. Bogoliubov and collaborators simplified the causality condition. They formulated it with the help of the so-called adiabatic switching with a test function. This tool must be used for mathematical reasons because the S-matrix is an operator-valued functional and not an operator, and also for physical reasons since the real asymptotic states are not simply generated by free fields. However, these authors failed at one essential point because they arrived at the usual formal expression for S as a time-ordered exponential (cf. (3.1.67) below). This expression is ill-defined and, therefore, has to be corrected by subsequent regularization. As mentioned in the preface, the program was successfully carried through by H. Epstein and V. Glaser in 1973. In their method (Sects. 3.1 and 3.2), the perturbation series is constructed inductively, order by order, by means of causality and translation invariance, unitarity is not used. All steps are under control, consequently, no ultraviolet divergences appear. The naive Feynman rules only hold for tree graphs, not for closed loops.

The most delicate step in this construction is the decomposition of distributions with causal support into retarded and advanced parts. If this distribution splitting is carried out without care by multiplication with step functions, then the usual ultraviolet divergences appear. But if it is carefully done by first multiplying with a C^∞ function and then performing the appropriate limit to the step function, everything is finite and well-defined. If the splitting is carried out in momentum space, it leads to dispersion integrals. The first who used dispersion techniques in QED was G. Källen (in *Handbuch der Physik, ed. by S. Flügge, Vol. V1 (1958)*). Our calculations in Sects. 3.6–9 are inspired by his work. The main difference is that Källen had to extract the input for the dispersion relations from some divergent expression, because he did not take causality into account from the very beginning. Finally, a comment on the infrared problem is indicated. As we point out at the beginning of the following Sect. 3.1 and discuss in more detail in Sect. 3.11 and 3.12, this is not really a problem, if the S-matrix is introduced by adiabatic switching with a test function $g(x)$. The limit $g \to 1$ exists, if the right physically measurable quantities (inclusive cross sections) are considered. There is no need to modify the definition of the S-matrix or to introduce a finite photon mass.

3.1 The Method of Epstein and Glaser

It is our aim to construct the S-matrix of full QED by perturbation theory. That is to say, we want to express S as a power series in the coupling constant $g = e$, where e is the unit of charge. We cannot make any statement about the (probably asymptotic) convergence of this series. The individual terms in the series are written down by utilizing the free field operators for the electron-positron and radiation fields. Since these operators are distributions, it is necessary to test the S-matrix with a C-number test function $g(x) \in \mathcal{S}(\mathbb{R}^4)$, assumed to be in Schwartz space, considering the limit $g \to 1$ at the end. We therefore start from the expression

$$S(g) = \mathbf{1} + \sum_{n=1}^{\infty} \frac{1}{n!} \int d^4x_1 \dots d^4x_n \, T_n(x_1, \dots x_n) g(x_1) \dots g(x_n)$$

$$\stackrel{\text{def}}{=} \mathbf{1} + T. \tag{3.1.1}$$

Since the test function $g(x)$ vanishes at infinity, it switches the interaction on and off in time. This process is completely unphysical in full QED. But it has a physical significance in the external field problem. For this reason, we have studied this subject in great detail in Chap. 2.

One problem in QED is the Coulomb interaction which is long range, due to the vanishing photon mass. This is the origin of infrared problems. For $g \in \mathcal{S}(\mathbb{R}^4)$, the long range part of the interaction is cut off. Accordingly, $S(g)$ (3.1.1) is free of infrared divergences. The latter appear in the so-called adiabatic limit $g \to 1$. However, the operator valued distributions $T_n(x_1, \dots, x_n)$ in (3.1.1) do not "know" what the test function g is, hence, they are also free from infrared divergences. These so-called n-point distributions $T_n(x_1, \dots, x_n)$ are the basic objects to be constructed, and this construction, therefore, cannot be plagued by infrared problems. This does not at all mean that such problems do not exist. They appear, however, at a much later stage, namely when one calculates observable quantities from the T_n for $g = 1$. This will be discussed in Sects. 3.9 and 3.10.

The method of Epstein and Glaser is a generalization of the simple inductive construction described at the end of Chap. 0 (see about (0.3.21) there) in the case of quantum mechanics. According to the definition (3.1.1), $T_n(x_1, \dots, x_n)$ is symmetric in $x_1, \dots, x_n$. It is therefore often convenient to consider the disordered set of n points in Minkowski space $\mathbb{M}$

$$X = \{x_j \in \mathbb{M} \,|\, j = 1, \dots n\} \tag{3.1.2}$$

as argument of T_n. Later on, we will expand $T_n(X)$ in terms of free field operators in normally ordered form

$$T_n(X) = \sum_k : \prod_j \overline{\psi}(x_j)\, t_n^k(x_1, \dots, x_n) \prod_l \psi(x_l) :: \prod_m A(x_m) :, \tag{3.1.3}$$

where t_n^k is a numerical distribution, the arguments of the field operators depend on the term k considered.

Like $S(g)$ (3.1.1), the inverse $S(g)^{-1}$ can be expressed by a perturbation series

$$S(g)^{-1} = \mathbf{1} + \sum_{n=1}^{\infty} \frac{1}{n!} \int d^4x_1 \dots d^4x_n\, \tilde{T}_n(x_1, \dots, x_n) g(x_1) \dots g(x_n) \tag{3.1.4}$$

$$= (\mathbf{1} + T)^{-1} = \mathbf{1} + \sum_{r=1}^{\infty} (-T)^r. \tag{3.1.5}$$

The corresponding n-point distributions $\tilde{T}_n$ follow from (3.1.5) as in the formal inversion of a power series (cf. (0.3.24))

$$\tilde{T}_n(X) = \sum_{r=1}^{n} (-)^r \sum_{P_r} T_{n_1}(X_1) \dots T_{n_r}(X_r), \tag{3.1.6}$$

where the second sum runs over all partitions P_r of X (3.1.2) into r disjoint subsets

$$X = X_1 \cup \dots \cup X_r, \quad X_j \neq \emptyset, \quad |X_j| = n_j.$$

All products of distributions in (3.1.6) are well-defined, because the arguments are disjoint sets of points such that the products are direct products of distributions.

Besides (3.1.6), there are two further relations between the T's which are important in the following. We write the product

$$\mathbf{1} = S(g) S(g)^{-1} = \mathbf{1} + \sum_{n=1}^{\infty} \sum_{n_1 + n_2 = n} \frac{1}{n_1! n_2!} \int T_{n_1}(x_1, \dots x_{n_1})$$

$$\times \tilde{T}_{n_2}(y_1,\dots y_{n_2})g(x_1)\dots g(x_{n_1})g(y_1)\dots g(y_{n_2})d^4x_1\dots d^4y_{n_2} \tag{3.1.7}$$

in symmetrical form by carrying out the $n!/n_1!n_2!$ permutations between the x and y variables

$$= \mathbf{1} + \sum_{n=1}^{\infty}\frac{1}{n!}\sum_{P_2^0}\int d^4x_1\dots d^4x_n\, T_{n_1}(X)\tilde{T}_{n_2}(Y)g(x_1)\dots g(x_n), \tag{3.1.8}$$

where P_2^0 are all partitions into two subsets

$$P_2^0 : \{x_1,\dots,x_n\} = X\cup Y, \quad |X| = n_1 \tag{3.1.9}$$

with empty sets X, $Y=\emptyset$ allowed. The trivial expressions

$$T_0(\emptyset) = \mathbf{1} = \tilde{T}_0(\emptyset) \tag{3.1.10}$$

have to be used. From (3.1.7, 8) we conclude that

$$\sum_{P_2^0} T_{n_1}(X)\tilde{T}_{n-n_1}(Z\setminus X) = 0, \tag{3.1.11}$$

for all Z with $|Z| = n \geq 1$, $|X| = n_1$ and

$$\sum_{P_2^0} T_{n-n_2}(Z\setminus Y)\tilde{T}_{n_2}(Y) = 0. \tag{3.1.12}$$

If we exchange S and S^{-1}, we obtain relations with T and $\tilde{T}$ interchanged, for example

$$\sum_{P_2^0} \tilde{T}_{n-n_1}(X)T_{n_1}(Z\setminus X) = 0. \tag{3.1.13}$$

We now proceed to discuss general properties which the S-matrix (3.1.1) should have. The first one is unitarity

$$S(g)^{-1} = S(g)^{+}. \tag{3.1.14}$$

Using (3.1.4), it can be expressed by means of the n-point distributions in the form

$$\tilde{T}_n(x_1,\dots,x_n) = T_n(x_1,\dots,x_n)^{+}.$$

We shall realize much later in Sect. 4.7 that unitarity is not as simple as this in QED. The second property is translation invariance: Let $U(a,\mathbf{1})$ be the unitary translation operator in the total Fock space $\mathcal{F}$

$$(U(a,\mathbf{1})\Phi)_j(x) = \Phi_j(x_1+a,\dots,x_j+a), \tag{3.1.15}$$

this transformation law is understood to hold in all sectors j of $\mathcal{F}$. Then one requires

$$U(a,\mathbf{1})S(g)U(a,\mathbf{1})^{-1} = S(g_a) \tag{3.1.16}$$

where

$$g_a(x) = g(x - a). \tag{3.1.17}$$

This implies a similar law for the T's

$$U(a, \mathbf{1})T_n(x_1, \ldots, x_n)U(a, \mathbf{1})^{-1} = T_n(x_1 + a, \ldots, x_n + a), \tag{3.1.18}$$

and the same must hold for $\tilde{T}_n$. The next property is Lorentz covariance. If $U(0, \Lambda)$ is the (pseudo-unitary) representation of the proper Lorentz group $\mathcal{L}_+^\uparrow$, defined by the free fields (cf. (2.2.65) and (2.11.39)), then we should have

$$U(0, \Lambda)S(g)U(0, \Lambda)^{-1} = S(g_\Lambda), \tag{3.1.19}$$

with

$$g_\Lambda(x) = g(\Lambda^{-1}x). \tag{3.1.20}$$

For the n-point distributions, this means

$$U(0, \Lambda)T_n(x_1, \ldots, x_n)U(0, \Lambda)^{-1} = T_n(\Lambda x_1, \ldots, \Lambda x_n). \tag{3.1.21}$$

We now come to the most important property which is causality. Let us suppose that there exists a frame of reference in which the test functions g_1, g_2 have disjoint supports in time

$$\mathrm{supp}\, g_1 \subset \{x \in \mathbb{M} \mid x^0 \in (-\infty, r)\} \quad , \quad \mathrm{supp}\, g_2 \subset \{x \in \mathbb{M} \mid x^0 \in (r, +\infty)\}. \tag{3.1.22}$$

Then we require

$$S(g_1 + g_2) = S(g_2)S(g_1). \tag{3.1.23}$$

In full QED, this causality condition does not have a direct physical foundation because, as discussed above, the switching on and off the interaction is unphysical. But the condition is physical in the external field problem and it works beautifully there, as we have found in the last sections of Chap. 2. This is our basis to take it over to full QED. In any case, the correctness of (3.1.23) can only be shown by working out its consequences. In an arbitrary Lorentz frame there exists a space-like plane which separates $\mathrm{supp}\, g_1$ and $\mathrm{supp}\, g_2$, such that $\mathrm{supp}\, g_2$ is later than $\mathrm{supp}\, g_1$. In this situation we shall write

$$\mathrm{supp}\, g_1 < \mathrm{supp}\, g_2. \tag{3.1.24}$$

For arbitrary sets X, Y of points in Minkowski space, the relation $X < Y$ means that all points $x \in X$ are earlier than all $y \in Y$ in some Lorentz frame.

We now proceed to investigate the consequences of (3.1.23) for the T's. The expression

$$S(g_1 + g_2) = \sum_n \frac{1}{n!} \int d^4x_1 \ldots d^4x_n\, T_n(x_1, \ldots, x_n)$$
$$\times (g_1(x_1) + g_2(x_1)) \ldots (g_1(x_n) + g_2(x_n)) \tag{3.1.25}$$

can be reordered by permutations of the integration variables x_j in the 2^n terms, to have them in the following form

$$g_2(x_1)\dots g_2(x_m)g_1(x_{m+1})\dots g_1(x_n).$$

Since there are

$$\frac{n!}{m!(n-m)!} \quad \text{permutations,} \quad \sum_{m=0}^{n} \frac{n!}{m!(n-m)!} = 2^n, \tag{3.1.26}$$

we arrive at

$$S(g_1+g_2) = \sum_{n=0}^{\infty}\sum_{m=0}^{n} \frac{1}{m!(n-m)!} \int d^4x_1 \dots d^4x_n$$

$$\times T_n(x_1,\dots,x_n)g_2(x_1)\dots g_2(x_m)g_1(x_{m+1})\dots g_1(x_n)$$

$$= S(g_2)S(g_1) = \sum_{n=0}^{\infty}\sum_{m=0}^{n} \frac{1}{m!(n-m)!} \int d^4x_1 \dots d^4x_n$$

$$\times T_m(x_1,\dots,x_m)T_{n-m}(x_{m+1},\dots,x_n)g_2(x_1)\dots g_2(x_m)g_1(x_{m+1})\dots g_1(x_n). \tag{3.1.27}$$

This leads to the condition

$$T_n(x_1,\dots,x_n) = T_m(x_1,\dots,x_m)T_{n-m}(x_{m+1},\dots,x_n) \tag{3.1.28}$$

if $\{x_1,\dots,x_m\} > \{x_{m+1},\dots,x_n\}$ (Problem 3.1). Similarly, the causality condition for $S^{-1}(g)$

$$S(g_1+g_2)^{-1} = S(g_1)^{-1}S(g_2)^{-1}$$

with g_1, g_2 satisfying (3.1.24), implies

$$\tilde{T}_n(x_1,\dots,x_n) = \tilde{T}_m(x_1,\dots,x_m)\tilde{T}_{n-m}(x_{m+1},\dots,x_n), \tag{3.1.29}$$

if $\{x_1,\dots,x_m\} < \{x_{m+1},\dots,x_n\}$.

The basic causality condition (3.1.28) shows that the T_n are time-ordered products. If all x_j have different temporal components x_j^0, the arguments of T_n can be permuted such that $x_1^0 > x_2^0 > \dots > x_n^0$, say, are ordered in time. Repeated application of (3.1.28) then gives

$$T_n(x_1,\dots,x_n) = T_1(x_1)T_2(x_2)\dots T_1(x_n). \tag{3.1.30}$$

The letter T has been chosen to denote the time-ordered products. They are the basic objects of the theory. For shortness the notion "n-point function" is also used for T_n.

Now we are ready to turn to the inductive construction of $T_n(x_1,\dots,x_n)$. The beginning of the induction and the idea of the construction have already been described at the end of Sect. 0.3 in Chap. 0 in the case of quantum mechanics. For QED, the first step, namely the construction of T_2 from T_1, will be carried out explicitly in the following sections. For this reason we treat here the general inductive step. Suppose all $T_m(x_1,\ldots,x_m)$ for $1\leq m\leq n-1$ are known and have the above properties. Then, according to (3.1.6), the $\tilde{T}_m(X)$ can be calculated for all $1\leq m=|X|\leq n-1$. From this it is possible to form the following distributions

$$A'_n(x_1,\ldots,x_n)=\sum_{P_2}\tilde{T}_{n_1}(X)T_{n-n_1}(Y,x_n) \tag{3.1.31}$$

$$R'_n(x_1,\ldots,x_n)=\sum_{P_2}T_{n-n_1}(Y,x_n)\tilde{T}_{n_1}(X), \tag{3.1.32}$$

where the sums run over all partitions

$$P_2:\ \{x_1,\ldots,x_{n-1}\}=X\cup Y,\quad X\neq\emptyset \tag{3.1.33}$$

into disjoint subsets with $|X|=n_1\geq 1$, $|Y|\leq n-2$. We also introduce

$$D_n(x_1,\ldots,x_n)=R'_n-A'_n. \tag{3.1.34}$$

If the sums are extended over all partitions P_2^0, including the empty set $X=\emptyset$, then we get the distributions

$$A_n(x_1,\ldots,x_n)=\sum_{P_2^0}\tilde{T}_{n_1}(X)T_{n-n_1}(Y,x_n)$$

$$=A'_n+T_n(x_1,\ldots,x_n), \tag{3.1.35}$$

$$R_n(x_1,\ldots,x_n)=\sum_{P_2^0}T_{n-n_1}(Y,x_n)\tilde{T}_{n_1}(X)$$

$$=R'_n+T_n(x_1,\ldots,x_n). \tag{3.1.36}$$

These two distribution are not known by the induction assumption because they contain the unknown $T_n(x_1,\ldots,x_n)$. Only the difference

$$D_n=R'_n-A'_n=R_n-A_n \tag{3.1.37}$$

is known (3.1.34). What remains to be done is to determine R_n (or A_n) in (3.1.37) separately. This is achieved by investigating the support properties of the various distributions.

Theorem 1.1. Let $Y=P\cup Q$, $P\neq\emptyset$, $P\cap Q=\emptyset$, $|Y|=n_1\leq n-1$ and $x\notin Y$. If $\{Q,\,x\}>P$, $|Q|=n_2$, then we have

$$R'_{n_1+1}(Y,\,x)=-T_{n_2+1}(Q,\,x)T_{n_1-n_2}(P). \tag{3.1.38}$$

If $\{Q,\,x\}<P$, then we have

$$A'_{n_1+1}(Y,\,x)=-T_{n_1-n_2}(P)T_{n_2+1}(Q,\,x). \tag{3.1.39}$$

Proof. We start from (3.1.32)

$$R'_{n_1+1}(Y,x)=\sum_{P_2}T_{n_1+1-n_3}(Y',x)\tilde{T}_{n_3}(X), \tag{3.1.40}$$

where P_2 are the partitions of Y

$$P_2:\ Y=X\cup Y'\,,\ |X|=n_3\neq 0. \tag{3.1.41}$$

Let

$$Y'=Y_1\cup Y_2\,,\quad Y_1=Y'\cap P\,,\quad Y_2=Y'\cap Q$$
$$X=X_1\cup X_2\,,\quad X_1=X\cap P\,,\quad X_2=X\cap Q, \tag{3.1.42}$$

then we obviously have

$$Y_1<Y_2\,,\quad X_1<X_2\,,\quad \{Y_2,x\}>Y_1. \tag{3.1.43}$$

Hence, causality implies

$$R'_{n_1+1}(Y,x)=\sum_{P_4^0}T(Y_2,x)T(Y_1)\tilde{T}(X_1)\tilde{T}(X_2), \tag{3.1.44}$$

where the subscripts of the T's have been dropped for simplicity, the latter are always equal to the number of points in the argument. P_4^0 are all partitions of the form

$$P_4^0:\ P=X_1\cup Y_1\,,\ Q=X_2\cup Y_2\,,\ X_1\cup X_2\neq\emptyset. \tag{3.1.45}$$

However, for $X_2\neq\emptyset$, we can have $X_1=\emptyset$. Then it follows from (3.1.12) that

$$\sum_{P_2^0}T(Y_1)\tilde{T}(X_1)=0. \tag{3.1.46}$$

Consequently, in (3.1.44) only the terms with $X_2=\emptyset$, $Y_2=Q$ remain

$$R'_{n_1+1}(Y,x)=T(Q,x)\sum_{P_2}T(Y_1)\tilde{T}(X_1). \tag{3.1.47}$$

The partitions P_2 are

$$P_2:\ P=X_1\cup Y_1\,,\ X_1\neq\emptyset,$$

with the empty set excluded. If one includes the empty set, one gets 0 according to (3.1.46). One may, therefore, rewrite (3.1.47) as

$$R'_{n_1+1}(Y,x)=-T(Q,x)T(P),$$

which proves (3.1.38). The proof of (3.1.39) is the same. □

By

$$\overline{V^+}(x)=\{y\,|\,(y-x)^2\geq 0\,,\ y^0\geq x^0\} \tag{3.1.48}$$

we denote the closed forward cone of x, and by

$$\overline{V^-}(x) = \{y \,|\, (y-x)^2 \geq 0\,,\; y^0 \leq x^0\} \tag{3.1.49}$$

the closed backward cone. The n-dimensional generalizations are

$$\Gamma_n^\pm(x) = \{(x_1,\dots,x_n) \,|\, x_j \in \overline{V^\pm}(x)\,,\; \forall j = 1,\dots n\}. \tag{3.1.50}$$

For $|Y| = n_1 \leq n-2$, it follows from (3.1.36) that

$$R_{n_1+1}(Y,\, x) = R'_{n_1+1}(Y,\, x) + T_{n_1+1}(P \cup Q,\, x)$$
$$= -T_{n_2+1}(Q,\, x)T_{n_1-n_2}(P) + T_{n_2+1}(Q,\, x)T_{n_1-n_2}(P) = 0, \tag{3.1.51}$$

where (3.1.38) and causality (3.1.28) was used. Since this holds for arbitrary Y as in Theorem 1.1, it follows that R (3.1.51) vanishes, if there exists a point earlier than x in some Lorentz frame. Consequently, R has a retarded support:

Corollary 1.2.

$$\operatorname{supp} R_{n_1+1}(Y,\, x) \subseteq \Gamma^+_{n_1+1}(x) \tag{3.1.52}$$

and similarly

$$\operatorname{supp} A_{n_1+1}(Y,\, x) \subseteq \Gamma^-_{n_1+1}(x). \tag{3.1.53}$$

Because of these support properties, R and A are called retarded and advanced distributions, respectively. The distribution D (3.1.37) then has a causal support:

Corollary 1.3. If $|Y| = n_1 \leq n-2$, then

$$\operatorname{supp} D_{n_1+1}(Y,\, x) \subseteq \Gamma^+_{n_1+1}(x) \cup \Gamma^-_{n_1+1}(x). \tag{3.1.54}$$

These support properties must be preserved in the step from $n-1$ to n. In particular, Corollary 1.3 must hold for $n_1 = n-1$, too. But $D_n(Y,\, x)$ is known according to (3.1.37) and the induction assumption. Hence, the support property (3.1.54) must follow without using causality of T_n.

Theorem 1.4. If $n \geq 3$, then

$$\operatorname{supp} D_n(x_1, \dots x_{n-1}, x_n) \subseteq \Gamma_n^+(x_n) \cup \Gamma_n^-(x_n). \tag{3.1.55}$$

Proof. We divide the proof into two parts.

1) According to Theorem 1.1, we have

$$R'_n(x_1,\dots,x_{n-1},x_n) = -T_{n_2+1}(Q,\, x_n)T_{n-n_2-1}(P), \tag{3.1.56}$$

if $\{Q, x_n\} > P$, $|Q| = n_2$, and

$$A'_n(x_1,\dots,x_{n-1},x_n) = -T_{n-n_2-1}(P)T_{n_2+1}(Q,x_n), \tag{3.1.57}$$

if $\{Q,x_n\} < P$. Let Ω be the set of all points $x = (x_1,\dots,x_n) \in \mathbb{M}^n$, such that in some Lorentz frame (which may depend on x) the n points of x can be decomposed as follows

$$\{x_1,\dots,x_n\} = P' \cup Q' \cup S\,,\ P',Q' \neq \emptyset \tag{3.1.58}$$

with

$$\begin{aligned} x_j^0 &> x_n^0 \quad \text{for} \quad \forall x_j \in P' \\ x_j^0 &< x_n^0 \quad \text{for} \quad \forall x_j \in Q' \\ x_j^0 &= x_n^0 \quad \text{for} \quad \forall x_j \in S. \end{aligned} \tag{3.1.59}$$

We obviously have $P' > Q'$. Then, by means of causality, we get from (3.1.56)

$$\begin{aligned} R'_n(x_1,\dots,x_{n-1},x_n) &= -T(P'\cup S)T(Q') \\ &= -T(P')T(S)T(Q'), \end{aligned} \tag{3.1.60}$$

and similarly for (3.1.57)

$$\begin{aligned} A'_n(x_1,\dots,x_{n-1},x_n) &= -T(P')T(Q'\cup S) \\ &= -T(P')T(S)T(Q'). \end{aligned} \tag{3.1.61}$$

Consequently,

$$D_n = R'_n - A'_n = 0 \tag{3.1.62}$$

vanishes in the open set Ω.

2) Suppose now

$$x = \{x_1,\dots,x_{n-1},x_n\} \notin \Gamma^+_{n-1}(x_n) \cup \Gamma^-_{n-1}(x_n) \tag{3.1.63}$$

is not in the support (3.1.55). This is possible in the following ways:

a) One point x_1 is in $\overline{V^+}(x_n)$ and another one, say x_2, is in $\overline{V^-}(x_n)$. Then $x \in \Omega$ and $D_n = 0$ according to (3.1.62).

b) One point, say x_1, is space-like with respect to x_n, $(x_1 - x_n)^2 < 0$. Then we choose a frame of reference such that $x_1^0 = x_n^0$. If there are two points x_j, x_k with $x_j^0 > x_n^0$ and $x_k^0 < x_n^0$, then x (3.1.63) is in Ω, hence, $D_n = 0$. We therefore assume

$$x_j^0 \geq x_n^0 \quad , \quad \forall j = 2,\dots,n-1.$$

The case $x_j^0 \leq x_n^0$ is similar. If there is a point, say x_2, with $x_2^0 > x_n^0$, then it is possible by a small Lorentz transformation to arrive at a situation with $x_1^0 < x_n^0$, but still $x_2^0 > x_n^0$. Once more, we find $x \in \Omega$, hence, $D_n = 0$.

c) There remains the case where all x_j are simultaneous

$$x_j^0 = x_n^0 \quad , \quad \forall j = 1,\dots,n-1. \tag{3.1.64}$$

Then we select a point, say x_1, with maximal spatial distance $|\boldsymbol{x}_1 - \boldsymbol{x}_n|$. Let P be the rest $\{x_2, \ldots, x_n\}$. It is now possible by small Lorentz transformations to get $x_1 > P$ or $x_1 < P$. In the first case, one finds according to (3.1.56)

$$R'_n(x_1, \ldots, x_n) = -T_1(x_1)T_{n-1}(P),$$

and in the second case from (3.1.57)

$$A'_n(x_1, \ldots, x_n) = -T_{n-1}(P)T_1(x_1).$$

From P we select another point, say x_2, which is space-like with respect to the rest P'. By suitable choice of Lorentz frames we can achieve that $x_2 > P'$ or $x_2 < P'$. Then causality of T_{n-1} implies

$$T_{n-1}(P) = T_1(x_2)T_{n-2}(P') = T_{n-2}(P')T_1(x_2).$$

For the same reason all three factors in

$$R'_n = -T_1(x_1)T_1(x_2)T_{n-2}(P')$$

commute which leads to $R'_n = A'_n$. Hence, $D_n = R'_n - A'_n = 0$, which completes the proof of Theorem 1.4. □

For $n \leq 2$, the support property (3.1.55) of D_n must be verified explicitly. If Theorem 1.4 would not be true, then the inductive construction of the T_n by means of causality would be impossible. Now we see this construction clearly before us: From the known $T_m(x_1, \ldots, x_m)$, $m \leq n-1$ one computes $A'_n(x_1, \ldots, x_n)$ (3.1.31) and $R'_n(x_1, \ldots, x_n)$ (3.1.32), and then $D_n = R'_n - A'_n$ (3.1.37). One decomposes D_n with respect to the supports (3.1.55)

$$D_n(x_1, \ldots, x_n) = R_n(x_1, \ldots, x_n) - A_n(x_1, \ldots, x_n), \tag{3.1.65}$$

$$\operatorname{supp} R_n \subseteq \Gamma^+_{n-1}(x_n) \quad , \quad \operatorname{supp} A_n \subseteq \Gamma^-_{n-1}(x_n).$$

Finally, T_n is found from (3.1.35) or (3.1.36)

$$T_n(x_1, \ldots, x_n) = R_n(x_1, \ldots, x_n) - R'_n(x_1, \ldots, x_n) \tag{3.1.66a}$$

$$= A_n(x_1, \ldots, x_n) - A'_n(x_1, \ldots, x_n). \tag{3.1.66b}$$

The only non-trivial step in this construction is the distribution splitting (3.1.65). This will be investigated in the following section.

To complete the inductive step we have to verify that T_n (3.1.66) satisfies all properties used in the inductive construction, in particular the causality condition (3.1.28). Assuming $\{x_1, \ldots, x_n\} = P \cup Q$ with $P < Q$, there are two cases to be examined: If $x_n \in P$, A_n vanishes and (3.1.66b) gives the causality condition

$$T_n = -A'_n = T(Q)T(P),$$

using (3.1.57). On the other hand, if $x_n \in Q$, R_n is zero and from (3.1.66a) and (3.1.56) we get the same result. Furthermore, T_n obeys translation invariance (3.1.18), because it cannot be destroyed in the distribution splitting

(3.1.65). Finally, if the distribution splitting is unique, $T_n(x_1,\dots,x_n)$ must also be totally symmetric in virtue of (3.1.36), although R'_n and R_n separately are not symmetrical with respect to x_n. We will learn in the next section that R_n is unique up to local terms. But the latter can be symmetrized, so that from an appropriate splitting solution R_n, we get a T_n with all desired properties.

At first sight, it seems to be a miracle that the first order $T_1(x)$ determines the whole S-matrix. However, as shown at the end of Sect. 0.3 in Chap. 0, this is a general feature of quantum theory. It was stressed there that the causality condition can indeed be used as a substitute for a dynamical equation. It is very instructive to follow the quantum mechanical analogy one step further. In (0.3.32) the splitting of D_2 into a retarded and advanced part was simply done by multiplication with a Θ-function $\Theta(t_1-t_2)$. This was possible there because $T_1(t)$ and all T_n are well-defined operators in Hilbert space. If we would proceed in the same way in (3.1.65), we would find the usual time-ordered exponential for the S-matrix (see (3.1.30))

$$S(g)=\sum_{n=0}^{\infty}\frac{1}{n!}\int d^4x_1\dots d^4x_n\,T\{T_1(x_1)\cdot\ldots\cdot T_1(x_n)\}g(x_1)\cdot\ldots\cdot g(x_n), \tag{3.1.67}$$

because the *formal* expression

$$T_n(x_1,\dots,x_n)=T\{T_1(x_1)\cdot\ldots\cdot T_1(x_n)\}$$

$$\stackrel{\text{def}}{=}\sum_{\Pi}\Theta(x^0_{\Pi 1}-x^0_{\Pi 2})\cdot\ldots\cdot\Theta(x^0_{\Pi(n-1)}-x^0_{\Pi n})T_1(x_{\Pi 1})\cdot\ldots\cdot T_1(x_{\Pi n}), \tag{3.1.68}$$

where the sum runs over all $n!$ permutations, obviously satisfies the basic causality condition (3.1.28). However, since the result (3.1.68) contains ultraviolet divergences, there must be an error in this simple procedure. The point is that the D_n (3.1.65) are operator-valued *distributions.* They cannot simply be multiplied by the discontinuous Θ-functions, in general. Distribution splitting is the crux of the matter. After correct splitting, the T_n (3.1.67) are the well-defined time-ordered products.

3.2 Splitting of Causal Distributions

The operator-valued distributions which we shall have to split later are expanded in terms of free fields, as in (3.1.3)

$$D_n(x_1,\dots,x_n)=\sum_k :\prod_j\overline{\psi}(x_j)d_n^k(x_1,\dots,x_n)\prod_l\psi(x_l)::\prod_m A(x_m):. \tag{3.2.1}$$

The numerical distributions d_n^k have causal support (3.1.55) and are assumed to be tempered $\in \mathcal{S}'(\mathbb{R}^{4n})$, because we will use Fourier transformation. There remains to split these numerical distributions as follows

$$d_n^k(x) = r_n(x) - a_n(x)$$

$$\operatorname{supp} r_n \subseteq \Gamma_{n-1}^+(x_n) \quad , \quad \operatorname{supp} a_n \subseteq \Gamma_{n-1}^-(x_n), \tag{3.2.2}$$

where $x = (x_1, \ldots, x_n)$. The simplest way of splitting would be

$$r_n(x) = \chi_n(x) d_n^k(x), \quad \text{with} \tag{3.2.3}$$

$$\chi_n(x) = \prod_{j=1}^{n-1} \Theta(x_j^0 - x_n^0). \tag{3.2.4}$$

The difficulty is that (3.2.4) is discontinuous, and then, if d_n^k is singular at $x = 0$, r_n (3.2.3) is generally not a tempered distribution. Because of translation invariance, it is sufficient to put $x_n = 0$ and to consider

$$d(x) \stackrel{\text{def}}{=} d_n^k(x_1, \ldots, x_{n-1}, 0) \in \mathcal{S}'(\mathbb{R}^m), \quad m = 4n - 4. \tag{3.2.5}$$

The intersection of the discontinuity surface of (3.2.4) and $\operatorname{supp} d$ is the origin $x = 0$. Therefore, the behaviour of $d(x)$ in the neighbourhood of $x = 0$ is essential for the splitting procedure. For this reason we introduce the following definition:

Definition 3.1. *The distribution $d(x) \in \mathcal{S}'(\mathbb{R}^m)$ has a quasi-asymptotics $d_0(x)$ at $x = 0$ with respect to a positive continuous function $\rho(\delta)$, $\delta > 0$, if the limit*

$$\lim_{\delta \to 0} \rho(\delta) \delta^m d(\delta x) = d_0(x) \not\equiv 0 \tag{3.2.6}$$

exists in $\mathcal{S}'(\mathbb{R}^m)$.

The quasi-asymptotics probes the vicinity of $x = 0$, only: If

$$d(x) = d_1(x) + d_2(x) \tag{3.2.7}$$

where d_1 has a compact support K_0 containing $x = 0$ and $\operatorname{supp} d_2$ is bounded away from 0, it follows that

$$\lim_{\delta \to 0} \rho(\delta) \delta^m \langle d_2(\delta x), \varphi_0 \rangle = \lim_{\delta \to 0} \rho(\delta) \left\langle d_2(x), \varphi_0\left(\frac{x}{\delta}\right) \right\rangle = 0 \tag{3.2.8}$$

for every $\varphi_0 \in C_0^\infty(\mathbb{R}^m)$. Since C_0^∞ is dense in $\mathcal{S}$, the distribution (3.2.8) vanishes on $\mathcal{S}$ also, hence

$$\lim_{\delta \to 0} \rho(\delta) \delta^m \langle d_1(\delta x), \varphi(x) \rangle = \langle d_0, \varphi \rangle \tag{3.2.9}$$

for all $\varphi \in \mathcal{S}$. In

$$\lim_{\delta\to 0}\rho(\delta)\Big\langle d(x),\varphi\Big(\frac{x}{\delta}\Big)\Big\rangle=\langle d_0,\varphi\rangle. \tag{3.2.10}$$

we go over to momentum space to find an equivalent condition for the Fourier transform $\hat{d}(p)$. Since

$$\Big\langle d(x),\varphi\Big(\frac{x}{\delta}\Big)\Big\rangle=\Big\langle \hat{d}(p),\Big(\varphi\Big(\frac{x}{\delta}\Big)\Big)^{\check{}}(p)\Big\rangle=\delta^m\langle\hat{d}(p),\check{\varphi}(\delta p)\rangle$$

$$=\Big\langle\hat{d}\Big(\frac{p}{\delta}\Big),\check{\varphi}(p)\Big\rangle, \tag{3.2.11}$$

where $\check{\varphi}$ denotes the inverse Fourier transform, we get the following equivalent definition:

Definition 3.2. *The distribution $\hat{d}(p)\in\mathcal{S}'(\mathbb{R}^{\mathbf{m}})$ has quasi-asymptotics $\hat{d}_0(p)$ at $p=\infty$ if*

$$\lim_{\delta\to 0}\rho(\delta)\Big\langle\hat{d}\Big(\frac{p}{\delta}\Big),\check{\varphi}(p)\Big\rangle=\langle\hat{d}_0,\check{\varphi}\rangle \tag{3.2.12}$$

exists for all $\check{\varphi}\in\mathcal{S}(\mathbb{R}^{\mathbf{m}})$.

In momentum space the quasi-asymptotics controls the ultraviolet behaviour of the distribution. Let us consider a scaling transformation

$$\lim_{\delta\to 0}\rho(\delta)\langle\hat{d}(\frac{p}{\delta}),\check{\varphi}(ap)\rangle=\langle\hat{d}_0(p),\check{\varphi}(ap)\rangle$$

$$=a^{-m}\lim_{\delta\to 0}\rho(\delta)\Big\langle\hat{d}\Big(\frac{p}{a\delta}\Big),\check{\varphi}(p)\Big\rangle=a^{-m}\lim_{\delta\to 0}\frac{\rho(\delta)}{\rho(a\delta)}\rho(a\delta)\Big\langle\hat{d}\Big(\frac{p}{a\delta}\Big),\check{\varphi}(p)\Big\rangle. \tag{3.2.13}$$

Since

$$\lim_{\delta\to 0}\rho(a\delta)\langle\hat{d}(\frac{p}{a\delta}),\check{\varphi}(p)\rangle=\langle\hat{d}_0(p),\check{\varphi}(p)\rangle$$

exists, we may conclude that the limit

$$\lim_{\delta\to 0}\frac{\rho(a\delta)}{\rho(\delta)}=a^{-m}\frac{\langle\hat{d}_0(p),\check{\varphi}(p)\rangle}{\langle\hat{d}_0(p),\check{\varphi}(ap)\rangle}\stackrel{\text{def}}{=}\rho_0(a) \tag{3.2.14}$$

exists, too, assuming that the denominator is different from 0. By another scaling transformation it follows

$$\rho_0(ab)=\rho_0(a)\rho_0(b), \tag{3.2.15}$$

which implies $\rho_0(a)=a^\omega$ with some real ω. We therefore call $\rho(\delta)$ the power-counting function.

Definition 3.3. *The distribution $d\in\mathcal{S}'(\mathbb{R}^{\mathbf{m}})$ is called singular of order ω, if it has a quasi-asymptotics $d_0(x)$ at $x=0$, or its Fourier transform has quasi-asymptotics $\hat{d}_0(p)$ at $p=\infty$, respectively, with power-counting function $\rho(\delta)$ satisfying*

$$\lim_{\delta\to 0}\frac{\rho(a\delta)}{\rho(\delta)}=a^\omega, \tag{3.2.16}$$

for each $a>0$.

Equation (3.2.14) implies

$$a^m\langle\hat{d}_0(p),\,\check{\varphi}(ap)\rangle=\langle\hat{d}_0(\frac{p}{a}),\,\check{\varphi}(p)\rangle=a^{-\omega}\langle\hat{d}_0(p),\,\check{\varphi}(p)\rangle$$

$$=\langle d_0(x),\,\varphi(\frac{x}{a})\rangle=a^m\langle d_0(ax),\,\varphi(x)\rangle=a^{-\omega}\langle d_0(x),\,\varphi(x)\rangle, \tag{3.2.17}$$

i.e. $\hat{d}_0$ is homogeneous of degree ω:

$$\hat{d}_0\left(\frac{p}{a}\right)=a^{-\omega}\hat{d}_0(p) \tag{3.2.18}$$

$$d_0(ax)=a^{-(m+\omega)}d_0(x). \tag{3.2.19}$$

This implies that d_0 has quasi-asymptotics $\rho(\delta)=\delta^\omega$ and the singular order ω, too. A positive measurable function $\rho(\delta)$, satisfying (3.2.16), is called regularly varying at zero by mathematicians (*E. Senata, Regularly Varying Functions, Lecture Notes in Mathematics 508, Springer-Verlag 1976*). We present the most important properties of those functions in Appendix B. In particular, we have the following estimates (B34-35): If $\varepsilon>0$ is an arbitrarily small number, then there exist constants C, C' and δ_0, such that

$$C\delta^{\omega+\varepsilon}\geq\rho(\delta)\geq C'\delta^{\omega-\varepsilon}. \tag{3.2.20}$$

We want to apply the definitions to the following examples:
1) $d=1$: From (3.2.6) we get $\rho(\delta)=\delta^{-m}$ and $\omega=-m$.
2) $d(x)=D^a\delta(x)$ where

$$D^a\stackrel{\text{def}}{=}\frac{\partial^{a_1+\ldots+a_m}}{\partial x_1^{a_1}\ldots\partial x_m^{a_m}}\quad,\quad |a|=a_1+\ldots+a_m.$$

Since

$$\hat{d}(p)=(2\pi)^{-m/2}(ip)^a,$$

we obtain $\rho(\delta)=\delta^{|a|}$ and $\omega=|a|$ from (3.2.12).
3) Let us consider the Jordan-Pauli distribution (2.3.18)

$$D(x)=\frac{\operatorname{sgn}x^0}{2\pi}\Big[\delta(x^2)-\frac{m}{2}\frac{\Theta(x^2)}{\sqrt{x^2}}J_1(m\sqrt{x^2})\Big]. \tag{3.2.21}$$

The one-dimensional δ-distribution satisfies

$$\delta(\delta^2x^2)=\frac{\delta(x^2)}{\delta^2},$$

whereas the term with the Bessel function stays bounded for $\delta\sqrt{x^2}\to 0$. Hence

$$\lim_{\delta\to 0}\delta^2 D(\delta x) = \frac{\operatorname{sgn} x^0}{2\pi}\delta(x^2) = D_0(x) \tag{3.2.22}$$

which is just the mass zero Jordan-Pauli distribution. This illustrates the general fact that the quasi-asymptotics d_0 is given by the corresponding mass zero distribution. Since the Jordan-Pauli distribution is considered in $\mathbb{R}^4$ ($m = 4$), we find $\rho(\delta) = \delta^{-2}$ and $\omega(D) = -2$.

4) The positive frequency part (2.3.39)

$$\hat{D}^{(+)}(p) = \frac{i}{2\pi}\Theta(p^0)\delta(p^2 - m^2) \tag{3.2.23}$$

is best considered in momentum space. Since

$$\int \Theta\Big(\frac{p_0}{\delta}\Big)\delta\Big(\frac{p^2}{\delta^2} - m^2\Big)\varphi(p)\,d^4p = \delta^2\int \Theta(p_0)\delta(p^2 - \delta^2 m^2)\varphi(p)\,d^4p$$

$$= \delta^2\int \frac{d^3p}{2\sqrt{\boldsymbol{p}^2 + \delta^2 m^2}}\varphi(\sqrt{\boldsymbol{p}^2 + \delta^2 m^2}, \boldsymbol{p}), \tag{3.2.24}$$

we find

$$\lim_{\delta\to 0}\delta^{-2}\hat{D}^{(+)}\Big(\frac{p}{\delta}\Big) = \hat{D}_0^{(+)}(p) \tag{3.2.25}$$

which implies $\omega(D^{(+)}) = -2$, in agreement with the foregoing example. We obviously have $\omega(D^{(-)}) = -2$, too.

We now turn to the splitting problem, where we have to distinguish two cases:

a) $\omega < 0$: In this case, the power-counting function goes to infinity

$$\rho(\delta) \to \infty \quad \text{for} \quad \delta \to 0. \tag{3.2.26}$$

This implies

$$\Big\langle d(x), \varphi\Big(\frac{x}{\delta}\Big)\Big\rangle \to \frac{\langle d_0, \varphi\rangle}{\rho(\delta)} \to 0. \tag{3.2.27}$$

We choose a monotonous C^∞-function χ_0 over $\mathbb{R}^1$ with

$$\chi_0(t) = \begin{cases} 0 & \text{for } t \le 0 \\ < 1 & \text{for } 0 < t < 1 \\ 1 & \text{for } t \ge 1. \end{cases} \tag{3.2.28}$$

In addition we choose a vector $v = (v_1, \dots v_{n-1}) \in \Gamma^+$, which means that all four-vectors v_j are inside the forward cone V^+. Then

$$v\cdot x = \sum_{j=1}^{n-1} v_j \cdot x_j = 0 \tag{3.2.29}$$

is a space-like hyperplane that separates the causal support: All products $v_j \cdot x_j$ are either ≥ 0 for $x \in \Gamma^+$ or ≤ 0 for $x \in \Gamma^-$ (see Fig. 2). It is our aim to prove that the limit

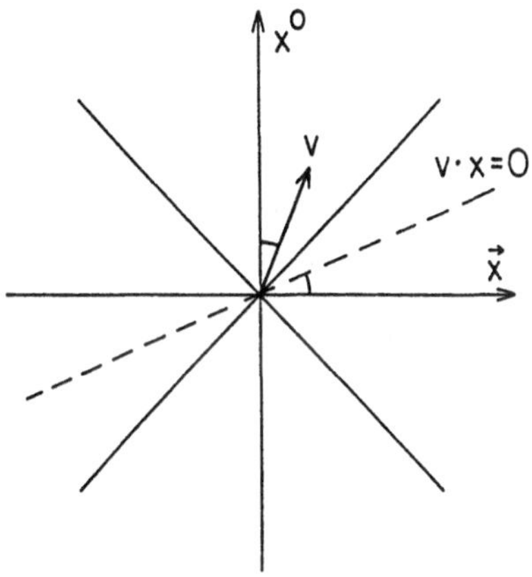

Fig. 2. Separation of the causal support into retarded and advanced parts

$$\lim_{\delta\to 0} \chi_0\Big(\frac{v\cdot x}{\delta}\Big) d(x) \stackrel{\text{def}}{=} \Theta(v\cdot x) d(x) = r(x) \tag{3.2.30}$$

exists.

To prove this we have to show that the difference

$$\Big[\chi_0\Big(a\frac{v\cdot x}{\delta}\Big) - \chi_0\Big(\frac{v\cdot x}{\delta}\Big)\Big] d(x) \stackrel{\text{def}}{=} \psi_0\Big(\frac{x}{\delta}\Big) d(x) \tag{3.2.31}$$

tends to 0 for $\delta \to 0$ uniformly in $a \geq a_1 > 1$. We choose an auxiliary real function $\psi_1(x) \in C^\infty(\mathbb{R}^{\mathbf{m}})$ that is $= 1$ in $\Gamma^+ \cup \Gamma^-$ and vanishes outside a certain neighbourhood of $\Gamma^+ \cup \Gamma^-$. Then, by means of the causal support of d, we have

$$\Big\langle \psi_0\Big(\frac{x}{\delta}\Big) d(x), \varphi(x) \Big\rangle = \Big\langle \psi_0\Big(\frac{x}{\delta}\Big) d(x), \psi_1\Big(\frac{x}{\delta}\Big)\varphi(x) \Big\rangle$$

$$= \Big\langle \varphi(x) d(x), \psi_0\Big(\frac{x}{\delta}\Big)\psi_1\Big(\frac{x}{\delta}\Big) \Big\rangle. \tag{3.2.32}$$

This vanishes for $\delta \to 0$ due to (3.2.27) for all $\varphi \in S(\mathbb{R}^{\mathbf{m}})$ because (i) $\psi_0(x)\psi_1(x)$ is a test function in $C_0^\infty(\mathbb{R}^{\mathbf{m}})$, (ii) if $d(x)$ has singular order < 0 then also $\varphi(x)d(x)$, which is a simple consequence of Def.3.1. The convergence is obviously uniform in $a_1 \geq a \geq 1$ for some finite fixed $a_1 > 1$. To show the uniformity in $a \geq a_1$, we consider instead of ψ_0 (3.2.31) the function

$$\chi_0\Big(a^n \frac{v\cdot x}{\delta}\Big) - \chi_0\Big(\frac{v\cdot x}{\delta}\Big) = \sum_{j=0}^{n-1} \psi_0\Big(a^j \frac{x}{\delta}\Big).$$

Since $d_0(x)$ in (3.2.27) is homogeneous of order ω, it follows

$$\Big\langle d_0, \sum_{j=0}^{n-1} \psi_0\Big(a^j\frac{x}{\delta}\Big)\psi_1\Big(a^j\frac{x}{\delta}\Big)\Big\rangle =$$

$$= \Big(\sum_{j=0}^{n-1} a^{j\omega}\Big)\Big\langle d_0,\, \psi_0\Big(\frac{x}{\delta}\Big)\psi_1\Big(\frac{x}{\delta}\Big)\Big\rangle.$$

The sum is bounded for $n \to \infty$ which proves the uniformity in $a \geq a_1$.

The same construction can be carried through with $\chi_0(-t)$ which leads to the advanced distribution

$$\lim_{\delta\to 0} \chi_0\Big(-\frac{v\cdot x}{\delta}\Big) d(x) \stackrel{\text{def}}{=} -a(x). \tag{3.2.33}$$

Furthermore, using $\tilde{\chi}_0(t) = 1 - \chi_0(-t)$ instead of $\chi_0(t)$, we get another retarded distribution $\tilde{r}(x)$. In fact, any point x with $v \cdot x < 0$ is cut off from $\operatorname{supp} \tilde{r}$ for sufficiently small δ. r and $\tilde{r}$ agree with d on $\Gamma^+ \setminus \{0\}$. Then $\tilde{r} - r$ must be a tempered distribution with point support $x = 0$:

$$\tilde{r} - r = \sum_a C_a D^a \delta(x). \tag{3.2.34}$$

It follows from example 2) above that $|a| \leq \omega$. For $\omega < 0$, which is the case here, all C_a must vanish, hence, $\tilde{r} = r$. Since

$$d = \tilde{r} - a = r - a, \tag{3.2.35}$$

we have a solution of the splitting problem. It follows as in (3.2.34) that this solution is unique, in particular, independent of the time-like vector v in (3.2.30) and independent of the choice of χ_0.

By construction we can apply r and a on discontinuous test functions

$$\langle r, \varphi\rangle = \langle r(x),\, \Theta(v\cdot x)\varphi(x)\rangle, \quad \langle r,\, (1-\Theta)\varphi\rangle = 0$$

$$\langle a, \varphi\rangle = \langle a,\, (1-\Theta)\varphi\rangle, \quad \langle a, \Theta\varphi\rangle = 0. \tag{3.2.36}$$

This allows us to extend d (3.2.35) to such test functions also:

$$\langle d,\, \Theta\varphi\rangle = \langle r,\, \Theta\varphi\rangle - \langle a,\, \Theta\varphi\rangle = \langle r,\, \varphi\rangle$$

$$\langle d,\, (1-\Theta)\varphi\rangle = -\langle a,\, \varphi\rangle. \tag{3.2.37}$$

The same is true for the quasi-asymptotics (3.2.6)

$$\langle d_0,\, \varphi\rangle = \lim_{\delta\to 0} \rho(\delta)\delta^m \langle d(\delta x),\, \varphi\rangle.$$

Choosing here $\Theta\varphi$ instead of φ it follows that

$$\langle d_0,\, \Theta\varphi\rangle = \lim_{\delta\to 0} \rho(\delta)\delta^m \langle r(\delta x),\, \varphi\rangle. \tag{3.2.38}$$

This shows that r and similarly a have the same singular order ω as d. We see in (3.2.37) that the case $\omega(d) < 0$ is the case of trivial splitting by multiplication with Θ-function.

The non-trivial case is

b) $\omega \geq 0$: Now the power-counting function satisfies

$$\frac{\rho(\delta)}{\delta^{\omega+1}} \to \infty \quad \text{for} \quad \delta \to 0. \tag{3.2.39}$$

To get a vanishing scaling limit as in (3.2.27) we choose a multi-index b with $|b| = \omega + 1$ and consider

$$\langle d(x)x^b, \, \psi(\frac{x}{\delta})\rangle = \langle d(\delta y)y^b, \psi(y)\rangle \delta^{m+\omega+1}$$

$$\to \langle d_0(y), \, y^b\psi\rangle \frac{\delta^{\omega+1}}{\rho(\delta)} \to 0. \tag{3.2.40}$$

If ω is not integer (which seems not to occur in Q.E.D.), we use the largest integer $[\omega] < \omega$ instead of ω. It follows that the splitting as in case a) is possible if the test function φ satisfies

$$D^a\varphi(0) = 0 \quad \text{for} \quad |a| \leq \omega. \tag{3.2.41}$$

To achieve that, we introduce an auxiliary function $w(x) \in \mathcal{S}(\mathbb{R}^{\mathbf{m}})$ with

$$w(0) = 1\,, \; D^a w(0) = 0 \quad \text{for} \quad 1 \leq |a| \leq \omega, \tag{3.2.42}$$

and define

$$(W\varphi)(x) \stackrel{\text{def}}{=} \varphi(x) - w(x) \sum_{|a|=0}^{\omega} \frac{x^a}{a!}(D^a\varphi)(0) \tag{3.2.43}$$

$$= \sum_{|b|=\omega+1} x^b \psi_b(x).$$

Now the decomposition according to a) (3.2.37) is possible

$$\langle r(x), \, \varphi\rangle \stackrel{\text{def}}{=} \langle d, \, \Theta(v \cdot x)W\varphi\rangle, \tag{3.2.44}$$

$$a(x) = r - d.$$

After construction $r(x)$ defines a tempered distribution with $\operatorname{supp} r \subseteq \Gamma^+(0)$. It agrees with $d(x)$ on $\Gamma^+(0) \setminus \{0\}$ in the sense of distributions, because a test function $\varphi \in \mathcal{S}$ with $\operatorname{supp} \varphi \subset \Gamma^+(0) \setminus \{0\}$ vanishes at $x = 0$, together with all its derivatives, so that the additional subtracted terms in (3.2.43) are 0. But without these terms, there is no splitting of $d(x)$ which makes sense for arbitrary $\varphi \in \mathcal{S}$, because the limit (3.2.30) exists on subtracted test functions only. *If one does the splitting incorrectly by simple multiplication with $\Theta(v \cdot x)$ as in a), one is punished by the well-known ultraviolet divergences in field theory.*

Again we have

$$\omega(r) = \omega(d) = \omega(a), \tag{3.2.45}$$

This is a direct consequence of the definitions (3.2.43-44), because the limit

$$\lim_{\delta\to 0} \rho(\delta)\Big\langle r(x), \, \varphi\Big(\frac{x}{\delta}\Big)\Big\rangle = \lim_{\delta\to 0} \rho(\delta)\Big\langle d(x), \, \Theta W\Big(\varphi\Big(\frac{x}{\delta}\Big)\Big)\Big\rangle$$

$$= \lim_{\delta\to 0} \rho(\delta)\langle d(x), (\Theta W\varphi)(\frac{x}{\delta})\rangle = \langle d_0(x), (\Theta W\varphi)(x)\rangle$$

exists with the same power counting function as $d(x)$. But in sharp contrast to case a), the splitting b) is not unique. If $\tilde{r}(x)$ is the retarded part of another decomposition, then the difference

$$\tilde{r} - r = \sum_{|a|=0}^{\omega} \tilde{C}_a D^a \delta(x) \tag{3.2.46}$$

is again a distribution with point support. Since $\omega > 0$, this time the splitting is only determined up to a finite sum of local terms (3.2.46). These undetermined local terms are not fixed by causality, additional physical normalization conditions are necessary to fix them.

All explicit calculations in QED are best done in momentum space. For this reason, we must investigate the splitting procedure in p-space. We need the distributional Fourier transforms

$$F^{-1}[\Theta(v\cdot x)] \stackrel{\text{def}}{=} \check{\chi}(k) \tag{3.2.47}$$

$$F^{-1}[x^a w](p) = (iD_p)^a \check{w}(p). \tag{3.2.48}$$

Since

$$(D^a\varphi)(0) = \langle (-)^a D^a\delta, \varphi\rangle = (-)^a \langle \widehat{D^a\delta}, \check{\varphi}\rangle$$

$$= (-)^a (2\pi)^{-m/2}\langle (-ip)^a, \check{\varphi}\rangle = (2\pi)^{-m/2}\langle (ip)^a, \check{\varphi}\rangle, \tag{3.2.49}$$

we conclude from (3.2.44) that

$$\langle \hat{r}, \check{\varphi}\rangle = \langle \hat{d}, (\Theta W\varphi)\check{}\,\rangle = (2\pi)^{-m/2}\Big\langle \hat{d}, \check{\chi}$$

$$* \Big[\check{\varphi} - \sum_{|a|=0}^{\omega} \frac{1}{a!}(iD_p)^a \check{w}(p)(2\pi)^{-m/2}\langle (ip')^a, \check{\varphi}\rangle\Big]\Big\rangle_p \tag{3.2.50}$$

$$= (2\pi)^{-m/2}\Big\langle \hat{\chi} * \hat{d}, \check{\varphi} - \sum_{|a|=0}^{\omega} \ldots \Big\rangle, \tag{3.2.51}$$

where the asterisk means convolution. We stress the fact that the convolution $\hat{\chi}*\hat{d}$ is only defined on subtracted test functions, not on $\check{\varphi}$ alone. Interchanging p' and p in the subtraction terms, we may write

$$\langle \hat{r}, \check{\varphi}\rangle = (2\pi)^{-m/2} \int dk\, \hat{\chi}(k)\Big\langle \hat{d}(p-k)$$

$$-(2\pi)^{-m/2} \sum_a \frac{(-)^a}{a!} p^a \int dp'\, \hat{d}(p'-k) D^a_{p'} \check{w}(p'), \check{\varphi}\Big\rangle_p . \tag{3.2.52}$$

After partial integration in the p'-integral this is equivalent to the following result for the retarded distribution

$$\hat{r}(p) = (2\pi)^{-m/2} \int dk\, \hat{\chi}(k) \Big[\hat{d}(p-k)$$

$$-(2\pi)^{-m/2} \sum_{|a|=0}^{\omega} \frac{p^a}{a!} \int dp' \, \big(D^a_{p'} \hat{d}(p'-k)\big) \check{w}(p') \Big]. \tag{3.2.53}$$

Here the k-integral is understood in the sense of distributions as in (3.2.52).

By Fourier transformation of (3.2.46) we see that $\hat{r}(p)$ is only determined up to a polynomial in p of degree ω. Consequently the general result for the retarded distribution reads

$$\tilde{r}(p) = \hat{r}(p) + \sum_{|a|=0}^{\omega} C_a p^a \tag{3.2.54}$$

with $\hat{r}(p)$ given by (3.2.53). We now assume that there exists a point $q \in \mathbb{R}^{\mathbf{m}}$ where the derivatives $D^b \hat{r}(q)$ exist in the usual sense of functions for all $|b| \leq \omega$. Let us define

$$\hat{r}_q(p) = \hat{r}(p) - \sum_{|b|=0}^{\omega} \frac{(p-q)^b}{b!} D^b \hat{r}(q). \tag{3.2.55}$$

This is another retarded distribution because we have only added a polynomial in p of degree ω. Furthermore, this solution of the splitting problem is *uniquely* specified by the normalization condition

$$D^b \hat{r}_q(q) = 0, \quad |b| \leq \omega. \tag{3.2.56}$$

We compute

$$D^b \hat{r}(q) = (2\pi)^{-m/2} \int dk\, \hat{\chi}(k) \Big[\big(D^b \hat{d}\big)(q-k)$$

$$-(2\pi)^{-m/2} \sum_{b \leq a} \frac{a! q^{a-b}}{(a-b)! a!} \int dp' \, \check{w}(p') D^a_{p'} \hat{d}(p'-k) \Big] \tag{3.2.57}$$

from (3.2.53) and substitute this into (3.2.55). Since

$$\sum_{b \leq a} \frac{(p-q)^b}{b!} \frac{q^{a-b}}{(a-b)!} = \frac{1}{a!} \sum_{b \leq a} \binom{a}{b} (p-q)^b q^{a-b} = \frac{p^a}{a!},$$

the subtracted terms in (3.2.53) drop out

$$\hat{r}_q(p) = (2\pi)^{-m/2} \int dk\, \hat{\chi}(k) \Big[\hat{d}(p-k) - \sum_{|b|=0}^{\omega} \frac{(p-q)^b}{b!} \big(D^b \hat{d}\big)(q-k) \Big]. \tag{3.2.58}$$

This is the splitting solution with normalization point q. It is uniquely specified by (3.2.56), that means it does not depend on the time-like vector v in (3.2.47). The subtracted terms are the beginning of the Taylor series at $p = q$.

This is an ultraviolet "regularization" in the usual terminology. It should be stressed, however, that here this is a consequence of the causal distribution splitting and not an ad hoc recipe.

In Sect. 2.8 it was found that causality is expressed in momentum space by dispersion relations. There must, therefore, exist a connection of the result (3.2.58) with dispersion relations. We take $q=0$ in (3.2.58), which is possible in QED with massive fermions, and consider time-like $p \in \Gamma^+$. We choose a special coordinate system such that $p=(p_1^0, \mathbf{0}, 0, \ldots)$. Note that this coordinate system is not obtained by a Lorentz transformation from the original one, but by an orthogonal transformation in $\mathbb{R}^{4m}$. Furthermore we take v parallel to p, i.e. $v=(1,\mathbf{0},0,\ldots)$. Then v varies with p, but this is admissible because (3.2.58) is actually independent of v. We now have $\Theta(v\cdot x)=\Theta(x_1^0)$ and the Fourier transform (3.2.47) is given by

$$\hat{\chi}(k)=(2\pi)^{m/2-1}\delta(\mathbf{k}_1, k_2, \ldots k_m)\frac{i}{k_1^0+i0}. \tag{3.2.59}$$

Using this in (3.2.58) we shall obtain

$$\hat{r}_0(p_1^0)=\frac{i}{2\pi}\int\limits_{-\infty}^{+\infty} dk_1^0\,\frac{1}{k_1^0+i0}\Bigg[\hat{d}(p_1^0-k_1^0,0,\ldots)$$

$$-\sum_{a=0}^{\omega}\frac{(p_1^0)^a}{a!}(-)^a D^a_{k_1^0}\hat{d}(q_1^0-k_1^0,0,\ldots)\bigg|_{q_1^0=0}\Bigg].$$

Integrating the last term by parts and using $k_1^0-p_1^0=k_0'$ as a new integration variable in the first term, we arrive at

$$\hat{r}_0(p_1^0)=\frac{i}{2\pi}\int\limits_{-\infty}^{+\infty} dk_0'\left[\frac{1}{p_1^0+k_0'+i0}-\sum_{a=0}^{\omega}\frac{(p_1^0)^a}{a!}\frac{\partial^a}{\partial k_0'^a}\frac{1}{k_0'+i0}\right]\hat{d}(-k_0').$$

The square bracket can be easily computed

$$\Big[\ldots\Big]=\left(-\frac{p_1^0}{k_0'+i0}\right)^{\omega+1}\frac{1}{p_1^0+k_0'+i0}\quad .$$

Changing the variable of integration from k_0' to $-k_0$, we obtain the following result

$$\hat{r}_0(p_1^0)=\frac{i}{2\pi}(p_1^0)^{\omega+1}\int\limits_{-\infty}^{+\infty} dk_0\,\frac{\hat{d}(k_0)}{(k_0-i0)^{\omega+1}(p_1^0-k_0+i0)}. \tag{3.2.60}$$

This is a subtracted dispersion relation like (2.8.52). We shall often apply it to the case of one four-momentum $p\in\mathbb{R}^4$. To write down the result for arbitrary $p\in\Gamma^+$, we use the variable of integration $t=k_0/p_1^0$ and arrive at

$$\hat{r}_0(p) = \frac{i}{2\pi} \int\limits_{-\infty}^{+\infty} dt \frac{\hat{d}(tp)}{(t - i0)^{\omega+1}(1 - t + i0)}. \tag{3.2.61}$$

For later reference we call this the central splitting solution, because it is normalized at the origin ($q = 0$ in (3.2.56)). The latter fact has two important consequences. (i) The central splitting solution does not introduce a mass scale into the theory. If $q \neq 0$, then $|q^2| = M^2$ defines such a scale. (ii) Most symmetry properties of $\hat{d}(p)$ are preserved under central splitting, as we will see later, because the origin $q = 0$ is a very symmetric point.

It is easy to verify that the dispersion integral (3.2.61) is convergent for $|t| \to \infty$. In fact, it follows from (3.2.12) that

$$\hat{d}(tp) \longrightarrow \frac{\hat{d}_0(p)}{\rho(1/t)}, \quad |t| \to \infty,$$

and since the power counting function ρ is bounded by

$$\frac{1}{\rho(1/t)} < \frac{|t|^{\omega+\varepsilon}}{C(\varepsilon)} \quad , \quad \varepsilon < 1,$$

the integral is absolutely convergent after smearing out with a test function $\varphi(p)$. But it would be ultraviolet divergent, if ω in (3.2.61) is chosen too small. Consequently, *the correct distribution splitting with the right singular order ω is terribly important. Incorrect distribution splitting leads to ultraviolet divergences.* This is the origin of the ultraviolet problem in field theory. Because of its central importance we want to illustrate this point with a classical 'comics' by Wilhelm Busch (Fig. 3, next page). The pictures (a) and (b) correspond, in a sense, to the cases a) and b) discussed above.

The dispersion relation (3.2.52) is only valid for $p \in \Gamma^+$. For arbitrary p we may either return to (3.2.49)

$$\hat{r}_q(p) = \frac{i}{2\pi} \int\limits_{-\infty}^{+\infty} \frac{dt}{t + i0} \Big[\hat{d}(p - tv) - \sum_{|a|=0}^{\omega} \frac{p^a}{a!} \big(D_q^a \hat{d}\big)(q - tv)\Big], \tag{3.2.62}$$

where $v \in \Gamma^+$ is fixed, or we may determine $\hat{r}(p)$ by analytic continuation. The latter method is based on the following well-known theorem (see e.g. *M. Reed, B. Simon, Methods of Modern Mathematical Physics, Vol. 2 (1978), p. 23 and Problem 23, p. 124*).

Theorem 3.4. The retarded distribution $\hat{r}(p)$ is the boundary value of an analytic function, regular in $\mathbb{R}^m + i\,\Gamma^+$.

We give the main idea of the proof in the simple case where $r(x)$ is an L^1-function. Then the ordinary Fourier integral

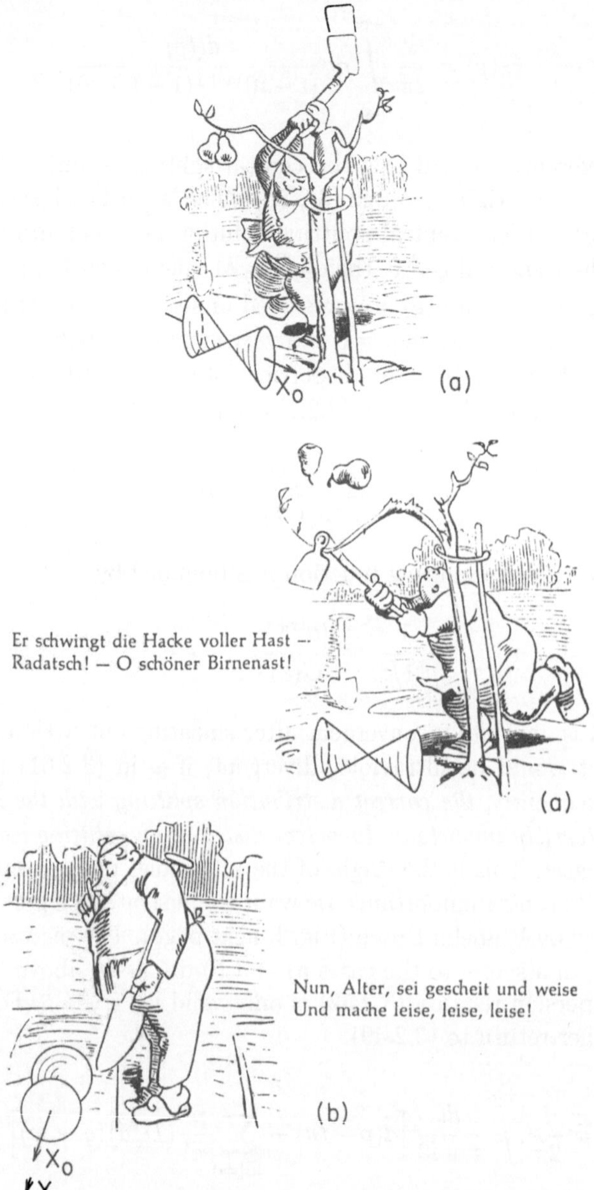

Fig. 3. Hasty (a) versus careful splitting (b) of a causal distribution. (Taken from: Humoristischer Hausschatz von Wilhelm Busch, Bassermann, München 1908)

$$\hat{r}(p) = (2\pi)^{-m/2} \int r(x)\, e^{ipx}\, d^m x \qquad (3.2.63)$$

is valid. Here $\operatorname{supp} r(x) \subseteq \Gamma^+$. Since $p \in \Gamma^+$ implies $p \cdot x \geq 0$ for all $x \in \Gamma^+$, it follows that

$$\operatorname{Re}(ipx) < 0 \quad \text{if} \quad p \in \mathbb{R}^{\mathbf{m}} + i\Gamma^+. \qquad (3.2.64)$$

Consequently, the integral (3.2.63) is exponentially convergent and, hence, analytic in p. Theorem 3.4 allows to continue $\hat{r}(p)$ from time-like to space-like p.

Finally we want to determine the analytic function in the tube $\Gamma^+ + i\Gamma^+$ from the dispersion integral (3.2.61). We take $p = \lambda e$, $e \in \Gamma^+$, $\lambda > 0$

$$\hat{r}_0(\lambda e) = \frac{i}{2\pi} \int\limits_{-\infty}^{+\infty} dt \, \frac{\hat{d}(t\lambda e)}{(t - i0)^{\omega+1}(1 - t + i0)},$$

and use $t\lambda = s$ as a new variable of integration

$$\hat{r}_0(\lambda e) = \frac{i}{2\pi} \lambda^{\omega+1} \int\limits_{-\infty}^{+\infty} ds \, \frac{\hat{d}(se)}{(s - i0)^{\omega+1}(\lambda - s + i0)}. \tag{3.2.65}$$

Here λ can be chosen complex with $\mathrm{Im}\,\lambda > 0$. In this case the $i0$ of the second factor in the denominator is superfluous.

3.3 Application to QED

The inductive construction is started by specifying the first order of the perturbation series. For QED, it is given by

$$T_1(x) = i\,e\, :\overline{\psi}(x)\gamma^\mu\psi(x):\, A_\mu(x) = -\tilde{T}_1(x), \tag{3.3.1}$$

where the last equality follows from (3.1.6). The coupling constant $e > 0$ is the unit of charge, another ("bare") charge appears nowhere. The field operators in (3.3.1) are free fields, the normal ordering is necessary to have a well-defined expression for the product of field operators at the same point (see Sect. 4.2)

$$j^\mu(x) = e\, :\overline{\psi}(x)\gamma^\mu\psi(x):\,. \tag{3.3.2}$$

Interacting fields are not used in the causal construction, but they can be derived from the S-matrix (see Sect. 4.9).

For going from $n = 1$ to $n = 2$, we first form

$$A_2'(x_1, x_2) = -T_1(x_1)T_1(x_2) \tag{3.3.3}$$

and

$$R_2'(x_1, x_2) = -T_1(x_2)T_1(x_1), \tag{3.3.4}$$

according to (3.1.31, 32), and then

$$D_2(x_1, x_2) = R_2' - A_2' = T_1(x_1)T_1(x_2) - T_1(x_2)T_1(x_1). \tag{3.3.5}$$

Here we will normally order the products, this is the technique which is now well known to the reader from Chap. 2. Normal ordering of various different field operators is best done by means of the

Theorem of Wick 3.1. A product of n field operators is normally ordered as follows:

$$A_1A_2\ldots A_n =: A_1A_2\ldots A_n : + : \overline{A_1A_2}\ldots A_n : + \text{permutations}$$

$$+ : \overline{A_1\overline{A_2\ldots A_j}}\ldots A_n : + \ldots + \overline{A_1A_2}\,\overline{A_3A_4}\ldots$$

$$+ \text{permutations,} \tag{3.3.6}$$

where the sum contains all normal products with all possible contractions (pairings).

The contractions are commutators (in case of Bose fields) or anticommutators (in case of Fermi fields) between the absorption and emission parts

$$\overline{A_jA_k} \stackrel{\text{def}}{=} [A_j^{(-)},\, A_k^{(+)}]_{\pm}. \tag{3.3.7}$$

Remember $A = A^{(-)} + A^{(+)}$, where $A^{(-)}$ contains absorption and $A^{(+)}$ emission operators, only. The contractions are assumed to be C-numbers. Therefore, they can be taken out of the normal products, but in the case of Fermi operators, the sign of the necessary permutation must be taken into account:

$$: A_1\ldots \overline{A_j\ldots A_k}\ldots A_n := (-)^P\, \overline{A_jA_k} : A_1\ldots \not{A}_j\ldots \not{A}_k\ldots A_n :, \tag{3.3.8}$$

where P is the permutation

$$P: \quad 1,\ldots n \to j,k,1\ldots \not{j}\ldots \not{k}\ldots n. \tag{3.3.9}$$

The proof of (3.3.6) is nothing but doing just the normal ordering: All absorption operators are commuted or anticommuted to the right. A detailed inductive proof is less evident than the theorem itself (see Problems 3.3, 3.4).

In QED, only the following three contractions appear:

$$\overline{\psi_a(x)\overline{\psi}_b(y)} \stackrel{\text{def}}{=} \{\psi_a^{(-)}(x),\, \overline{\psi}_b^{(+)}(y)\} = \frac{1}{i}S_{ab}^{(+)}(x-y) \tag{3.3.10}$$

$$\overline{\overline{\psi}_a(x)\psi_b(y)} \stackrel{\text{def}}{=} \{\overline{\psi}_a^{(-)}(x),\, \psi_b^{(+)}(y)\} = \frac{1}{i}S_{ba}^{(-)}(y-x) \tag{3.3.11}$$

$$\overline{A_\mu(x)A_\nu(y)} \stackrel{\text{def}}{=} [A_\mu^{(-)}(x),\, A_\nu^{(+)}(y)] = g_{\mu\nu}iD_0^{(+)}(x-y). \tag{3.3.12}$$

These results follow from (2.2.37, 38) and (2.11.23).

We are now ready to calculate the advanced function (3.3.3)

$$A_2'(x_1,x_2) = e^2\gamma_{ab}^{\mu}\gamma_{cd}^{\nu} : \overline{\psi}_a(x_1)\psi_b(x_1) :: \overline{\psi}_c(x_2)\psi_d(x_2) :$$

$$\times A_\mu(x_1)A_\nu(x_2). \tag{3.3.13}$$

Because of the normal orderings, only contractions between ψ and $\overline{\psi}$ with different coordinates must be taken into account:

$$= e^2\gamma^\mu_{ab}\gamma^\nu_{cd}\Big[:\overline{\psi}_a(x_1)\psi_b(x_1)\overline{\psi}_c(x_2)\psi_d(x_2):$$

$$+ :\overline{\psi}_a(x_1)\psi_d(x_2): \frac{1}{i}S^{(+)}_{bc}(x_1-x_2) + :\psi_b(x_1)\overline{\psi}_c(x_2): \frac{1}{i}S^{(-)}_{da}(x_2-x_1)$$

$$-S^{(+)}_{bc}(x_1-x_2)S^{(-)}_{da}(x_2-x_1)\Big]$$

$$\times\Big[:A_\mu(x_1)A_\nu(x_2): +g_{\mu\nu}iD^{(+)}_0(x_1-x_2)\Big]. \tag{3.3.14}$$

The retarded function (3.3.4) is obtained by interchanging $x_1 \longleftrightarrow x_2$ and by replacing the indices of summation $\mu \longleftrightarrow \nu$, $a \longleftrightarrow c$, $b \longleftrightarrow d$:

$$R'_2(x_1,x_2) = e^2\gamma^\mu_{ab}\gamma^\nu_{cd}\Big[:\overline{\psi}_a(x_1)\psi_b(x_1)\overline{\psi}_c(x_2)\psi_d(x_2):$$

$$- :\psi_b(x_1)\overline{\psi}_c(x_2): \frac{1}{i}S^{(+)}_{da}(x_2-x_1) - :\overline{\psi}_a(x_1)\psi_d(x_2): \frac{1}{i}S^{(-)}_{bc}(x_1-x_2)$$

$$-S^{(+)}_{da}(x_2-x_1)S^{(-)}_{bc}(x_1-x_2)\Big]$$

$$\times\Big[:A_\mu(x_1)A_\nu(x_2): +g_{\mu\nu}iD^{(+)}_0(x_2-x_1)\Big]. \tag{3.3.15}$$

From these expressions, we obtain $D_2(x_1,x_2)$ (3.3.5)

$$D_2(x_1,x_2) = R'_2 - A'_2 = e^2\gamma^\mu_{ab}\gamma^\nu_{cd} \tag{3.3.16}$$

$$\Big\{:\overline{\psi}_a(x_1)\psi_b(x_1)\overline{\psi}_c(x_2)\psi_d(x_2): g_{\mu\nu}i\Big[D^{(+)}_0(x_2-x_1) - D^{(+)}_0(x_1-x_2)\Big] \quad (1)$$

$$- :\psi_b(x_1)\overline{\psi}_c(x_2)::A_\mu(x_1)A_\nu(x_2): \frac{1}{i}\Big[S^{(+)}_{da}(x_2-x_1) + S^{(-)}_{da}(x_2-x_1)\Big] \quad (2)$$

$$- :\psi_b(x_1)\overline{\psi}_c(x_2): \Big[S^{(+)}_{da}(x_2-x_1)D^{(+)}_0(x_2-x_1)$$

$$+S^{(-)}_{da}(x_2-x_1)D^{(+)}_0(x_1-x_2)\Big]g_{\mu\nu} \quad (3)$$

$$- :\overline{\psi}_a(x_1)\psi_d(x_2)::A_\mu(x_1)A_\nu(x_2): \frac{1}{i}\Big[S^{(-)}_{bc}(x_1-x_2) + S^{(+)}_{bc}(x_1-x_2)\Big] \quad (4)$$

$$- :\overline{\psi}_a(x_1)\psi_d(x_2): \Big[S^{(-)}_{bc}(x_1-x_2)D^{(+)}_0(x_2-x_1)+$$

$$+S^{(+)}_{bc}(x_1-x_2)D^{(+)}_0(x_1-x_2)\Big]g_{\mu\nu} \quad (5)$$

$$+ :A_\mu(x_1)A_\nu(x_2): \Big[-S^{(-)}_{bc}(x_1-x_2)S^{(+)}_{da}(x_2-x_1)$$

$$+S^{(+)}_{bc}(x_1-x_2)S^{(-)}_{da}(x_2-x_1)\Big] \quad (6)$$

$$+\frac{1}{i}g_{\mu\nu}\Big[S_{bc}^{(-)}(x_1-x_2)S_{da}^{(+)}(x_2-x_1)D_0^{(+)}(x_2-x_1)$$

$$-S_{bc}^{(+)}(x_1-x_2)S_{da}^{(-)}(x_2-x_1)D_0^{(+)}(x_1-x_2)\Big]\Big\}. \qquad (7)$$

This large formula is the starting point for many applications. In the later calculations of S-matrix elements, the field operators in the various terms of (3.3.16) create or annihilate the external particles. The first term (1) describes two incoming and two outgoing electrons (or positrons). This is electron scattering which is considered in the following section. The terms (2) and (4) correspond to electron-photon scattering which is treated in Sect. 3.5. The remaining terms are the interesting loop graphs, because they contain products of propagators S, D. We emphasize the fact that in contrast to ordinary Feynman rules, the products in (3.3.16) are all well-defined distributions. Although we do not have Feynman rules, it is still very convenient to represent the terms in (3.3.16) by graphs, called Feynman diagrams. The graphs are depicted as follows (see Fig. 4): External fermions are represented by normal lines and the photons by wavy lines. The same lines are used for the fermion (S) and photon propagators (D), with the only difference that the propagator lines go from one vertex to another, where two fermion lines and one photon line come together, whereas the external lines have one open end. The graphs corresponding to (1) till (7) are shown in Fig. 4. The terms (3) and (5) describe the so-called self-energy which is discussed in Sect. 3.7. The vacuum polarization (6), which has already been discussed in Sect. 2.10, is considered once more in Sect. 3.6 and the vacuum graph (7) in Sect. 4.1.

It is important to note that normal ordering of the causal distribution D_n does not obscure the causal support property, on the contrary, it fully develops it. In fact, since there is no compensation between terms with different normal products of field operators, those terms must separately have causal supports. Therefore, the distribution splitting can be done for every scattering process separately.

3.4 Electron Scattering (Moeller Scattering)

Electron scattering comes from expression (1) in (3.3.16) or from graph (1) in Fig. 4,

$$D_2^{(1)}(x_1,x_2)=ie^2\ :\overline{\psi}(x_1)\gamma^\mu\psi(x_1)\overline{\psi}(x_2)\gamma_\mu\psi(x_2):$$

$$\times\big[D_0^{(+)}(x_2-x_1)-D_0^{(+)}(x_1-x_2)\big]. \qquad (3.4.1)$$

Since

$$D_0^{(+)}(x_2-x_1)=-D_0^{(-)}(x_1-x_2), \qquad (3.4.2)$$

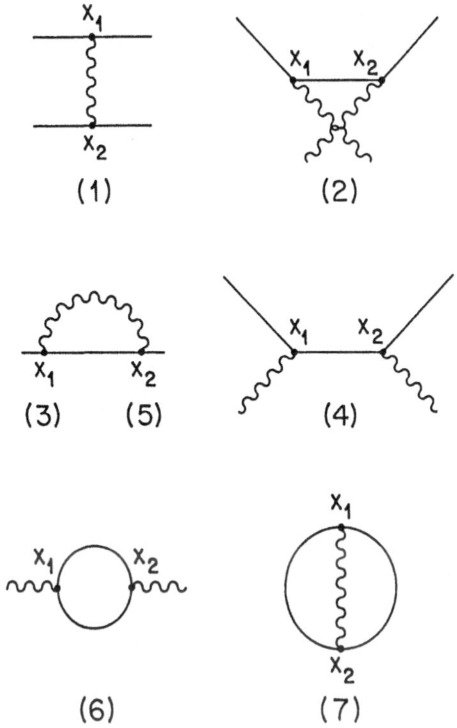

Fig. 4. Second order graphs corresponding to the terms (1–7) in (3.3.16)

the square bracket in (3.4.1) is equal to the Jordan-Pauli function $-D_0(x_1 - x_2)$ for mass 0 (see (2.3.7)). This distribution has a causal support, as required. The singular order is $\omega = -2$ (3.2.9), so that the splitting is trivial (see (3.2.13))

$$D_0(x_1 - x_2) = D_0^{\rm ret}(x_1 - x_2) - D_0^{\rm av}(x_1 - x_2). \tag{3.4.3}$$

We then get the retarded function

$$R_2^{(1)}(x_1, x_2) = -ie^2 \ :\ldots: \ D_0^{\rm ret}(x_1 - x_2). \tag{3.4.4}$$

The first term in the square bracket in (3.4.1) comes from R_2' (3.3.15), therefore, according to (3.1.66), the second order two-point function is given by

$$T_2^{(1)}(x_1, x_2) = R_2^{(1)} - R_2'^{(1)}$$

$$= -ie^2 \ :\ldots: \left[D_0^{\rm ret}(x_1 - x_2) + D_0^{(+)}(x_2 - x_1)\right]$$

$$= -ie^2 \ :\overline{\psi}(x_1)\gamma^\mu\psi(x_1)\overline{\psi}(x_2)\gamma_\mu\psi(x_2): \ D_0^F(x_1 - x_2), \tag{3.4.5}$$

where again (3.4.2) and (3.2.14) have been used. We see that the Feynman propagator describes the photon exchange in a tree graph like (1). More generally, any internal photon line which is not part of a loop is represented by the Feynman propagator D_0^F (see also (5.2.9)). This is due to the trivial distribution splitting. In fact, according to (3.1.68) we have then to consider a contraction in a usual time ordered product

$$T\{\overline{A_\mu(x_1)A_\nu(x_2)}\} = \Theta(x_1^0 - x_2^0)\big[A_\mu^{(-)}(x_1),\, A_\nu^{(+)}(x_2)\big]$$
$$+\Theta(x_2^0 - x_1^0)\big[A_\nu^{(-)}(x_2),\, A_\nu^{(+)}(x_1)\big]$$
$$= g_{\mu\nu} i\{\Theta(x_1^0 - x_2^0)D_0^{(+)}(x_1 - x_2) + \Theta(x_2^0 - x_1^0)D_0^{(+)}(x_2 - x_1)\}$$
$$= g_{\mu\nu} i\, D_0^{\mathrm{F}}(x_1 - x_2). \tag{3.4.6}$$

We now turn to the external fermion lines, that is to say, we want to calculate S-matrix elements $(\psi_f,\, S\psi_i)$ for in- and outgoing electrons

$$\psi_i = b_s^+(\boldsymbol{p})\Omega$$

and similarly for ψ_f. The emission operator in ψ_i is contracted with the absorption operator

$$\psi^{(-)}(x) = (2\pi)^{-3/2} \int d^3p_1\, b_{s_1}(\boldsymbol{p}_1)u_{s_1}(\boldsymbol{p}_1)e^{-ip_1 x}, \tag{3.4.7}$$

appearing in the normal product expansion of S: Anticommuting the absorption operator in (3.4.7) to the right, it gives 0 on the vacuum; from the anticommutator

$$\{b_{s_1}(\boldsymbol{p}_1),\, b_s^+(\boldsymbol{p})\} = \delta_{s_1 s}\delta(\boldsymbol{p}_1 - \boldsymbol{p}) \tag{3.4.8}$$

there remains

$$\psi^{(-)}(x)b_s^+(\boldsymbol{p})\Omega = (2\pi)^{-3/2}u_s(\boldsymbol{p})e^{-ipx}\Omega. \tag{3.4.9}$$

Similarly, the emission operator in the final state ψ_f is contracted with $\overline{\psi}^{(+)}(x)$ in the normal product:

$$(\psi_f,\, \overline{\psi}^{(+)}(x)\ldots = (\Omega,\, b_s(\boldsymbol{p})(2\pi)^{-3/2} \int d^3p_1 b_{s_1}(\boldsymbol{p}_1)^+\overline{u}_{s_1}(\boldsymbol{p}_1)e^{ip_1 x}\ldots$$
$$= (2\pi)^{-3/2}\overline{u}_s(\boldsymbol{p})e^{ipx}(\Omega,\ldots \tag{3.4.10}$$

Since the second order S-matrix (with $g = 1$) is given by

$$S_2 = \frac{1}{2!}\int d^4x_1 d^4x_2 T_2(x_1, x_2), \tag{3.4.11}$$

we have the following expression for the desired matrix element

$$S_{fi} = (b_{s_f}(\boldsymbol{p}_f)^+ b_{\sigma_f}(\boldsymbol{q}_f)^+\Omega,\, S_2^{(1)}\, b_{s_i}(\boldsymbol{p}_i)^+ b_{\sigma_i}(\boldsymbol{q}_i)^+\Omega)$$
$$= -\frac{i}{2}e^2 \int d^4x_1 d^4x_2 D_0^F(x_1 - x_2)\times$$

$$\times\left(\Omega,\, b_{\sigma_f}(\boldsymbol{q}_f) b_{s_f}(\boldsymbol{p}_f) : \overline{\psi}(x_1)\gamma^\mu\psi(x_1)\overline{\psi}(x_2)\gamma_\mu\psi(x_2) : b_{s_i}(\boldsymbol{p}_i)^+\, b_{\sigma_i}(\boldsymbol{q}_i)^+\Omega\right). \tag{3.4.12}$$

The b's must be contracted with $\overline{\psi}$ according to (3.4.10) and the b^+'s with ψ according to (3.4.9). The sign is determined by the permutation which must be carried out in order to bring the contracted Fermi operators together (3.3.8). This sign is easily found by means of the following rule: If the contracted operators are joined by lines, the sign is positive if these lines cross an even number of times; the sign is negative if the lines cross an odd number of times (as in (3.4.12) upstairs). There are four possibilities to contract the operators:

$$S_{fi} = \frac{i}{2}e^2(2\pi)^{-6}\int d^4x_1 d^4x_2\, D_0^F(x_1-x_2)$$
$$\times\big[\overline{u}_{s_f}(\boldsymbol{p}_f)\gamma^\mu u_{\sigma_i}(\boldsymbol{q}_i)\overline{u}_{\sigma_f}(\boldsymbol{q}_f)\gamma_\mu u_{s_i}(\boldsymbol{p}_i)e^{i(p_f-q_i)x_1+i(q_f-p_i)x_2}$$
$$-\overline{u}_{s_f}(\boldsymbol{p}_f)\gamma^\mu u_{s_i}(\boldsymbol{p}_i)\overline{u}_{\sigma_f}(\boldsymbol{q}_f)\gamma_\mu u_{\sigma_i}(\boldsymbol{q}_i)e^{i(p_f-p_i)x_1+i(q_f-q_i)x_2}$$
$$-\overline{u}_{\sigma_f}(\boldsymbol{q}_f)\gamma^\mu u_{\sigma_i}(\boldsymbol{q}_i)\overline{u}_{s_f}(\boldsymbol{p}_f)\gamma_\mu u_{s_i}(\boldsymbol{p}_i)e^{i(q_f-q_i)x_1+i(p_f-p_i)x_2}$$
$$+\overline{u}_{\sigma_f}(\boldsymbol{q}_f)\gamma^\mu u_{s_i}(\boldsymbol{p}_i)\overline{u}_{s_f}(\boldsymbol{p}_f)\gamma_\mu u_{\sigma_i}(\boldsymbol{q}_i)e^{i(q_f-p_i)x_1+i(p_f-q_i)x_2}\big]. \tag{3.4.13}$$

We first integrate over x_1, x_2, observing that the terms in (3.4.13) are equal in pairs. The following distributional Fourier transform must be computed:

$$\int d^4x_1 d^4x_2\, D_0^F(x_1-x_2)e^{iPx_1+iQx_2} = \dots$$

Substituting

$$x_1 = x + \frac{1}{2}y \quad , \quad x_2 = x - \frac{1}{2}y,$$

we get

$$\dots = \int d^4x\, e^{i(P+Q)x}\int d^4y\, D_0^F(y)e^{i(P-Q)y/2}$$
$$= (2\pi)^4\delta(P+Q)\tilde{D}_0^F\Big(\frac{P-Q}{2}\Big) = (2\pi)^4\delta(P+Q)\tilde{D}_0^F(P), \tag{3.4.14}$$

where

$$\tilde{D}_0^F(P) = \int d^4y\, D_0^F(y)e^{iPy} = -\frac{1}{P^2+i0}, \tag{3.4.15}$$

according to (2.3.44). The argument $P+Q$ of the δ-distribution in (3.4.14) is the same for all terms in (3.4.13) and expresses energy-momentum conservation:

$$S_{fi} = \delta(p_f+q_f-p_i-q_i)M \tag{3.4.16}$$

with

$$M = ie^2(2\pi)^{-2}\Big[-\frac{1}{(p_f-q_i)^2+i0}\overline{u}_{s_f}(\boldsymbol{p}_f)\gamma^\mu u_{\sigma_i}(\boldsymbol{q}_i)\overline{u}_{\sigma_f}(\boldsymbol{q}_f)\gamma_\mu u_{s_i}(\boldsymbol{p}_i)+$$

$$+\frac{1}{(p_f-p_i)^2+i0}\overline{u}_{s_f}(\boldsymbol{p}_f)\gamma^\mu u_{s_i}(\boldsymbol{p}_i)\overline{u}_{\sigma_f}(\boldsymbol{q}_f)\gamma_\mu u_{\sigma_i}(\boldsymbol{q}_i)\Big]. \tag{3.4.17}$$

The S-matrix element S_{fi} (3.4.16) is a distribution $S_{fi}(\boldsymbol{p}_i, \boldsymbol{q}_i, \boldsymbol{p}_f, \boldsymbol{q}_f)_{s_i\sigma_i s_f\sigma_f}$. For this reason, it is not possible to compute $|S_{fi}|^2$ directly. As in Sect. 2.6, we must calculate with wave packets

$$\psi_i = \int d^3p_1 d^3q_1\, \Phi_1(\boldsymbol{p}_1)\Phi_2(\boldsymbol{q}_1)b_{s_i}(\boldsymbol{p}_1)^+ b_{\sigma_i}(\boldsymbol{q}_1)^+\Omega, \tag{3.4.18}$$

and similarly for ψ_f. Because of the Pauli principle, Φ_1 must be orthogonal to Φ_2. If both wave functions are normalized, then (3.4.18) is also normalized. We now have

$$S_{fi} = \int d^3p_1 d^3q_1 d^3p_2 d^3q_2\, \psi_f(\boldsymbol{p}_2, \boldsymbol{q}_2)^* S_{fi}(\boldsymbol{p}_1, \boldsymbol{q}_1, \boldsymbol{p}_2, \boldsymbol{q}_2)_{s_i\sigma_i s_f\sigma_f}$$
$$\times \Phi_1(\boldsymbol{p}_1)\Phi_2(\boldsymbol{q}_1), \tag{3.4.19}$$

and the transition probability makes sense

$$p_{fi} \stackrel{\text{def}}{=} |S_{fi}|^2. \tag{3.4.20}$$

The following steps are the same as in Sect. 2.6 (2.6.21): Summing over a complete system of two-particle final states by means of the completeness relation

$$\sum_f \psi_f(\boldsymbol{p}_2, \boldsymbol{q}_2)^*\psi_f(\boldsymbol{p}_2', \boldsymbol{q}_2') = \delta(\boldsymbol{p}_2-\boldsymbol{p}_2')\delta(\boldsymbol{q}_2-\boldsymbol{q}_2'),$$

we shall obtain

$$\sum_f p_{fi} = \sum_{s_f\sigma_f}\int d^3p_1 d^3q_1 d^3p_1' d^3q_1' d^3p_2 d^3q_2$$
$$\times M_{s_i\sigma_i s_f\sigma_f}(p_1, q_1, p_2, q_2)\delta(p_1+q_1-p_2-q_2)$$
$$\times M_{s_i\sigma_i s_f\sigma_f}(p_1', q_1', p_2, q_2)^*\delta(p_1'+q_1'-p_2-q_2)$$
$$\times \Phi_1(\boldsymbol{p}_1)\Phi_2(\boldsymbol{q}_1)\Phi_1(\boldsymbol{p}_1')^*\Phi_2(\boldsymbol{q}_1')^*. \tag{3.4.21}$$

Assuming the wave functions Φ_1, Φ_2 to be sharply peaked around $\boldsymbol{p}_i$, $\boldsymbol{q}_i$, respectively, compared with the distance of variation of M, (3.4.21) may be simplified as follows

$$\sum_f p_{fi} = \sum_{s_f\sigma_f}\int d^3p_2 d^3q_2\, |M_{s_i\sigma_i s_f\sigma_f}(\boldsymbol{p}_i, \boldsymbol{q}_i, \boldsymbol{p}_2, \boldsymbol{q}_2)|^2$$
$$\times \int d^3p_1 d^3q_1 d^3p_1' d^3q_1' \delta(p_1+q_1-p_2-q_2)\delta(p_1'+q_1'-p_2-q_2)$$
$$\times \Phi_1(\boldsymbol{p}_1)\Phi_2(\boldsymbol{q}_1)\Phi_1(\boldsymbol{p}_1')^*\Phi_2(\boldsymbol{q}_1')^*. \tag{3.4.22}$$

The second integral in (3.4.22) only depends on the initial state. Denoting it by $F(p)$, $p_2 + q_2 = p$, we proceed similarly to (2.6.26): Substituting

$$\delta(p) = (2\pi)^{-4} \int e^{\pm ipx} d^4x,$$

we write the last integral in (3.4.22) in the form

$$F(p) = (2\pi)^{-8} \int d^4x_1 d^4x_2 \int d^3p_1 \dots d^3q_1' e^{-i(p_1+q_1-p)x_1}$$

$$\times e^{i(p_1'+q_1'-p)x_2} \Phi_1(\boldsymbol{p}_1)\Phi_2(\boldsymbol{q}_1)\Phi_1(\boldsymbol{p}_1')^* \Phi_2(\boldsymbol{q}_1')^*. \tag{3.4.23}$$

Let

$$\tilde{\Phi}(x) = (2\pi)^{-3/2} \int d^3p \, e^{-ipx} \Phi(\boldsymbol{p}) \tag{3.4.24}$$

be the free wave packet in x-space, then (omitting the tilde in the following)

$$F(p) = (2\pi)^{-2} \int d^4x_1 d^4x_2 \, \Phi_1(x_1)\Phi_2(x_1)\Phi_1(x_2)^*\Phi_2(x_2)^* e^{ip(x_1-x_2)}. \tag{3.4.25}$$

The positive function $F(p)$ is normalized according to

$$\int d^4p \, F(p) = (2\pi)^2 \int d^4x \, |\Phi_1(x)|^2 |\Phi_2(x)|^2, \tag{3.4.26}$$

and it is concentrated around $p = p_2 + q_2 = p_1 + q_1 \approx p_i + q_i$. In the limit of infinitely sharp wave packets we, therefore, may represent it by

$$F(p) = \delta(p - p_i - q_i)(2\pi)^2 \int d^4x \, |\Phi_1(x)|^2 |\Phi_2(x)|^2. \tag{3.4.27}$$

Then we get the following result for (3.4.22)

$$\sum_f p_{fi} = \sum_{s_f \sigma_f} \int d^3p_2 d^3q_2 \, |M|^2 \delta(p_2 + q_2 - p_i - q_i)$$

$$\times (2\pi)^2 \int d^4x \, |\Phi_1(x)|^2 |\Phi_2(x)|^2. \tag{3.4.28}$$

The spreading of the wave packets in the course of time can be neglected, if they are sharply concentrated in momentum space. The free wave packet is then shifted with the velocity $\boldsymbol{v}$ of the particle without change of the shape

$$\Phi_1(t, \boldsymbol{x}) = \varphi_1(\boldsymbol{x} + \boldsymbol{x}_1 + \boldsymbol{v}t). \tag{3.4.29}$$

Particle 2 is assumed to be the target which is at rest

$$\Phi_2(x) = \varphi_2(\boldsymbol{x}). \tag{3.4.30}$$

Scattering is only possible, if we consider a beam of incoming particles. Therefore, the expression (3.4.28) must be averaged over a cylinder of radius R parallel to $\boldsymbol{v}$

$$\sum_f p_{fi}(R) = \frac{1}{\pi R^2} \int\limits_{|\boldsymbol{x}_{1\perp}| \le R} d^2x_{1\perp} \int d^4x \, |\varphi_1(\boldsymbol{x}+\boldsymbol{x}_1+\boldsymbol{v}t)|^2 |\varphi_2(\boldsymbol{x})|^2 \ldots . \tag{3.4.31}$$

The scattering cross section in the laboratory frame is then given by

$$\sigma = \lim_{R\to\infty} \pi R^2 \sum_f p_{fi}(R). \tag{3.4.32}$$

Since

$$\int d^2x_{1\perp} \int dt \int d^3x \, |\varphi_1(\boldsymbol{x}+\boldsymbol{x}_1+\boldsymbol{v}t)|^2 |\varphi_2(\boldsymbol{x})|^2$$

$$= \frac{1}{|\boldsymbol{v}|} \int d^3x_1 \int d^3x \, |\varphi_1(\boldsymbol{x}+\boldsymbol{x}_1)|^2 |\varphi_2(\boldsymbol{x})|^2 = \frac{1}{|\boldsymbol{v}|} = \frac{E_i}{|\boldsymbol{p}_i|},$$

we arrive at

$$\sigma = (2\pi)^2 \frac{E_i}{|\boldsymbol{p}_i|} \sum_{s_f \sigma_f} \int d^3p_2 d^3q_2 \, \delta(p_2+q_2-p_i-q_i) |M|^2. \tag{3.4.33}$$

This result can be written in Lorentz invariant form by substituting

$$|\boldsymbol{p}_i| = \sqrt{E_i^2 - m_1^2} = \sqrt{(p_iq_i)^2 \frac{1}{m_2^2} - m_1^2}$$

$$= \frac{1}{m_2} \sqrt{(p_iq_i)^2 - m_1^2 m_2^2},$$

and $m_2 = E(q_i)$:

$$\sigma = (2\pi)^2 \frac{E(p_i)E(q_i)}{\sqrt{(p_iq_i)^2 - m_1^2 m_2^2}} \sum_{s_f \sigma_f} \int d^3p_f d^3q_f$$

$$\times \, \delta(p_f+q_f-p_i-q_i) |M|^2, \tag{3.4.34}$$

where p_f has been written instead of p_2.We want to specialize this to the center-of-mass system, defined by

$$\boldsymbol{q}_i = -\boldsymbol{p}_i \quad , \quad \boldsymbol{q}_f = -\boldsymbol{p}_f, \tag{3.4.35}$$

taking $m_1 = m_2 = m$ for simplicity. This finally yields

$$\sigma_{\text{c.m.}} = (2\pi)^2 \frac{E}{2\sqrt{E^2-m^2}} \int d^3p_f d^3q_f \delta^3(\boldsymbol{p}_f + \boldsymbol{q}_f - \boldsymbol{p}_i - \boldsymbol{q}_i)$$

$$\times\, \delta(2E(p_f) - 2E) \sum_{s_f \sigma_f} |M|^2$$

$$= (2\pi)^2 \frac{E}{2\sqrt{E^2 - m^2}} \int d^3p_f \frac{1}{2} \delta(E(p_f) - E) \sum |M|^2$$

$$= (2\pi)^2 \frac{E}{4\sqrt{E^2 - m^2}} E\sqrt{E^2 - m^2} \int d\Omega \sum_{s_f \sigma_f} |M|^2, \tag{3.4.36}$$

where we have written $E = E(p_i)$ and the integral $\int d^3p_f$ has been transformed by means of

$$\boldsymbol{p}_f^2 d|\boldsymbol{p}_f| d\Omega = dE(p_f) E(p_f) |\boldsymbol{p}_f| d\Omega. \tag{3.4.37}$$

This leads to the simple result for the differential cross section

$$\left(\frac{d\sigma}{d\Omega}\right)_{\text{c.m.}} = (2\pi)^2 \frac{E^2}{4} |M|^2. \tag{3.4.38}$$

For simplicity, we do not consider the polarizations of the incoming and outgoing electrons: We sum over s_f, σ_f and average over s_i, σ_i. We find with the aid of (3.4.17)

$$\frac{1}{4} \sum_{s_i \sigma_i s_f \sigma_f} |M|^2 = \frac{e^4}{4(2\pi)^4} \sum_{s_i \sigma_i s_f \sigma_f} \left[\frac{1}{(p_f - q_i)^4} u_{s_i}(\boldsymbol{p}_i)^+ \gamma^0 \gamma_\mu \gamma^0 \gamma^0 u_{\sigma_f}(\boldsymbol{q}_f) \right.$$

$$\times\, u_{\sigma_i}(\boldsymbol{q}_i)^+ \gamma^0 \gamma^\mu \gamma^0 \gamma^0 u_{s_f}(\boldsymbol{p}_f) \overline{u}_{s_f}(\boldsymbol{p}_f) \gamma^\nu u_{\sigma_i}(\boldsymbol{q}_i) \overline{u}_{\sigma_f}(\boldsymbol{q}_f) \gamma_\nu u_{s_i}(\boldsymbol{p}_i)$$

$$\left. + \, 3 \text{ terms} \right]. \tag{3.4.39}$$

Using $\gamma^{\mu *} = \gamma^0 \gamma_\mu \gamma^0$ and (1.4.43), we can express the spin sums by traces

$$\frac{1}{4} \sum |M|^2 = \frac{e^4}{64(2\pi)^4 E(p_i) E(q_i) E(p_f) E(q_f)}$$

$$\times \left[\frac{1}{(p_f - q_i)^4} \operatorname{tr}_1[\gamma^\mu (\not{p}_f + m) \gamma^\nu (\not{q}_i + m)] \operatorname{tr}_2[\gamma_\mu (\not{q}_f + m) \gamma_\nu (\not{p}_i + m)] \right.$$

$$- \frac{1}{(p_f - q_i)^2 (p_f - p_i)^2} \operatorname{tr}_3[\gamma^\mu (\not{p}_f + m) \gamma^\nu (\not{q}_i + m) \gamma_\mu (\not{q}_f + m) \gamma_\nu (\not{p}_i + m)]$$

$$\left. + \, q_i \longleftrightarrow p_i \right]. \tag{3.4.40}$$

The three traces have been numbered. The first two can be calculated immediately by means of (2.6.43)

$$\operatorname{tr}_1 = 4(p_f^\mu q_i^\nu + p_f^\nu q_i^\mu - g^{\mu\nu} p_f \cdot q_i + m^2 g^{\mu\nu})$$

$$\operatorname{tr}_2 = 4(q_{f\mu} p_{i\nu} + q_{f\nu} p_{i\mu} - g_{\mu\nu} q_f \cdot p_i + m^2 g_{\mu\nu}). \tag{3.4.41}$$

In the third trace, we use the formulae

$$\gamma_\mu \not{p} \gamma^\mu = -2\not{p} \quad , \quad \gamma_\mu \not{p}\not{q}\not{r} \gamma^\mu = -2\not{r}\not{q}\not{p}, \tag{3.4.42}$$

and obtain

$$\mathrm{tr}\,_3 = 16(-2p_i \cdot q_i p_f \cdot q_f + m^2 p_i \cdot p_f + m^2 p_i \cdot q_f + m^2 p_i \cdot q_i$$
$$+ m^2 p_f \cdot q_f + m^2 p_f \cdot q_i + m^2 q_i \cdot q_f - 2m^4). \tag{3.4.43}$$

The results are best expressed in terms of the so-called Mandelstam variables (*S. Mandelstam, Phys. Rev. 112 (1958) 1344*)

$$s = (p_i + q_i)^2 = (p_f + q_f)^2 \quad , \quad u = (p_f - q_i)^2 = (p_i - q_f)^2$$
$$t = (p_i - p_f)^2 = (q_i - q_f)^2 \quad , \quad s + t + u = 4m^2. \tag{3.4.44}$$

Then evidently

$$\mathrm{tr}\,_1 \cdot \mathrm{tr}\,_2 = 8(s^2 + t^2 + 8m^2 u - 8m^4)$$
$$\mathrm{tr}\,_3 = 8[-s^2 + 4m^2 s - 4m^4 - 2m^2(t+u) + 2m^2 s],$$

which yields

$$\frac{1}{4}\sum |M|^2 = \frac{e^4}{8(2\pi)^4 E(p_i)E(q_i)E(p_f)E(q_f)} F(s,t,u), \tag{3.4.45}$$

with

$$F(s,t,u) = \frac{1}{t^2}(s^2 + u^2 + 8m^2 t - 8m^4) + \frac{1}{u^2}(s^2 + t^2 + 8m^2 u - 8m^4)$$
$$+ \frac{1}{tu}(2s^2 - 16m^2 s + 24m^4). \tag{3.4.46}$$

The total cross section (3.4.34) now becomes

$$\sigma = \frac{e^4}{(2\pi)^2\sqrt{s(s-4m^2)}} \int \frac{d^3p_f}{2E(p_f)} \frac{d^3q_f}{2E(q_f)} \delta(p_f + q_f - p_i - q_i) F, \tag{3.4.47}$$

which is obviously Lorentz invariant. For the differential cross section (3.4.38) in the center of mass system, we get

$$\left(\frac{d\sigma}{d\Omega}\right)_{\mathrm{c.m.}} = \frac{e^4}{32(2\pi)^2 E^2} F(s,t,u). \tag{3.4.48}$$

As a check, let us consider the non-relativistic limit $|\boldsymbol{p}_i| \ll m$, $\boldsymbol{p}_i = \boldsymbol{p}$. If $\vartheta = \angle(\boldsymbol{p}_i, \boldsymbol{p}_f)$ is the scattering angle in the center-of-mass system, then the Mandelstam variables (3.4.44) assume the following values

$$s = 4m^2 + 4\boldsymbol{p}^2 \quad , \quad u = -4\boldsymbol{p}^2 \sin^2 \frac{\vartheta}{2}$$

$$t = -4\boldsymbol{p}^2 \cos^2 \frac{\vartheta}{2}. \tag{3.4.49}$$

This shows that in the non-relativistic limit, we may neglect t, u in the nominators in (3.4.46) and put $s \approx 4m^2$

$$F \approx \frac{8m^4}{16\boldsymbol{p}^4}\left(\frac{1}{\sin^4 \frac{\vartheta}{2}} + \frac{1}{\cos^4 \frac{\vartheta}{2}} - \frac{1}{\sin^2 \frac{\vartheta}{2} \cos^2 \frac{\vartheta}{2}}\right).$$

With $E \approx m$, we finally obtain

$$\left(\frac{d\sigma}{d\Omega}\right)_{\text{c.m.}} = \frac{e^4 m^2}{64(2\pi)^2 \boldsymbol{p}^4}\left(\frac{1}{\sin^4 \frac{\vartheta}{2}} + \frac{1}{\cos^4 \frac{\vartheta}{2}} - \frac{1}{\sin^2 \frac{\vartheta}{2} \cos^2 \frac{\vartheta}{2}}\right). \tag{3.4.50}$$

Introducing the fine-structure constant

$$\alpha = \frac{e^2}{4\pi} \tag{3.4.51}$$

and $\boldsymbol{p} = m\boldsymbol{v}$, this is the Rutherford scattering formula with exchange

$$\left(\frac{d\sigma}{d\Omega}\right)_{\text{c.m.}} = \frac{\alpha^2}{16m^2 v^4}\left(\frac{1}{\sin^4 \frac{\vartheta}{2}} + \frac{1}{\cos^4 \frac{\vartheta}{2}} - \frac{1}{\sin^2 \frac{\vartheta}{2} \cos^2 \frac{\vartheta}{2}}\right). \tag{3.4.52}$$

The first two terms in (3.4.52) are due to the impossibility to distinguish between the scattered and the recoil particles. The third term is an interference term, which is typical for exchange.

3.5 Electron-Photon Scattering (Compton Scattering)

Compton scattering comes from the graphs (4) and (2) in Fig. 3 or from the corresponding terms in (3.3.16). Let us first consider the term (4) in (3.3.16)

$$D_2^{(4)}(x_1, x_2) = ie^2 :\overline{\psi}(x_1)\gamma^\mu \left[S^{(-)}(x_1 - x_2) + S^{(+)}(x_1 - x_2)\right]$$
$$\times \gamma^\nu \psi(x_2) :: A_\mu(x_1) A_\nu(x_2) : . \tag{3.5.1}$$

The square bracket is just the function S (2.3.6) which has causal support, as required by Theorem 1.4. The distribution splitting is again trivial

$$S(x) = (i\partial\!\!\!/ + m)(D^{\text{ret}} - D^{\text{av}}) = S^{\text{ret}} - S^{\text{av}}. \tag{3.5.2}$$

The retarded part $R_2^{(4)}$ contains S^{ret} because we must have

$$x_1^0 > x_2^0.$$

The first term in the square bracket in (3.5.1) comes from R_2' (3.3.15). Therefore, the two-point function

$$T_2^{(4)} = R_2^{(4)} - R_2'^{(4)} \sim S^{\rm ret} - S^{(-)} = -S^F \tag{3.5.3}$$

contains the Feynman propagator. We see that the Feynman propagator describes the inner fermion lines in a tree graph like (4), as in Chap. 2 in the external field problem. More generally, any internal fermion line which is not part of a loop is represented by the Feynman propagator S^F. This is again a consequence of the trivial distribution splitting: According to (3.1.68) we have to consider a fermionic contraction in a usual time ordered product

$$T\big[\overbrace{\psi(x_1)\overline{\psi}(x_2)}\big] = \Theta(x_1^0 - x_2^0)\{\psi^{(-)}(x_1), \overline{\psi}^{(+)}(x_2)\}$$

$$-\Theta(x_2^0 - x_1^0)\{\overline{\psi}^{(-)}(x_2), \psi^{(+)}(x_1)\}$$

$$= \frac{1}{i}\big[\Theta(x_1^0 - x_2^0)S^{(+)}(x_1 - x_2) - \Theta(x_2^0 - x_1^0)S^{(-)}(x_1 - x_2)\big]$$

$$= i\, S^{\rm F}(x_1 - x_2). \tag{3.5.4}$$

Thus we have from (3.5.1) and (3.5.3)

$$T_2^{(4)}(x_1, x_2) = -ie^2 \; :\overline{\psi}(x_1)\gamma^\mu S^F(x_1 - x_2)\gamma^\nu\psi(x_2): \\ \times :A_\mu(x_1)A_\nu(x_2):. \tag{3.5.5}$$

There is a second term in (3.3.16) which contributes to electron-photon scattering, namely (2):

$$D_2^{(2)}(x_1, x_2) = -ie^2 \; :\overline{\psi}(x_2)\gamma^\nu\big[S^{(+)}(x_2 - x_1) + S^{(-)}(x_2 - x_1)\big] \\ \times \gamma^\mu\psi(x_1) :: A_\mu(x_1)A_\nu(x_2):. \tag{3.5.6}$$

This is the so-called crossed graph (2) in Fig. 3 with the photon lines crossing each other. According to (3.5.2), the square bracket is equal to

$$S(x_2 - x_1) = S^{\rm ret}(x_2 - x_1) - S^{\rm av}(x_2 - x_1).$$

However, since $x_1^0 > x_2^0$ (3.5.3), here $-S^{\rm av}(x_2 - x_1)$ contributes to the retarded function

$$R_2^{(2)} \sim -S^{\rm av}.$$

Again the first term in the square bracket in (3.5.6) comes from $R_2'^{(2)}$, consequently,

$$T_2^{(2)} = R_2^{(2)} - R_2'^{(2)} \sim -S^{\rm av} - S^{(+)} = S^F(x_2 - x_1)$$

which gives

$$T_2^{(2)}(x_1, x_2) = -ie^2 \; :\overline{\psi}(x_2)\gamma^\nu S^F(x_2 - x_1)\gamma^\mu\psi(x_1): \\ \times :A_\mu(x_1)A_\nu(x_2):. \tag{3.5.7}$$

Comparing this result with (3.5.5), we see that

$$T_2^{(2)}(x_2, x_1) = T_2^{(4)}(x_1, x_2). \tag{3.5.8}$$

This means that the two terms together automatically lead to a symmetric two-point function, as required.

We now turn to electron-photon scattering. The external photons are transversal. It is convenient to introduce real polarization vectors ε_ν as in (2.11.32, 33):

$$\varepsilon_\nu = (0, \boldsymbol{\varepsilon}) \quad , \quad \boldsymbol{k} \cdot \boldsymbol{\varepsilon}(\boldsymbol{k}) = 0 \quad , \quad \boldsymbol{\varepsilon}^2 = 1. \tag{3.5.9}$$

The initial and final photon states are then given by

$$\varphi_{\substack{i\\f}} = \varepsilon_{\substack{i\\f}\nu} a_\nu(\boldsymbol{k})^+ \Omega.$$

Remember that the two indices ν are written downstairs because the sum over ν is not a Minkowski scalar product. The emission operator in φ_i must be contracted with an absorption operator in the normal product of the A's in (3.5.5) as follows:

$$\begin{aligned} A_\mu^{(-)}(x)\varphi_i &= (2\pi)^{-3/2} \int \frac{d^3k}{\sqrt{2\omega}}\, a_\mu(\boldsymbol{k}) e^{-ikx} \varepsilon_{i\nu}(\boldsymbol{k}_i) a_\nu(\boldsymbol{k}_i)^+ \Omega \\ &= (2\pi)^{-3/2} \frac{\varepsilon_{i\mu}(\boldsymbol{k}_i)}{\sqrt{2\omega_i}} e^{-ik_i x} \Omega. \end{aligned} \tag{3.5.10}$$

The outgoing photon in φ_f is treated in a similar manner

$$(\varphi_f, A^{(+)\mu}(x) \ldots) = (2\pi)^{-3/2} \frac{\varepsilon_f^\mu(\boldsymbol{k}_f)}{\sqrt{2\omega_f}} e^{ik_f x} (\Omega, \ldots) \quad . \tag{3.5.11}$$

We have to compute the S-matrix element

$$\begin{aligned} S_{fi} = -\frac{i}{2} e^2 \int d^4x_1 d^4x_2 \, &\big[(\psi_f, :\overline{\psi}(x_1)\gamma^\mu S^F(x_1 - x_2)\gamma^\nu \psi(x_2): \psi_i) \\ &\times (\varphi_f, :A_\mu(x_1) A_\nu(x_2): \varphi_i) + x_1 \longleftrightarrow x_2\big]. \end{aligned} \tag{3.5.12}$$

Since the term with x_1, x_2 interchanged gives the same contribution, we find with the aid of (3.5.10, 11)

$$\begin{aligned} S_{fi} = -ie^2 (2\pi)^{-6} \int d^4x_1 d^4x_2 \, &\overline{u}_{s_f}(\boldsymbol{p}_f) \gamma^\mu S^F(x_1 - x_2) \gamma^\nu u_{s_i}(\boldsymbol{p}_i) \\ \times e^{i(p_f x_1 - p_i x_2)} &\left[\frac{\varepsilon_{f\mu}(\boldsymbol{k}_f)}{\sqrt{2\omega_f}} \frac{\varepsilon_{i\nu}(\boldsymbol{k}_i)}{\sqrt{2\omega_i}} e^{i(k_f x_1 - k_i x_2)} \right. \\ &\left. + \frac{\varepsilon_{f\nu}(\boldsymbol{k}_f)}{\sqrt{2\omega_f}} \frac{\varepsilon_{i\mu}(\boldsymbol{k}_i)}{\sqrt{2\omega_i}} e^{i(k_f x_2 - k_i x_1)} \right]. \end{aligned} \tag{3.5.13}$$

Here the x_1, x_2-integrals are carried out as in (3.4.14)

$$\int d^4x_1 d^4x_2\, S^F(x_1-x_2)e^{iPx_1+iQx_2} = (2\pi)^4\delta(P+Q)\frac{\not{P}+m}{P^2-m^2+i0}. \qquad (3.5.14)$$

The δ-distribution again expresses energy-momentum conservation

$$S_{fi} = \delta(p_f + k_f - p_i - k_i)M, \qquad (3.5.15)$$

where

$$M = -ie^2(2\pi)^{-2}\Big[\overline{u}_{s_f}(\boldsymbol{p}_f)\frac{\not{\varepsilon}_f}{\sqrt{2\omega_f}}\frac{\not{p}_i+\not{k}_i+m}{(p_i+k_i)^2-m^2+i0}\frac{\not{\varepsilon}_i}{\sqrt{2\omega_i}}u_{s_i}(\boldsymbol{p}_i)$$

$$+\overline{u}_{s_f}(\boldsymbol{p}_f)\frac{\not{\varepsilon}_i}{\sqrt{2\omega_i}}\frac{\not{p}_i-\not{k}_f+m}{(p_i-k_f)^2-m^2+i0}\frac{\not{\varepsilon}_f}{\sqrt{2\omega_f}}u_{s_i}(\boldsymbol{p}_i)\Big]. \qquad (3.5.16)$$

We restrict ourselves to the situation where the electron is initially at rest, $p_i = (m, \mathbf{0})$. Then, since $k_i^2 = k_f^2 = 0$, the denominators in (3.5.16) simply become

$$(p_i + k_i)^2 - m^2 = 2p_ik_i = 2m\omega_i$$

$$(p_i - k_f)^2 - m^2 = -2p_ik_f = -2m\omega_f.$$

This yields

$$M = -i\frac{e^2}{(4\pi)^2}\frac{1}{m(\omega_i\omega_f)^{3/2}}\overline{u}_{s_f}(\boldsymbol{p}_f)Au_{s_i}(\boldsymbol{p}_i), \qquad (3.5.17)$$

with

$$A = \omega_f\not{\varepsilon}_f(\not{p}_i + \not{k}_i + m)\not{\varepsilon}_i - \omega_i\not{\varepsilon}_i(\not{p}_i - \not{k}_f + m)\not{\varepsilon}_f. \qquad (3.5.18)$$

With the aid of the relation

$$\not{p}\not{q} = -\not{q}\not{p} + 2pq, \qquad (3.5.19)$$

we anticommute $\not{p}_i$ to the right, where it gives m due to the Dirac equation $\not{p}u(p) = mu(p)$. Taking into account that

$$p_i\varepsilon_i = 0 = p_i\varepsilon_f \quad , \quad \varepsilon_ik_i = \varepsilon_fk_f = 0, \qquad (3.5.20)$$

where the last equality is a consequence of the transversal polarization vectors (3.5.9), we end up with

$$M_{s_fs_i} = -i\frac{e^2}{(4\pi)^2}\frac{1}{m(\omega_i\omega_f)^{3/2}}\overline{u}_{s_f}(\boldsymbol{p}_f)A'u_{s_i}(\boldsymbol{p}_i), \qquad (3.5.21)$$

where

$$A' = -\omega_f\not{\varepsilon}_f\not{\varepsilon}_i\not{k}_i - \omega_i\not{\varepsilon}_i\not{\varepsilon}_f\not{k}_f. \qquad (3.5.22)$$

In order to compute the cross-section, one has to calculate with wave packets as in the last section in the case of electron scattering. The final formula (3.4.33) can immediately be taken over

$$\sigma = (2\pi)^2 \frac{1}{|\boldsymbol{v}|} \sum_f \int d^3p_f d^3k_f \, \delta(p_f + k_f - p_i - k_i)|M|^2, \tag{3.5.23}$$

where the relative velocity of the two incoming particles is now equal to the velocity of light (=1). We are not interested in observing the recoil electron, therefore, the integration $\int d^3p_f$ is carried out

$$\sigma = (2\pi)^2 \sum_f \int d^3k_f \, \delta(E_f + \omega_f - E_i - \omega_i)|M|^2. \tag{3.5.24}$$

Transforming the remaining integral by means of

$$d^3k_f = \boldsymbol{k}_f^2 d|\boldsymbol{k}_f| d\Omega = \omega_f^2 d\omega_f d\Omega,$$

we obtain the differential cross section for the scattered photons

$$\frac{d\sigma}{d\Omega} = (2\pi)^2 \frac{\omega_f^2}{|g'(\omega_f)|} |M|^2. \tag{3.5.25}$$

Here, $g(\omega_f)$ is the argument of the δ-distribution in (3.5.24):

$$\begin{aligned} g(\omega_f) &= E_f + \omega_f - E_i - \omega_i \\ &= \sqrt{m^2 + (\boldsymbol{k}_i - \boldsymbol{k}_f)^2} + \omega_f - m - \omega_i \\ &= \sqrt{m^2 + \omega_i^2 + \omega_f^2 - 2\omega_i\omega_f \cos\vartheta} + \omega_f - m - \omega_i, \end{aligned} \tag{3.5.26}$$

where ϑ is the scattering angle in the laboratory frame. The derivative

$$g'(\omega_f) = \frac{2\omega_f - 2\omega_i \cos\vartheta}{2E_f} + 1$$

greatly simplifies by back substitution of $\cos\vartheta$ (see (3.5.26))

$$g'(\omega_f) = \frac{m\omega_i}{E_f \omega_f}. \tag{3.5.27}$$

Thus we have

$$\frac{d\sigma}{d\Omega} = (2\pi)^2 \frac{\omega_f^3 E_f}{m\omega_i} |M|^2. \tag{3.5.28}$$

For simplicity, we again do not consider polarization of the electrons: We sum over s_f and average over s_i

$$\begin{aligned} \frac{1}{2} \sum_{s_i s_f} M_{s_f s_i} M^*_{s_f s_i} &= \frac{e^4}{2(4\pi)^4 m^2 (\omega_i \omega_f)^3} \\ \times \sum_{s_i s_f} \overline{u}_{s_f}(\boldsymbol{p}_f) A' u_{s_i}(\boldsymbol{p}_i) \overline{u}_{s_i}(\boldsymbol{p}_i) \overline{A'} u_{s_f}(\boldsymbol{p}_f) &= \frac{e^4}{8(4\pi)^4 (m\omega_i\omega_f)^3 E_f} \operatorname{tr} B, \end{aligned} \tag{3.5.29}$$

where $\overline{A'} = \gamma^0 A'^+ \gamma^0$ and

$$B = (\not{p}_f + m)A'(\not{p}_i + m)\overline{A'}. \tag{3.5.30}$$

There remain the four different traces to be computed. We show the method in some detail. Using the relation

$$\overline{\not{p}\not{a}\not{k}} = \not{k}\not{a}\not{p}, \tag{3.5.31}$$

we have to calculate

$$\mathrm{tr}\, B_1 \stackrel{\mathrm{def}}{=} \mathrm{tr}\,(\not{p}_f + m)\not{\varepsilon}_i\not{\varepsilon}_f\not{k}_f(\not{p}_i + m)\not{k}_f\not{\varepsilon}_f\not{\varepsilon}_i.$$

$$= \mathrm{tr}\,\not{p}_f\not{\varepsilon}_i\not{\varepsilon}_f\not{k}_f\not{p}_i\not{k}_f\not{\varepsilon}_f\not{\varepsilon}_i + m^2\mathrm{tr}\,\not{\varepsilon}_i\not{\varepsilon}_f\not{k}_f\not{k}_f\not{\varepsilon}_f\not{\varepsilon}_i.$$

The second term vanishes because $k_f^2 = 0$. In the first term we bring together the two $\not{k}_f$ by means of (3.5.19)

$$= 2p_ik_f\mathrm{tr}\,\not{p}_f\not{\varepsilon}_i\not{\varepsilon}_f\not{k}_f\not{\varepsilon}_f\not{\varepsilon}_i.$$

Since $k_f\varepsilon_f = 0$, we can bring together the two $\not{\varepsilon}_f$ and use $\not{\varepsilon}_f\not{\varepsilon}_f = -1$:

$$= 2p_ik_f\mathrm{tr}\,\not{p}_f\not{\varepsilon}_i\not{k}_f\not{\varepsilon}_i.$$

Finally, we bring together the $\not{\varepsilon}_i$ and compute the remaining traces of two γ's by means of (2.6.43)

$$= 2p_ik_f(2k_f\varepsilon_i\mathrm{tr}\,\not{p}_f\not{\varepsilon}_i + \mathrm{tr}\,\not{p}_f\not{k}_f) = 8p_ik_f(2k_f\cdot\varepsilon_ip_f\cdot\varepsilon_i + p_fk_f). \tag{3.5.32}$$

Energy-momentum conservation

$$p_i + k_i = p_f + k_f \tag{3.5.33}$$

implies the relations

$$p_ik_i = p_fk_f = m\omega_i \quad , \quad p_ik_f = p_fk_i = m\omega_f \tag{3.5.34}$$

and with (3.5.20)

$$p_f\varepsilon_i = -k_f\varepsilon_i. \tag{3.5.35}$$

The trace (3.5.32) then becomes

$$\mathrm{tr}\, B_1 = 8m\omega_f[m\omega_i - 2(k_f\varepsilon_i)^2]. \tag{3.5.36}$$

Likewise we shall obtain the other three traces

$$\mathrm{tr}\, B_2 \stackrel{\mathrm{def}}{=} \mathrm{tr}\,(\not{p}_f + m)\not{\varepsilon}_i\not{\varepsilon}_f\not{k}_f(\not{p}_i + m)\not{k}_i\not{\varepsilon}_i\not{\varepsilon}_f$$

$$= 8m^2\omega_i\omega_f[2(\varepsilon_i\varepsilon_f)^2 - 1] + 8m\omega_i(k_f\varepsilon_i)^2 - 8m\omega_f(k_i\varepsilon_f)^2, \tag{3.5.37}$$

$$\mathrm{tr}\, B_3 \stackrel{\mathrm{def}}{=} \mathrm{tr}\,(\not{p}_f + m)\not{\varepsilon}_f\not{\varepsilon}_i\not{k}_i(\not{p}_i + m)\not{k}_f\not{\varepsilon}_f\not{\varepsilon}_i$$

$$= 16m^2\omega_i\omega_f(\varepsilon_i\varepsilon_f)^2 - 8m^2\omega_i\omega_f - 8m\omega_f(k_i\varepsilon_f)^2 + 8m\omega_i(k_f\varepsilon_i)^2, \tag{3.5.38}$$

$$\operatorname{tr} B_4 \stackrel{\text{def}}{=} \operatorname{tr}(\not{p}_f + m)\not{\varepsilon}_f \not{\varepsilon}_i \not{k}_i (\not{p}_i + m)\not{k}_i \not{\varepsilon}_i \not{\varepsilon}_f$$
$$= 8m\omega_i [m\omega_f + 2(k_i\varepsilon_f)^2]. \tag{3.5.39}$$

This leads to the total trace

$$\operatorname{tr} B = 8m^2\omega_i^2\omega_f^2 \left[\frac{\omega_i}{\omega_f} + \frac{\omega_f}{\omega_i} - 2 + 4(\varepsilon_i\varepsilon_f)^2\right]. \tag{3.5.40}$$

Substituting this into (3.5.29), we obtain from (3.5.28)

$$\frac{d\sigma}{d\Omega} = \frac{e^4\omega_f^2}{64\pi^2 m^2\omega_i^2}\left[\frac{\omega_i}{\omega_f} + \frac{\omega_f}{\omega_i} - 2 + 4(\varepsilon_i\varepsilon_f)^2\right], \tag{3.5.41}$$

which is the Klein-Nishina formula. The dimensional factor in front is conveniently expressed by the classical electron radius

$$r_0 = \frac{\alpha}{m} = \frac{e^2}{4\pi mc^2} \tag{3.5.42}$$

as follows

$$\frac{d\sigma}{d\Omega} = \frac{r_0^2}{4}\frac{\omega_f^2}{\omega_i^2}\left[\frac{\omega_i}{\omega_f} + \frac{\omega_f}{\omega_i} - 2 + 4(\varepsilon_i\varepsilon_f)^2\right]. \tag{3.5.43}$$

The energy ω_f of the scattered photon is given by ω_i and the scattering angle $\vartheta = \angle(\boldsymbol{k}_i, \boldsymbol{k}_f)$ through energy conservation

$$\sqrt{m^2 + (\boldsymbol{k}_i - \boldsymbol{k}_f)^2} = m + \omega_i - \omega_f .$$

Squaring this equation and solving for ω_f, we find

$$\omega_f = \frac{\omega_i}{1 + \frac{\omega_i}{m}(1 - \cos\vartheta)}. \tag{3.5.44}$$

The product of the polarization vectors in (3.5.43) is $\varepsilon_i\varepsilon_f = -\cos\Theta$, where Θ is the angle between the directions of polarizations of the incoming and outgoing photons. If these polarizations are not observed, one must average over $\boldsymbol{\varepsilon}_i$ and sum over $\boldsymbol{\varepsilon}_f$. Since there are two transversal polarizations, we have to calculate

$$\frac{1}{2}\sum_{n_1,n_2=1}^{2}\sum_{j=1}^{3}\left(\varepsilon_{ij}^{(n_1)}\varepsilon_{fj}^{(n_2)}\right)^2 = \frac{1}{2}\sum \varepsilon_{ij}^{(n_1)}\varepsilon_{fj}^{(n_2)}\varepsilon_{im}^{(n_1)}\varepsilon_{fm}^{(n_2)} . \tag{3.5.45}$$

We use the completeness relation for the 3-dimensional basis $\boldsymbol{\varepsilon}_i^{(1)}, \boldsymbol{\varepsilon}_i^{(2)}, \boldsymbol{k}_i/|\boldsymbol{k}_i|$

$$\sum_{n_1=1}^{2}\varepsilon_{ij}^{(n_1)}\varepsilon_{im}^{(n_1)} + \frac{k_{ij}k_{im}}{|\boldsymbol{k}_i|^2} = \delta_{jm}. \tag{3.5.46}$$

We then get for (3.5.45)

$$= \frac{1}{2}\sum_{jm}\Big(\delta_{jm} - \frac{k_{ij}k_{im}}{|\boldsymbol{k}_i|^2}\Big)\Big(\delta_{jm} - \frac{k_{fj}k_{fm}}{|\boldsymbol{k}_f|^2}\Big)$$

$$= \frac{1}{2}\Big(3 - 1 - 1 + \frac{(\boldsymbol{k}_i \cdot \boldsymbol{k}_f)^2}{|\boldsymbol{k}_i|^2|\boldsymbol{k}_f|^2}\Big) = \frac{1}{2}(1 + \cos^2\vartheta). \tag{3.5.47}$$

The other terms in (3.5.43) are multiplied by 2

$$\frac{d\sigma}{d\Omega} = \frac{r_0^2}{4}\frac{\omega_f^2}{\omega_i^2}\Big(2\frac{\omega_i}{\omega_f} + 2\frac{\omega_f}{\omega_i} - 4 + 2 + 2\cos^2\vartheta\Big)$$

$$= \frac{r_0^2}{2}\frac{\omega_f^2}{\omega_i^2}\Big(\frac{\omega_i}{\omega_f} + \frac{\omega_f}{\omega_i} - \sin^2\vartheta\Big). \tag{3.5.48}$$

In the non-relativistic limit $\omega_i \ll m$, it follows from (3.5.44) that $\omega_f \approx \omega_i$, thus

$$\frac{d\sigma}{d\Omega} = \frac{r_0^2}{2}(2 - \sin^2\vartheta) = \frac{r_0^2}{2}(1 + \cos^2\vartheta). \tag{3.5.49}$$

This is the classical Thomson formula. Note that the Compton cross section goes with the classical electron radius squared and not with the Compton wave length.

3.6 Vacuum Polarization

The first loop graph, we are going to consider, is (6) in Fig. 3. The corresponding term (6) in (3.3.16) reads

$$D_2^{(6)}(x_1, x_2) = -e^2 \mathrm{tr}\,\big[\gamma^\mu S^{(-)}(y)\gamma^\nu S^{(+)}(-y)$$

$$-\gamma^\mu S^{(+)}(y)\gamma^\nu S^{(-)}(-y)\big] \, :A_\mu(x_1)A_\nu(x_2): , \tag{3.6.1}$$

with

$$y = x_1 - x_2. \tag{3.6.2}$$

As will be seen below, this term leads to vacuum polarization which was already treated in Sect. 2.10. The comparison with the results obtained there is the first non-trivial test of the construction of the S-matrix of full QED by causal perturbation theory. In particular, this will lead to modifications of the naive Feynman rules in the case of loop graphs. Due to the invariance of the trace under cyclic permutations, the two terms in the square bracket in (3.6.1) are little different

$$D_2^{(6)}(x_1, x_2) = [P_1^{\mu\nu}(y) - P_1^{\nu\mu}(-y)] \, :A_\mu(x_1)A_\nu(x_2): , \tag{3.6.3}$$

where

$$P_1^{\mu\nu}(y) = e^2 \mathrm{tr}\,\gamma^\mu S^{(+)}(y)\gamma^\nu S^{(-)}(-y). \tag{3.6.4}$$

First we have to check the causal support of (3.6.1):

$$\gamma^\mu\left[S^{(-)}(y)\gamma^\nu S^{(+)}(-y)-S^{(+)}(y)\gamma^\nu S^{(-)}(-y)\right]$$

$$=\gamma^\mu\left[S(y)\gamma^\nu S^{(+)}(-y)-S^{(+)}(y)\gamma^\nu S(-y)\right].$$

Since both terms contain the full S, the support is causal with respect to x_2 in agreement with Theorem 1.4.

The further calculations are carried out in momentum space. The Fourier transform of (3.6.4) is given by

$$\hat{P}_1^{\mu\nu}(k)=e^2(2\pi)^{-2}\mathrm{tr}\,\gamma^\mu\int dy\, S^{(+)}(y)\gamma^\nu S^{(-)}(-y)e^{iky}$$

$$=e^2(2\pi)^{-2}\mathrm{tr}\int dq_1dq_2\,\gamma^\mu(\not{q}_1+m)\gamma^\nu(\not{q}_2+m)$$

$$\times\frac{i}{2\pi}\Theta(q_1^0)\delta(q_1^2-m^2)\Big(-\frac{i}{2\pi}\Big)\Theta(-q_2^0)\delta(q_2^2-m^2)\delta(q_1-q_2-k). \qquad (3.6.5)$$

Using $p=q_1$, $q=-q_2$ as new integration variables, we arrive at

$$\hat{P}_1^{\mu\nu}(k)=-e^2(2\pi)^{-4}\int dpdq\,\mathrm{tr}\,\gamma^\mu(\not{p}+m)\gamma^\nu(\not{q}-m)$$

$$\times\,\Theta(p^0)\delta(p^2-m^2)\Theta(q^0)\delta(q^2-m^2)\delta(p+q-k)$$

$$\stackrel{\mathrm{def}}{=}-e^2(2\pi)^{-4}T^{\nu\mu}(k). \qquad (3.6.6)$$

This second rank tensor is known to us from pair creation (2.7.31). According to (2.8.33), we can immediately write down the result

$$\hat{P}_1^{\mu\nu}(k)=-e^2(2\pi)^{-4}\Big(\frac{k^\mu k^\nu}{k^2}-g^{\mu\nu}\Big)\hat{d}_1(k), \qquad (3.6.7)$$

where

$$\hat{d}_1(k)=\frac{2\pi}{3}(k^2+2m^2)\sqrt{1-\frac{4m^2}{k^2}}\Theta(k^2-4m^2)\Theta(k_0). \qquad (3.6.8)$$

According to (3.6.3), we must split the distribution

$$d(y)=d_1(y)-d_1(-y)\quad,\quad\text{or} \qquad (3.6.9)$$

$$\hat{d}(k)=\hat{d}_1(k)-\hat{d}_1(-k). \qquad (3.6.10)$$

In fact, it follows from the central splitting solution (3.2.61) that the tensorial factor in (3.6.7) is not involved in the splitting. As discussed at the end of Sect. 3.2, it is sufficient to consider time-like k. The general result is then obtained by analytic continuation. We may use a Lorentz frame such that $k=(k_0,\mathbf{0})$

$$\hat{d}(k_0) = \frac{2\pi}{3}(k_0^2 + 2m^2)\sqrt{1 - \frac{4m^2}{k_0^2}}\Theta(k_0^2 - 4m^2)\text{sgn}\, k_0. \tag{3.6.11}$$

This distribution behaves as $\sim k_0^2$ for $|k_0| \to \infty$, consequently, the singular order is $\omega = 2$ (3.2.60). We wish to emphasize that the distribution splitting is non-trivial ($\omega > -1$). For this reason, *the ordinary Feynman rules do not hold for this loop graph. They would lead to ultraviolet divergent expressions.* Applying (3.2.60) with $\omega = 2$, we shall obtain

$$\hat{r}(k_0) = \frac{i}{3}k_0^3 \int\limits_{-\infty}^{+\infty} dk_0' \frac{k_0'^2 + 2m^2}{k_0'^3(k_0 - k_0' + i0)}\sqrt{1 - \frac{4m^2}{k_0'^2}}\Theta(k_0'^2 - 4m^2)\text{sgn}\, k_0$$

$$= \frac{2}{3}ik_0^4 \int\limits_0^{\infty} \frac{dk_0'}{k_0'^3} \frac{k_0'^2 + 2m^2}{(k_0 + i0)^2 - k_0'^2}\sqrt{1 - \frac{4m^2}{k_0'^2}}\Theta(k_0'^2 - 4m^2). \tag{3.6.12}$$

Introducing $s = k_0'^2$ as the new variable of integration, we may write

$$\hat{r}(k_0) = \frac{i}{3}k_0^4 \int\limits_{4m^2}^{\infty} ds \frac{s + 2m^2}{s^2(k_0^2 - s + ik_00)}\sqrt{1 - \frac{4m^2}{s}}. \tag{3.6.13}$$

We decompose the integral (3.6.13) into real and imaginary part

$$\hat{r}(k_0) = \frac{i}{3}k_0^4\Bigg[\text{P}\int \ldots - i\pi[\Theta(k_0) - \Theta(-k_0)]\frac{k_0^2 + 2m^2}{k_0^4}\sqrt{1 - \frac{4m^2}{k_0^2}}$$

$$\times\, \Theta(k_0^2 - 4m^2)\Bigg], \tag{3.6.14}$$

to combine it with the retarded function r', which is given by the first term in (3.6.1) or the second in (3.6.3):

$$\widehat{r'}(k_0) = -\hat{d}_1(-k_0)$$

$$= -\frac{2\pi}{3}(k_0^2 + 2m^2)\sqrt{1 - \frac{4m^2}{k_0^2}}\Theta(k_0^2 - 4m^2)\Theta(-k_0). \tag{3.6.15}$$

The two-point function is now given by (3.1.66)

$$\hat{t}(k_0) = \hat{r}(k_0) - \widehat{r'}(k_0)$$

$$= \frac{i}{3}k_0^4\Bigg[\text{P}\int\limits_{4m^2}^{\infty} \ldots - i\pi \frac{k_0^2 + 2m^2}{k_0^4}\sqrt{1 - \frac{4m^2}{k_0^2}}\Theta(k_0^2 - 4m^2)\Bigg]$$

$$= \frac{i}{3} k_0^4 \int\limits_{4m^2}^{\infty} ds \, \frac{s + 2m^2}{s^2(k_0^2 - s + i0)} \sqrt{1 - \frac{4m^2}{s}}. \tag{3.6.16}$$

This result agrees with (2.8.69) up to an overall factor. This is the desired test of the method. The two-point function in x-space may be represented in the form

$$T_2^{(6)}(x_1, x_2) = -i \, : A_\mu(x_1) \Pi^{\mu\nu}(x_1 - x_2) A_\nu(x_2) : , \tag{3.6.17}$$

where

$$\hat{\Pi}^{\mu\nu}(k) = (2\pi)^{-4} \Big(\frac{k^\mu k^\nu}{k^2} - g^{\mu\nu} \Big) \hat{\Pi}(k), \tag{3.6.18}$$

$$\hat{\Pi}(k) = \frac{e^2}{3} k^4 \int\limits_{4m^2}^{\infty} ds \, \frac{s + 2m^2}{s^2(k^2 - s + i0)} \sqrt{1 - \frac{4m^2}{s}}. \tag{3.6.19}$$

The dispersion integral (3.6.19) is calculated by means of the following Euler substitution

$$\frac{s}{m^2} = -\frac{(1-\eta)^2}{\eta} \quad , \quad -1 < \eta < 0. \tag{3.6.20}$$

At first we get the result for time-like momenta. As discussed at the end of Sect. 3.2, the result for space-like k can be obtained by analytic continuation. In this case we use the following variables

$$k^2 = t < 0 \quad , \quad \frac{t}{m^2} = -\frac{(1-\xi)^2}{\xi} \quad , \quad 0 < \xi < 1. \tag{3.6.21}$$

Since

$$\frac{ds}{m^2} = \frac{1-\eta^2}{\eta^2} d\eta \quad , \quad 1 - \frac{4m^2}{s} = \frac{(1+\eta)^2}{(1-\eta)^2},$$

we find

$$\hat{\Pi}(k) = \frac{e^2 m^2}{3} \frac{(1-\xi)^4}{\xi^2} \int\limits_{-1}^{0} d\eta \, \frac{(1+\eta)^2(\eta^2 - 4\eta + 1)}{(\xi - \eta)(\eta - 1/\xi)(\eta - 1)^4}$$

$$= \frac{e^2 m^2}{3} \Big[\frac{1+\xi}{1-\xi} \Big(\xi - 4 + \frac{1}{\xi} \Big) \log \xi + \frac{5}{3} \xi + \frac{5}{3\xi} - \frac{22}{3} \Big]. \tag{3.6.22}$$

For time-like $0 < t < 4m^2$, we put

$$\xi = e^{i\varphi} \quad , \quad \log \xi = i\varphi, \tag{3.6.23}$$

and get

$$\hat{\Pi}(k) = \frac{e^2 m^2}{3} \Big[2(1 + 2\sin^2 \frac{\varphi}{2}) \varphi \cot \frac{\varphi}{2} - \frac{20}{3} \sin^2 \frac{\varphi}{2} - 4 \Big]. \tag{3.6.24}$$

For $t > 4m^2$, we can use ξ (3.6.21) again which becomes negative, however. Then the logarithm in (3.6.22) gets complex $\log\xi = \log|\xi| + i\pi$. The sign of the imaginary part follows from $+i0$ in the denominator of (3.6.19).

It was discussed in Sect. 3.2 (3.2.54) that the general solution of the distribution splitting contains an undetermined polynomial of degree ω. Here we have $\omega = 2$, consequently, instead of (3.6.19) we have to consider the general expression

$$\tilde{\Pi}(k) = \hat{\Pi}(k) + C_0 + c_\mu k^\mu + C_2 k^2. \tag{3.6.25}$$

The constants C_0, c_μ, C_2 are not fixed by causality. In perturbation theory, additional physical conditions are necessary to fix them. By parity, c_μ must vanish.

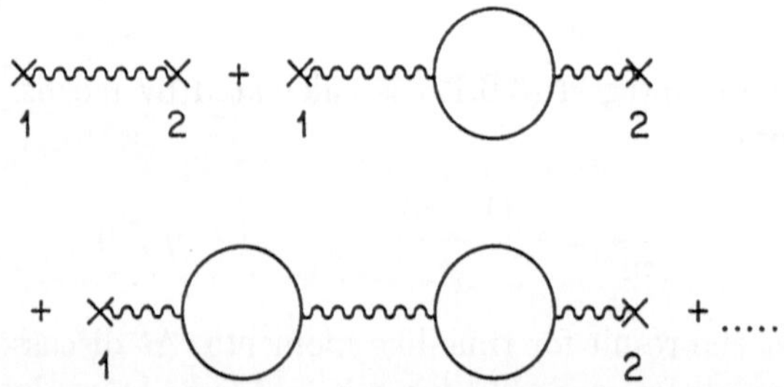

Fig. 5. Proper vacuum polarization insertions

To get a condition for C_0 and C_2, we consider the sum of vacuum polarization diagrams in Fig. 5, which are called proper vacuum polarization insertions. According to (3.4.5) the Feynman propagator $D_0^F(x_1 - x_2)$ describes an inner photon line in a tree graph. By (3.6.17) the sum of loops in Fig. 5 is then equal to

$$g^{\mu\nu} D_0^F(x_1 - x_2) + \int D_0^F(x_1 - x_1')\Pi^{\mu\nu}(x_1' - x_2) D_0^F(x_2' - x_2)\, dx_1' dx_2'$$

$$+ \int D_0^F(x_1 - x_1')\Pi^{\mu\lambda}(x_1' - x_2') D_0^F(x_2' - x_3')\Pi^{\lambda\nu}(x_3' - x_2)\, dx_1' dx_2' dx_3' + \dots \tag{3.6.26}$$

Here the integrations over the inner variables $x_1', x_2', \dots$ have been carried out in the adiabatic limit $g(x') = 1$. After Fourier transformation, the convolutions go over into products

$$g^{\mu\nu}\hat{D}_0^F(k) + \hat{D}_0^F(k)\tilde{\Pi}^{\mu\nu}(k)\hat{D}_0^F(k)$$

$$+\hat{D}_0^F(k)\tilde{\Pi}^{\mu\lambda}(k)\hat{D}_0^F(k)\tilde{\Pi}^{\lambda\nu}(k)\hat{D}_0^F(k)+\ldots$$

$$\stackrel{\text{def}}{=} \hat{D}^{\mu\nu}_{\text{tot}}(k)=\hat{D}_0^F(g^{\mu\nu}+\tilde{\Pi}^{\mu}{}_{\lambda}\hat{D}^{\lambda\nu}_{\text{tot}}), \tag{3.6.27}$$

where

$$\tilde{\Pi}^{\mu\nu}=(2\pi)^4\hat{\Pi}^{\mu\nu}.$$

This implies

$$(g^{\mu}{}_{\lambda}-\hat{D}_0^F\tilde{\Pi}^{\mu}{}_{\lambda})\hat{D}^{\lambda\nu}_{\text{tot}}=g^{\mu\nu}\hat{D}_0^F$$

and

$$(\hat{D}^{-1}_{\text{tot}})^{\mu\nu}=(\hat{D}_0^F)^{-1}g^{\mu\nu}-\tilde{\Pi}^{\mu\nu}=(2\pi)^2\Big[-g^{\mu\nu}k^2-\Big(\frac{k^\mu k^\nu}{k^2}-g^{\mu\nu}\Big)\Pi\Big] \tag{3.6.28}$$

with

$$\Pi(k)=(2\pi)^2\hat{\Pi}(k). \tag{3.6.29}$$

To calculate the inverse, we write the right-hand side as follows

$$(\hat{D}^{-1}_{\text{tot}})^{\mu\nu}(k)=(2\pi)^2\Big[\Big(g^{\mu\nu}-\frac{k^\mu k^\nu}{k^2}\Big)(\Pi-k^2)-\frac{k^\mu k^\nu}{k^2}k^2\Big]. \tag{3.6.30}$$

The two factors

$$P_1=g^{\mu\nu}-\frac{k^\mu k^\nu}{k^2},\quad P_2=\frac{k^\mu k^\nu}{k^2} \tag{3.6.31}$$

are orthogonal projection operators, satisfying

$$P_jP_k=\delta_{jk}P_k,\quad j,k=1,2. \tag{3.6.32}$$

The inverse of (3.6.30) is then given by

$$(2\pi)^2\hat{D}^{\mu\nu}_{\text{tot}}(k)=\Big(g^{\mu\nu}-\frac{k^\mu k^\nu}{k^2}\Big)\frac{1}{\Pi-k^2-i0}-\frac{k^\mu k^\nu}{k^2}\frac{1}{k^2+i0}. \tag{3.6.33}$$

The last term is unimportant, because it vanishes between transversal photon operators.

We are now able to discuss the consequences of a change of normalization (3.6.25). The constant C_0 must vanish because it would give a mass to the photon. The constant C_2 gives rise to a factor in the total photon propagator which changes the physical charge (charge renormalization). We want to exclude this by requiring $C_2=0$. Then the vacuum polarization function is normalized by the two conditions

$$\hat{\Pi}(0)=0 \tag{3.6.34}$$

$$\frac{\hat{\Pi}(k)}{k^2}\Big|_{k^2=0}=0. \tag{3.6.35}$$

This normalization is in accordance with the one found in the discussion of the causal phase (2.9.46).

3.7 Self-Energy

The next loop graph to be discussed is (3) (5) in Fig. 4. We turn to the corresponding term (5) in (3.3.16)

$$D_2^{(5)}(x_1,x_2) =: \overline{\psi}(x_1)d^{(5)}(y)\psi(x_2):, \tag{3.7.1}$$

where $y = x_1 - x_2$ and

$$d^{(5)}(y) = -e^2\gamma^\mu\left[S^{(-)}(y)D_0^{(+)}(-y) + S^{(+)}(y)D_0^{(+)}(y)\right]\gamma_\mu. \tag{3.7.2}$$

First we have to check the causal support of (3.7.2)

$$d^{(5)}(y) = -e^2\gamma^\mu\left[S(y)D_0^{(+)}(-y) + S^{(+)}(y)D_0(y)\right]\gamma_\mu. \tag{3.7.3}$$

Since S and D_0 have causal support, so has (3.7.3). Let us denote the two terms in (3.7.2) by

$$d_-(y) \stackrel{\text{def}}{=} S^{(-)}(y)D_0^{(+)}(-y) = -S^{(-)}(y)D_0^{(-)}(y) \tag{3.7.4}$$

$$d_+(y) \stackrel{\text{def}}{=} S^{(+)}(y)D_0^{(+)}(y). \tag{3.7.5}$$

The Fourier transform of (3.7.4) is given by

$$\hat{d}_-(p) = -(2\pi)^{-2}\int d^4q\,\hat{D}_0^{(-)}(p-q)(\not{q}+m)\hat{D}_m^{(-)}(q)$$

$$= (2\pi)^{-4}\int d^4q\,\Theta(q^0-p^0)\delta((p-q)^2)(\not{q}+m)\Theta(-q^0)\delta(q^2-m^2). \tag{3.7.6}$$

Since q lies on the mass shell $q^2 = m^2$, $q^0 < 0$ and $p - q$ on the backward light-cone $(p-q)^2 = 0$, $p^0 - q^0 < 0$, it follows that $p = p - q + q$ is time-like. For the further calculations, it is therefore convenient to use a Lorentz frame such that $p = (p_0, \mathbf{0})$.

First let us compute the term proportional to m in (3.7.6). Since

$$(p-q)^2 = (p_0 - q_0)^2 - \boldsymbol{q}^2, \tag{3.7.7}$$

it is equal to

$$I_1 \stackrel{\text{def}}{=} \int d^4q\,\Theta(q^0-p^0)\Theta(-q^0)\delta(p_0^2 - 2p_0q_0 + m^2)\delta(q^2-m^2). \tag{3.7.8}$$

Integrating out the last δ-distribution which is equal to $\delta(q_0 + E_q)/2E_q$, we arrive at

$$I_1 = \int \frac{d^3q}{2E_q}\delta(p_0^2 + 2p_0E_q + m^2)\Theta(-E_q - p_0). \tag{3.7.9}$$

The remaining δ-distribution and Θ-function imply

$$p_0 = -E_q - \sqrt{E_q^2 - m^2} = -E_q - |\boldsymbol{q}|.$$

From $E_q^2 = \boldsymbol{q}^2 + m^2 = (p_0 + ||\boldsymbol{q}|)^2$, it follows

$$|\boldsymbol{q}| = \frac{m^2 - p_0^2}{2p_0}, \qquad (3.7.10)$$

which, because of $p_0 < 0$, yields $p_0^2 \geq m^2$. The angular integrations in (3.7.9) are trivial. Thus we have

$$\begin{aligned} I_1 &= \frac{4\pi}{2}\Theta(p_0^2 - m^2)\Theta(-p_0) \int_0^\infty d|\boldsymbol{q}| \, \frac{|\boldsymbol{q}|^2}{E_q 2|p_0|}\delta\left(E_q - \frac{p_0^2 + m^2}{2|p_0|}\right) \\ &= 2\pi\Theta(p_0^2 - m^2)\Theta(-p_0) \int_m^\infty dE \, \frac{|\boldsymbol{q}|}{2|p_0|}\delta\left(E - \frac{p_0^2 + m^2}{2|p_0|}\right) \\ &= \frac{\pi}{2}\left(1 - \frac{m^2}{p_0^2}\right)\Theta(p_0^2 - m^2)\Theta(-p_0). \end{aligned} \qquad (3.7.11)$$

The second integral in (3.7.6) to be computed is

$$I_{2\nu} = \int d^4q \, q_\nu \Theta(q_0 - p_0)\Theta(-q_0)\delta(p_0^2 - 2p_0 q_0 + m^2)\delta(q^2 - m^2). \quad (3.7.12)$$

It obviously vanishes for $\nu \neq 0$ for symmetry reasons. The integral I_{20} can be calculated like I_1

$$\begin{aligned} I_{20} &= \int d^4q \, q_0 \Theta(q_0 - p_0)\delta(p_0^2 - 2p_0 q_0 + m^2)\frac{\delta(q_0 + E_q)}{2E_q} \\ &= \int \frac{d^3q}{2E_q}(-E_q)\Theta(-E_q - p_0)\delta\left(2p_0(E_q + \frac{p_0^2 + m^2}{2p_0})\right) \\ &= -\frac{\pi}{|p_0|}\Theta(p_0^2 - m^2)\Theta(-p_0) \int_m^\infty dE \, E|\boldsymbol{q}|\Theta(-E - p_0)\delta\left(E - \frac{p_0^2 + m^2}{2|p_0|}\right) \\ &= \frac{\pi}{4}\Theta(p_0^2 - m^2)\Theta(-p_0)\left(1 + \frac{m^2}{p_0^2}\right)\left(1 - \frac{m^2}{p_0^2}\right)p_0. \end{aligned} \qquad (3.7.13)$$

According to (3.7.6), this has to be multiplied by γ^0 which leads to $p_0\gamma^0 = \not{p}$. The final result for (3.7.6) in an arbitrary Lorentz frame is then given by

$$\begin{aligned} \hat{d}_-(p) = (2\pi)^{-4}\frac{\pi}{2}\Theta(p^2 - m^2)\Theta(-p_0)\left(1 - \frac{m^2}{p^2}\right) \\ \times \left[m + \frac{\not{p}}{2}\left(1 + \frac{m^2}{p^2}\right)\right]. \end{aligned} \qquad (3.7.14)$$

The total contribution to (3.7.3) becomes

$$\hat{d}_-^{(5)}(p) = -e^2\gamma^\mu \hat{d}_-(p)\gamma_\mu. \qquad (3.7.15)$$

Using the formulae

$$\gamma^\mu \not{p} \gamma_\mu = -2\not{p} \quad , \quad \gamma^\mu \gamma_\mu = 4, \tag{3.7.16}$$

this may be written as

$$\hat{d}_-^{(5)}(p) = -e^2(2\pi)^{-3}\Theta(p^2 - m^2)\Theta(-p_0)\Big(1 - \frac{m^2}{p^2}\Big)$$

$$\times \Big[m - \frac{\not{p}}{4}\Big(1 + \frac{m^2}{p^2}\Big)\Big] = r'^{(5)}(p). \tag{3.7.17}$$

This is also the contribution to the retarded function r' (3.3.15). The second term (3.7.5) is calculated in the same way

$$\hat{d}_+^{(5)}(p) = -e^2\gamma^\mu\big[S^{(+)}(y)D_0^{(+)}(y)\big](p)\gamma_\mu$$

$$= e^2(2\pi)^{-3}\Theta(p^2 - m^2)\Theta(p_0)\Big(1 - \frac{m^2}{p^2}\Big)$$

$$\times \Big[m - \frac{\not{p}}{4}\Big(1 + \frac{m^2}{p^2}\Big)\Big]. \tag{3.7.18}$$

Both terms together give the final real result

$$d^{(5)}(p) = e^2(2\pi)^{-3}\Theta(p^2 - m^2)\mathrm{sgn}\, p_0$$

$$\times \Big(1 - \frac{m^2}{p^2}\Big)\Big[m - \frac{\not{p}}{4}\Big(1 + \frac{m^2}{p^2}\Big)\Big]. \tag{3.7.19}$$

We note that the dependence on p_0 is through the sign $\mathrm{sgn}\, p_0$ as before in (3.6.11).

The most delicate step in the calculation is the splitting of the distribution (3.7.19). Again it is sufficient to consider time-like p. We return to the special frame with $p = (p_0, \mathbf{0})$

$$d^{(5)}(k_0) = e^2(2\pi)^{-3}\Theta(k_0^2 - m^2)\mathrm{sgn}\, k_0$$

$$\times \Big(1 - \frac{m^2}{k_0^2}\Big)\Big[m - \frac{1}{4}k_0\gamma^0\Big(1 + \frac{m^2}{k_0^2}\Big)\Big]. \tag{3.7.20}$$

Since this is $\sim |k_0|$ for $|k_0| \to \infty$, the singular order is $\omega = 1$ in (3.2.34). *Again the distribution splitting is non-trivial and the usual Feynman rules are not valid.* According to (3.2.34), there remains to calculate the following dispersion integral

$$\hat{r}(p_0) = \frac{i}{2\pi}p_0^2 \int\limits_{-\infty}^{+\infty} dk_0 \frac{d^{(5)}(k_0)}{(k_0 - i0)^2(p_0 - k_0 + i0)}. \tag{3.7.21}$$

Substituting (3.7.20) in this, we abbreviate the term coming from m in the square bracket in (3.7.20) by r_1 and the rest by r_2.

We show the calculation of

$$r_1(p_0) = e^2(2\pi)^{-4} i p_0^2 \int_{-\infty}^{+\infty} dk_0 \, \frac{m}{k_0^2(p_0 - k_0 + i0)} \Theta(k_0^2 - m^2)$$

$$\times \operatorname{sgn} k_0 \left(1 - \frac{m^2}{k_0^2}\right) \tag{3.7.22}$$

in some detail. This integral is further decomposed into two integrals r_{11} and r_{12} according to the last bracket in (3.7.22). We have

$$r_{11} = e^2(2\pi)^{-4} i p_0^2 m \int_0^{\infty} dk_0 \, \frac{\Theta(k_0^2 - m^2)}{k_0^2} \left(\frac{1}{p_0 - k_0 + i0} - \frac{1}{p_0 + k_0 + i0}\right)$$

$$= e^2(2\pi)^{-4} i p_0^2 m \int_{m^2}^{\infty} \frac{ds}{s(p_0^2 - s + i p_0 0)}, \tag{3.7.23}$$

where the new integration variable $s = k_0^2$ was introduced. We decompose this integral into real and imaginary parts

$$r_{11} = e^2(2\pi)^{-4} i p_0^2 m \left[\frac{1}{p_0^2} \mathrm{P} \int_{m^2}^{\infty} ds \left(\frac{1}{s} + \frac{1}{p_0^2 - s}\right)\right.$$

$$\left. - i\pi \operatorname{sgn} p_0 \int_{m^2}^{\infty} ds \, \frac{\delta(s - p_0^2)}{s}\right],$$

and obtain

$$r_{11} = e^2(2\pi)^{-4} i m \left[\log \frac{|p_0^2 - m^2|}{m^2} - i\pi \operatorname{sgn} p_0 \Theta(p_0^2 - m^2)\right]. \tag{3.7.24}$$

In the same way, we get for the second integral in (3.7.22)

$$r_{12} = e^2(2\pi)^{-4} i m^3 \left[\frac{1}{p_0^2}\left(\log \frac{m^2}{|p_0^2 - m^2|} + i\pi \operatorname{sgn} p_0 \Theta(p_0^2 - m^2)\right) - \frac{1}{m^2}\right]. \tag{3.7.25}$$

Both terms together lead to

$$r_1(p_0) = e^2(2\pi)^{-4} i m \left[\left(1 - \frac{m^2}{p_0^2}\right)\left(\log \frac{|p_0^2 - m^2|}{m^2} - i\pi \operatorname{sgn} p_0 \Theta(p_0^2 - m^2)\right) - 1\right]. \tag{3.7.26}$$

There remains to compute the integral r_2 coming from the last term in (3.7.20)

$$r_2(p_0) = -e^2(2\pi)^{-4} \frac{i}{4} \gamma^0 p_0^2 \int_{-\infty}^{+\infty} dk_0 \, \frac{k_0}{k_0^2(p_0 - k_0 + i0)}$$

$$\times\,\Theta(k_0^2-m^2)\mathrm{sgn}\,k_0\Big(1-\frac{m^4}{k_0^4}\Big). \tag{3.7.27}$$

Again, according to the last bracket, it splits into two integrals r_{21}, r_{22} which are easily calculated as above:

$$r_{21}=-e^2(2\pi)^{-4}\frac{i}{4}\gamma^0p_0\Big(\log\frac{|p_0^2-m^2|}{m^2}-i\pi\,\mathrm{sgn}\,p_0\Theta(p_0^2-m^2)\Big) \tag{3.7.28}$$

$$r_{22}=e^2(2\pi)^{-4}\frac{i}{4}\gamma^0p_0\Big[\frac{m^4}{p_0^4}\Big(\log\frac{|p_0^2-m^2|}{m^2}-i\pi\,\mathrm{sgn}\,p_0\Theta(p_0^2-m^2)\Big)+\frac{m^2}{p_0^2}+\frac{1}{2}\Big]. \tag{3.7.29}$$

Both terms together yield

$$r_2(p_0)=e^2(2\pi)^{-4}\frac{i}{4}\gamma^0p_0\Big[-\Big(1-\frac{m^4}{p_0^4}\Big)\Big(\log\frac{|p_0^2-m^2|}{m^2}-i\pi\,\mathrm{sgn}\,p_0\Theta(p_0^2-m^2)\Big)$$

$$+\frac{m^2}{p_0^2}+\frac{1}{2}\Big]. \tag{3.7.30}$$

From (3.7.26) and (3.7.30), we now obtain the total function (3.7.21) which we just write down in an arbitrary Lorentz frame

$$\hat{r}(p)=e^2(2\pi)^{-4}i\Big\{(\log|1-b^2|-i\pi\,\mathrm{sgn}\,p_0\Theta(p^2-m^2))$$

$$\times\Big[m(1-\frac{1}{b^2})-\frac{\not p}{4}(1-\frac{1}{b^4})\Big]+\frac{\not p}{4b^2}-m+\frac{\not p}{8}\Big\}, \tag{3.7.31}$$

where we have introduced the dimensionless parameter

$$b^2=\frac{p^2}{m^2}.$$

According to Theorem 3.4 (3.2.64), this retarded function is the boundary value of an analytic function of complex momentum $p+i\eta$, $\eta=(\varepsilon,\mathbf{0})$, $\varepsilon>0$. In fact, since

$$\log|1-b^2|-i\pi\mathrm{sgn}\,p_0\,\Theta(b^2-1)=\log(1-b^2-i0p_0)$$

and $(p+i\eta)^2=p^2+2i\varepsilon p_0-\varepsilon^2$, the analytic function is simply given by

$$\hat{r}_{\rm an}(p)=e^2(2\pi)^{-4}i\Big\{\log(1-b^2)\Big[m\Big(1-\frac{1}{b^2}\Big)-\frac{\not p}{4}\Big(1-\frac{1}{b^4}\Big)\Big]$$

$$+\frac{\not p}{4b^2}-m+\frac{\not p}{8}\Big\}. \tag{3.7.32}$$

It has to be used if the self-energy for space-like p is needed.

The result (3.7.31) must be combined with (3.7.17)

$$\widehat{r'}(p)=e^2(2\pi)^{-4}i(2\pi i)\Theta(-p_0)\Theta(p^2-m^2)$$

$$\times \Big[m\Big(1-\frac{1}{b^2}\Big)-\frac{\not p}{4}\Big(1-\frac{1}{b^4}\Big)\Big],$$

according to the general formula (3.1.66), to obtain the two-point function

$$T_2^{(5)}(x_1,x_2) \stackrel{\text{def}}{=} i:\overline{\psi}(x_1)\Sigma(x_1-x_2)\psi(x_2): . \tag{3.7.33}$$

We note that in

$$\hat{\Sigma}(p) = -i(\hat{r}(p)-\widehat{r'}(p))$$

the $\operatorname{sgn} p_0$ of (3.7.31) is changed into a factor 1. The same compensation has taken place in the case of vacuum polarization (3.6.14–16). We end up with the following result for the self-energy

$$\hat{\Sigma}(p) = e^2(2\pi)^{-4}\Big\{\big(\log|1-b^2| - i\pi\Theta(p^2-m^2)\big)$$

$$\times \Big[m(1-\frac{1}{b^2})-\frac{\not p}{4}(1-\frac{1}{b^4})\Big]+\frac{\not p}{4b^2}-m+\frac{\not p}{8}\Big\}. \tag{3.7.34}$$

It is a remarkable feature of this calculation that the result (3.7.34) is free of any infrared divergence. The reason for this rather good behavior is that $\hat{\Sigma}(p)$ has been normalized at $p=0$, because we have used the central splitting solution. The often used normalization on the mass shell is problematic because many distributions in QED are ill-defined there (see Sect. 3.11).

The distribution (3.7.19), which we had to split, has singular order $\omega=1$. Consequently, the general solution of the problem contains an undetermined linear polynomial in p:

$$\tilde{\Sigma}(p) = e^2(2\pi)^{-4}\Big\{\big[(\log|1-b^2| - i\pi\Theta(p^2-m^2)\big]$$

$$\times \Big[m-\frac{\not p}{4}(1+\frac{1}{b^2})\Big](1-\frac{1}{b^2})+\frac{\not p}{4b^2}+C_0+C_1\not p\Big\}. \tag{3.7.35}$$

The two constants C_0, C_1 must be determined by additional physical conditions. As in the discussion of vacuum polarization (3.6.26), we consider radiative corrections by photon self-energy loops of the following form:

$$S^F(x_1-x_2)+\int S^F(x_1-x_1')\Sigma(x_1'-x_2')S^F(x_2'-x_2)dx_1'dx_2'+\dots . \tag{3.7.36}$$

After Fourier transformation, the convolutions go over into products

$$\hat{S}^F(p)+\hat{S}^F(p)\hat{\Sigma}(p)\hat{S}^F(p)+\hat{S}^F(p)\hat{\Sigma}(p)\hat{S}^F(p)\hat{\Sigma}(p)\hat{S}^F(p)$$

$$+\dots \stackrel{\text{def}}{=} \hat{S}_{\text{tot}}(p). \tag{3.7.37}$$

Writing the Feynman propagator as an inverse matrix

$$\hat{S}^F(p) = \frac{1}{\not p - m + i0},$$

enables us to sum the geometric series (3.7.37)

$$\hat{S}_{\mathrm{tot}}(p) = \frac{1}{\not{p} - m + i0 - \hat{\Sigma}(p)}. \tag{3.7.38}$$

Here, $\hat{\Sigma}(p)$ appears like the mass. This is the reason why it is called self-mass or self-energy.

We want m to be the physical mass of the electron. This implies the normalization condition

$$\hat{\Sigma}(\not{p} = m) = 0, \tag{3.7.39}$$

which restricts the constant C_0, C_1 in (3.7.35) according to

$$C_0 = -m\Big(C_1 + \frac{1}{4}\Big). \tag{3.7.40}$$

A further reason for this normalization and, even a more fundamental one, comes from the discussion of the adiabatic limit (see (3.11.19)). In fact, this limit only exists, if the condition (3.7.39) is satisfied. Eliminating C_0 by (3.7.40) and taking the analytic continuation (3.7.32) into account, we get the following result for the self-energy, valid for arbitrary p:

$$\Sigma(p) = e^2(2\pi)^{-4}\Big\{\big[\log|1-b^2| - i\pi\Theta(p^2-m^2)\big]$$
$$\times\Big[m - \frac{\not{p}}{4}\Big(1+\frac{1}{b^2}\Big)\Big]\Big(1-\frac{1}{b^2}\Big) + \frac{\not{p}}{4b^2} - \frac{m}{4} + C_1(\not{p} - m)\Big\}. \tag{3.7.41}$$

We emphasize that this is no longer the central splitting solution, because both constants C_0, C_1 cannot be 0 due to (3.7.40). The remaining constant C_1 will be discussed in Sect. 3.13.

3.8 Vertex Function: Causal Distribution

In the preceding sections we have discussed all terms appearing in (3.3.16), which contribute to the two-point function in second order. We go one step further and consider a particularly interesting term of the three-point function, the so-called vertex function. We start from the retarded and advanced distributions (3.1.31–32)

$$R_3'(x_1,x_2,x_3) = \sum_{P_2} T(Y,x_3)\tilde{T}(X)$$

$$= \underbrace{T_2(x_1,x_3)\,\tilde{T}_1(x_2)}_{R'_{31}} + \underbrace{T_2(x_2,x_3)\,\tilde{T}_1(x_1)}_{R'_{32}} + \underbrace{T_1(x_3)\,\tilde{T}_2(x_1,x_2)}_{R'_{33}}, \tag{3.8.1}$$

$$A_3'(x_1,x_2,x_3) = \sum_{P_2} \tilde{T}(X)T(Y,x_3)$$

$$= \underset{A'_{31}}{\tilde{T}_1(x_2)\, T_2(x_1,x_3)} + \underset{A'_{32}}{\tilde{T}_1(x_1)\, T_2(x_2,x_3)} + \underset{A'_{33}}{\tilde{T}_2(x_1,x_2)\, T_1(x_3)}. \tag{3.8.2}$$

The abbreviations below the terms will be used in the following. Every term in (3.8.1, 2) corresponds to a certain decomposition of the vertex diagram into a single vertex ($T_1 = -\tilde{T}_1$) and a second order diagram (T_2, $\tilde{T}_2$) (see Fig. 6). The latter are of two types, the first is the two-point function for Moeller scattering (3.4.5)

$$T_2^{(1)}(x_1,x_2) = -ie^2 :\overline{\psi}(x_1)\gamma^\mu\psi(x_1)\overline{\psi}(x_2)\gamma_\mu\psi(x_2): D_0^F(x_1-x_2), \tag{3.8.3}$$

and the second one comes from Compton scattering (3.5.5, 7)

$$T_2^{(2)}(x_1,x_3) = -ie^2 :\overline{\psi}(x_1)\gamma^\mu S_F(x_1-x_3)\gamma^\nu\psi(x_3)::A_\mu(x_1)A_\nu(x_3):$$
$$-ie^2 :\overline{\psi}(x_3)\gamma^\nu S_F(x_3-x_1)\gamma^\mu\psi(x_1)::A_\mu(x_1)A_\nu(x_3):. \tag{3.8.4}$$

The inverse two-point functions are given by (3.1.6)

$$\tilde{T}_2(x_1,x_2) = -T_2(x_1,x_2) + T_1(x_1)T_1(x_2) + T_1(x_2)T_1(x_1). \tag{3.8.5}$$

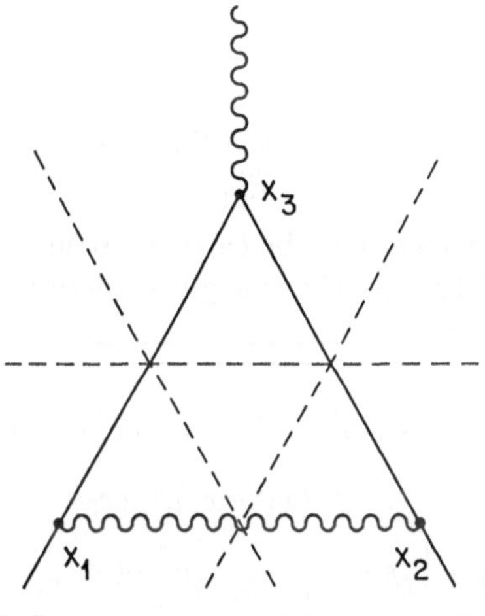

Fig. 6. Decompositions of the vertex diagram

In dispersion-theoretical calculations of the vertex function, one sometimes restricts oneself to the horizontal cut in Fig. 6 without taking the other two decompositions into account. This is not correct if the external electrons are off the mass-shell, i.e. $p^2, q^2 \neq m^2$.

For the first term in (3.8.1) we obtain

$$R'_{31} = -T_2^{(2)}(x_1,x_3)T_1(x_2)$$

$$= ie^2\big[:\overline{\psi}(x_1)\gamma^\mu S_F(x_1-x_3)\gamma^\nu\overbrace{\psi(x_3): ie :\overline{\psi}}(x_2)\gamma^\lambda\psi(x_2):$$

$$+ :\overline{\psi}(x_3)\gamma^\nu S_F(x_3-x_1)\gamma^\mu\psi(x_1): ie :\overline{\psi}(x_2)\gamma^\lambda\psi(x_2):]$$

$$\times :A_\mu(x_1)A_\nu(x_3):A_\lambda(x_2). \tag{3.8.6}$$

We only look for terms where x_1 corresponds to an outgoing electron, x_2 to an incoming electron and x_3 to the external photon. Then the field operators $\overline{\psi}(x_1), \psi(x_2)$ and $A(x_3)$ must survive in the expression, all other field operators are contracted in pairs. We obviously get

$$R'_{31} = -e^3 :\overline{\psi}(x_1)\gamma^\mu S_F(x_1-x_3)\gamma^\nu S^{(+)}(x_3-x_2)\gamma^\lambda\psi(x_2):$$

$$\times g_{\mu\lambda}D_0^{(+)}(x_1-x_2)A_\nu(x_3), \tag{3.8.7}$$

and similarly

$$R'_{32} = -T_2^{(2)}(x_2,x_3)T_1(x_1)$$

$$= e^3 :\overline{\psi}(x_1)\gamma_\mu S^{(-)}(x_1-x_3)\gamma^\nu S_F(x_3-x_2)\gamma^\mu\psi(x_2):$$

$$\times D_0^{(+)}(x_2-x_1)A_\nu(x_3). \tag{3.8.8}$$

Here the anticommutation of $\overline{\psi}(x_1)$ to the left gives rise to a minus sign. The last term in (3.8.1) requires a little more work:

$$R'_{33} = T_1(x_3)\tilde{T}_2^{(1)}(x_1,x_2)$$

$$= \underbrace{-T_1(x_3)T_2^{(1)}(x_1,x_2)}_{R'_1} + \underbrace{T_1(x_3)T_1(x_1)\,T_1(x_2)}_{R'_{21}} + \underbrace{T_1(x_3)T_1(x_2)\,T_1(x_1)}_{R'_{22}}. \tag{3.8.9}$$

The two contractions which must be carried out in the first term give rise to a minus sign (see (3.4.12) and the rule given there):

$$R'_1 = ie :\overline{\psi}(x_3)\gamma^\nu\psi(x_3): ie^2 :\overline{\psi}(x_1)\gamma^\mu\psi(x_1)\overline{\psi}(x_2)\gamma_\mu\psi(x_2):$$

$$\times D_0^F(x_1-x_2)A_\nu(x_3)$$

$$= e^3 :\overline{\psi}(x_1)\gamma^\mu\frac{1}{i}S^{(-)}(x_1-x_3)\gamma^\nu\frac{1}{i}S^{(+)}(x_3-x_2)\gamma_\mu\psi(x_2):$$

$$\times D_0^F(x_1-x_2)A_\nu(x_3). \tag{3.8.10}$$

Similarly we compute the other two terms

$$R'_{21} = -ie^3 :\overline{\psi}(x_3)\gamma^\nu\psi(x_3)::\overline{\psi}(x_1)\gamma^\mu\psi(x_1)::\overline{\psi}(x_2)\gamma^\lambda\psi(x_2):$$

$$\times A_\nu(x_3)A_\mu(x_1)A_\lambda(x_2)$$

$$= e^3 :\overline{\psi}(x_1)\gamma^\mu S^{(-)}(x_1-x_3)\gamma^\nu S^{(+)}(x_3-x_2)\gamma^\lambda\psi(x_2):$$

$$\times g_{\mu\lambda}D_0^{(+)}(x_1-x_2)A_\nu(x_3), \tag{3.8.11}$$

$$
\begin{aligned}
R'_{22} = -ie^3 \ &:\overline{\psi}(x_3)\gamma^\nu\psi(x_3)::\overline{\psi}(x_2)\gamma^\mu\psi(x_2)::\overline{\psi}(x_1)\gamma^\lambda\psi(x_1): \\
&\times A_\nu(x_3)A_\mu(x_2)A_\lambda(x_1) \\
= e^3 \ &:\overline{\psi}(x_1)\gamma^\lambda S^{(-)}(x_1-x_3)\gamma^\nu S^{(+)}(x_3-x_2)\gamma^\mu\psi(x_2): \\
&\times g_{\mu\lambda}D_0^{(+)}(x_2-x_1)A_\nu(x_3).
\end{aligned} \tag{3.8.12}
$$

The advanced function (3.8.2) is computed in a similar manner

$$
\begin{aligned}
A'_{31} &= -T_1(x_2)T_2^{(2)}(x_1,x_3) \\
&= e^3 \ :\overline{\psi}(x_1)\gamma^\mu S^F(x_1-x_3)\gamma^\nu S^{(-)}(x_3-x_2)\gamma^\lambda\psi(x_2): \\
&\quad \times g_{\lambda\mu}D_0^{(+)}(x_2-x_1)A_\nu(x_3),
\end{aligned} \tag{3.8.13}
$$

$$
\begin{aligned}
A'_{32} &= -T_1(x_1)T_2^{(2)}(x_2,x_3) \\
&= -e^3 \ :\overline{\psi}(x_1)\gamma^\lambda S^{(+)}(x_1-x_3)\gamma^\nu S^F(x_3-x_2)\gamma^\mu\psi(x_2): \\
&\quad \times g_{\lambda\mu}D_0^{(+)}(x_1-x_2)A_\nu(x_3).
\end{aligned} \tag{3.8.14}
$$

The last term in (3.8.2) splits into three terms

$$
\begin{aligned}
A'_{33} &= \tilde{T}_2(x_1,x_2)T_1(x_3) \\
&= \underbrace{-T_2^{(1)}(x_1,x_2)T_1(x_3)}_{A'_1} + \underbrace{T_1(x_1)T_1(x_2)\,T_1(x_3)}_{A'_{21}} \\
&\quad \underbrace{+T_1(x_2)T_1(x_1)\,T_1(x_3)}_{A'_{22}},
\end{aligned} \tag{3.8.15}
$$

with the following results:

$$
\begin{aligned}
A'_1 = -e^3 \ &:\overline{\psi}(x_1)\gamma^\mu S^{(+)}(x_1-x_3)\gamma^\nu S^{(-)}(x_3-x_2)\gamma_\mu\psi(x_2): \\
&\times D_0^F(x_1-x_2)A_\nu(x_3),
\end{aligned} \tag{3.8.16}
$$

$$
\begin{aligned}
A'_{21} = e^3 \ &:\overline{\psi}(x_1)\gamma^\mu S^{(+)}(x_1-x_3)\gamma^\nu S^{(-)}(x_3-x_2)\gamma^\lambda\psi(x_2): \\
&\times g_{\mu\lambda}D_0^{(+)}(x_1-x_2)A_\nu(x_3),
\end{aligned} \tag{3.8.17}
$$

$$
\begin{aligned}
A'_{22} = e^3 \ &:\overline{\psi}(x_1)\gamma^\lambda S^{(+)}(x_1-x_3)\gamma^\nu S^{(-)}(x_3-x_2)\gamma^\mu\psi(x_2): \\
&\times g_{\mu\lambda}D_0^{(+)}(x_2-x_1)A_\nu(x_3).
\end{aligned} \tag{3.8.18}
$$

Now we are ready to write down the difference

$$
D = R' - A' \stackrel{\mathrm{def}}{=} -e^3 \ :\overline{\psi}(x_1)D^\nu(x_1,x_2,x_3)\psi(x_2):\ A_\nu(x_3), \tag{3.8.19}
$$

where the 10 terms contributing to D^ν all have the same matrix structure. The following pairs of terms combine: R'_{31} and R'_{21}, R'_{32} and R'_{22}, $-A'_{31}$ and

$-A'_{22}$, $-A'_{32}$ and $-A'_{21}$. Then all spinor Feynman propagators are converted to retarded or advanced propagators

$$\begin{aligned}D^\nu =&\gamma^\mu S^{(-)}(x_1-x_3)\gamma^\nu S^{(+)}(x_3-x_2)\gamma_\mu D_0^F(x_1-x_2)\\ &-\gamma^\mu S^{(+)}(x_1-x_3)\gamma^\nu S^{(-)}(x_3-x_2)\gamma_\mu D_0^F(x_1-x_2)\\ &-\gamma^\mu S^{\rm ret}(x_1-x_3)\gamma^\nu S^{(+)}(x_3-x_2)\gamma_\mu D_0^{(+)}(x_1-x_2)\\ &+\gamma^\mu S^{(-)}(x_1-x_3)\gamma^\nu S^{\rm av}(x_3-x_2)\gamma_\mu D_0^{(+)}(x_2-x_1)\\ &-\gamma^\mu S^{\rm av}(x_1-x_3)\gamma^\nu S^{(-)}(x_3-x_2)\gamma_\mu D_0^{(+)}(x_2-x_1)\\ &+\gamma^\mu S^{(+)}(x_1-x_3)\gamma^\nu S^{\rm ret}(x_3-x_2)\gamma_\mu D_0^{(+)}(x_1-x_2).\end{aligned} \tag{3.8.20}$$

It is not necessary to check the causal support of D^ν

$$\operatorname{supp} D^\nu(x_1,x_2,x_3)\subseteq \Gamma_2^+(x_3)\cup\Gamma_2^-(x_3) \tag{3.8.21}$$

because Theorem 1.4 is now applicable. Nevertheless, the direct verification is useful, in particular for eliminating a possible sign error in (3.8.20).

The further calculations are carried out in p-space

$$\hat D^\nu(p,q)=(2\pi)^{-4}\int dy_1dy_2\, D^\nu(x_1,x_2,x_3)e^{ipy_1+iqy_2}, \tag{3.8.22}$$

where

$$y_1=x_1-x_3\quad ,\quad y_2=x_3-x_2. \tag{3.8.23}$$

Substituting (3.8.20) in here, we arrive at

$$\begin{aligned}\hat D^\nu(p,q)=&(2\pi)^{-2}\int dk\\ \times\Big[&\gamma^\mu \hat S^{(-)}(p-k)\gamma^\nu \hat S^{(+)}(q-k)\gamma_\mu \hat D_0^F(k)\\ &-\gamma^\mu \hat S^{(+)}(p-k)\gamma^\nu \hat S^{(-)}(q-k)\gamma_\mu \hat D_0^F(k)\\ &-\gamma^\mu \hat S^{\rm ret}(p-k)\gamma^\nu \hat S^{(+)}(q-k)\gamma_\mu \hat D_0^{(+)}(k)\\ &+\gamma^\mu \hat S^{(-)}(p-k)\gamma^\nu \hat S^{\rm av}(q-k)\gamma_\mu \hat D_0^{(+)}(-k)\\ &-\gamma^\mu \hat S^{\rm av}(p-k)\gamma^\nu \hat S^{(-)}(q-k)\gamma_\mu \hat D_0^{(+)}(-k)\\ &+\gamma^\mu \hat S^{(+)}(p-k)\gamma^\nu \hat S^{\rm ret}(q-k)\gamma_\mu \hat D_0^{(+)}(k)\Big].\end{aligned} \tag{3.8.24}$$

We first compute the matrix part, say G, in (3.8.24). Using the formulae

$$\begin{aligned}&\gamma^\mu \not p\not k\not q\gamma_\mu=-2\not q\not k\not p\\ &\gamma^\mu\not p\not q\gamma_\mu=4p\cdot q\quad ,\quad \gamma^\mu\gamma^\nu\gamma_\mu=-2\gamma^\nu,\end{aligned} \tag{3.8.25}$$

we get

$$\hat D^\nu(p,q)=(2\pi)^{-2}\int d^4k\, G^\nu\times$$

$$\Big[\hat{D}^{(-)}(p-k)\hat{D}^{(+)}(q-k)\hat{D}_0^F(k)-\hat{D}^{\rm ret}(p-k)\hat{D}^{(+)}(q-k)\hat{D}_0^{(+)}(k)$$

$$+\hat{D}^{(-)}(p-k)\hat{D}^{\rm av}(q-k)\hat{D}_0^{(+)}(-k)-\hat{D}^{(+)}(p-k)\hat{D}^{(-)}(q-k)\hat{D}_0^F(k)$$

$$+\hat{D}^{(+)}(p-k)\hat{D}^{\rm ret}(q-k)\hat{D}_0^{(+)}(k)-\hat{D}^{\rm av}(p-k)\hat{D}^{(-)}(q-k)\hat{D}_0^{(+)}(-k)\Big] \quad (3.8.26)$$

with

$$G^\nu=-2(\not q-\not k)\gamma^\nu(\not p-\not k)+4m(p^\nu-k^\nu)+4m(q^\nu-k^\nu)-2m^2\gamma^\nu. \quad (3.8.27)$$

The six terms in (3.8.26) are in a new order, where the first three terms are equal to R' whereas the last three ones give A'. The k-dependence of G^ν (3.8.27) shows that we have to compute scalar, vector and tensor integrals.

We now turn to the scalar integral

$$I_D(p,q)=(2\pi)^{-2}\int d^4k\,(2\pi)^{-4}$$

$$\times\Big[-\Theta(k_0-p_0)\delta((p-k)^2-m^2)\Theta(q_0-k_0)\delta((q-k)^2-m^2)\frac{1}{k^2+i0}$$

$$-\frac{1}{(p-k)^2-m^2+i0(p_0-k_0)}\Theta(q_0-k_0)\delta((q-k)^2-m^2)\Theta(k_0)\delta(k^2)$$

$$-\Theta(k_0-p_0)\delta((p-k)^2-m^2)\frac{1}{(q-k)^2-m^2-i0(q_0-k_0)}\Theta(-k_0)\delta(k^2)$$

$$+p\leftrightarrow q\Big], \quad (3.8.28)$$

where the last three terms (not written down) have all plus signs and are obtained from the first three terms by interchanging p and q. The following symmetry relations are direct consequences of (3.8.28):

$$I_D(q,p)=-I_D(p,q)=I_D(-p,-q) \quad (3.8.29)$$

$$I_D(-p,q)=-I_D(q,-p)=I_D(-q,p). \quad (3.8.30)$$

Let us start with the computation of the first integral in (3.8.28)

$$I_1(p,q)\stackrel{\rm def}{=}\int\frac{dk}{k^2+i0}\Theta(k_0-p_0)\Theta(q_0-k_0)\delta((p-k)^2-m^2)\delta((q-k)^2-m^2). \quad (3.8.31)$$

Introducing

$$P=p-q \quad (3.8.32)$$

and using the new integration variable $k'=k-p$, we obtain

$$I_1=\int\frac{dk'}{(k'+p)^2+i0}\Theta(k'_0)\delta(k'^2-m^2)\Theta(-P_0-k'_0)\delta((P+k')^2-m^2). \quad (3.8.33)$$

In the following we write k instead of k' for the integration variable. The first Θ and δ functions together give $\delta(k_0-E_k)/2E_k$ which implies

$$k_0 = E_k = \sqrt{\boldsymbol{k}^2 + m^2} \quad , \quad P_0 < -k_0 < -m. \tag{3.8.34}$$

For space-like P there exists a Lorentz frame with $P_0 = 0$ such that I_1 vanishes. Consequently, P must be time-like. Then there exists a frame with $\boldsymbol{P} = \boldsymbol{0} = \boldsymbol{p} - \boldsymbol{q}$ (center of mass system) such that

$$\boldsymbol{p} = \boldsymbol{q} \quad .$$

The last δ-distribution in (3.8.33) implies

$$(P_0 + k_0)^2 - \boldsymbol{k}^2 - m^2 = 0, \tag{3.8.35}$$

which gets specialized to

$$P_0^2 = -2E_k P_0, \quad E_k = -\frac{P_0}{2} = -\frac{1}{2}(p_0 - q_0), \quad k_0 = -\frac{1}{2}P_0. \tag{3.8.36}$$

There remains to calculate the integral

$$I_1(p,q) = \Theta(-P_0) \int \frac{d^3k}{2E_k} \frac{1}{(k+p)^2 + i0} \frac{\delta(E_k - \frac{|P_0|}{2})}{2|P_0|}, \tag{3.8.37}$$

k_0 in the denominator is given by (3.8.36).

Let us now use $E_k = E$ as integration variable

$$\int d^3k = \int\limits_m^\infty dE\, E\sqrt{E^2 - m^2} \int d\Omega, \tag{3.8.38}$$

where the E-integral becomes trivial due to the δ-distribution in (3.8.37). The denominator can be written as follows

$$(k+p)^2 = (k_0 + p_0)^2 - (\boldsymbol{k} + \boldsymbol{p})^2$$

$$= \lambda^2 - \boldsymbol{k}^2 - \boldsymbol{p}^2 - 2|\boldsymbol{k}||\boldsymbol{p}| \cos\vartheta,$$

where $\lambda = k_0 + p_0 = -\frac{1}{2}P_0 + p_0$ and $|\boldsymbol{k}|$ is given by (3.8.35-36)

$$|\boldsymbol{k}| = \sqrt{\frac{P_0^2}{4} - m^2}. \tag{3.8.39}$$

Only the $\cos\vartheta$-integration remains to be done

$$I_1 = \frac{1}{4}\Theta(-P_0)\Theta(P_0^2 - 4m^2)\sqrt{\frac{1}{4} - \frac{m^2}{P_0^2}}\, 2\pi \int\limits_{-1}^{1} \frac{d\cos\vartheta}{\lambda^2 - \boldsymbol{k}^2 - \boldsymbol{p}^2 - 2|\boldsymbol{k}||\boldsymbol{p}|\cos\vartheta + i0}$$

$$= -\frac{\pi}{4} \frac{\Theta(-P_0)\Theta(P_0^2 - 4m^2)}{|P_0||\boldsymbol{p}|} \log \frac{\lambda^2 - (|\boldsymbol{k}| + |\boldsymbol{p}|)^2 + i0}{\lambda^2 - (|\boldsymbol{k}| - |\boldsymbol{p}|)^2 + i0}$$

$$= -\frac{\pi}{4}\frac{\Theta(-P_0)\Theta(P_0^2-4m^2)}{|P_0||\boldsymbol{p}|}\log\frac{p^2+m^2-P_0p_0-|\boldsymbol{p}|\sqrt{P_0^2-4m^2}+i0}{p^2+m^2-P_0p_0+|\boldsymbol{p}|\sqrt{P_0^2-4m^2}+i0}. \tag{3.8.40}$$

It is now not evident how the result looks like in an arbitrary Lorentz frame. We therefore consider the Lorentz boost transforming $P=(P_0,\mathbf{0})$ into a general four-vector $P'=(P_0',\boldsymbol{P}')$

$$\boldsymbol{P}'=-P_0\frac{\boldsymbol{v}}{\sqrt{1-\boldsymbol{v}^2}} \tag{3.8.41}$$

$$P_0'=\frac{P_0-\boldsymbol{v}\cdot\boldsymbol{P}}{\sqrt{1-\boldsymbol{v}^2}}=\frac{P_0}{1-\boldsymbol{v}^2}, \tag{3.8.42}$$

with velocity

$$\boldsymbol{v}=-\frac{\boldsymbol{P}'}{P_0'}. \tag{3.8.43}$$

This enables us to determine the covariant expressions which must be substituted into (3.8.40). Since the boost from the primed to the unprimed system has relative velocity $-\boldsymbol{v}$, we have (see Problem 0.5)

$$p_0=\frac{p_0'+\boldsymbol{v}\cdot\boldsymbol{p}'}{\sqrt{1-\boldsymbol{v}^2}}=\frac{p_0'P_0'}{P_0}-\frac{\boldsymbol{P}'\cdot\boldsymbol{p}'}{P_0}.$$

Using the obvious result $P_0=\sqrt{P^2}=\sqrt{P'^2}$, we get

$$p_0=\frac{p'P'}{\sqrt{P'^2}}. \tag{3.8.44}$$

The spatial part follows from

$$\boldsymbol{p}=\boldsymbol{p}'+\frac{\boldsymbol{v}}{\boldsymbol{v}^2}\Big[\boldsymbol{v}\cdot\boldsymbol{p}'\Big(\frac{1}{\sqrt{1-\boldsymbol{v}^2}}-1\Big)+\frac{p_0'\boldsymbol{v}^2}{\sqrt{1-\boldsymbol{v}^2}}\Big]$$

$$=\boldsymbol{p}'+\boldsymbol{P}'\frac{\boldsymbol{P}'\cdot\boldsymbol{p}'}{\boldsymbol{P}'^2}\Big(\frac{P_0'}{\sqrt{P'^2}}-1\Big)-\boldsymbol{P}'\frac{p_0'}{\sqrt{P'^2}}. \tag{3.8.45}$$

Taking the square

$$\boldsymbol{p}^2=\boldsymbol{p}'^2+\frac{(\boldsymbol{P}'\cdot\boldsymbol{p}')^2-2\boldsymbol{P}'\cdot\boldsymbol{p}'P_0'p_0'+\boldsymbol{P}'^2p_0'^2}{P'^2}$$

$$=-p'^2+\frac{1}{P'^2}(\boldsymbol{P}'\cdot\boldsymbol{p}'-P_0'p_0')^2=-p'^2+\frac{(P'p')^2}{P'^2},$$

we obtain the following covariant result

$$|\boldsymbol{p}|\sqrt{P'^2}=\sqrt{(P'p')^2-P'^2p'^2}. \tag{3.8.46}$$

It can be written in different forms

$$N\stackrel{\text{def}}{=}(Pp)^2-P^2p^2=(Pq)^2-P^2q^2=(pq)^2-p^2q^2, \tag{3.8.47}$$

omitting the primes from now on. The general result for I_1 (3.8.40) now reads

$$I_1(p,q) = -\frac{\pi}{4}\frac{\Theta(-P_0)\Theta(P^2-4m^2)}{\sqrt{N}}\log\frac{p^2+m^2-Pp-\sqrt{N}\sqrt{1-\frac{4m^2}{P^2}}+i0}{p^2+m^2-Pp+\sqrt{N}\sqrt{1-\frac{4m^2}{P^2}}+i0}.$$

Using the relation

$$\log\frac{a+ib0}{c+ib0} = \log\Big(\frac{a}{c}+ib(c-a)0\Big), \tag{3.8.48}$$

which holds for real a, b, c, the logarithm can be written as follows

$$-\log_1 \stackrel{\rm def}{=} -\log\left(\frac{pq+m^2+\sqrt{N}\sqrt{1-\frac{4m^2}{P^2}}}{pq+m^2-\sqrt{N}\sqrt{1-\frac{4m^2}{P^2}}}-i0\right). \tag{3.8.49}$$

Next we turn to the second integral in (3.8.28)

$$I_2(p,q) = \int\frac{dk}{(p-k)^2-m^2+i(p_0-k_0)0}\Theta(q_0-k_0)\delta((q-k)^2-m^2)$$
$$\Theta(k_0)\delta(k^2). \tag{3.8.50}$$

Since k lies on the forward light-cone ($k_0=|\boldsymbol{k}|$) and $k-q$ on the backward mass hyperboloid, it follows that q must be inside the forward cone $q\in V^+$. We choose a Lorentz frame such that $\boldsymbol{q}=\boldsymbol{0}$. Then the first δ-distribution implies

$$(q-k)^2 = (q_0-|\boldsymbol{k}|)^2-(\boldsymbol{q}-\boldsymbol{k})^2 = q_0^2-2q_0|\boldsymbol{k}| = m^2,$$

$$|\boldsymbol{k}| = \frac{q_0^2-m^2}{2q_0}. \tag{3.8.51}$$

Since $q_0>0$ and $|\boldsymbol{k}|>0$, we must have $q_0^2-m^2\ge 0$. Then the argument of $\Theta(q_0-k_0)$ is automatically positive. Integrating out the two δ-distributions we arrive at

$$I_2 = \Theta(q_0)\Theta(q_0^2-m^2)\frac{1}{4q_0}\int\frac{d\cos\vartheta\,d\varphi\,|\boldsymbol{k}|}{(p_0-k_0)^2-(\boldsymbol{p}-\boldsymbol{k})^2-m^2+i\alpha 0},$$

where $\alpha = p_0-|\boldsymbol{k}|$. Taking the polar axis along $\boldsymbol{p}$ and using (3.8.51), the denominator becomes

$$p^2-m^2-\frac{p_0}{q_0}(q_0^2-m^2)+2|\boldsymbol{k}||\boldsymbol{p}|\cos\vartheta+i\alpha 0$$
$$\stackrel{\rm def}{=} \lambda+2|\boldsymbol{k}||\boldsymbol{p}|\cos\vartheta+i\alpha 0,$$

hence

$$I_2 = \frac{\pi}{4} \frac{\Theta(q_0)\Theta(q_0^2 - m^2)}{q_0|\boldsymbol{p}|} \log \frac{\lambda + \frac{|\boldsymbol{p}|}{q_0}(q_0^2 - m^2) + i\alpha 0}{\lambda - \frac{|\boldsymbol{p}|}{q_0}(q_0^2 - m^2) + i\alpha 0}. \tag{3.8.52}$$

The logarithm can be written as follows

$$\log_2 \stackrel{\text{def}}{=} \log \frac{p^2 - m^2 + (|\boldsymbol{p}|q_0 - p_0 q_0)(1 - \frac{m^2}{q_0^2}) + i\alpha 0}{p^2 - m^2 - (|\boldsymbol{p}|q_0 - p_0 q_0)(1 - \frac{m^2}{q^2}) + i\alpha 0}, \tag{3.8.53}$$

with

$$\alpha = p_0 - \frac{q_0^2 - m^2}{2q_0} = \frac{2p_0 q_0 - q_0^2 + m^2}{2q_0}. \tag{3.8.54}$$

Here the denominator $2q_0$ can be dropped because it is positive. The transformation to an arbitrary Lorentz frame is carried out as above (3.8.44-46) with P substituted by q:

$$q_0 \to \sqrt{q^2}, \quad p_0 \to \frac{pq}{\sqrt{q^2}}, \quad |\boldsymbol{p}|q_0 \to \sqrt{N}. \tag{3.8.55}$$

With (3.8.48) this leads to the final result

$$I_2(p,q) = \frac{\pi}{4} \frac{\Theta(q_0)\Theta(q^2 - m^2)}{\sqrt{N}} \log_2, \tag{3.8.56}$$

$$\log_2 = \log\left(\frac{p^2 - m^2 - pq(1 - \frac{m^2}{q^2}) + \sqrt{N}(1 - \frac{m^2}{q^2})}{p^2 - m^2 - pq(1 - \frac{m^2}{q^2}) - \sqrt{N}(1 - \frac{m^2}{q^2})} + i(q^2 - m^2 - 2pq)0\right). \tag{3.8.57}$$

It is easy to see that the imaginary part can also be written as $i0(m^2 - p^2)$.

The third integral in (3.8.28) is obtained from the second by interchanging $p \leftrightarrow q$ and $p_0 \to -p_0$, $q_0 \to -q_0$:

$$I_3(p,q) = \frac{\pi}{4} \frac{\Theta(-p_0)\Theta(p^2 - m^2)}{\sqrt{N}} \log_3 \tag{3.8.58}$$

$$\log_3 = \log\left(\frac{q^2 - m^2 - pq(1 - \frac{m^2}{p^2}) + \sqrt{N}(1 - \frac{m^2}{p^2})}{q^2 - m^2 - pq(1 - \frac{m^2}{p^2}) - \sqrt{N}(1 - \frac{m^2}{p^2})} - i(q^2 - m^2)0\right). \tag{3.8.59}$$

Summing up, the scalar integral for R' (i.e. the first three terms in (3.8.28)) is equal to

$$I_{R'}(p,q) = \frac{\pi}{4(2\pi)^6}\left\{-\frac{\Theta(-P_0)\Theta(P^2 - 4m^2)}{\sqrt{N}} \log_1\right.$$

$$\left.-\frac{\Theta(q_0)\Theta(q^2 - m^2)}{\sqrt{N}} \log_2 - \frac{\Theta(-p_0)\Theta(p^2 - m^2)}{\sqrt{N}} \log_3\right\}. \tag{3.8.60}$$

To get the causal scalar integral (3.8.28) we have to subtract the same expression with p and q interchanged ($I_{A'}$). This alters the Θ-functions into sign-functions:

$$I_D(p,q)=\frac{\pi}{4(2\pi)^6}\Big\{\frac{\mathrm{sgn}\,(P_0)\Theta(P^2-4m^2)}{\sqrt{N}}\log_1$$

$$-\frac{\mathrm{sgn}\,q_0\Theta(q^2-m^2)}{\sqrt{N}}\log_2+\frac{\mathrm{sgn}\,p_0\Theta(p^2-m^2)}{\sqrt{N}}\log_3\Big\}. \tag{3.8.61}$$

It is important to notice that the imaginary parts cancel out in I_D. This follows most easily from the original expression (3.8.28). We can therefore take the absolute values under the logarithms. However, the imaginary parts do not cancel in $I_{R'}$ alone:

$$\mathrm{Im}\, I_{R'}(p,q)=\frac{\pi^2}{4(2\pi)^6\sqrt{N}}\Theta(p^2-m^2)\Theta(q^2-m^2)\Theta\Big(2pq-p^2\frac{q^2-m^2}{p^2-m^2}$$

$$-q^2\frac{p^2-m^2}{q^2-m^2}\Big)=\mathrm{Im}\, I_{A'}(p,q). \tag{3.8.62}$$

This result can be obtained more directly by taking the imaginary part in the original definition of $I_{R'}$ (analogous to (3.8.28)).

We now turn to the vectorial integral

$$I_D^\nu(p,q)=(2\pi)^{-6}\int d^4k\,k^\nu[(3.8.28)],$$

where in the bracket all previous terms appear. We note the relations

$$I_D^\nu(-p,-q)=I_D^\nu(p,q)=-I_D^\nu(q,p)$$

$$I_D^\nu(-p,q)=-I_D^\nu(q,-p)=-I_D^\nu(-q,p). \tag{3.8.63}$$

The most economic way to compute this is, perhaps, to generate it as derivative of certain other scalar integrals. For example

$$I_1^\nu(p,q)=\int\frac{d^4k\,k^\nu}{k^2+i0}\Theta(k_0-p_0)\Theta(q_0-k_0)\delta((p-k)^2-m^2)\delta((q-k)^2-m^2)$$

$$=\tfrac{1}{2}\int d^4k[\partial^\nu\log(k^2+i0)]\Theta\ldots$$

$$=-\tfrac{1}{2}\int d^4k\,\log(k^2+i0)\partial_k^\nu[\Theta\ldots]$$

$$=\tfrac{1}{2}\int d^4k\,\log(k^2+i0)(\partial_p^\nu+\partial_q^\nu)[\Theta\ldots]$$

$$=\tfrac{1}{2}(\partial_p^\nu+\partial_q^\nu)J_1(p,q), \tag{3.8.64}$$

with

$$J_1(p,q) = \int d^4k \, \log(k^2+i0)\Theta(k_0-p_0)\delta((p-k)^2-m^2)$$
$$\times\,\Theta(q_0-k_0)\delta((q-k)^2-m^2). \tag{3.8.65}$$

This integral can be calculated as I_1 (3.8.32) above:

$$J_1(p,q) = \frac{\pi}{4}\frac{\Theta(-P_0)\Theta(P^2-4m^2)}{\sqrt{N}}\Big\{(pq+m^2)\log_1 + \sqrt{N}\sqrt{1-\frac{4m^2}{P^2}}$$
$$\times\log\Big[(pq+m^2)^2 - N\big(1-\frac{4m^2}{P^2}\big)\Big] - 2\sqrt{N}\sqrt{1-\frac{4m^2}{P^2}}\Big\}. \tag{3.8.66}$$

Similarly we write

$$I_2^\nu(p,q) = \int \frac{d^4k\,(k^\nu-p^\nu+p^\nu)}{(p-k)^2-m^2+i0(p_0-k_0)}\Theta(q_0-k_0)\delta((q-k)^2-m^2)$$
$$\times\,\Theta(k_0)\delta(k^2) = p^\nu I_2 - \tfrac{1}{2}\partial_p^\nu J_2 \tag{3.8.67}$$

with

$$J_2(p,q) = \int d^4k \, \log\Big[(p-k)^2-m^2+i0\Big]\Theta(q_0-k_0)\delta((q-k)^2-m^2)$$
$$\times\,\Theta(k_0)\delta(k^2). \tag{3.8.68}$$

This integral is calculated as I_2 (3.8.50)

$$J_2(p,q) = \frac{\pi}{4}\frac{\Theta(q_0)\Theta(q^2-m^2)}{\sqrt{N}}\Big\{A_1\log A_1 - A_2\log A_2 + A_2 - A_1\Big\}, \tag{3.8.69}$$

where

$$A_{1,2} = p^2-m^2-\Big(1-\frac{m^2}{q^2}\Big)(pq\mp\sqrt{N}) + i0(p^2-m^2). \tag{3.8.70}$$

The differentiations are tedious but straightforward and lead to the following result for the vector integrals:

$$I_D^\nu(p,q) = \frac{I_D(p,q)}{2N}\Big[p^\nu\Big(pq(q^2-m^2)-q^2(p^2-m^2)\Big)+q^\nu\Big(pq(p^2-m^2)$$
$$-p^2(q^2-m^2)\Big)\Big] + \frac{\pi}{4(2\pi)^6N}\Big[\mathrm{sgn}\,p_0\Theta(p^2-m^2)\Big(1-\frac{m^2}{p^2}\Big)(q^\nu p^2-p^\nu pq)$$
$$-\mathrm{sgn}\,q_0\Theta(q^2-m^2)\Big(1-\frac{m^2}{q^2}\Big)(p^\nu q^2-q^\nu pq)$$
$$+\mathrm{sgn}\,P_0\Theta(P^2-4m^2)\Big(p^\nu(pq-q^2)+q^\nu(pq-p^2)\Big)\sqrt{1-\frac{4m^2}{P^2}}\Big]. \tag{3.8.71}$$

$I_{R'}^\nu$ is obtained by replacing I_D by $I_{R'}$ and

$$\operatorname{sgn} p_0 \to -\Theta(-p_0), \quad \operatorname{sgn} q_0 \to \Theta(q_0), \quad \operatorname{sgn} P_0 \to -\Theta(-P_0). \tag{3.8.72}$$

Finally we turn to the tensorial integral

$$I_D^{\mu\nu}(p,q) = (2\pi)^{-6} \int d^4k k^\mu k^\nu [(3.8.28)]. \tag{3.8.73}$$

The symmetry relations read

$$I^{\mu\nu}(-p,-q) = -I^{\mu\nu}(p,q) = I^{\mu\nu}(q,p)$$

$$I^{\mu\nu}(-p,q) = -I^{\mu\nu}(q,-p) = I^{\mu\nu}(-q,p). \tag{3.8.74}$$

Again we generate the integral

$$I_1^{\mu\nu}(p,q) = \int \frac{d^4k\, k^\mu k^\nu}{k^2+i0} \Theta(k_0-p_0)\Theta(q_0-k_0)\delta((p-k)^2-m^2)\delta((q-k)^2-m^2) \tag{3.8.75}$$

as derivative by means of

$$\frac{k^\mu k^\nu}{k^2+i0} = \frac{1}{2}\partial^\nu [k^\mu \log(k^2+i0)] - \frac{1}{2} g^{\mu\nu} \log(k^2+i0).$$

This leads to

$$I_1^{\mu\nu} = -\tfrac{1}{2} g^{\mu\nu} J_1 + \tfrac{1}{4} \int d^4k\, \partial_k^\mu \partial_k^\nu [k^2 \log(k^2+i0) - k^2] \ldots$$

$$= -\tfrac{1}{2} g^{\mu\nu} J_1 + \tfrac{1}{4} (\partial_p^\mu + \partial_q^\mu)(\partial_p^\nu + \partial_q^\nu) K_1, \tag{3.8.76}$$

where J_1 is given by (3.8.65) and

$$K_1(p,q) = \int d^4k\, [k^2 \log(k^2+i0) - k^2] \Theta(k_0-p_0)\delta((p-k)^2-m^2)$$

$$\times\, \Theta(q_0-k_0)\delta((q-k)^2-m^2) \tag{3.8.77}$$

can be calculated as before

$$K_1(p,q) = \frac{\pi}{4} \frac{\Theta(-P_0)\Theta(P^2-4m^2)}{\sqrt{N}} \Bigl\{ \Bigl[(pq+m^2)^2 + (Pp)^2 \bigl(1-\frac{4m^2}{P^2}\bigr) + p^2(4m^2$$

$$-P^2) \Bigr] \log_1 - (pq+m^2)\sqrt{N}\sqrt{1-\frac{4m^2}{P^2}} \Bigl[\log\bigl((pq+m^2)^2 - N(1-\frac{4m^2}{P^2})\bigr) + 3 \Bigr] \Bigr\}. \tag{3.8.78}$$

The second tensor integral is obtained similarly to (3.8.66):

$$I_2^{\mu\nu}(p,q) = \int d^4k \frac{(k^\mu - p^\mu + p^\mu)(k^\nu - p^\nu + p^\nu)}{(k-p)^2 - m^2 + i0(p_0-k_0)} \Theta(q_0-k_0)\delta((q-k)^2-m^2)$$

$$\times\, \Theta(k_0)\delta(k^2) = \int d^4k \frac{(k^\mu-p^\mu)(k^\nu-p^\nu)}{\ldots} \ldots$$

$$+p^\nu (I_2^\mu - p^\mu I_2) + p^\mu (I_2^\nu - p^\nu I_2) - p^\mu p^\nu I_2, \tag{3.8.79}$$

where the first integral, say $J_2^{\mu\nu}$, can be expressed as (3.8.67) by derivatives

$$J_2^{\mu\nu} = \tfrac{1}{4}\partial_p^\mu \partial_p^\nu K_2 - \tfrac{1}{2} g^{\mu\nu} J_2, \tag{3.8.80}$$

with

$$K_2(p,q) = \int d^4k\, [(k-p)^2 - m^2]\{\log[(k-p)^2 - m^2 + i0(p_0 - k_0)] - 1\}$$

$$\times\, \Theta(q_0 - k_0)\delta((q-k)^2 - m^2)\Theta(k_0)\delta(k^2)$$

$$= \frac{\pi}{4}\frac{\Theta(q_0)\Theta(q^2 - m^2)}{\sqrt{N}}[F(A_1) - F(A_2)], \tag{3.8.81}$$

$$F(A) = \frac{A^2}{2}\log A - \frac{3}{4}A^2, \tag{3.8.82}$$

where $A_{1,2}$ are given by (3.8.70). As before we write the final result by means of the scalar integrals $I_{R'}$, I_D:

$$I_D^{\mu\nu}(p,q) = \frac{I_D(p,q)}{N}\Bigg\{-q^\mu q^\nu\Big[-\frac{3}{8N}\Big((q^2-m^2)p^2-pq(p^2-m^2)\Big)^2+\frac{1}{8}(p^2-m^2)^2\Big]$$

$$-(p^\mu q^\nu + q^\mu p^\nu)\frac{1}{8}\Big[2(p^2-m^2)(q^2-m^2) + \frac{3pq}{N}\Big((q^2-m^2)^2p^2$$

$$-2(q^2-m^2)(p^2-m^2)pq + (p^2-m^2)^2q^2\Big)\Big]$$

$$-p^\mu p^\nu\Big[-\frac{3}{8N}\Big((p^2-m^2)q^2 - pq(q^2-m^2)\Big)^2 + \frac{1}{8}(q^2-m^2)^2\Big]$$

$$+g^{\mu\nu}\frac{1}{8}\Big[p^2(q^2-m^2)^2 - 2pq(q^2-m^2)(p^2-m^2) + q^2(p^2-m^2)^2\Big]\Bigg\}$$

$$+\frac{\pi}{4(2\pi)^6N}\Bigg\{\mathrm{sgn}\,p_0\Theta(p^2-m^2)\Big(1-\frac{m^2}{p^2}\Big)\Big\{q^\mu q^\nu\frac{3p^2}{4N}\Big[pq(p^2-m^2)-p^2(q^2-m^2)\Big]$$

$$+p^\mu p^\nu\frac{1}{4N}\Big[pq\Big(1-\frac{m^2}{p^2}\Big)\Big)(5p^2q^2-2(pq)^2) - (q^2-m^2)(2(pq)^2+p^2q^2)\Big]$$

$$+(p^\mu q^\nu + p^\nu q^\mu)\frac{p^2}{4N}\Big[3pq(q^2-m^2) - \Big(1-\frac{m^2}{p^2}\Big)((pq)^2+2p^2q^2)\Big]$$

$$+g^{\mu\nu}\frac{p^2}{4}\Big[pq\Big(1-\frac{m^2}{p^2}\Big)-(q^2-m^2)\Big]\Big\}$$

$$-\mathrm{sgn}\,q_0\Theta(q^2-m^2)\Big(1-\frac{m^2}{q^2}\Big)\Big\{p\leftrightarrow q\Big\}$$

$$+\mathrm{sgn}\,P_0\Theta(P^2-4m^2)\frac{1}{4}\sqrt{1-\frac{4m^2}{P^2}}\Big\{p^\mu p^\nu\Big[5pq - 4q^2 + m^2$$

$$-3\frac{(pq+m^2)(pq-q^2)^2}{N}\Bigg]+q^\mu q^\nu\Bigg[5pq-4p^2+m^2-3\frac{(pq+m^2)(pq-p^2)^2}{N}\Bigg]$$

$$+(p^\mu q^\nu+p^\nu q^\mu)\Bigg[3pq-2p^2-2q^2-m^2-3\frac{(pq+m^2)(pq-q^2)(pq-p^2)}{N}\Bigg]$$

$$-g^{\mu\nu}\Big[P^2(pq+m^2)\Big]\Bigg]\Bigg\}. \tag{3.8.83}$$

$I^{\mu\nu}_{R'}$ is again obtained by the replacements (3.8.72). Although these are rather complicated formulae, one should keep in mind that the only non-trivial operation in their computation, namely integration, has been very simple: only 4-dimensional integrals with two 1-dimensional δ-distributions had to be computed.

The total causal distribution $\hat{D}^\nu(p,q)$ (3.8.26) is now given in terms of the I_D-integrals by

$$\hat{D}^\nu(p,q)=-2\not{q}\gamma^\nu\not{p}I+2\not{q}\gamma^\nu\gamma^\mu I_\mu+2I_\mu\gamma^\mu\gamma^\nu\not{p}$$

$$+2I^\mu_\mu\gamma^\nu-4I^\nu_\mu\gamma^\mu+4m(p^\nu+q^\nu)I-8mI^\nu-2m^2\gamma^\nu I, \tag{3.8.84}$$

due to (3.8.27) (omitting the subscripts D).

3.9 Vertex Function: Retarded Distribution

The distribution D^ν (3.8.84) has a causal support in x-space. It must now be decomposed into retarded and advanced parts. The retarded part R^ν is given in momentum space by the dispersion integral (3.2.61)

$$R^\nu(p,q)=\frac{i}{2\pi}\int\limits_{-\infty}^{+\infty}dt\,\frac{\hat{D}^\nu(tp,tq)}{(t-i0)^{\omega+1}(1-t+i0)}. \tag{3.9.1}$$

where ω is the singular order of the distributions which is $=0$ in case of the vertex function. If $\omega\le -1$, one has to take $\omega=-1$ in (3.9.1). p is assumed to be in the forward light-cone V^+ and q in the backward light-cone V^-. The result for arbitrary p,q can be obtained by analytic continuation (3.2.64): $R^\nu(p,q)$ is the boundary value of an analytic function, regular in $p\in\mathbf{R^4}+iV^+$, $q\in\mathbb{R}^4+iV^-$. This central splitting solution is normalized at $p=0=q$

$$(D^a\,R^\nu)(0,0)=0\quad\forall|a|\le\omega. \tag{3.9.2}$$

The general solution is then given by

$$\tilde{R}^\nu(p,q)=R^\nu(p,q)+C\gamma^\nu \tag{3.9.3}$$

with an arbitrary finite constant C. The splitting of (3.8.84) can be achieved by splitting the scalar, vector and tensor integrals $I^{\cdots}$ separately with their individual ω's. In fact, the integrals $I^{\cdots}$ separately have causal supports. The multiplications by p and q in (3.8.84) correspond to derivatives in x-space which do not disturb the causal structure.

We now turn to the splitting of the scalar integral which has $\omega = -2$

$$I_{\rm ret}(p,q) = \frac{i}{2\pi} \int\limits_{-\infty}^{+\infty} \frac{I(tp,tq)}{1-t+i0}\, dt. \tag{3.9.4}$$

Using the expression (3.8.61), we have first to compute

$$J_0 = \int\limits_{-\infty}^{+\infty} dt \, \frac{\operatorname{sgn} t\, \Theta(t^2P^2 - 4m^2)}{t^2(1-t+i0)} \log \frac{t^2pq + m^2 + t^2\sqrt{N}\sqrt{1-\frac{4m^2}{t^2P^2}}}{t^2pq + m^2 - t^2\sqrt{N}\sqrt{1-\frac{4m^2}{t^2P^2}}}, \tag{3.9.5}$$

where $P = p - q$. This integral contains a non-elementary part which can be expressed in terms of the so-called Spence function

$$L(z) = \int_0^z \frac{\log(1-t)}{t}\, dt. \tag{3.9.6}$$

This function often occurs in exact calculations of radiative corrections. Its most important properties are given in Appendix C.

Calculating with Spence functions is tricky because they obey very many identities. For this reason we give the substitutions which have to be done in the integral (3.9.5): First set $t^2 = s$ and then

$$\sqrt{s^2 - 4bs} = sx, \quad \text{where} \quad b = \frac{m^2}{P^2},$$

which leads to

$$J_0 = \int_0^1 dx \, \frac{2x}{1 - x^2 - 4b + i0} \log \frac{4bpq + m^2(1-x^2) + 4b\sqrt{N}x}{4bpq + m^2(1-x^2) - 4b\sqrt{N}x}. \tag{3.9.7}$$

This integral defines an analytic function of p, q, because it can be differentiated under the integral sign, where $i0$ may be omitted if b is complex. Therefore it can be used for analytic continuation to arbitrary real p, q. According to Theorem 3.4 (3.2.64), the continuation has to be done via complex momenta by means of the substitutions $p \to p + i\eta$, $q \to q - i\eta$ with $\eta = (\varepsilon, 0, 0, 0)$, $\varepsilon > 0$. Then we find $p - q \to P + 2i\eta$ and $(p-q)^2 \to P^2 + 4iP\eta \to P^2 + i0P_0$ for the boundary value for real momenta. This leads to the substitution $-4b + i0 \to -4b + i0P_0$ in the denominator of (3.9.7). The boundary value for arbitrary real p and q is therefore given by

$$J_0 = \int\limits_0^1 dx \, \frac{2x}{1 - x^2 - 4b + i0P_0} \log \ldots \,. \tag{3.9.8}$$

Since the logarithmic singularities are locally integrable, (3.9.8) is a well defined distribution. The imaginary parts are always uniquely determined by the boundary values from complex $p \in \mathbb{R}^4 + iV^+$, $q \in \mathbb{R}^4 + iV^-$.

We next write the quadratic polynomials under the logarithm in product form

$$J_0 = \int_0^1 dx \, \frac{2x}{1 - x^2 - 4b + i0P_0} \log \frac{(x - x_1)(x - x_2)}{(x + x_1)(x + x_2)}, \tag{3.9.9}$$

$$\text{with} \quad x_{1,2} = \frac{2}{P^2}\sqrt{N} \pm \left| \frac{p^2 - q^2}{P^2} \right| \quad \text{and} \quad N = (pq)^2 - p^2 q^2. \tag{3.9.10}$$

The integral (3.9.9) can now be easily expressed by Spence functions. Since it is regular on the mass shell

$$p^2 = m^2, \quad q^2 = m^2, \tag{3.9.11}$$

we give the final result only in this case. Furthermore, we have calculated it for space-like $P^2 < 0$ for later applications and, therefore have no contribution from $i0P_0$ in the denominator of (3.9.9),

$$J_0 = 2L\Big(\frac{|x_1| + 1}{2|x_1|}\Big) - 2L\Big(\frac{|x_1| - 1}{2|x_1|}\Big) - \log^2\Big(\frac{|x_1| + 1}{2|x_1|}\Big) + \log^2\Big(\frac{|x_1| - 1}{2|x_1|}\Big), \tag{3.9.12}$$

$$\text{with} \quad |x_1| = \sqrt{1 - \frac{4m^2}{P^2}} \tag{3.9.13}$$

and $\log^2 x = (\log x)^2$. J_0 (3.9.12) is real in the considered region.

The second integral in $I_{\rm ret}$ is

$$J_1 = \int_{-\infty}^{+\infty} dt \, \frac{\operatorname{sgn} t \, \Theta(t^2 p^2 - m^2)}{t^2(1 - t + i0)} \log \frac{a_0 + b_0 t^2 + a_1 + b_1 t^2}{a_0 + b_0 t^2 - a_1 - b_1 t^2} \tag{3.9.14}$$

where

$$a_0 = \frac{m^2}{p^2} pq - m^2 \qquad b_0 = q^2 - pq \tag{3.9.15}$$

$$a_1 = -\frac{m^2}{p^2}\sqrt{N} \qquad b_1 = \sqrt{N}. \tag{3.9.16}$$

The corresponding analytic function, defined for $\tilde{p} \in \mathbb{R}^4 + iV^+$, $\tilde{q} \in \mathbb{R}^4 + iV^-$ is given by

$$J_1^{\rm an}(\tilde{p}, \tilde{q}) = \int_{m^2}^{\infty} \frac{d\tau}{\tilde{p}^2 - \tau} \log \frac{\tilde{a}_0 \tilde{p}^2 + \tilde{b}_0 \tau + \tilde{a}_1 \tilde{p}^2 + \tilde{b}_1 \tau}{\tilde{a}_0 \tilde{p}^2 + \tilde{b}_0 \tau - \tilde{a}_1 \tilde{p}^2 - \tilde{b}_1 \tau}. \tag{3.9.17}$$

Taking the boundary value for $(p, q) \in \mathbb{R}^8$, the denominator $\tilde{p}^2 - \tau$ goes over into $p^2 - \tau + i0p^0$. Transforming to Spence functions we obtain for its principle value part

$$J_1^P(p,q) = \log\frac{a_0+a_1+b_0+b_1}{a_0-a_1+b_0-b_1}\log\frac{m^2-p^2}{p^2} + \log\frac{a_0+a_1}{a_0-a_1}\log\frac{p^2}{m^2}$$

$$-\frac{1}{2}\log^2\frac{a_1-a_0}{b_0-b_1} + \frac{1}{2}\log^2\frac{-a_0-a_1}{b_0+b_1} + \frac{1}{2}\log^2\frac{a_0-a_1+b_0-b_1}{b_0-b_1}$$

$$-\frac{1}{2}\log^2\frac{a_0+a_1+b_0+b_1}{b_0+b_1} + L\Big(\frac{m^2}{p^2}\frac{b_0-b_1}{a_0-a_1}\Big)$$

$$-L\Big(\frac{m^2}{p^2}\frac{b_0+b_1}{a_0+a_1}\Big) - L\Big(\frac{m^2-p^2}{p^2}\frac{b_0-b_1}{t_1}\Big) + L\Big(\frac{m^2-p^2}{p^2}\frac{b_0+b_1}{t_2}\Big), \qquad (3.9.18)$$

$$\text{where} \quad t_{1,2} = -a_0 \pm a_1 - b_0 \pm b_1.$$

Here the first term has a logarithmic singularity on the mass shell, which will be discussed in Sect. 3.11, all other singularities cancel out in the following (i.e. in (3.9.20, 21)). We shall use (3.9.18) only for the imaginary part of $I_{\rm ret}$, i.e. only $\mathrm{Re}\, J_1^P$ is of interest here, $\mathrm{Re}\, I_{\rm ret}$ will be obtained in a more general way. Therefore, we don't have to specify the imaginary parts of the logarithms in (3.9.18). But $\mathrm{Re}\, J_1$ gets an additional contribution J_1^δ coming from $-i\pi \mathrm{sgn}\, p^0 \delta(p^2-\tau) i\mathrm{Im}\,(\log\ldots)$ in the boundary value of (3.9.17). For $p \in V^+$, $q \in V^+$ we obtain near the mass shell ($|p^2-m^2| \ll m^2$, $|q^2-m^2| \ll m^2$)

$$J_1^\delta(p,q) = -i\pi\Theta(p^2-m^2) i\mathrm{Im}\left[\log\frac{(m^2-p^2)(pq-\sqrt{N})+(q^2-m^2)p^2-i0}{(m^2-p^2)(pq+\sqrt{N})+(q^2-m^2)p^2-i0}\right]$$

$$= \pi^2\Theta(p^2-m^2)\Theta(q^2-m^2)\Theta(pq)\Theta(2pq - p^2\frac{q^2-m^2}{p^2-m^2} - q^2\frac{p^2-m^2}{q^2-m^2}), \qquad (3.9.19)$$

which is symmetrical in p and q. Hence, these $\delta \cdot \mathrm{Im}\,(\log)$ - terms double in the total scalar integral

$$\mathrm{Im}\,(I_{\rm ret}) = \frac{1}{(2\pi)^6 8\sqrt{N}}\mathrm{Re}\,[J_0(p,q)+J_1(p,q)+J_1(q,p)], \qquad (3.9.20)$$

they don't compensate each other as the imaginary parts of the logarithms in the D-distribution (3.8.61). Fortunately, the $\hat{R}$-distribution (3.9.20) is not the whole story: In $\hat{T} = \hat{R} - \hat{R}'$ the terms (3.9.19) cancel against the imaginary part of $\hat{R}'$ (3.8.62). Then, for the imaginary part of the $\hat{T}$-distribution of the total scalar integral we end up with

$$\mathrm{Im}\,(I_T) = \frac{1}{(2\pi)^6 8\sqrt{N}}[J_0(p,q) + \mathrm{Re}\, J_1^P(p,q) + \mathrm{Re}\, J_1^P(q,p)] \qquad (3.9.21)$$

for $p, q \in V^+$ near the mass shell.

Now we turn to the real part of $I_{\rm ret}$ resp. I_T. The analytic function associated with $I_{\rm ret}$ (analogous to (3.9.17)) is given by

$$I^{\mathrm{an}}(\tilde{p},\tilde{q})=\frac{i}{2\pi}\frac{\pi}{4(2\pi)^6\sqrt{N}}\Bigg[\int\limits_{m^2}^{\infty}\frac{d\tau}{\tilde{p}^2-\tau}\log_1(\tau;\tilde{p},\tilde{q})$$

$$+\int\limits_{m^2}^{\infty}\frac{d\tau}{\tilde{q}^2-\tau}\log_2(\tau;\tilde{p},\tilde{q})+\int\limits_{4m^2}^{\infty}\frac{d\tau}{\tilde{P}^2-\tau}\log_3(\tau;\tilde{p},\tilde{q})\Bigg], \tag{3.9.22}$$

where $\tilde{p}\in\mathbb{R}^4+iV^+$, $\tilde{q}\in\mathbb{R}^4+iV^-$ and $\log_1(\tau;\tilde{p},\tilde{q})$ is the logarithm in (3.9.17). The function $I^{\mathrm{an}}(\tilde{p},\tilde{q})$ can be analytically continued to $\tilde{p}\in\mathbb{R}^4+iV^-$, $\tilde{q}\in\mathbb{R}^4+iV^+$ and the boundary value from this (opposite) side to real p, q gives the *advanced* distribution $I_{\mathrm{av}}(p,q)$. The imaginary parts of the logarithms in (3.9.22) get the opposite sign in this case. Calculating $I=I_{\mathrm{ret}}-I_{\mathrm{av}}$ as the difference of these two boundary values of $I^{\mathrm{an}}(\tilde{p},\tilde{q})$ we obtain

$$I(p,q)=\frac{i}{4(2\pi)^6\sqrt{N}}\Bigg[P\int\limits_{m^2}^{\infty}\frac{d\tau}{p^2-\tau}\,i\,\mathrm{Im}\,\log_1(\tau;p,q)$$

$$+P\int\limits_{m^2}^{\infty}\frac{d\tau}{q^2-\tau}\,i\,\mathrm{Im}\,\log_2(\tau;p,q)+P\int\limits_{4m^2}^{\infty}\frac{d\tau}{P^2-\tau}\,i\,\mathrm{Im}\,\log_3(\tau;p,q)\Bigg] \tag{3.9.23}$$

$$+\frac{\pi}{4(2\pi)^6\sqrt{N}}\Big[\mathrm{sgn}\,p^0\Theta(p^2-m^2)\mathrm{Re}\,\log_1(p^2;p,q)$$

$$-\mathrm{sgn}\,q_0\Theta(q^2-m^2)\mathrm{Re}\,\log_2(q^2;p,q)+\mathrm{sgn}\,P^0\Theta(P^2-m^2)\mathrm{Re}\,\log_3(P^2;p,q)\Big], \tag{3.9.24}$$

where we have decomposed the denominators in (3.9.22) into principle value (3.9.23) and δ-terms (3.9.24). We have used the convention to give the imaginary parts of the logarithms the signs of I_{ret}. Comparing the result with (3.8.61) we see that (3.9.23) must vanish. This proves again that the imaginary parts of the logarithms in (3.8.61) drop out.

Concerning the real part of I_{ret}, we see that it is given by the contribution of I_{ret} to (3.9.23), that is $\frac{1}{2}I(p,q)=\frac{1}{2}\mathrm{Re}\,(I_{R'}(p,q)-I_{A'}(p,q))$. Including $\mathrm{Re}\,\hat{R}'$ we finally get for the T-distribution

$$I_T(p,q)=-\frac{1}{2}\mathrm{Re}\,\Big(I_{R'}(p,q)+I_{A'}(p,q)\Big)+i\,\mathrm{Im}\,I_T(p,q), \tag{3.9.25}$$

where $I_{R'}$ and $I_{A'}$ are given by (3.8.60).

We now turn to the vector and tensor integrals. Since the logarithms appear here in the same combination as in the scalar integral, the compensation of the $\delta-\mathrm{Im}\,(\log)$-terms of $\hat{R}$ with the imaginary part of $\hat{R}'$ works for those integrals too. The above reasoning (3.9.22–25) concerning the real part of I_T can be extended in an obvious way to the vector and tensor integrals. Moreover, these integrals are regular on the mass shell (3.9.11). We therefore give the final results only in this case (i.e. after analytic continuation and inclusion of R'), which saves us from writing down long formulas as (3.8.71) and (3.8.83). The singular orders are $\omega = -1$ and 0, respectively. To get the vector integral, we have therefore to compute

$$I^{\nu}_{\rm ret}(p,q) = \frac{i}{2\pi}\int\limits_{-\infty}^{+\infty} dt\, \frac{I^{\nu}(tp,tq)}{1-t+i0}. \tag{3.9.26}$$

The tensor integral is split according to

$$I^{\mu\nu}_{\rm ret}(p,q) = \frac{i}{2\pi}\int\limits_{-\infty}^{+\infty} dt\, \frac{I^{\mu\nu}(tp,tq)}{t(1-t+i0)}. \tag{3.9.27}$$

The integrals needed for the splitting of I^{ν} and $I^{\mu\nu}$ are all of the following form:

$$K_n \stackrel{\rm def}{=} \int\limits_{-\infty}^{+\infty} dt\, \frac{\operatorname{sgn} t\,\Theta(t^2-a^2)}{t^{n+1}(1-t+i0)} f(t^2). \tag{3.9.28}$$

Note the relation

$$K_{2n} = K_{2n+1}. \tag{3.9.29}$$

For the vector integral (3.9.26) we especially need

$$K_2' \stackrel{\rm def}{=} K_2 - K_0 = \int\limits_{a^2}^{\infty} \frac{ds}{s^2} f(s) \tag{3.9.30}$$

and for the tensor integral (3.9.27)

$$K_4' \stackrel{\rm def}{=} K_4 - 2K_2 + K_0 = -K_2' + \int\limits_{a^2}^{\infty} \frac{ds}{s^3} f(s). \tag{3.9.31}$$

The splitting formula (3.9.28) holds only for $p \in V^+$, $q \in V^-$. As in (3.9.17-19) we go over to the corresponding analytic function $K_n^{\rm an}(\tilde{p},\tilde{q})$, regular in $\tilde{p} \in \mathbb{R}^4 + iV^+$, $\tilde{q} \in \mathbb{R}^4 + iV^-$. Then we take the boundary value for arbitrary p, $q \in \mathbb{R}^4$. We only need the real part of the principle value contribution in the following, which we denote by $\operatorname{Re} K_n^P$.

We note that in K_2' (3.9.30) and K_4' the complex factor in the denominator drops out. The following special cases in (3.9.28) arise:

$$(i)\quad a^2 = \frac{m^2}{p^2},\qquad f(t^2) = 1$$

$$K_2' = \operatorname{Re} K_2' = \frac{p^2}{m^2},\qquad K_4' = \operatorname{Re} K_4' = \frac{p^4}{2m^4} - \frac{p^2}{m^2} \tag{3.9.32}$$

$$(ii)\quad a^2=\frac{4m^2}{P^2},\quad f(t^2)=\sqrt{1-\frac{4m^2}{t^2P^2}}$$

$$\mathrm{Re}\,K_0^P=-2-B\log\left|\frac{B-1}{B+1}\right|,\quad K_2'=\mathrm{Re}\,K_2'=\frac{P^2}{6m^2},\tag{3.9.33}$$

$$\text{where}\quad B=\sqrt{1-\frac{4m^2}{P^2+iP_00}}.\tag{3.9.34}$$

For simplicity we give the other integrals, needed to split I^ν and $I^{\mu\nu}$, only on the mass shell after analytic continuation to $q^0>0$ (and therefore $P^2<0$):

$$(iii)\quad a^2=\frac{4m^2}{P^2},\quad f(t^2)=\log_1(tp,tq)\quad (\text{see } (3.8.49))$$

$$\mathrm{Re}\,K_2'=-\frac{P^2}{m^2}|x_1|+2\log\frac{|x_1|-1}{|x_1|+1},\tag{3.9.35}$$

$$\mathrm{Re}\,K_4'=\left(\frac{P^2}{2m^2}-\frac{P^4}{12m^4}\right)|x_1|-\log\frac{|x_1|-1}{|x_1|+1}.\tag{3.9.36}$$

$$(iv)\quad a^2=\frac{m^2}{p^2},\quad f(t^2)=\log_3(tp,tq)\quad (\text{see } (3.8.59))$$

$$\mathrm{Re}\,K_2'=\log\frac{|x_1|+1}{|x_1|-1},\quad \mathrm{Re}\,K_4'=\frac{1}{2}\log\frac{|x_1|-1}{|x_1|+1},\tag{3.9.37}$$

where $|x_1|$ is given in (3.9.13). J_1 (3.9.14-18) is the only integral in the splitting of the vertex D-distribution which has a singularity on the mass shell, besides $\log_1(p,q)$ (3.8.49) in the real part of the scalar integral (3.9.25).

Inserting all these integrals into (3.9.26), (3.9.27), one finds that I^ν_{ret} and $I^{\mu\nu}_{\mathrm{ret}}$ consist mainly of the integral K_0 in (3.9.33) (in some terms only its logarithmic part appears), since the other contributions compensate each other. As we see, no Spence functions appear. We shall obtain after analytic continuation to $q^0>0$, $P^2<0$ and with the inclusion of R'

$$I_T^\nu=\frac{i}{(2\pi)^6}\frac{p^\nu+q^\nu}{4P^2|x_1|}\log\frac{|x_1|-1}{|x_1|+1}-\frac{1}{2}\mathrm{Re}\Big(I^\nu_{R'}(p,q)+I^\nu_{A'}(p,q)\Big).\tag{3.9.38}$$

For the tensor integral we find the following final result on the mass shell

$$I_T^{\mu\nu}=\frac{i}{8(2\pi)^6}\Bigg\{g^{\mu\nu}\left[-1-\frac{|x_1|}{2}\log\Big(\frac{|x_1|-1}{|x_1|+1}\Big)-\frac{1}{4}+C\right]$$

$$+\frac{P^\mu P^\nu}{P^2}\left[1+\frac{|x_1|}{2}\log\Big(\frac{|x_1|-1}{|x_1|+1}\Big)\right]+(p^\mu+q^\mu)(p^\nu+q^\nu)\frac{1}{2P^2|x_1|}\log\Big(\frac{|x_1|-1}{|x_1|+1}\Big)\Bigg\}$$

$$-\frac{1}{2}\mathrm{Re}\left[I^{\mu\nu}_{R'}(p,q)+I^{\mu\nu}_{A'}(p,q)\right],\tag{3.9.39}$$

where the normalization constant C is 0 for the central splitting solution (3.9.1) of the tensor integral. These results agree with the finite part of the corresponding Feynman integrals (*A.I. Akhiezer, V.B. Berestetski, Quantum Electrodynamics, Wiley-Interscience, New York (1965)*) up to finite normalization terms. However, as we will see in the next chapter (Sect. 4.6), also such finite normalization terms must be computed correctly in the vertex function in order to get the right gauge-invariant results.

Since we have splitted the scalar integral (3.9.4) and the vector integral (3.9.26) with $\omega=-1$, $I_{\rm ret}(0,0)$ and $I^\nu_{\rm ret}(0,0)$ eventually do not vanish. For the vector integral this is not true, because it has the following tensor structure

$$I^\nu_{\rm ret}=f_1(p,q)p^\nu+f_2(p,q)q^\nu, \tag{3.9.40}$$

(see (3.8.71)). In order to obtain $I_{\rm ret}(0,0)$ we consider

$$I^{(0)}_{\rm ret}(p,q)\stackrel{\rm def}{=}I_{\rm ret}(p,q)-I_{\rm ret}(0,0)=\frac{i}{2\pi}\int\frac{I(tp,tq)}{t(1-t+i0)}\,dt, \tag{3.9.41}$$

which is a subtracted dispersion integral. We already know the integrals appearing in

$$I_{\rm ret}(0,0)=\frac{i}{2\pi}\int dt\,\frac{I(tp,tq)}{1-t+i0}\Big(1-\frac{1}{t}\Big) \tag{3.9.42}$$

on the mass shell from the integrals K_2' in the above computation. Inserting them into (3.9.42) we get the result

$$I_{\rm ret}(0,0)=\frac{i}{4(2\pi)^6m^2}. \tag{3.9.43}$$

Substituting $I_{\rm ret}$, $I^\nu_{\rm ret}$ and $I^{\mu\nu}_{{\rm ret},C=0}$ for I, I^ν, $I^{\mu\nu}$ into (3.8.54) and (3.9.1), we see that we have to add $2m^2\gamma^\nu I_{\rm ret}(0,0)$ in order to get the central splitting solution $\hat{R}^\nu(p,q)$ (3.9.2) of the total vertex function. This addition can be absorbed in the normalization constant C in (3.9.39) by choosing

$$C=1. \tag{3.9.44}$$

As we shall see in Sect. 3.13, this is the right choice to get a gauge invariant result.

The total vertex function $\hat{T}^\nu(p,q)$ is obtained by substituting the T-integrals into (3.8.84). In later applications, p and q are momenta of external electrons and the γ-matrices in (3.8.84) appear between Dirac spinors $\bar{u}(p), u(q)$. Then the expression can be simplified by anticommuting $\not p$ to the left and $\not q$ to the right and using the Dirac equations

$$\overline{u}(p)\not p=\overline{u}(p)m \quad , \quad \not q u(q)=mu(q). \tag{3.9.45}$$

Instead of (3.8.84) the following matrix structure appears:

$$\hat{\Lambda}^\nu(p,q)=4pq\gamma^\nu I_T-4mI^\nu_T-4\gamma^\nu(p^\mu+q^\mu)I_{T\mu}+4(p^\nu+q^\nu)\gamma^\mu I_{T\mu}$$
$$+2I^\mu_{T\mu}\gamma^\nu-4I^\nu_{T\mu}\gamma^\mu. \tag{3.9.46}$$

Furthermore, $p^\nu + q^\nu$ can be expressed by means of the Gordon decomposition (1.4.68)

$$\overline{u}_s(p)(p^\nu + q^\nu)u_\sigma(q) = 2m\,\overline{u}_s(p)\gamma^\nu u_\sigma(q) - iP_\mu \overline{u}_s(p)\sigma^{\nu\mu}u_\sigma(q). \qquad (3.9.47)$$

Then the vertex function can be written in the form

$$\hat{\Lambda}^\nu(p,q) = if(P^2)\gamma^\nu + i\sigma^{\nu\mu}P_\mu \frac{i}{2m} g(P^2). \qquad (3.9.48)$$

The quantity $f(P^2)$ is the so-called electric form factor and $g(P^2)$ the magnetic form factor. The reason for this terminology is discussed in the next section. The notions Dirac and Pauli form factors are also in use.

3.10 Form Factors

It is often said that the electron is a point particle without structure in contrast to the proton, for example. We will see in this section that this is not true. The electromagnetic structure of the electron is contained in the form factors, introduced at the end of the last section. The point is that this structure is calculable with extremely high precision by QED, which is not the case for the proton at present. It is fascinating to realize how the simple electromagnetic coupling (3.3.1)

$$T_1(x) = ie : \overline{\psi}(x)\gamma^\nu\psi(x) : A_\nu(x) \qquad (3.10.1)$$

together with causality and Poincaré covariance successively determines finer and finer details of the structure of the leptons.

To see this, we start from the contribution of the vertex function in third order

$$T_3(x_1,x_2,x_3) = -e^3 : \overline{\psi}(x_1)T_3^\nu(x_1,x_2,x_3)\psi(x_2) : A_\nu(x_3) \qquad (3.10.2)$$

to the S-matrix element for electron scattering

$$(b_s^+(\mathbf{p})\Omega,\, S_3 b_\sigma^+(\mathbf{q})\Omega) = -e^3(2\pi)^{-3}\int d^4x_1 d^4x_2 d^4x_3$$

$$\times\, \overline{u}_s(p)T_3^\nu(x_1,x_2,x_3)u_\sigma(q)e^{ipx_1 - iqx_2}A_\nu(x_3). \qquad (3.10.3)$$

A factor 1/3! does not appear because there are 3! terms with permuted coordinates x_1, x_2, x_3. Using relative coordinates $y_1 = x_1 - x_3$, $y_2 = x_3 - x_2$ again, we get

$$= -e^3(2\pi)^{-3}\int dy_1 dy_2\, \overline{u}_s(p)T_3^\nu u_\sigma(q)e^{ipy_1 + iqy_2}$$

$$\times \int dx_3\, e^{i(p-q)x_3}A_\nu(x_3) = -e^3(2\pi)^3\overline{u}_s(p)\hat{\Lambda}^\nu(p,q)u_\sigma(q)\hat{A}_\nu(P) \qquad (3.10.4)$$

with $P = p - q$. From (3.10.1) we obtain for the first order S-matrix element

$$(b_s^+(\boldsymbol{p})\Omega,\, S_1 b_\sigma^+(\boldsymbol{q})\Omega) = ie(2\pi)^{-3}\int dx\, \overline{u}_s(p)\gamma^\nu u_\sigma(q) e^{ipx-iqx} A_\nu(x)$$

$$= ie(2\pi)^{-1}\overline{u}_s(p)\gamma^\nu u_\sigma(q)\hat{A}_\nu(p-q). \tag{3.10.5}$$

Taking (3.9.48) into account, we find for the total S-matrix element

$$S_{fi} = (b_s^+(\boldsymbol{p})\Omega,\, (S_1+S_3)b_\sigma^+(\boldsymbol{q})\Omega) = ie(2\pi)^{-1}\overline{u}_s(p)$$

$$\times\Big[\gamma^\nu(1-e^2(2\pi)^4 f(P^2)) - i\frac{e^2}{2m}(2\pi)^4 g(P^2)\sigma^{\nu\mu}P_\mu\Big]u_\sigma(q)\hat{A}_\nu(p-q). \tag{3.10.6}$$

In order to identify electric and magnetic quantities, we have to select a special frame of reference. Let us consider electron scattering at an external static (C-number) field at low energies. We calculate in the standard representation (1.4.22)

$$u_s(\boldsymbol{p}) = \sqrt{\frac{E+m}{2E}}\begin{pmatrix} \chi_s \\ \dfrac{\boldsymbol{\sigma}\cdot\boldsymbol{p}}{E+m}\chi_s \end{pmatrix}, \tag{3.10.7}$$

because we are interested in the non-relativistic limit $|\boldsymbol{p}| \ll m$. Using

$$\gamma^0 = \begin{pmatrix} 1 & 0 \\ 0 & -1 \end{pmatrix} \qquad \gamma^j = \begin{pmatrix} 0 & \sigma_j \\ -\sigma_j & 0 \end{pmatrix} \qquad \sigma^{kl} = \varepsilon^{klm}\begin{pmatrix} \sigma_m & 0 \\ 0 & \sigma_m \end{pmatrix}, \tag{3.10.8}$$

we find

$$\overline{u}_s(\boldsymbol{p})\gamma^0 u_\sigma(\boldsymbol{q}) = (\chi_s^+, \chi_\sigma)$$

$$\overline{u}_s(\boldsymbol{p})\gamma^j u_\sigma(\boldsymbol{q}) = \frac{1}{2m}\Big(\chi_s^+, (\boldsymbol{p}+\boldsymbol{q}+i\boldsymbol{\sigma}\wedge(\boldsymbol{p}-\boldsymbol{q}))^j\chi_\sigma\Big), \tag{3.10.9}$$

to leading order in $|\boldsymbol{p}|/m$. Substituting this into (3.10.6), we finally get

$$S_{fi} = ie(2\pi)^{-1}\Big\{(1-e^2(2\pi)^4 f(P^2))\chi_s^+\chi_\sigma\Big(\hat{A}_0(\boldsymbol{P}) - \frac{\boldsymbol{p}+\boldsymbol{q}}{2m}\boldsymbol{A}(\boldsymbol{P})\Big)$$

$$-\frac{1}{2m}\Big[1-e^2(2\pi)^4(f(P^2)+g(P^2))\Big]\chi_s^+\boldsymbol{\sigma}\chi_\sigma i\boldsymbol{P}\wedge\hat{\boldsymbol{A}}(\boldsymbol{P})\Big\}. \tag{3.10.10}$$

The last term is proportional to the magnetic field $\hat{\boldsymbol{B}}(\boldsymbol{P}) = i\boldsymbol{P}\wedge\hat{\boldsymbol{A}}(\boldsymbol{P})$, so that we can identify the magnetic moment

$$\mu = \frac{e}{2m}\Big[1-e^2(2\pi)^4(f(0)+g(0))\Big]. \tag{3.10.11}$$

Hence, we find a small change $O(e^2)$ of the magnetic moment. The value $P = 0$ is approached from space-like momenta, $P^2 < 0$, because we have $P = (0, \boldsymbol{p}-\boldsymbol{q})$ for a static potential. In discussing the infrared problem, we shall see in Sect. 3.13 (3.13.15) that only the magnetic form factor $g(0)$

contributes to the anomalous magnetic moment. This is the reason for the name and for the factor $1/2m$ in the definition (3.9.48).

To calculate $g(P^2)$, we return to (3.9.46). The terms $\sim \gamma^\nu$ contribute to the electric form factor, so that only the second, forth and last term involving the vector and tensor integrals must be considered. The vector integral follows from (3.9.38), using (3.8.71, 72) for the last real part:

$$I^\nu_{R'} + I^\nu_{A'} = -\frac{\pi}{4(2\pi)^6 N}\Theta(P^2 - 4m^2)(p^\nu + q^\nu)(pq - m^2)\sqrt{1 - \frac{4m^2}{P^2}}. \quad (3.10.12)$$

Since the Θ-function vanishes for $P^2 < 0$, this gives no contribution. The second and forth term in (3.9.46) then yield

$$T^\nu_V = \frac{i}{(2\pi)^6}\frac{p^\nu + q^\nu}{4P^2|x_1|}\log\frac{|x_1|-1}{|x_1|+1}\Big[-4m + 4(\not p + \not q)\Big]$$

$$= \frac{im}{(2\pi)^6}\frac{p^\nu + q^\nu}{P^2|x_1|}\log\frac{|x_1|-1}{|x_1|+1}, \quad (3.10.13)$$

where we have substituted $\not p$ and $\not q$ by m between the Dirac spinors.

We finally turn to the contribution $-4I^\nu_\mu\gamma^\mu$ from the tensor integral (3.9.39). Again, the last real part vanishes for $P^2 < 0$. The first term $\sim g^{\mu\nu}$ becomes proportional to γ^ν and, hence, contributes to the electric form factor. The second term vanishes between Dirac spinors, because $P_\mu\gamma^\mu = \not p - \not q$, so that there only remains the third term

$$T^\nu_T = -4\frac{i}{8(2\pi)^6}(p^\nu + q^\nu)(\not p + \not q)\frac{1}{2P^2|x_1|}\log\frac{|x_1|-1}{|x_1|+1}$$

$$= -\frac{im}{(2\pi)^6}\frac{p^\nu + q^\nu}{2P^2|x_1|}\log\frac{|x_1|-1}{|x_1|+1}. \quad (3.10.14)$$

Totally we have

$$\hat{\Lambda}^\nu = \frac{im}{(2\pi)^6}\frac{p^\nu + q^\nu}{2P^2|x_1|}\log\frac{|x_1|-1}{|x_1|+1}. \quad (3.10.15)$$

Using the Gordon decomposition (3.9.47), this is equivalent to

$$\hat{\Lambda}^\nu = \frac{m}{(2\pi)^6}\sigma^{\nu\mu}P_\mu\frac{1}{2P^2|x_1|}\log\frac{|x_1|-1}{|x_1|+1}. \quad (3.10.16)$$

Comparing this with the definition (3.9.48) of the form factors, we get the magnetic form factor

$$g(P^2) = -\frac{m^2}{(2\pi)^6}\frac{1}{P^2|x_1|}\log\frac{|x_1|-1}{|x_1|+1}. \quad (3.10.17)$$

To calculate the anomalous magnetic moment, we have to expand this for $P^2 \uparrow 0$. Using (3.9.13)

$$|x_1| = \sqrt{1 - \frac{4m^2}{P^2}},$$

we find

$$g(0) = -(2\pi)^{-6}\frac{1}{2}. \tag{3.10.18}$$

This gives the following result for the total magnetic moment (3.10.11)

$$\mu \stackrel{\text{def}}{=} \frac{e}{2m}(1+a) = \frac{e}{2m}\left(1 + \frac{e^2}{8\pi^2}\right) = \frac{e}{2m}\left(1 + \frac{\alpha}{2\pi}\right). \tag{3.10.19}$$

The so-called anomaly a, also called $g-2$ (where g stands for the gyromagnetic ratio), has been computed to three further orders in the fine structure constant α (*T. Kinoshita, Metrologia 25, 233 (1988)*):

$$a = \frac{\alpha}{2\pi} - 0.328\,478\,965\left(\frac{\alpha}{\pi}\right)^2 + (1.175\,62 \pm (56))\left(\frac{\alpha}{\pi}\right)^3 - (1.472 \pm (152))\left(\frac{\alpha}{\pi}\right)^4$$

$$= (1\,159\,652\,164 \pm 108) \cdot 10^{-12}, \tag{3.10.20}$$

with excellent agreement with the experimental values for electron and positron (*R.S. Van Dyck, Jr., P.B. Schwinberg, H.G. Dehmelt, Phys. Rev. Lett. 59, p. 26 (1987)*):

$$a_{\text{exp}} = (1\,159\,652\,188.4 \pm 4.3) \cdot 10^{-12}.$$

The uncertainty in (3.10.20) is due to the uncertainty in α. The magnetic moment of the electron seems to be the most accurately calculated and measured quantity in physics.

3.11 Adiabatic Limit

In all application so far, the adiabatic limit where the switching function $g(x)$ goes to 1, could be carried out without problem in the S-matrix elements. This is somewhat surprising for the following reason: The adiabatic switching of the interaction has the unphysical consequence that it produces uncharged electrons in the asymptotic region. Indeed, our asymptotic electronic states are solutions of the free Dirac equation which describes *neutral* particles. But real electrons are charged and carry their Coulomb field. The latter is switched off by $g(x)$ in a "gedanken-experiment", and, therefore, it should appear as an outgoing electromagnetic field. In sharp contrast to the photon states, this decoupled Coulomb field is not transversal, but consists of scalar and longitudinal photons. In reality this field is confined to the electrons and makes them charged. Hence, there is already a confinement problem in QED, and the adiabatic limit is the tool to study it.

The adiabatic limit will be carried out in the following way: We choose a fixed test function $g_0(x) \in \mathcal{S}(\mathbb{R}^4)$ with $g_0(0) = 1$ and perform the scaling limit

$$g(x) = g_0(\epsilon x) \quad \text{where} \quad \epsilon \to 0. \tag{3.11.1}$$

This scaling limit is very useful, because any spurious effect in the asymptotic region vanishes in the limit. Then the Fourier transform

$$\hat{g}(k) = (2\pi)^{-2} \int g_0(\epsilon x) e^{-ikx} d^4x = \frac{1}{\epsilon^4} \hat{g}_0\left(\frac{k}{\epsilon}\right) \tag{3.11.2}$$

goes to $(2\pi)^2 \delta(k)$ and

$$\int d^4k\, \hat{g}_0(k) = (2\pi)^2 g_0(0) = (2\pi)^2. \tag{3.11.3}$$

For later convenience we have changed the convention of the Fourier transformation in (3.11.2); all other Fourier transforms are defined with the usual exponential factor $\exp(ikx)$.

We have found in Sect. 3.9 that the vertex function contains terms which become singular on the mass shell. We are now interested in the consequences of these infrared singularities. To discuss the adiabatic limit in the vertex function we consider its contribution to electron scattering on an external (C-number) potential $A^{\text{ext}}(x)$. This is called a radiative correction to electron scattering. Let $f^i(\boldsymbol{p})$ and $f^f(\boldsymbol{p})$ be the wave functions of the initial and final electrons, then the transition amplitude is given in p-space by

$$S^V_{fi}(g) = -\frac{ie^3}{(2\pi)^3} \sum_{s_i s_f} \int \frac{d^3p_i}{2E_i} \frac{d^3p_f}{2E_f}\, f^{f*}_{s_f}(\boldsymbol{p}_f) f^i_{s_i}(\boldsymbol{p}_i)$$

$$\times \frac{1}{\epsilon^{12}} \int dk_1\, dk_2\, dk_3\, \hat{g}_0\left(\frac{k_1}{\epsilon}\right) \hat{g}_0\left(\frac{k_2}{\epsilon}\right) \hat{g}_0\left(\frac{k_3}{\epsilon}\right)$$

$$\overline{u}_{s_f}(\boldsymbol{p}_f) \hat{\Lambda}^\nu (p_f - k_1, p_i + k_2) u_{s_i}(\boldsymbol{p}_i) \hat{A}^{\text{ext}}_\nu (p_f - p_i - k_1 - k_2 - k_3), \tag{3.11.4}$$

where V stands for vertex. After transformation of variables in the k-integrals we shall obtain

$$S^{(1)}(g) = \ldots \int dk_1\, dk_2\, dk_3\, \hat{g}_0(k_1) \hat{g}_0(k_2) \hat{g}_0(k_3)$$

$$\overline{u}_{s_f}(\boldsymbol{p}_f) \hat{\Lambda}^\nu (p_f - \epsilon k_1, p_i + \epsilon k_2) u_{s_i}(\boldsymbol{p}_i) \hat{A}^{\text{ext}}_\nu (p_f - p_i - \epsilon(k_1 + k_2 + k_3)). \tag{3.11.5}$$

Here the limit $\epsilon \to 0$ can immediately be performed in the argument of $\hat{A}^{\text{ext}}_\nu$. But in the vertex function there is a divergent part $\hat{\Lambda}^\nu_{\text{div}}$ that comes from the first term in (3.9.18), that means from the last two terms in (3.9.20). The singularity of $\Theta(P^2 - 4m^2) \log_1(p, q)$ in $\operatorname{Re} I_T$ (3.9.25) (see (3.8.60)) gives no contribution here because $P^2 < 0$. We finally get

$$\hat{\Lambda}^\nu_{\text{div}}(p, q) = 4i\, pq \gamma^\nu (\tilde{J}_{\text{div}}(p, q) + \tilde{J}_{\text{div}}(q, p)), \tag{3.11.6}$$

where

$$\tilde{J}_{\text{div}}(p,q) = (2\pi)^{-6}\frac{i}{8\sqrt{N}}\log|A|\log\left|\frac{m^2-p^2}{p^2}\right|, \tag{3.11.7}$$

$$\text{with} \quad A = \frac{\frac{q^2-m^2}{p^2-m^2}p^2 - pq - \sqrt{N}}{\frac{q^2-m^2}{p^2-m^2}p^2 - pq + \sqrt{N}}, \tag{3.11.8}$$

according to (3.9.18). The simple pre-factor in (3.11.6) follows from (3.9.2) using

$$-2\not{q}\gamma^\nu\not{p} = 2\not{p}\gamma^\nu\not{q} - 4p^\nu\not{q} - 4q^\nu\not{p} + 4pq\gamma^\nu \tag{3.11.9}$$

and the fact that $\hat{\Lambda}^\nu$ is sandwiched between Dirac spinors $\overline{u}(p)$ and $u(q)$.

Let us now specialize to the values near the mass shell:

$$p^2 - m^2 = (p_f - \epsilon k_1)^2 - m^2 = -2\epsilon p_f k_1 + O(\epsilon^2)$$

$$q^2 - m^2 = (p_i + \epsilon k_2)^2 - m^2 = 2\epsilon p_i k_2 + O(\epsilon^2)$$

$$pq = p_f p_i + O(\epsilon) = m^2 - \frac{P^2}{2} + O(\epsilon)$$

$$\sqrt{N} = \frac{|P^2|}{2}\sqrt{1-\frac{4m^2}{P^2}} + O(\epsilon), \tag{3.11.10}$$

where $P = p_f - p_i$. The last logarithm in (3.11.7) gives rise to the well-known infrared singularity $\log|\epsilon|$. The other logarithm is rewritten as follows

$$\log|A| = \log\left|\frac{-m^2x - m^2 + \frac{P^2}{2} - \frac{|P^2|}{2}|x_1|}{-m^2x - m^2 + \frac{P^2}{2} + \frac{|P^2|}{2}|x_1|}\right|, \tag{3.11.11}$$

where

$$x = \frac{p_i k_2}{p_f k_1} \tag{3.11.12}$$

and $|x_1|$ is given by (3.9.13). We separate the x-dependence

$$\log|A| = \log\left|\frac{1+|x_1|}{1-|x_1|}\right| + \log\left|\frac{1-x+(1+x)|x_1|}{1-x-(1+x)|x_1|}\right|. \tag{3.11.13}$$

The second divergent term in (3.11.6) gives a contribution with p and q interchanged, that means we must substitute x (3.11.12) by $1/x$ in the second term of (3.11.13):

$$\log|A(q,p)| = \log\left|\frac{1+|x_1|}{1-|x_1|}\right| + \log\left|\frac{1-\frac{1}{x}+(1+\frac{1}{x})|x_1|}{1-\frac{1}{x}-(1+\frac{1}{x})|x_1|}\right|. \tag{3.11.14}$$

Adding (3.11.14) to (3.11.13), the two logarithms depending on x cancel each other and the integrals over k_1, k_2 and k_3 in (3.11.5) become trivial

$$\int \hat{\Lambda}^\nu_{\text{div}} = 2i(2m^2 - P^2)\gamma^\nu \frac{i}{4|P^2||x_1|} 2\log\left|\frac{1+|x_1|}{1-|x_1|}\right| \log|\epsilon|. \tag{3.11.15}$$

We set

$$b = \frac{\sqrt{|P^2|}}{2m} \tag{3.11.16}$$

and finally obtain

$$\int \hat{\Lambda}^{\nu}_{\rm div} = -\frac{1+2b^2}{b\sqrt{1+b^2}}\gamma^\nu \log(b+\sqrt{1+b^2})\log|\epsilon|$$

$$\stackrel{\rm def}{=} -F(b)\log|\epsilon|\gamma^\nu. \tag{3.11.17}$$

This result differs slightly from the one in other books, because the vertex function is here normalized at $p = 0 = q$ (3.9.2), whereas it is usually normalized on the mass shell. The latter procedure is dangerous because of the infrared singularities. With our normalization we get also contributions from the two self-energy graphs of third order (Fig. 7), which we are now going to compute.

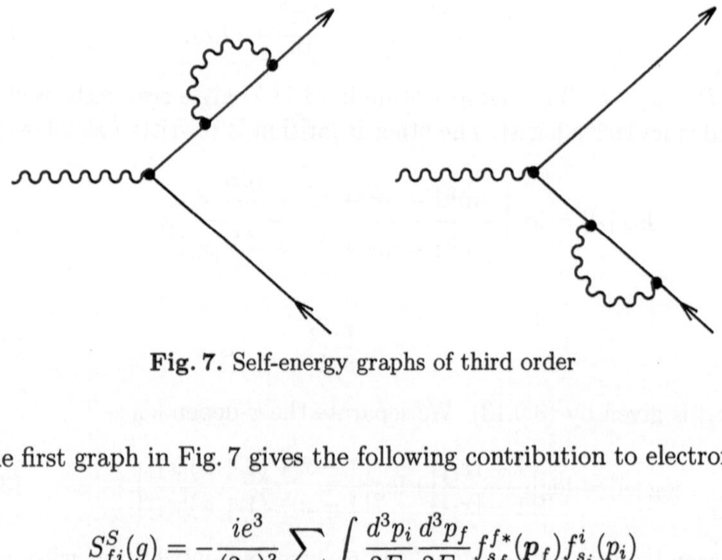

Fig. 7. Self-energy graphs of third order

The first graph in Fig. 7 gives the following contribution to electron scattering

$$S^S_{fi}(g) = -\frac{ie^3}{(2\pi)^3}\sum_{s_i s_f}\int \frac{d^3p_i}{2E_i}\frac{d^3p_f}{2E_f} f^{f*}_{s_f}(\boldsymbol{p}_f) f^i_{s_i}(\boldsymbol{p}_i)$$

$$\times \int d^4k_1\, d^4k_2\, d^4k_3\, \hat{g}_0(k_1)\hat{g}_0(k_2)\hat{g}_0(k_3)\overline{u}_{s_f}(\boldsymbol{p}_f)$$

$$\hat{\Sigma}(p_f-\epsilon k_1)\hat{S}^F(p_f-\epsilon k_1-\epsilon k_2)\hat{A\!\!\!/}_{\rm ext}(p_f-p_i-\epsilon(k_1+k_2+k_3))u_{s_i}(\boldsymbol{p}_i). \tag{3.11.18}$$

The self-energy function is given by Eq. (3.7.41)

$$\hat{\Sigma}(p) = (2\pi)^{-4}\Big\{\Big[\log\Big|\frac{m^2-p^2}{m^2}\Big| - i\pi\Theta(p^2-m^2)\Big]\Big[m - \frac{p\!\!\!/}{4}\Big(1+\frac{m^2}{p^2}\Big)\Big]\frac{p^2-m^2}{p^2}$$

$$+\frac{m^2}{4}\frac{\not p}{p^2}-\frac{m}{4}+C_1(\not p-m)\Big\}. \tag{3.11.19}$$

Although this function vanishes on the mass shell, it gives a divergent contribution to (3.11.18) because the Feynman propagator

$$\hat S^F(p_f-\epsilon(k_1+k_2))=\frac{1}{(2\pi)^2}\frac{\not p_f+m-\epsilon(\not k_1+\not k_2)}{-2\epsilon p_f(k_1+k_2)+i0} \tag{3.11.20}$$

cancels a factor ϵ in the numerator. Concentrating again on the divergent term, there remains to calculate the integral

$$\int d^4k_1\,d^4k_2\,\hat g_0(k_1)\hat g_0(k_2)\log\Big|2\epsilon\frac{p_fk_1}{m^2}\Big|\frac{p_fk_1}{p_f(k_1+k_2)-i0}$$

$$=\log|\epsilon|\int d^4k_1\,d^4k_2\,\hat g_0(k_1)\hat g_0(k_2)\frac{p_fk_1}{p_f(k_1+k_2)-i0}+O(1).$$

Interchanging the integration variables k_1 and k_2, the integral is also equal to

$$=\log|\epsilon|\int\cdots\frac{p_fk_2}{p_f(k_1+k_2)-i0}. \tag{3.11.21}$$

Since the sum of the two integrals is trivial (3.11.3), each of them must be equal to

$$=\log|\epsilon|\frac{(2\pi)^4}{2}. \tag{3.11.22}$$

The second self-energy diagram gives the same result. Without the normalization condition (3.7.39), $\hat\Sigma(p)$ would not vanish on the mass shell. Then a $1/\epsilon$-singularity would remain and the adiabatic limit does not exist. The existence of the limit forbids mass renormalization!

The total infrared divergent self-energy contribution is therefore given by

$$S^S_{fi}=-\frac{ie^3}{(2\pi)^3}\log|\epsilon|\sum_{s_is_f}\int\frac{d^3p_i}{2E_i}\frac{d^3p_f}{2E_f}f^{f*}_{s_f}(\boldsymbol p_f)f^i_{s_i}(\boldsymbol p_i)$$

$$\times\overline u_{s_f}(\boldsymbol p_f)\hat{\not A}_{\rm ext}(p_f-p_i)u_{s_i}(\boldsymbol p_i).$$

Together with the contribution (3.11.17) of the vertex function we shall obtain

$$S^V_{fi}+S^S_{fi}=-\frac{ie^3}{(2\pi)^3}\log|\epsilon|\sum_{s_is_f}\int\frac{d^3p_i}{2E_i}\frac{d^3p_f}{2E_f}f^{f*}_{s_f}(\boldsymbol p_f)f^i_{s_i}(\boldsymbol p_i)(1-F(b))$$

$$\times\overline u_{s_f}(\boldsymbol p_f)\hat{\not A}_{\rm ext}(p_f-p_i)u_{s_i}(\boldsymbol p_i)\stackrel{\rm def}{=}h(b,\varepsilon)S^{(1.\rm order)}_{fi},\quad b=\frac{1}{2m}|\boldsymbol p_i-\boldsymbol p_f|. \tag{3.11.23}$$

The factorization of the first order S-matrix element would be impossible if we had a contribution of the vector or tensor integral in $S^{(1)}_{fi}$ (see (3.8.84)). Note that in the computation of the differential cross section in forth order

$$\left(\frac{\partial\sigma}{\partial\Omega}\right)_{(4.\mathrm{order})} \sim |S_{fi}|^2_{(4.\mathrm{order})} = |S_{fi(2.\mathrm{order})}|^2 + |S_{fi(1.\mathrm{order})}|^2 2\mathrm{Re}\, h(b,\varepsilon) \tag{3.11.23a}$$

only the real part of $h(b,\epsilon)$ contributes. This means that $\mathrm{Re}\, I_T(p,q)$ (3.9.25) drops out. Summing up, the contribution of $S^{(1)}_{fi} + S^{(2)}_{fi}$ (3.11.23) to the differential cross section (3.11.23a) is equal to

$$\frac{d\sigma^{V+S}}{d\Omega} = \frac{d\sigma^{(1)}}{d\Omega}\left[1 - 2\frac{e^2}{(2\pi)^2}(1 - F(b))\log|\epsilon|\right], \tag{3.11.24}$$

where the factor in front is the first order electron scattering cross section and $F(b)$ is defined in (3.11.17). This result alone is physically meaningless, not only because it diverges for $\epsilon \to 0$. Since $1 - F(b)$ is negative, the cross section would even become negative for $|\epsilon|$ small enough.

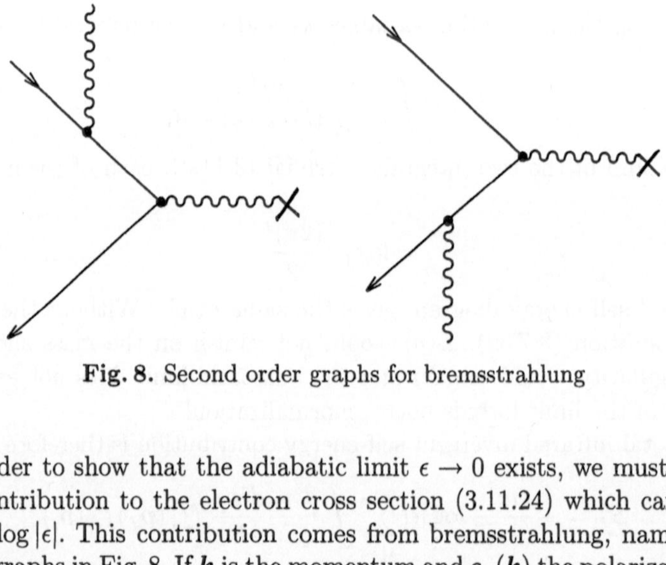

Fig. 8. Second order graphs for bremsstrahlung

In order to show that the adiabatic limit $\epsilon \to 0$ exists, we must find another contribution to the electron cross section (3.11.24) which cancels the term $\sim \log|\epsilon|$. This contribution comes from bremsstrahlung, namely from the two graphs in Fig. 8. If $\boldsymbol{k}$ is the momentum and $\varepsilon_\nu(\boldsymbol{k})$ the polarization vector of the bremsstrahlungs-photon, the corresponding transition amplitude is given by

$$S_{fi} = -ie^2(2\pi)^{-13/2}\int dk_1\, dk_2\, \hat{g}(k_1)\hat{g}(k_2)\overline{u}_s(p)$$

$$\left[\gamma^\mu \frac{\not{q} - \not{k} + \not{k}_2 + m}{(q-k+k_2)^2 - m^2 + i0}\gamma^\nu + \gamma^\nu \frac{\not{p} + \not{k} - \not{k}_2 + m}{(p+k-k_2)^2 - m^2 + i0}\gamma^\mu\right]$$

$$u_\sigma(q)\hat{A}^{\mathrm{ext}}_\mu(p - q + k - k_1 - k_2)\frac{\varepsilon_\nu(\boldsymbol{k})}{\sqrt{2\omega}}. \tag{3.11.25}$$

Here we set $k_1 = 0$ because the adiabatic limit in this variable is trivial. Furthermore, we only consider soft photons and neglect all small quantities in the numerators

$$S_{fi}=-ie^2(2\pi)^{-9/2}\int dk_2\,\hat{g}(k_2)\overline{u}_s(p)\Big[\gamma^\mu\frac{\not{q}+m}{-2q(k-k_2)}\frac{\not{\varepsilon}(\boldsymbol{k})}{\sqrt{2\omega}}$$

$$+\frac{\not{\varepsilon}(\boldsymbol{k})}{\sqrt{2\omega}}\frac{\not{p}+m}{2p(k-k_2)}\gamma^\mu\Big]u_\sigma(q)\hat{A}_\mu^{\rm ext}(p-q). \tag{3.11.26}$$

Anticommuting $\not{q}$ to the right and $\not{p}$ to the left and using the Dirac equation, we arrive at

$$S_{fi}=\frac{-ie^2}{\sqrt{2\omega}}(2\pi)^{-9/2}\overline{u}_s(p)\hat{\not{A}}^{\rm ext}(p-q)u_\sigma(q)\int dk_2\Big(\frac{p\varepsilon}{p(k-k_2)}$$

$$-\frac{q\varepsilon}{q(k-k_2)}\Big)\hat{g}(k_2)=S_{fi}^{(1)}\frac{e}{\sqrt{2\omega}}(2\pi)^{-7/2}\int dk_2\Big(\frac{p\varepsilon}{p(k-k_2)}-\frac{q\varepsilon}{q(k-k_2)}\Big)\hat{g}(k_2), \tag{3.11.27}$$

where $S_{fi}^{(1)}$ is the first order S-matrix element. This gives the following cross section

$$\frac{d\sigma(k)}{d\Omega}=\frac{d\sigma^{(1)}}{d\Omega}\frac{e^2}{(2\pi)^7}\frac{1}{2|\boldsymbol{k}|}\Big|\int dk_2\Big(\frac{p\varepsilon}{p(k-k_2)}-\frac{q\varepsilon}{q(k-k_2)}\Big)\hat{g}(k_2)\Big|^2.$$

As discussed in detail in the following section, we have to sum over *four* polarizations, in order to get a *unique* adiabatic limit. For the cancellation of the infrared divergence $\sim\log|\epsilon|$, it is sufficient to use the two transversal polarizations only (Problem 3.15). However, the contributions $O(1)$ depending on the switching function g_0 in (3.11.1) must also cancel out. As discussed in the next section (3.12.10), this can only be achieved by summing over four polarizations, using the covariant expression

$$\sum_{n=1}^{3}\varepsilon_\mu^{(n)*}\varepsilon_\nu^{(n)}-\varepsilon_\mu^{(0)*}\varepsilon_\nu^{(0)}=-g_{\mu\nu}. \tag{3.11.28}$$

Then there remains to calculate

$$J=\int dk_1dk_2\Big(-\frac{p^2}{p(k-k_1)p(k-k_2)}+\frac{pq}{p(k-k_1)q(k-k_2)}$$

$$+\frac{pq}{q(k-k_1)p(k-k_2)}-\frac{q^2}{q(k-k_1)q(k-k_2)}\Big)\hat{g}(k_1)^*\hat{g}(k_2). \tag{3.11.29}$$

It is our goal to compute the inclusive cross section

$$\frac{d\sigma^B}{d\Omega}=\int\limits_{|\boldsymbol{k}|\le\omega_0}d^3k\,\frac{d\sigma(k)}{d\Omega}, \tag{3.11.30}$$

where ω_0 is some finite energy resolution of the photon detector. We perform the integral $\int d^3k$ using a trick due to Feynman. It allows to replace the products in the denominator by a square by means of the following identity

$$\frac{1}{\alpha\beta} = \int_0^1 \frac{dx}{[\alpha x + \beta(1-x)]^2}.$$

For the first integral in (3.11.29) we must compute

$$I_1 \stackrel{\text{def}}{=} \int \frac{d\Omega_k}{[p(k-k_1)x + p(k-k_2)(1-x)]^2}$$

$$= 2\pi \int_{-1}^{1} \frac{d\cos\vartheta}{[p_0|\boldsymbol{k}| - |\boldsymbol{p}||\boldsymbol{k}|\cos\vartheta - pk_1x - pk_2(1-x)]^2}$$

$$= \frac{4\pi}{a + b|\boldsymbol{k}| + c\boldsymbol{k}^2}, \tag{3.11.31}$$

where

$$a = [p(k_1x + k_2(1-x))]^2 \tag{3.11.32}$$

$$b = -2p_0 p \cdot (k_1x + k_2(1-x)) \tag{3.11.33}$$

$$c = p^2.$$

We now perform the integral over $|\boldsymbol{k}|$:

$$I_2 = \int_0^{\omega_0} \frac{d|\boldsymbol{k}||\boldsymbol{k}|}{a + b|\boldsymbol{k}| + c\boldsymbol{k}^2} = -\frac{\log p(k_1x + k_2(1-x))}{p^2}$$

$$+\frac{1}{2p^2}\Big(1 - \frac{p_0}{|\boldsymbol{p}|}\Big)\log(\omega_0^2p^2) - \frac{p_0}{p^2|\boldsymbol{p}|}\log\frac{\omega_0 p^2}{-p_0 - |\boldsymbol{p}|}. \tag{3.11.34}$$

The final nontrivial integrations over k_1 and k_2 are of the following form:

$$\int d^4k_1\, d^4k_2\, \hat{g}(k_1)^*\hat{g}(k_2)\log|pk_1x + qk_2(1-x)|$$

$$= \int d^4k_1'\, d^4k_2'\, \hat{g}_0(k_1')^*\hat{g}_0(k_2')\log|pk_1'\epsilon x + qk_2'\epsilon(1-x)| = (2\pi)^4\log|\epsilon| + O(1). \tag{3.11.35}$$

Then there remains to compute the x-integrals, for example

$$I_3 \stackrel{\text{def}}{=} \int_0^1 \frac{dx}{x^2(p^2 - 2pq + q^2) + 2xq(p-q) + q^2}. \tag{3.11.36}$$

In the case of elastic scattering which is only of interest to us, we express all quantities by $\boldsymbol{P} = \boldsymbol{p} - \boldsymbol{q}$,

$$p \cdot q = E^2 - \boldsymbol{p}\cdot\boldsymbol{q} = m^2 + \frac{1}{2}\boldsymbol{P}^2.$$

Then the integral (3.11.36) can be calculated as follows

$$I_3 = -\frac{2}{\boldsymbol{P}^2} \int\limits_0^{1/2} \frac{dy}{y^2 - a}$$

$$= -\frac{1}{\boldsymbol{P}^2} \frac{|\boldsymbol{P}|}{m\sqrt{1+b^2}} \log \frac{\left|\frac{1}{2} - \frac{m}{|\boldsymbol{P}|}\sqrt{1+b^2}\right|}{\frac{1}{2} + \frac{m}{|\boldsymbol{P}|}\sqrt{1+b^2}}, \tag{3.11.37}$$

where

$$a = \frac{m^2}{\boldsymbol{P}^2}(1+b^2) \quad , \quad b = \frac{|\boldsymbol{P}|}{2m}. \tag{3.11.38}$$

Writing the result in terms of b, we end up with

$$I_3 = \frac{1}{m^2 b\sqrt{1+b^2}} \log(b + \sqrt{1+b^2}). \tag{3.11.39}$$

Now we are ready to compute the infrared divergent part of the cross section (3.11.30) with bremsstrahlung of soft photons included:

$$\frac{d\sigma^B}{d\Omega} = \frac{d\sigma^{(1)}}{d\Omega} \frac{e^2}{(2\pi)^2} \log \frac{|\epsilon|}{\omega_0^2} (2 - 2pqI_3)$$

$$= \frac{d\sigma^{(1)}}{d\Omega} \frac{2}{\pi} \frac{e^2}{4\pi} \log \frac{|\epsilon|}{\omega_0^2} \left[1 - \frac{1+2b^2}{b\sqrt{1+b^2}} \log(b + \sqrt{1+b^2})\right]$$

$$= \frac{d\sigma^{(1)}}{d\Omega} \frac{2e^2}{(2\pi)^2} [1 - F(b)] \log \frac{|\epsilon|}{\omega_0^2}. \tag{3.11.40}$$

This exactly cancels against (3.11.24) which shows the existence of the adiabatic limit. To prove the uniqueness of the limit requires more work, because one must calculate the finite part and show that it is independent of the switching function $\hat{g}_0(k)$. We consider this problem in the next section.

3.12 Charged Particles in Perturbative QED

In proving the existence of the adiabatic limit in the last section, we have summed over *four* polarizations of soft photons (see (3.11.28)). Seemingly, this is in contradiction to the fact that soft photons of bremsstrahlung are transversal, like any other photon. But we would not get a unique adiabatic limit with the two physical polarizations only, as we will realize below. This forces us to think in more detail about the physical meaning of the adiabatic switching of the interaction by means of the function $g(x) \in \mathcal{S}(\mathbb{R}^4)$.

This $g(x)$ which was essential for elaborating the causal structure of the theory is also important as a natural infrared regulator, as we have seen in the last section. Physically speaking, the electric charge is switched off in the asymptotic region which then leads to the free uncharged asymptotic spinor fields. Although this switching-off is only possible in a gedanken-experiment, we have it under control because the adiabatic limit $g(x) \to 1$ where the switching is removed, is independent of the switching function in the following sense: We again take $g(x) = g_0(\epsilon x)$ with $g_0(0) = 1$ and consider the limit $\epsilon \to 0$. Then, we shall see that the finite results for inclusive cross sections are independent of g_0. So far so good, but now comes the peculiarity: Since the switching annihilates electric charge, charge conservation is violated. This manifests itself in non-transversal polarizations of the asymptotic electromagnetic field. In the gedanken-experiment the Coulomb field is separated from the outgoing charged particles which become neutral, while the electromagnetic fields propagates as a free but non-transversal field. This is the reason for the occurrence of the unphysical polarizations. At the end of this section we shall illustrate this mechanism in the framework of classical electrodynamics. Of course, in the adiabatic limit $\epsilon \to 0$ charge conservation and gauge invariance are restored, because the switching-off is moved to infinity. As in the last section, we consider electron scattering on an external (C-number) potential $A^{\rm ext}(x)$, where we have to consider self-energy, vertex and bremsstrahlung contributions. The divergent contributions $\sim \log|\epsilon|$ have been calculated in the last section, now we are concerned with the finite contributions.

Let us first consider the vertex function (3.11.5) where we have to calculate the integral

$$\int dk_1 dk_2\, \hat{g}_0(k_1)\hat{g}_0(k_2)\hat{\Lambda}^\nu(p_f - \epsilon k_1, p_i + \epsilon k_2). \tag{3.12.1}$$

p_i and p_f are initial and final momenta of the scattered electrons. We symmetrize in k_1, k_2, because an antisymmetric contribution drops out in (3.12.1), and introduce the variables $P = p_f - p_i$

$$|x_1| = \sqrt{1 - \frac{4m^2}{P^2}}, \quad x = \frac{p_i k_2}{p_f k_1}, \quad y = \frac{p_i k_1}{p_f k_2}. \tag{3.12.2}$$

Only the scalar integral $I_T(p,q)$ (3.9.25) gives g_0-dependent contributions to the vertex function. Since its real part $\mathrm{Re}\, I_T(p,q)$ drops out in the cross section (3.11.23a), the k_1, k_2-dependent logarithms $\log_1$, $\log_2$ (3.8.49, 53) must not be taken into account. Furthermore, only J_1 (3.9.14) must be considered, the g_0-dependent contributions in J_0 (3.9.8) compensate, as can be shown by detailed calculation. From (3.9.18) we find the symmetrical expression

$$\mathrm{Re}\Big(J_1^P(p,q) + J_1^P(q,p)\Big) \overset{\text{def}}{=} \frac{1}{2}V$$

$$= \frac{1}{2}\Big\{\log\Big|\frac{1-|x_1|}{1+|x_1|}\Big|\Big[4\log|\epsilon| + \log 16 + \log\Big|\frac{p_f k_1}{m^2}\Big| + \log\Big|\frac{p_f k_2}{m^2}\Big|$$

$$+\log\Big|\frac{p_i k_1}{m^2}\Big| + \log\Big|\frac{p_i k_2}{m^2}\Big|\Big] - \log|x|\log\Big|\frac{1-x-(1+x)|x_1|}{1-x+(1+x)|x_1|}\Big|$$

$$-\log|y|\log\Big|\frac{1-y-(1+y)|x_1|}{1-y+(1+y)|x_1|}\Big|$$

$$+L\Big(\frac{1-|x_1|}{1-|x_1|-\frac{2m^2}{P^2}(1+x)}\Big) - L\Big(\frac{1+|x_1|}{1+|x_1|-\frac{2m^2}{P^2}(1+x)}\Big)$$

$$+L\Big(\frac{1-|x_1|}{1-|x_1|-\frac{2m^2}{P^2}(1+\frac{1}{x})}\Big) - L\Big(\frac{1+|x_1|}{1+|x_1|-\frac{2m^2}{P^2}(1+\frac{1}{x})}\Big)$$

$$+L\Big(\frac{1-|x_1|}{1-|x_1|-\frac{2m^2}{P^2}(1+y)}\Big) - L\Big(\frac{1+|x_1|}{1+|x_1|-\frac{2m^2}{P^2}(1+y)}\Big)$$

$$+L\Big(\frac{1-|x_1|}{1-|x_1|-\frac{2m^2}{P^2}(1+\frac{1}{y})}\Big) - L\Big(\frac{1+|x_1|}{1+|x_1|-\frac{2m^2}{P^2}(1+\frac{1}{y})}\Big)\Big\}, \tag{3.12.3}$$

where $p = p_f - \epsilon k_1$, $q = p_i + \epsilon k_2$, plus terms independent of k_1, k_2 for $\epsilon \to 0$. The logarithmic terms in the first two lines come from the first term in J_1^P (3.9.18), while the Spence functions in (3.12.3) result from the last two Spence functions in (3.9.18).

Next we turn to the self-energy contribution. It follows from the expression (3.11.19) for the self-energy $\hat{\Sigma}(p)$ that

$$\hat{\Sigma}(p_f - \epsilon k_1) = (2\pi)^{-4}\Big\{-\frac{\epsilon p_f k_1}{m}\log\Big|\frac{2\epsilon p_f k_1}{m^2}\Big| - \epsilon \not{k}_1\Big(C_1 + \frac{1}{4}\Big)$$

$$+\frac{\epsilon}{2}\frac{p_f k_1}{m} + O(\epsilon^2 \log|\epsilon|)\Big\}. \tag{3.12.4}$$

This must be multiplied by the electron propagator

$$\hat{S}^F(p_f - \epsilon(k_1+k_2)) = -\frac{1}{(2\pi)^2}\frac{\not{p}_f + m}{2\epsilon p_f(k_1+k_2)} + O(1) \tag{3.12.5}$$

and sandwiched between Dirac spinors $\bar{u}(p_f)$, $u(p_i)$. It is easy to see that only the first term in (3.12.4) gives a k_1, k_2-dependent contribution, if one symmetrizes in k_1, k_2. Taking both self-energy graphs together one arrives at

$$S = \log|\epsilon| + \frac{p_f k_1 \log|\frac{p_f k_1}{m^2}| + p_f k_2 \log|\frac{p_f k_2}{m^2}|}{2p_f(k_1+k_2)} + \frac{p_i k_1 \log|\frac{p_i k_1}{m^2}| + p_i k_2 \log|\frac{p_i k_2}{m^2}|}{2p_i(k_1+k_2)}. \tag{3.12.6}$$

Summing up the vertex and self-energy contributions to the forth order differential cross section are given by

$$\frac{d\sigma^{V+S}}{d\Omega} = -\frac{d\sigma^{(1)}}{d\Omega}\frac{e^2}{(2\pi)^6}\int dk_1 dk_2\, \hat{g}_0(k_1)\hat{g}_0(k_2)$$
$$\times 2\left[\frac{m^2-\frac{1}{2}P^2}{2|x_1||P^2|}V + S\right] + \cdots \tag{3.12.7}$$

where $\sigma^{(1)}$ is the lowest order cross section. The dots in (3.12.7) represent terms independent of $\hat{g}_0$ and terms that vanish for $\epsilon \to 0$.

Finally we must calculate the inclusive cross section for soft bremsstrahlung (3.11.30)

$$\frac{d\sigma^B}{d\Omega} = \int\limits_{|k|\le\omega_0} d^3k\, \frac{d\sigma(k)}{d\Omega}. \tag{3.12.8}$$

It is our goal to show that the adiabatic limit of

$$\frac{d\sigma}{d\Omega} = \lim_{\epsilon\to 0}\left[\frac{d\sigma^{V+S}}{d\Omega} + \frac{d\sigma^B}{d\Omega}\right] \tag{3.12.9}$$

is independent of $\hat{g}_0$. This result holds true only if we take *four* polarizations of bremsstrahlung into account by means of the covariant polarization sum

$$\sum_{n=1}^{3}\varepsilon_\nu^{(n)}\varepsilon_\lambda^{(n)} - \varepsilon_\nu^{(0)}\varepsilon_\lambda^{(0)} = -g_{\nu\lambda}. \tag{3.12.10}$$

Here we have assumed real polarization vectors, an explicit choice is given below (3.12.42). Although the signs in (3.12.10) are obvious for reasons of Lorentz covariance, we must be aware that we herewith extend the statistical interpretation of the S-matrix to the unphysical scalar ($n=0$) and longitudinal ($n=3$ in (3.12.43)) photon states. Since the S-matrix is not unitary on the big Fock space $\mathcal{F}$ (Sects. 2.11 and 4.7), it is not clear how this has to be done. In Sect. 4.7 we shall learn that $S(g)$ is pseudo-unitary in $\mathcal{F}$:

$$S(g)^K S(g) = \eta S(g)^+ \eta S(g) = 1, \quad \eta = (-1)^{\mathbf{N}_0}.$$

The conjugation K was defined in (2.11.43), where the sign η depends on the number of scalar photons. If Φ_j is a complete system of normalized states in $\mathcal{F}$, then pseudo-unitarity implies

$$\sum_j (\Phi_j, \eta S^K \Psi)^* (\Phi_j, \eta S\Psi) = \sum_j (\Phi_j, S\eta\Psi)^* (\Phi_j, \eta S\Psi)$$

$$= \sum_j (S\eta\Psi, \Phi_j)(\Phi_j, \eta S\Psi) = (S\eta\Psi, \eta S\Psi) = 1. \tag{3.12.11}$$

Now we consider an initial state $\Psi \in \mathcal{H}_{\rm phys}$ without scalar photons. Then $S\Psi$ does contain scalar photons, and the corresponding contribution in (3.12.11) with one scalar photon of bremsstrahlung in Φ_j gets a minus sign. This explains the minus sign in (3.12.10). We postpone the further discussion of the unphysical states to the end of this section.

In (3.12.8) we must substitute the previous integral J (3.11.29). The resulting integral is absolutely convergent; using Fubini's theorem we do the $\int d^3k$ integral first:

$$\int\limits_{|\boldsymbol{k}|\le\omega_0} \frac{d^3k}{|\boldsymbol{k}|}\Bigl(\frac{-p_i^2}{p_i(k-\epsilon k_1)p_i(k-\epsilon k_2)} + \frac{p_ip_f}{p_i(k-\epsilon k_1)p_f(k-\epsilon k_2)}$$

$$+\frac{p_ip_f}{p_f(k-\epsilon k_1)p_i(k-\epsilon k_2)} - \frac{p_f^2}{p_f(k-\epsilon k_1)p_f(k-\epsilon k_2)}\Bigr) \stackrel{\rm def}{=} I + III + IV + II.$$

Since in the cross sections the complex conjugate $\hat g_0^*(k_1) = \hat g_0(-k_1)$ appears, we substitute k_1 by $-k_1$ in order to compare with (3.12.7). The integrals are calculated as before (3.11.29) with the result

$$I = 4\pi\Bigl(\log|\epsilon| + \frac{p_ik_1\log|p_ik_1| + p_ik_2\log|p_ik_2|}{p_ik_1+p_ik_2}\Bigr) \tag{3.12.12}$$

$$II = 4\pi\bigl(\log|\epsilon| + p_i \longrightarrow p_f\bigr) \tag{3.12.13}$$

$$III = \frac{4\pi(\frac{1}{2}P^2-m^2)}{|x_1|P^2}\Bigl\{L\Bigl(-\frac{(1+x)(1-|x_1|}{1+|x_1|-x(1-|x_1|)}\Bigr)$$

$$-L\Bigl(\frac{(1+x)(1+|x_1|)}{1+|x_1|-x(1-|x_1|)}\Bigr)+L\Bigl(\frac{(1+x)(1-|x_1|)}{1-|x_1|-x(1+|x_1|)}\Bigr)$$

$$-L\Bigl(-\frac{(1+x)(1+|x_1|)}{1-|x_1|-x(1+|x_1|)}\Bigr)+\log\Bigl|\frac{1-|x_1|}{1+|x_1|}\Bigr|\Bigl[2\log|\epsilon|$$

$$+\log|p_fk_1|+\log|1-x+|x_1|(1+x)|+\log|p_ik_2|+\log\Bigl|1-\frac{1}{x}+|x_1|(1+\frac{1}{x})\Bigr|\Bigr]\Bigr\} \tag{3.12.14}$$

$$IV = \frac{4\pi(\frac{1}{2}P^2-m^2)}{|x_1|P^2}\Bigl\{L\Bigl(-\frac{(1+y)(1-|x_1|)}{1+|x_1|-y(1-|x_1|)}\Bigr)$$

$$-L\Bigl(\frac{(1+y)(1+|x_1|)}{1+|x_1|-y(1-|x_1|)}\Bigr)+L\Bigl(\frac{(1+y)(1-|x_1|)}{1-|x_1|-y(1+|x_1|)}\Bigr)$$

$$-L\Bigl(-\frac{(1+y)(1+|x_1|)}{1-|x_1|-y(1+|x_1|)}\Bigr)+\log\Bigl|\frac{1-|x_1|}{1+|x_1|}\Bigr|\Bigl[2\log|\epsilon|+\log|p_fk_2|$$

$$+\log|1-y+|x_1|(1+y)|+\log|p_ik_1|+\log\Big|1-\frac{1}{y}+|x_1|(1+\frac{1}{y})\Big|\Big]\Big\}, \quad (3.12.15)$$

up to terms independent of k_1, k_2. (3.12.12, 13) agree with the self-energy contribution (3.12.6) up to k_1, k_2 - independent terms. In (3.12.14) and (3.12.15) Spence functions appear which complicate the comparison with the vertex contribution V in (3.12.7).

We use the identity (C.10) of Appendix C

$$L(z)=-L(1-z)+\log|z|\log|1-z|-\frac{\pi^2}{6}. \quad (3.12.16)$$

Then (3.12.14) may be written as follows

$$III=\frac{4\pi(m^2-\frac{1}{2}P^2)}{|x_1||P^2|}\Big\{L\Big(\frac{1-|x_1|}{1-|x_1|-\frac{2m^2}{P^2}(1+x)}\Big)$$

$$-L\Big(\frac{1+|x_1|}{1+|x_1|-\frac{2m^2}{P^2}(1+x)}\Big)+L\Big(\frac{1-|x_1|}{1-|x_1|-\frac{2m^2}{P^2}(1+\frac{1}{x})}\Big)$$

$$-L\Big(\frac{1+|x_1|}{1+|x_1|-\frac{2m^2}{P^2}(1+\frac{1}{x})}\Big) \quad (3.12.17a)$$

$$+\log\Big|\frac{1-|x_1|}{1+|x_1|}\Big|\Big[2\log|\epsilon|+\log|p_fk_1|+\log|1-x+|x_1|(1+x)|$$

$$+\log|p_ik_2|+\log\Big|1-\frac{1}{x}+|x_1|(1+\frac{1}{x})\Big|\Big] \quad (3.12.17b)$$

$$+\log\Big|\frac{(1+x)(1-|x_1|)}{1-x+|x_1|(1+x)}\Big|\log\Big|\frac{2}{1-x+|x_1|(1+x)}\Big| \quad (3.12.17c)$$

$$+\log\Big|\frac{(1+x)(1-|x_1|)}{x-1+|x_1|(1+x)}\Big|\log\Big|\frac{2x}{x-1+|x_1|(1+x)}\Big| \quad (3.12.17d)$$

$$-\log\Big|\frac{(1+x)(1+|x_1|)}{1-x+|x_1|(1+x)}\Big|\log\Big|\frac{2x}{1-x+|x_1|(1+x)}\Big| \quad (3.12.17e)$$

$$-\log\Big|\frac{(1+x)(1+|x_1|)}{x-1+|x_1|(1+x)}\Big|\log\Big|\frac{2}{x-1+|x_1|(1+x)}\Big|\Big\}. \quad (3.12.17f)$$

The terms (3.12.17a) and the $(\log|p_fk_1|+\log|p_ik_2|)$ - term in (3.12.17b) now agree with the corresponding terms in the vertex V (3.12.3). The remaining logarithms in (3.12.17b) are equal to

$$\log\Big|\frac{1-|x_1|}{1+|x_1|}\Big|\log\Big|-\frac{4m^2}{P^2}\Big(x+\frac{1}{x}\Big)+2\Big(2-\frac{4m^2}{P^2}\Big)\Big|. \quad (3.12.18)$$

The other logarithms (3.12.17c-f) can be transformed as follows

$$\log|x|\log\Big|\frac{1-x+|x_1|(1+x)}{1-x-|x_1|(1+x)}\Big| \quad (3.12.19a)$$

$$+\log\left|\frac{1-|x_1|}{1+|x_1|}\right|\log\left|\frac{4x}{(1-x+|x_1|(1+x))(x-1+|x_1|(1+x))}\right|. \tag{3.12.19b}$$

Here (3.12.19a) appears in the vertex V (3.12.3), the term (3.12.19b) cancels against (3.12.18) up to $\log 4\log|1-|x_1||/|1+|x_1||$ which is independent of k_1, k_2. Summing up we find

$$III=\frac{4\pi(m^2-\frac{1}{2}P^2)}{|x_1||P^2|}\Big\{L(z_1)-L(z_2)+L(z_3)-L(z_4)$$

$$+\log\left|\frac{1-|x_1|}{1+|x_1|}\right|\Big[2\log|\epsilon|+\log|p_fk_1|+\log|p_ik_2|\Big]$$

$$+\log|x|\log\left|\frac{1-x+|x_1|(1+x)}{1-x-|x_1|(1+x)}\right|\Big\}, \tag{3.12.20}$$

where $z_1,\cdots z_4$ are the arguments of the Spence functions in (3.12.17a).

The term IV (3.12.15) is obtained from III by interchangeing $k_1\leftrightarrow k_2$, that means $x\leftrightarrow y$. Then it follows that

$$III+IV=\frac{4\pi(m^2-\frac{1}{2}P^2)}{|x_1||P^2|}V \tag{3.12.21}$$

is essentially the vertex contribution (3.12.3). On the other hand $I+II$ (3.12.12) (3.12.13) give just the self-energy (3.12.6)

$$I+II=8\pi S. \tag{3.12.22}$$

Hence, the $\hat{g}_0$ - dependent terms of the bremsstrahlung cross section of forth order are equal to

$$\frac{d\sigma^B}{d\Omega}=\frac{d\sigma^{(1)}}{d\Omega}\frac{e^2}{(2\pi)^7}\int dk_1dk_2\,\hat{g}_0(k_1)\hat{g}_0(k_2)\frac{1}{2}[I+II+III+IV]$$

$$=\frac{d\sigma^{(1)}}{d\Omega}\frac{e^2}{(2\pi)^6}\int dk_1dk_2\,\hat{g}_0(k_1)\hat{g}_0(k_2)2\left[\frac{m^2-\frac{1}{2}P^2}{2|x_1||P^2|}V+S\right]+\cdots$$

$$=-\frac{d\sigma^{V+S}}{d\Omega}. \tag{3.12.23}$$

Consequently, in the sum of bremsstrahlung, vertex and self-energy cross sections the divergent terms $\sim\log|\epsilon|$ and the finite g_0-dependent contributions cancel out. This shows existence and uniqueness of the adiabatic limit. The g_0-independent contributions are not considered here. If we had summed over two polarizations in (3.12.10) only, then additional non-covariant terms would survive, so that the limit is no longer unique.

Finally we want to illuminate the picture sketched at the beginning of this section. In the process of adiabatic switching, the charge of the outgoing particles is annihilated, their Coulomb field is liberated and the resulting free field is no longer transversal because current conservation is violated.

Since we are only interested in the infrared limit in this situation, we will study classical electrodynamics *without current conservation.* In this case the basic equations are not the inhomogeneous Maxwell equations but the inhomogeneous wave equation

$$\partial_\mu \partial^\mu A^\nu(x) = j^\nu(x). \tag{3.12.24}$$

The corresponding Lagrangian density is

$$L = L_0 + j^\nu A_\nu, \tag{3.12.25}$$

where

$$L_0 = \tfrac{1}{2} \sum_{\mu,\nu} (\partial_\mu A_\nu) \partial^\mu A^\nu \tag{3.12.26}$$

agrees with the Lagrangian of four independent free, massless scalar fields; in fact

$$\partial_\mu \frac{\partial L}{\partial(A_{\nu,\mu})} - \frac{\partial L}{\partial A_\nu} = \square A^\nu - j^\nu = 0. \tag{3.12.27}$$

To find the energy momentum tensor, let us express the translation invariance of L: L depends on x only through the fields and not explicitly

$$\partial_\nu L = \frac{\partial L}{\partial A^\lambda} \frac{\partial A^\lambda}{\partial x^\nu} + \frac{\partial L}{\partial (A^\lambda_{,\mu})} A^\lambda_{,\mu,\nu}$$

$$= \Big(\frac{\partial}{\partial x^\mu} \frac{\partial L}{\partial A^\lambda_{,\mu}} \Big) \frac{\partial A^\lambda}{\partial x^\nu} + \frac{\partial L}{\partial (A^\lambda_{,\mu})} A^\lambda_{,\mu,\nu} = \frac{\partial}{\partial x^\mu} \Big(\frac{\partial L}{\partial A^\lambda_{,\mu}} A^\lambda_{,\nu} \Big) \tag{3.12.28}$$

where (3.12.27) has been used. Hence,

$$\partial_\mu \Big(g^\mu_\nu L - \frac{\partial L}{\partial A^\lambda_{,\mu}} A^\lambda_{,\nu} \Big) = 0 \tag{3.12.29}$$

and the bracket is the energy momentum tensor. Substituting L_0 (3.12.26) we get the energy momentum tensor of the electromagnetic field alone

$$\tilde{T}^\mu{}_\nu = -\partial^\mu A_\lambda A^\lambda_{,\nu} + \frac{1}{2} g^\mu_\nu \sum_{\alpha,\beta} (\partial_\alpha A_\beta) \partial^\alpha A^\beta. \tag{3.12.30}$$

Its divergence

$$\partial_\mu \tilde{T}^\mu{}_\nu = -(\square A_\lambda) A^\lambda_{,\nu} - A_{\lambda,\mu} \partial^\mu A^\lambda_{,\nu} + A_{\beta,\alpha} \partial^\alpha A^\beta_{,\nu}$$

$$= -j_\lambda A^\lambda_{,\nu} \tag{3.12.31}$$

is the analogue of the Lorentz force. Integrating this over the four-dimensional volume $\mathbb{R}^3 \times [-T, T]$, we obtain for the change of the energy momentum of the field

$$P_\nu(T) - P_\nu(-T) = -\int_{-T}^{T} d^4x\, j_\lambda A^\lambda_{,\nu}. \tag{3.12.32}$$

Let us now consider a current density generated by a moving point charge that is switched on and off by our switching function $g(x)$

$$j_\lambda(x) = eg(x)\int d\tau\, \frac{dx_\lambda}{d\tau}\delta^4(x - x(\tau)), \tag{3.12.33}$$

where $x(\tau)$ is the trajectory of the point charge in space-time. We have $\partial^\lambda j_\lambda \neq 0$ because of the switching. Since we are interested in radiation of low frequencies, it is sufficient to describe the path of the moving particle by a sudden change between initial and final momentum

$$x^\lambda(\tau) = \frac{1}{m}\begin{cases} p_i^\lambda \tau, & \tau < 0 \\ p_f^\lambda \tau, & \tau > 0. \end{cases} \tag{3.12.34}$$

Without switching this leads to the following current density in momentum space

$$\begin{aligned} \tilde{j}^\lambda(k) &= e(2\pi)^{-2}\int_{-\infty}^{+\infty} d\tau\, \frac{dx^\lambda}{d\tau} e^{ikx(\tau)} \\ &= \frac{e}{m}(2\pi)^{-2}\left[\frac{-ip_i^\lambda}{p_i k - i0} + \frac{ip_f^\lambda}{p_f k + i0}\right]. \end{aligned} \tag{3.12.35}$$

The current density (3.12.33) is then given by

$$\begin{aligned} \hat{j}^\lambda(k) &= (2\pi)^{-2}\int d^4k'\, \tilde{j}^\lambda(k-k')\hat{g}(k') \\ &= \frac{-ie}{(2\pi)^4}\int d^4k' \left(\frac{p_i^\lambda}{p_i(k-k') - i0} - \frac{p_f^\lambda}{p_f(k-k') + i0}\right)\hat{g}(k') \end{aligned} \tag{3.12.36}$$

in momentum space. According to (3.12.24) it emits an electromagnetic field

$$A^\lambda(x) = \int d^4y\, D_0^{\mathrm{ret}}(x-y) j^\lambda(y), \tag{3.12.37}$$

that is not transversal.

We insert (3.12.37) into (3.12.32) to calculate the emitted energy - momentum in the limit $T \to \infty$

$$\begin{aligned} P_\nu(\infty) - P_\nu(-\infty) &= -\int d^4x\, d^4y\, j_\lambda(x)\partial_\nu^x D_0^{\mathrm{ret}}(x-y) j^\lambda(y) \\ &= -\frac{1}{2}\int d^4x\, d^4y\, j_\lambda(x)\Big[\partial_\nu^x D_0^{\mathrm{ret}}(x-y) + \partial_\nu^y D_0^{\mathrm{ret}}(y-x)\Big] j^\lambda(y). \end{aligned}$$

Writing the square bracket as

$$\partial_\nu^x[D_0^{\rm ret}(x-y)-D_0^{\rm av}(x-y)]=\partial_\nu^x D_0(x-y)$$

we arrive at

$$P_\nu(\infty)-P_\nu(-\infty)=-\frac{1}{2}\int d^4x\,d^4y\,j_\lambda(x)j^\lambda(y)\partial_\nu^x D_0(x-y). \tag{3.12.38}$$

In momentum space this becomes

$$P_\nu(\infty)-P_\nu(-\infty)=-\frac{(2\pi)^2}{2}\int d^4k\,\hat{j}_\lambda(k)\hat{j}^\lambda(k)^*(-ik_\nu)\hat{D}_0(k)$$

$$=-2\pi\int d^4k\,\hat{j}_\lambda(k)\hat{j}^\lambda(k)^*k_\nu\Theta(k^0)\delta(k^2). \tag{3.12.39}$$

For $\nu=0$ we get the total emitted energy

$$E=-2\pi\tfrac{1}{2}\int d^3k\,\hat{j}_\lambda(k)\hat{j}^\lambda(k)^*. \tag{3.12.40}$$

Using (3.12.36) and dividing by $\omega=|\boldsymbol{k}|$, we find the number of emitted photons with momentum k:

$$dN(k)=\frac{e^2}{(2\pi)^7}\frac{1}{2|\boldsymbol{k}|}\int d^4k_1\int d^4k_2\Bigg(-\frac{p_i^2}{p_i(k-k_1)p_i(k-k_2)}$$

$$+\frac{p_ip_f}{p_i(k-k_1)p_f(k-k_2)}+\frac{p_ip_f}{p_f(k-k_1)p_i(k-k_2)}$$

$$-\frac{p_f^2}{p_f(k-k_1)p_f(k-k_2)}\Bigg)\hat{g}(k_1)^*\hat{g}(k_2)d^3k. \tag{3.12.41}$$

This agrees precisely with the QED result for soft bremsstrahlung (3.11.29), and justifies that we have to work with four polarizations there, and also shows that the extension of the statistical interpretation of the S-matrix is correct. The modified classical electrodynamics gives the correct low energy limit of QED. The gedanken experiment of switching the electric charge in the asymtotic region is controlled by classical electrodynamics and, hence, is without any mystery. In this way charged scattering states can be simulated by neutral fermions plus their adiabatically detached electromagnetic field. This procedure is justified because the resulting inclusive cross sections have a general meaning independent of the switching function.

To isolate the detached self-field we introduce two transverse orthonormal and real polarization vectors

$$\varepsilon_n^\mu(\boldsymbol{k})=(0,\boldsymbol{\varepsilon}_n)\quad,\quad \boldsymbol{k}\cdot\boldsymbol{\varepsilon}_n(\boldsymbol{k})=0\quad,\quad n=1,2$$

$$\boldsymbol{\varepsilon}_n\cdot\boldsymbol{\varepsilon}_m=\delta_{nm}. \tag{3.12.42}$$

Together with the two vectors

$$\eta^\mu = (1,0,0,0) \quad , \quad \hat{k}^\mu = \frac{k^\mu - (k^\nu \eta_\nu)\eta^\mu}{|\boldsymbol{k}|} \tag{3.12.43}$$

they form a basis in Minkowski space. Then we have

$$g^{\lambda\mu} = \eta^\lambda \eta^\mu - \sum_{n=1}^{2} \varepsilon_n^\lambda(\boldsymbol{k})\varepsilon_n^\mu(\boldsymbol{k}) - \hat{k}^\lambda \hat{k}^\mu \tag{3.12.44}$$

$$= \eta^\lambda \eta^\mu - \sum_{n=1}^{2} \varepsilon_n^\lambda(\boldsymbol{k})\varepsilon_n^\mu(\boldsymbol{k})$$

$$-\frac{1}{\boldsymbol{k}^2}\Big[k_0^2 \eta^\lambda \eta^\mu + k^\lambda k^\mu - k_0(k^\lambda \eta^\mu + k^\mu \eta^\lambda)\Big].$$

This identity which has already been used above, can be applied to decompose the emitted electromagnetic field (3.12.37). In momentum space we obtain

$$\hat{A}^\lambda(k) = -\frac{1}{k^2 + i0k_0} g^{\lambda\mu} \hat{j}_\mu(k)$$

$$= -\frac{1}{k^2 + i0k_0}\Big\{\hat{j}_\perp^\lambda + \eta^\lambda \hat{j}_0$$

$$-\frac{1}{\boldsymbol{k}^2}\Big[k_0^2 \eta^\lambda \hat{j}_0 + k^\lambda k^\mu \hat{j}_\mu - k_0(k^\lambda \hat{j}_0 + \eta^\lambda k^\mu \hat{j}_\mu)\Big]\Big\}$$

$$= \hat{A}_\perp^\lambda + \frac{\hat{j}_0}{\boldsymbol{k}^2}\eta^\lambda - \frac{1}{(k^2 + i0k_0)\boldsymbol{k}^2}\Big[k_0 k^\lambda \hat{j}_0$$

$$+(k_0 \eta^\lambda - k^\lambda) k^\mu \hat{j}_\mu)\Big], \tag{3.12.45}$$

where the transverse part $A_\perp$ comes from the two transverse polarization vectors $\varepsilon_1, \varepsilon_2$ in (3.12.44). The last term in (3.12.45) is $O(\epsilon)$, because the current (3.12.35) is conserved

$$k_\lambda \tilde{j}^\lambda(k) = \frac{e}{m}(2\pi)^{-2}\left[\frac{-ikp_i}{p_i k - i0} + \frac{ikp_f}{p_f k + i0}\right] = 0. \tag{3.12.46}$$

The rest consists of "scalar photons"

$$\hat{A}^0(k) = \hat{j}_0\left(\frac{1}{\boldsymbol{k}^2} - \frac{k_0^2}{(k^2 + i0k_0)\boldsymbol{k}^2}\right) = -\frac{\hat{j}_0}{k^2 + i0k_0} \tag{3.12.47}$$

and the longitudinal part

$$\hat{A}_\parallel^l(k) = -\hat{j}_0 \frac{k_0 k^l}{(k^2 + i0k_0)\boldsymbol{k}^2}$$

$$= -\frac{1}{k^2 + i0k_0}\frac{k^i}{|\boldsymbol{k}|}\hat{j}^i \frac{k^l}{|\boldsymbol{k}|} + O(\epsilon). \tag{3.12.48}$$

These two contributions represent the self-field of the charged in- and outgoing particles.

3.13 Charge Normalization

Charge normalization is very different from mass normalization. The latter amounts to fixing one of the two free normalization constants in the free energy (3.7.39). We have found after (3.11.22) that this fixing is unique due to the existence of the adiabatic limit. But the other constant C_1 is not unique, because it is related to the constant C (3.9.3) in the vertex function by a Ward identity (see (4.6.49)). Since the vertex function contains infrared singularities, we must use the adiabatic limit again, in order to investigate the significance of those constants.

If one makes a renormalization of the vertex function

$$\tilde{\Lambda}^\nu = \hat{\Lambda}^\nu + C\gamma^\nu \tag{3.13.1}$$

with a finite constant C, the contribution to electron scattering is

$$S^{(3)} = e^3 (2\pi)^4 C \overline{u}(p)\gamma^\nu u(q)\hat{A}_\nu^{\rm ext}(p-q), \tag{3.13.2}$$

where the factor $(2\pi)^4$ comes from the trivial adiabatic limit. Writing the self-energy (3.11.19) in the form

$$\hat{\Sigma}(p) = (\not{p} - m)\hat{\Sigma}_1(p^2) + (p^2 - m^2)\hat{\Sigma}_2(p^2), \tag{3.13.3}$$

a finite renormalization without change of the mass normalization is given by

$$\tilde{\Sigma}_1 = \hat{\Sigma}_1 + C', \tag{3.13.4}$$

i.e. by changing C_1 in (3.11.19). However, the constant C' is related to C in (3.13.1) by the Ward identity (4.6.49)

$$C + \frac{C'}{(2\pi)^2} = 0. \tag{3.13.5}$$

The contribution to the first self-energy graph in Fig. 7 must again be calculated in the adiabatic limit as in (3.11.18):

$$e^3 C' \int d^4k_1\, d^4k_2\, \hat{g}_0(k_1)\hat{g}_0(k_2)\overline{u}(p)(\not{p} - \epsilon\not{k}_1 - m)$$

$$\times \frac{1}{(2\pi)^2}\frac{\not{p} + m - \epsilon(\not{k}_1 + \not{k}_2)}{(p - \epsilon(k_1+k_2))^2 - m^2 + i0}\gamma^\nu u(q)\hat{A}_\nu^{\rm ext}(p-q)$$

$$= e^3 \frac{C'}{(2\pi)^2}\overline{u}(p)\int d^4k_1\, d^4k_2\, \hat{g}_0(k_1)\hat{g}_0(k_2)\frac{-\epsilon\not{k}_1(\not{p}+m)}{-2\epsilon p(k_1+k_2) + i0}$$

$$\gamma^\nu u(q)\hat{A}_\nu^{\rm ext}(p-q) + O(\epsilon). \tag{3.13.6}$$

Anticommuting $\not{p}$ to the left and operating on $\overline{u}(p)$, we finally obtain

$$= e^3 \frac{C'}{(2\pi)^2}\overline{u}(p)\int d^4k_1\, d^4k_2\, \hat{g}_0(k_1)\hat{g}_0(k_2)\frac{2\epsilon p k_1}{2\epsilon p(k_1+k_2) - i0} \times$$

$$\times \gamma^\nu u(q) \hat{A}_\nu^{\rm ext}(p-q). \tag{3.13.7}$$

This integral is of the same form as (3.11.21), so that the final result is given by

$$= \frac{e^3}{2}(2\pi)^4 \frac{C'}{(2\pi)^2} \overline{u}(p)\gamma^\nu u(q) \hat{A}_\nu^{\rm ext}(p-q). \tag{3.13.8}$$

Since the second self-energy graph gives the same contribution, they exactly cancel the vertex contribution (3.13.2). Hence, the constants C and C' do not influence the charge normalization, they do not contain any physics. If the vacuum polarization is fixed by other conditions (3.6.34, 35), charge normalization is unambiguously specified in finite QED.

There arises the question whether there is a finite charge renormalization by radiative corrections. We hold the view that the physical charge is defined in electron scattering at low energies. To investigate this problem it is sufficient to consider the limit of vanishing momentum transfer $p_f = p_i = p$. Then we obtain from the self-energy and vertex functions

$$\overline{u}_s(p) \Big[\hat{\Lambda}^\nu(p-\epsilon k_1, p+\epsilon k_2) - \hat{\Sigma}(p-\epsilon k_1) \hat{S}_F(p-\epsilon k_1 - \epsilon k_2)\gamma^\nu$$

$$-\gamma^\nu \hat{S}_F(p+\epsilon k_1 + \epsilon k_2)\hat{\Sigma}(p+\epsilon k_2)\Big] u_\sigma(p)$$

$$= \overline{u}_s(p) \left[\frac{p^\nu}{m} F(p, k_1, k_2; \epsilon) - \frac{1}{(2\pi)^2} \frac{p^\nu}{\epsilon p(k_1+k_2)} \Big(\hat{\Sigma}(p+\epsilon k_2) - \hat{\Sigma}(p-\epsilon k_1) \Big) \right]$$

$$\times u_\sigma(p) + O(\epsilon \log|\epsilon|). \tag{3.13.9}$$

The precise form of F may be deduced from the previous sections, but this is not necessary in the following. From the Ward identity (4.6.48)

$$\frac{1}{(2\pi)^2} \Big[\hat{\Sigma}(\tilde{p}) - \hat{\Sigma}(q) \Big] = (\tilde{p}-q)_\mu \hat{\Lambda}^\mu(\tilde{p}, q), \tag{3.13.10}$$

with $\tilde{p} = p - \epsilon k_1$, $q = p + \epsilon k_2$, $\not{p} = m$, which is fulfilled in all orders of ϵ, in particular $O(\epsilon \log|\epsilon|)$ and $O(\epsilon)$, we see that the leading contributions $O(\log|\epsilon|)$ and $O(1)$ in (3.13.9) drop out so that (3.13.9) is of order $O(\epsilon \log|\epsilon|)$. Hence, there is no contribution in the adiabatic limit $\epsilon \to 0$.

There remains bremsstrahlung to be considered. Its contribution simplifies for zero momentum transfer as follows

$$S_{fi}^{(n)} = -ie^2(2\pi)^{-9/2} \int dk_2\, \hat{g}(k_2) \overline{u}_s(p) \Big[\gamma^\mu (\not{k} - \not{k}_2) \not{\varepsilon}^{(n)}$$

$$+ \not{\varepsilon}^{(n)} (\not{k} - \not{k}_2)\gamma^\mu \Big] u_\sigma(p) \frac{1}{2p(k-k_2)+i0} \hat{A}_\mu^{\rm ext}(k-k_2) \frac{1}{\sqrt{2\omega(k)}}, \tag{3.13.11}$$

where $\varepsilon^{(n)}$ is the polarization vector of the photon of bremsstrahlung. In the inclusive cross section the following k-integral occurs

$$I_{\alpha\beta}(p,k_1,k_2)=\int\limits_{|\boldsymbol{k}|\le\omega_0}\frac{d^3k}{2|\boldsymbol{k}|}\frac{(k-\epsilon k_2)_\alpha(k-\epsilon k_1)_\beta}{p(k-\epsilon k_2)p(k-\epsilon k_1)}. \tag{3.13.12}$$

After a tedious but straight forward computation one finds that the result is of the following form

$$I_{\alpha\beta}=\omega_0^2 f_{\alpha\beta}(p)+O(\epsilon). \tag{3.13.13}$$

The first term independent of k_1, k_2 comes from the upper limit of the $|\boldsymbol{k}|$-integration:

$$I_{\alpha\beta}=\int\limits_0^{\omega_0} d|\boldsymbol{k}|\,\frac{|\boldsymbol{k}|^2}{2|\boldsymbol{k}|}\int d\Omega_k\,\frac{(k-\epsilon k_2)_\alpha(k-\epsilon k_1)_\beta}{p(k-\epsilon k_2)p(k-\epsilon k_1)}$$

$$\sim\int\limits_0^{\omega_0} d|\boldsymbol{k}|\,|\boldsymbol{k}|\sim\omega_0^2. \tag{3.13.14}$$

In the adiabatic limit $\epsilon\to 0$ the result (3.13.13) only reflects the dependence of the cross section on the energy resolution ω_0 of the photon detector. There is no constant contribution independent of ω_0 and, hence, no finite renormalization of the charge. Up to forth order perturbation theory the physical charge agrees with the coupling constant appearing in the first order S-matrix. Consequently, the electric form factor $f(P^2)$ (3.9.48) satisfies

$$f(0)=0. \tag{3.13.15}$$

It should be remembereded that the vacuum polarization tensor, for other reasons, is normalized by the conditions (3.6.34, 35)

$$\Pi(0)=0,\qquad \frac{\Pi(k)}{k^2}\bigg|_{k^2=0}=0. \tag{3.13.16}$$

Then, it gives no contribution to charge normalization, too. If one assumes a different normalization of $\Pi(k)$, then the coupling constant in $T_1(x)$ and the physical charge are no longer equal. This is the starting point for the renormalization group. This subject will be discussed in Sect. 4.8.

Until now we have assumed that a gauge invariant normalization of the vertex function and self-energy exists and have investigated finite renormalizations. There remains to show how this gauge invariant normalization can actually be achieved. Our starting point is the Ward identity (3.13.10), which in the limit $q\to\tilde{p}=p$ assumes the following form

$$\frac{\partial}{\partial p_\mu}\hat{\Sigma}(p)=(2\pi)^2\hat{\Lambda}^\mu(p,p). \tag{3.13.17}$$

To avoid any infrared problem, we consider this identity at $p=0$ and choose the central splitting solution $\hat{\Lambda}(0,0)=0$ for the vertex function (see (3.9.44)). Then the self-energy must be normalized according to

$$\frac{\partial}{\partial p_\mu}\hat{\Sigma}(p)\Big|_{p=0} = 0. \tag{3.13.18}$$

Expanding our previous result (3.7.41) for $p \to 0$, we find

$$\hat{\Sigma}(p) = e^2(2\pi)^{-4}\Big\{\frac{3}{4}m - \frac{\not p}{8} + O(p^2) + C_1(\not p - m)\Big\}.$$

The normalization condition (3.13.18) now leads to

$$C_1 = \frac{1}{8}. \tag{3.13.19}$$

Curiously enough, the same value for C_1 is obtained from the "on-shell" condition

$$\Big(\frac{\partial}{\partial \not p}\hat{\Sigma}\Big)(\not p = m) = 0. \tag{3.13.20}$$

But the latter cannot directly be used in (3.13.17), because of the singular behaviour of the vertex function on the mass shell.

3.14 Problems

3.1. Prove the perturbative causality condition (3.1.28) from the global condition (3.1.27), using the symmetry under permutations.

3.2. What is the singular order of $\log(p^2/m^2) \in \mathcal{S}'(\mathbb{R}^4)$?

3.3. Prove the following identity for normal products by induction

$$:A_1 \ldots A_n : B = : A_1 \ldots A_{n-1} : A_n B$$
$$+ \sum_{k=1}^{n} : A_1 \ldots \overline{A_k \ldots A_n B} : . \tag{3.14.1}$$

3.4. Give an inductive proof of Wick's theorem by means of (3.14.1).

3.5. Write the differential cross section for electron scattering in Lorentz invariant form: $d\sigma/dt = ?$, $t = (p_i - q_f)^2$.

3.6. In electron-positron colliders e^+e^--scattering (Bbabba - scattering) is measured in order to determine the flux of the machine. How can the results for this process be obtained from electron scattering? Calculate the center-of-mass cross section. Show that, in the non-relativistic limit, it is equal to the first term in (3.4.52).

3.7. Calculate the differential cross section for the process $e^- + e^+ \to \mu^- + \mu^+$. Result:

$$\frac{d\sigma}{dt} = \frac{e^4}{4\pi s^6}\Big[\frac{t^2+u^2}{2} + (m_e^2 + m_\mu^2)(2s - m_e^2 - m_\mu^2)\Big], \tag{3.14.2}$$

where

$$s = (p_i + q_i)^2 = (p_f + q_f)^2, \quad t = (p_f - p_i)^2, \quad u = (p_i - q_f)^2. \tag{3.14.3}$$

3.8. Show that for high energies $E \gg m_\mu$, (3.14.2) implies

$$\left(\frac{d\sigma}{d\Omega}\right)_{\text{c.m.}} = \frac{e^4}{64\pi^2 s}(1 + \cos^2 \vartheta) \tag{3.14.4}$$

in the center-of-mass frame.

3.9. Write the differential cross section (3.5.48) for Compton scattering without polarizations in Lorentz invariant form, using the variables

$$s = (p_i + k_i)^2, \quad t = (p_i - p_f)^2, \quad u = (p_i - k_f)^2. \tag{3.14.5}$$

Result:

$$\frac{d\sigma}{dt} = \frac{8\pi e^4}{(s-m)^2}\Big[\Big(\frac{m^2}{s-m^2} + \frac{m^2}{u-m^2}\Big)^2 + \Big(\frac{m^2}{s-m^2} + \frac{m^2}{u-m^2}\Big)$$
$$-\frac{1}{4}\Big(\frac{s-m^2}{u-m^2} + \frac{u-m^2}{s-m^2}\Big)\Big]. \tag{3.14.6}$$

3.10. Show that the S-matrix element for pair annihilation $e^- + e^+ \to 2\gamma$ can be obtained from Compton scattering by exchanging particles in the initial state by antiparticles in the final state with reversed momentum (crossing symmetry).

3.11. Determine the differential cross section for pair annihilation by means of problem 3.9. Result:

$$\frac{d\sigma}{dt} = \frac{8\pi e^4}{t(t-4m^2)}\Big[\Big(\frac{m^2}{s-m^2} + \frac{m^2}{u-m^2}\Big)^2 + \Big(\frac{m^2}{s-m^2} + \frac{m^2}{u-m^2}\Big)$$
$$-\frac{1}{4}\Big(\frac{s-m^2}{u-m^2} + \frac{u-m^2}{s-m^2}\Big)\Big], \tag{3.14.7}$$

with

$$s = (p_i - k_f)^2, \quad t = (p_i + q_i)^2, \quad u = (p_i - k_f')^2. \tag{3.14.8}$$

3.12. Write the causal distribution (3.8.20) for the vertex function in a manifestly causal form.

3.13. Show that naive distribution splitting of the result of problem 3.12 would formally lead to the usual Feynman integral for $T^\nu = R^\nu - R'^\nu$. Where is the error in this procedure?

3.14. Verify that the imaginary part in I_D (3.8.61) vanishes.

3.15. Consider the inclusive cross section (3.11.30) with only two transversal polarizations of soft bremsstrahlung. Show that the divergent contribution $\sim \log|\epsilon|$ agrees with (3.11.40).

4. Properties of the S-Matrix

The inductive construction of causal perturbation theory is very well suited to derive properties of the S-matrix that are valid in all orders. One has only to show that a property is true in first order and that it is preserved in the inductive step. In this way we get comparably simple proofs of normalizability, various symmetries, gauge invariance and unitarity. Before coming to these themes we will analyse vacuum graphs and show that the perturbative S-matrix $S(g)$ is a well-defined operator in Fock space. The chapter closes with some more sophisticated techniques, namely the renormalization group and interacting fields. The latter are widely used in Lagrangian field theory. However, we shall obtain different results: the interacting fields must depend not only on the space-time argument, but also on the switching function $g(x)$. The limit $g \to 1$ is impossible in general. Nevertheless, these fields fulfill suitably defined field equations which are similar to the basic equations of classical Lagrangian field theory.

4.1 Vacuum Graphs

In finite QED even vacuum graphs have to be pretty finite, because there is no illegitimate step in their derivations. Let us return to the second order causal distribution (3.3.16 (7)):

$$D^{(7)}(x_1,x_2) = -ie^2 \mathrm{tr}\left[\gamma^\mu S^{(-)}(x_1-x_2)\gamma_\mu S^{(+)}(x_2-x_1)\right] D_0^{(+)}(x_2-x_1)$$

$$+ie^2 \mathrm{tr}\left[\gamma^\mu S^{(-)}(x_2-x_1)\gamma_\mu S^{(+)}(x_1-x_2)\right] D_0^{(+)}(x_1-x_2). \tag{4.1.1}$$

Using the relative coordinate $y = x_1 - x_2$, This can be written as follows:

$$D^{(7)}(x_1,x_2) = -ie^2[d(y)-d(-y)]. \tag{4.1.2}$$

Here the first term is the R'-distribution (3.1.37), which we have to know separately for the calculation of the T-distribution:

$$R_2^{(7)\prime}(x_1,x_2) = -ie^2 d(y), \tag{4.1.3}$$

with

$$d(y) = \mathrm{tr}\,[\gamma^\mu S^{(-)}(y)\gamma_\mu S^{(+)}(-y)] D_0^{(+)}(-y). \tag{4.1.4}$$

After Fourier transformation we get

$$\hat{d}(p) = i(2\pi)^{-7}\int dq_1\, dq_2\, \mathrm{tr}\,[\gamma^\mu(\not{q}_1 + m)\gamma_\mu(\not{q}_2 + m)]$$

$$\delta(q_1^2 - m^2)\Theta(q_1^0)\delta(q_2^2 - m^2)\Theta(-q_2^0)\delta((q_2 - q_1 - p)^2)\Theta(q_2^0 - q_1^0 - p^0). \quad (4.1.5)$$

Here we make contact with the calculation of vacuum polarization in Sect. 3.6. Inserting a trivial integration

$$1 = \int d^4k\, \delta(q_1 - q_2 - k) \quad (4.1.6)$$

into (4.1.5), we arrive at

$$\hat{d}(p) = i(2\pi)^{-7}\int dq_1\, dq_2\, dk\, \delta(q_1 - q_2 - k)\mathrm{tr}\,[\gamma^\mu(\not{q}_1 + m)\gamma_\mu(\not{q}_2 + m)]$$

$$\delta(q_1^2 - m^2)\Theta(q_1^0)\delta(q_2^2 - m^2)\Theta(-q_2^0)\delta((-k - p)^2)\Theta(-k^0 - p^0). \quad (4.1.7)$$

Here it is possible to use the associative law for the convolution

$$\hat{d}(p) = i(2\pi)^{-7}\int dk\, \delta((-k - p)^2)\Theta(-k^0 - p^0)$$

$$\Big\{\int dq_1\, dq_2 \mathrm{tr}\,[\gamma^\mu(\not{q}_1 + m)\gamma_\mu(\not{q}_2 + m)]$$

$$\delta(q_1^2 - m^2)\Theta(q_1^0)\delta(q_2^2 - m^2)\Theta(-q_2^0)\delta(q_1 - q_2 - k)\Big\}. \quad (4.1.8)$$

The curly bracket agrees with the vacuum polarization function (3.6.5)

$$\{\ldots\} = e^{-2}(2\pi)^4 \hat{P}^\mu_{1\mu}(k). \quad (4.1.9)$$

Using the previous result (3.6.7, 8), we obtain

$$\hat{d}(p) = i(2\pi)^{-6}\int dk\, k^2\Big(1 + \frac{2m^2}{k^2}\Big)\sqrt{1 - \frac{4m^2}{k^2}}$$

$$\times\, \Theta(k^2 - 4m^2)\Theta(k_0)\Theta(-k_0 - p_0)\delta((-k - p)^2). \quad (4.1.10)$$

The last integral is calculated in a special Lorentz frame. It is easy to see that p must be time-like with $p_0 < 0$. We therefore may take $p = (p_0, \mathbf{0})$. Then we get

$$\hat{d}(p_0) = i(2\pi)^{-6}\int dk\, k^2\Big(1 + \frac{2m^2}{k^2}\Big)\sqrt{1 - \frac{4m^2}{k^2}}$$

$$\times\, \Theta(k^2 - 4m^2)\Theta(k_0)\Theta(-k_0 - p_0)\delta((k_0 + p_0)^2 - |\boldsymbol{k}|^2)$$

$$= -i(2\pi)^{-5}\int dk_0\, (k_0 + p_0)(k_0^2 - (k_0 + p_0)^2)$$

$$\left(1+\frac{2m^2}{k_0^2-(k_0+p_0)^2}\right)\sqrt{1-\frac{4m^2}{k_0^2-(k_0+p_0)^2}}$$

$$\Theta(k_0)\Theta(k_0^2-(k_0+p_0)^2-4m^2)\Theta(-k_0-p_0). \tag{4.1.11}$$

Introducing the new integration variable s

$$s^2=k_0^2-(k_0+p_0)^2=-p_0^2-2p_0k_0, \tag{4.1.12}$$

we arrive at

$$\hat{d}(p_0)=i(2\pi)^{-5}\Theta(p_0^2-4m^2)\Theta(-p_0)\frac{m}{p_0^2}$$

$$\int\limits_{2m}^{-p_0} ds\,(p_0^2-s^2)(s^2+2m^2)\sqrt{\frac{s^2}{4m^2}-1}. \tag{4.1.13}$$

The remaining integral is elementary, the result in an arbitrary Lorentz system is given by

$$\hat{d}(p)=i(2\pi)^{-5}\Theta(p^2-4m^2)\Theta(-p_0)\left[\left(\frac{p^4}{24}+\frac{m^2}{12}p^2+m^4\right)\sqrt{1-\frac{4m^2}{p^2}}\right.$$

$$\left.+\frac{m^4}{p^2}(4m^2-3p^2)\log\left(\sqrt{\frac{p^2}{4m^2}}+\sqrt{\frac{p^2}{4m^2}-1}\right)\right] \tag{4.1.14}$$

$$=ie^{-2}R^{(7)\prime}(p).$$

The D-distribution (4.1.3) is now equal to

$$\hat{D}^{(7)}(p)=-e^2(2\pi)^{-5}\Theta(p^2-4m^2)\mathrm{sgn}\,p_0 f(p^2), \tag{4.1.15}$$

where

$$f(p^2)=\left(\frac{p^4}{24}+\frac{m^2}{12}p^2+m^4\right)\sqrt{1-\frac{4m^2}{p^2}} \quad : \quad I_1$$

$$+\frac{m^4}{p^2}(4m^2-3p^2)\log\left(\sqrt{\frac{p^2}{4m^2}}+\sqrt{\frac{p^2}{4m^2}-1}\right) \quad : \quad I_2. \tag{4.1.16}$$

Since $\hat{d}(p)=O(p^4)$ for $p\to\infty$, we conclude that the singular order is $\omega=4$. In Sect. 4.3 we shall derive the general expression for the singular order of an arbitrary diagram. The retarded distribution is obtained by central splitting (3.2.61) of (4.1.2) with $\omega=4$. For time-like p, $p^2>0$, $p_0>0$ it is given by

$$\hat{R}^{(7)}(p)=\frac{i}{2\pi}\int\limits_{-\infty}^{+\infty} dt\,\frac{\hat{D}^{(7)}(tp)}{(t-i0)^5(1-t+i0)} \tag{4.1.17}$$

$$= \frac{i}{2\pi} \mathrm{P} \int_{-\infty}^{+\infty} dt \frac{\hat{D}^{(7)}(tp)}{(t-i0)^5(1-t)} + \frac{1}{2}\hat{D}^{(7)}(p). \tag{4.1.18}$$

Let us consider the integral

$$r_7 = \left[\int_0^\infty \frac{dt}{t^5(1-t)} - \int_{-\infty}^0 \frac{dt}{t^5(1-t)} \right] \Theta(t^2p^2 - 4m^2) f(t^2p^2)$$

$$= p^6 \int_{4m^2}^\infty \frac{ds}{s^3} \frac{f(s)}{p^2 - s}. \tag{4.1.19}$$

Inserting (4.1.16), we have to calculate first

$$I_1(p^2) = \int_{4m^2}^\infty \frac{ds}{s^3} \frac{1}{p^2 - s} \left(\frac{s^2}{24} + m^2 \frac{s}{12} + m^4 \right) \sqrt{1 - \frac{4m^2}{s}}$$

$$= \frac{-1}{p^6} \left\{ \frac{19}{360} p^4 + 2m^4 + \left(\frac{p^4}{24} + m^2 \frac{p^2}{12} + m^4 \right) \right.$$

$$\times \left[\Theta(p^2 - 4m^2) \sqrt{1 - \frac{4m^2}{p^2}} \log \left| \frac{1 - \sqrt{1 - \frac{4m^2}{p^2}}}{1 + \sqrt{1 - \frac{4m^2}{p^2}}} \right| \right.$$

$$\left. \left. - \Theta(4m^2 - p^2) 2 \sqrt{\frac{4m^2}{p^2} - 1} \arctan \frac{1}{\sqrt{\frac{4m^2}{p^2} - 1}} \right] \right\}. \tag{4.1.20}$$

Since this is part of the retarded distribution, it must be the boundary value of an analytic function (theorem 3.3 (3.2.63)). Indeed, it can be easily verified that this analytic function is given by

$$I_1^{\mathrm{an}}(p) = \frac{-1}{p^6} \left[\frac{19}{360} p^4 + 2m^4 + \left(\frac{p^4}{24} + m^2 \frac{p^2}{12} + m^4 \right) \sqrt{1 - \frac{4m^2}{p^2}} \right.$$

$$\left. \times \log \frac{\sqrt{1 - \frac{4m^2}{p^2}} - 1}{\sqrt{1 - \frac{4m^2}{p^2}} + 1} \right]. \tag{4.1.21}$$

The second integral in (4.1.16) can be computed by means of the Euler substitutions (3.6.20). We immediately write down the result for the analytic function

$$I_2^{\mathrm{an}}(p) = \frac{-1}{p^6} \left[\frac{37}{720} \frac{p^4}{m^4} + \frac{2}{3} \frac{p^2}{m^2} - 1 + \left(3 - \frac{4m^2}{p^2} \right) \log^2 \left(\sqrt{\frac{-p^2}{4m^2}} + \sqrt{1 - \frac{p^2}{4m^2}} \right) \right]. \tag{4.1.22}$$

Substituting these results into (4.1.17), we get the total retarded distribution

$$R^{(7)}(p)=ie^2(2\pi)^{-6}m^4\Big[\frac{5}{3}a^4-\frac{8}{3}a^2+1+\Big(3+\frac{1}{a^2}\Big)\log^2\Big(\sqrt{a^2}+\sqrt{1+a^2}\Big)$$

$$+\Big(\frac{2}{3}a^4-\frac{a^2}{3}+1\Big)\sqrt{1+a^{-2}}\log\frac{\sqrt{1+a^{-2}}-1}{\sqrt{1+a^{-2}}+1}\Big], \tag{4.1.23}$$

where

$$a^2=\frac{-p^2}{4m^2}. \tag{4.1.24}$$

This can be analytically continued to arbitrary p. The T-distribution is obtained by subtraction of

$$R_2^{(7)\prime}(p)=e^2(2\pi)^{-5}\Theta(p^2-4m^2)\Theta(-p_0)$$

$$\times\Big[\Big(\frac{p^4}{24}+\frac{m^2}{12}p^2+m^4\Big)\sqrt{1-\frac{4m^2}{p^2}}$$

$$+\frac{m^4}{p^2}(4m^2-3p^2)\log\Big(\sqrt{\frac{p^2}{4m^2}}+\sqrt{\frac{p^2}{4m^2}-1}\Big)\Big]. \tag{4.1.25}$$

The last result does not contribute in the limit $p\to 0$. Consequently, we find from (4.1.23)

$$\hat{T}^{(7)}(p)=ie^2(2\pi)^{-6}\frac{4m^4}{15}\Big(\frac{p^2}{4m^2}\Big)^3+\dots \quad \text{for} \quad p\to 0. \tag{4.1.26}$$

The general solution for the T-distribution contains three free constants

$$\tilde{T}^{(7)}(p)=\hat{T}^{(7)}(p)+C_0+C_2p^2+C_4p^4, \tag{4.1.27}$$

because $\omega=4$. It certainly contributes to the vacuum-to-vacuum transition amplitude

$$\lim_{g\to 1}(\Omega,S_2(g)\Omega)$$

$$=\frac{1}{2}\lim_{g\to 1}\int dy\,g(y)\int dp\,\tilde{T}^{(7)}(p)e^{ipy}\hat{g}(-p). \tag{4.1.28}$$

We will shortly see that we must again be careful with the adiabatic limit. As before (3.11.1–3), we perform this limit in scaling form (3.11.1)

$$=\frac{1}{2}\lim_{\varepsilon\to 0}\int dy\,g_0(\varepsilon y)\int dp\,(\hat{T}^{(7)}(p)+C_0+C_2p^2+C_4p^4)$$

$$\times e^{ipy}\frac{1}{\varepsilon^4}\hat{g}_0\Big(\frac{-p}{\varepsilon}\Big) \tag{4.1.29}$$

$$=\frac{1}{2}(2\pi)^2\lim_{\varepsilon\to 0}\frac{1}{\varepsilon^4}\int dp\,\Big[\hat{T}^{(7)}(\varepsilon p)+C_0+C_2\varepsilon^2p^2+C_4\varepsilon^4p^4\Big]\hat{g}_0(p)\hat{g}(-p). \tag{4.1.30}$$

For $C_0, C_2\neq 0$, the limit does not exist, so that we must require the normalization

$$C_0 = 0 = C_2. \tag{4.1.31}$$

Now, by (4.1.26), the limit exists and is independent of g_0, if we choose $C_4 = 0$.

The vanishing of the vacuum-to-vacuum amplitude can be traced back to the existence of the central solution (4.1.17–19) which implies the vanishing at $p = 0$. The same is true in higher orders. It follows that

$$\lim_{g\to 1} (\Omega, S(g)\Omega) = 1. \tag{4.1.32}$$

This means that the free vacuum is stable, it is not modified by the interaction. One might think that there is no other place where vacuum graphs play a rôle. This in not quite true, because in

$$T_n = R_n - R'_n \tag{4.1.33}$$

also disconnected graphs contribute to R'_n! The latter may contain vacuum graphs. However, their contribution vanishes in the adiabatic limit for the same reason as above, and since this is a factor, it follows that a disconnected graph with a vacuum subgraph also vanishes.

4.2 Operator Character of the S-Matrix

The reader may wonder why we have calculated the vacuum graph in detail, if it has no observable consequences. The reason is that we will now use these results in the study of the operator nature of the S-matrix. Since scattering processes are described by the S-matrix elements, which are scalar products in Fock space, it is necessary that the S-matrix is a bilinear form. The situation in QED is better: $S(g)$ is even an operator in Fock space for all $g \in \mathcal{S}(\mathbb{R}^4)$.

We now show this in detail for the first order $S_1(g)$ of perturbation theory. To do so, we must estimate the norm

$$\|S_1(g)\Phi\|^2 = (\Phi,\, S_1^+(g)S_1(g)\Phi) \tag{4.2.1}$$

for vectors Φ in a certain dense domain. Since

$$S_1^+(g)S_1(g) = -\int dx_1 dx_2\, g(x_1)g(x_2)T_1(x_2)T_1(x_1)$$

$$= \int dx_1 dx_1\, g(x_1)g(x_2)R'_2(x_1, x_2), \tag{4.2.2}$$

we have

$$\|S_1(g)\Phi\|^2 = \int dx_1 dx_2\, g(x_1)g(x_2)(\Phi,\, R'_2(x_1, x_2)\Phi), \tag{4.2.3}$$

where R'_2 is given by (3.3.15):

$$R_2'(x_1,x_2)=e^2:\overline{\psi}(x_1)\gamma^\mu\psi(x_1)\overline{\psi}(x_2)\gamma^\nu\psi(x_2)::A_\mu(x_1)A_\nu(x_2):\qquad(1)$$

$$+ie^2[:\overline{\psi}(x_1)\gamma^\mu S^{(-)}(x_1-x_2)\gamma^\nu\psi(x_2):-:\overline{\psi}(x_2)\gamma^\nu S^{(+)}(x_2-x_1)\gamma^\mu\psi(x_1):]$$

$$\times:A_\mu(x_1)A_\nu(x_2):\qquad(2)$$

$$-e^2\mathrm{tr}\,[S^{(+)}(x_2-x_1)\gamma^\mu S^{(-)}(x_1-x_2)\gamma^\nu]:A_\mu(x_1)A_\nu(x_2):\qquad(3)$$

$$+ie^2:\overline{\psi}(x_1)\gamma^\mu\psi(x_1)\overline{\psi}(x_2)\gamma_\mu\psi(x_2):D_0^{(+)}(x_2-x_1)\qquad(4)$$

$$-e^2[:\overline{\psi}(x_1)\gamma^\mu S^{(-)}(x_1-x_2)\gamma_\mu\psi(x_2):-:\overline{\psi}(x_2)\gamma^\mu S^{(+)}(x_2-x_1)\gamma_\mu\psi(x_1):]$$

$$\times D_0^{(+)}(x_2-x_1)\qquad(5)$$

$$-ie^2\mathrm{tr}\,[S^{(+)}(x_2-x_1)\gamma^\mu S^{(-)}(x_1-x_2)\gamma^\mu]D_0^{(+)}(x_2-x_1).\quad(6)\qquad(4.2.4)$$

To show that (4.2.3) is bounded for all $g\in\mathcal{S}(\mathbb{R}^4)$, it is sufficient to verify that $(\Phi,R_2'(x_1,x_2)\Phi)$ is a tempered distribution $\in\mathcal{S}'(\mathbb{R}^4)$. There are only a few cases for Φ to be examined.

If Φ is the vacuum Ω, only the last term (6) in (4.2.4) contributes. This contribution is given by $R_2^{(7)\prime}$ (4.1.3) in the last section. From its computation in momentum space (4.1.25), it follows that this is a tempered distribution. Note that there is no singularity at $p^2=0$, because of the Θ-function.

The second case is a one-electron state

$$\Phi=\int d^3p\,\varphi(\boldsymbol{p})b_s^+(\boldsymbol{p})\Omega.\qquad(4.2.5)$$

Beside the vacuum contribution (6), which obviously gives a tempered distribution, we have to concentrate on (5) in (4.2.4):

$$(\Phi,\,(5)\,\Phi)=-e^2D_0^{(+)}(x_2-x_1)[\bar{\varphi}(x_1)\gamma^\mu S^{(-)}(x_1-x_2)\gamma_\mu\varphi(x_2)$$

$$-\bar{\varphi}(x_2)\gamma^\mu S^{(+)}(x_2-x_1)\gamma_\mu\varphi(x_1)].\qquad(4.2.6)$$

Such an expression appeared in the self-energy calculation (3.7.2):

$$=-e^2\bar{\varphi}(x_1)\gamma^\mu[d_-^{(5)}(x_1-x_2)-d_+^{(5)}(x_1-x_2)]\gamma_\mu\varphi(x_2).\qquad(4.2.7)$$

It follows from (3.7.14, 18) that the square bracket is a tempered distribution. Choosing the electron wave function φ with compact support in momentum space, the multiplication in (4.2.7) does not destroy the distributional character. This is shown in detail below (4.2.12). The third case is a two electron state

$$\Phi=\int d^3p\int d^3q\,\varphi_s(\boldsymbol{p})\psi_\sigma(\boldsymbol{q})b_s^+(\boldsymbol{p})b_\sigma^+(\boldsymbol{q})\Omega.\qquad(4.2.8)$$

Now the term (4) in (4.2.4) gives a contribution which is known from electron scattering (3.4.13). With smooth wave functions, we get again a tempered distribution.

We now turn to a photon state (2.11.33)

$$\Phi = \int d^3k\, f(\boldsymbol{k})\varepsilon_\mu(\boldsymbol{k})a_\mu(\boldsymbol{k})^+\Omega. \tag{4.2.9}$$

In this case (3) in (4.2.4) leads to a contribution which can be expressed by the vacuum polarization tensor $P_1^{\nu\mu}$ (3.6.5)

$$(\Phi,\, R_2'(x_1,x_2)\Phi) = -(2\pi)^{-3}P_1^{\nu\mu}(x_2-x_1)\int d^3kd^3h\, f^*(\boldsymbol{k})f(\boldsymbol{h})$$

$$\left[\frac{\varepsilon_\mu(\boldsymbol{k})\varepsilon_\nu(\boldsymbol{h})}{\sqrt{4\omega(\boldsymbol{k})\omega(\boldsymbol{h})}}e^{ikx_1-ihx_2}+\frac{\varepsilon_\nu(\boldsymbol{k})\varepsilon_\mu(\boldsymbol{h})}{\sqrt{4\omega(\boldsymbol{k})\omega(\boldsymbol{h})}}e^{ikx_2-ihx_1}\right]. \tag{4.2.10}$$

The vacuum polarization tensor is a tempered distribution, so that there remains the function

$$E_\mu(x) = (2\pi)^{-3/2}\int\frac{d^3k}{\sqrt{2\omega}}\, f(\boldsymbol{k})\varepsilon_\mu(\boldsymbol{k})e^{-ikx}. \tag{4.2.11}$$

to be considered. It can be written as a four-dimensional distributional Fourier transform

$$E_\mu(x) = (2\pi)^{-3/2}\int d^4k\,\delta(k^2)\Theta(k^0)f(\boldsymbol{k})\sqrt{2\omega}\varepsilon_\mu(\boldsymbol{k})e^{-ikx}. \tag{4.2.12}$$

If $f(\boldsymbol{k})$ has compact support in (3-dimensional) momentum space, $E_\mu(x)$ is the Fourier transform of a distribution with compact support in $\mathbb{R}^4$. It is therefore an entire analytic function and thus C^∞ in x-space. On the other hand, assuming $f(\boldsymbol{k})\in C_0^\infty(\mathbb{R}^3)$, then $E_\mu(x)$ is a smooth solution of the wave equation equation, a so-called regular solution (*M. Reed, B. Simon, Methods of Modern Mathematical Physics, Vol. III, Theorem XI.19*). Such a solution is polynomially bounded, in fact

$$|E_\mu(t,\boldsymbol{x})| \le \text{const.}\cdot(1+|t|)^{-1} \tag{4.2.13}$$

for all $\boldsymbol{x}$ and t. Consequently, by multiplying the tempered distribution $P_1^{\nu\mu}(x_2-x_1)$ by $E_\mu(x_1)E_\nu(x_2)$ one gets again a tempered distribution.

The next case is an electron-photon state

$$\Phi = \int d^3k\, f(\boldsymbol{k})\varepsilon_\mu(\boldsymbol{k})a_\mu(\boldsymbol{k})^+\cdot\int d^3p\,\psi_s(\boldsymbol{p})b_s^+(\boldsymbol{p})\Omega). \tag{4.2.14}$$

A non-vanishing contribution to (4.2.3) now comes from (2) in (4.2.4). Choosing the wave functions in $C_0^\infty(\mathbb{R}^3)$ again, this contribution is a product of $(S^{(-)}-S^{(+)})(x_1-x_2)$ times smooth solutions of the wave and Dirac equations (Problem 4.1). This gives again a tempered distribution. In the last case

$$\Phi = \int d^3k\, f(\boldsymbol{k})\varepsilon_\mu(\boldsymbol{k})a_\mu(\boldsymbol{k})^+\cdot\int d^3p\,\psi_s(\boldsymbol{p})b_s^+(\boldsymbol{p})$$

$$\times\int d^3q\,\varphi_\sigma(\boldsymbol{q})b_\sigma(\boldsymbol{q})^+\Omega \tag{4.2.15}$$

the result is a sum of products of only smooth solutions of the wave and Dirac equations, which is a tempered distribution, of course.

Summing up, we have shown that the first order S-matrix

$$S_1(g) = \int d^4x\, T_1(x)g(x) \tag{4.2.16}$$

is an operator in Fock space with a dense domain. It is clear from the general nature of the arguments, that the same is true in higher orders. This result is very desirable because it explains why the often used products $T_n(X)T_m(Y)$ are well defined: Considered as distributions, they are direct products, but as operators in Fock space, after smearing out, they are ordinary products.

4.3 Normalizability of QED

The subject of this section is called renormalizability in other textbooks. The reader will agree that the prefix "re" is of no use here. By renormalization we always mean finite renormalization of an already normalized T-distribution, as discussed in Sect. 3.13, for example. We have found that for every graph with singular order $\omega \geq 0$, a polynomial of degree ω remains undetermined in momentum space. The coefficients of this polynomial must be fixed by further physical normalization conditions. In every order n of perturbation theory, there is then a finite number of free parameters. Considering the inductive procedure in n, there are two possibilities: (i) The number of free parameters increases with n without bound; then the theory is called non-normalizable. (ii) The total number of free parameters appearing in all orders is finite; then the theory is normalizable. (iii) There is only a finite number of low-order graphs with $\omega \geq 0$, then the theory is called super-normalizable. In case (i) the theory has a weaker predictive power, but it is still well defined. We are now going to show that QED belongs to the normalizable case (ii).

In order to achieve the desired result, we must determine the singular order ω of arbitrary graphs. We consider a graph g of the following form

$$T_n^g(x_1,\ldots,x_n) =: \prod_{j=1}^{n_g} \overline{\psi}(x_{k_j}) t_g(x_1,\ldots,x_n) \prod_{j=1}^{n_g} \psi(x_{n_j}) :: \prod_{j=1}^{m_g} A(x_{m_j}) : . \tag{4.3.1}$$

Here m_g is the number of external photons and n_g the total number of external electrons and positrons divided by two. The main result of this section is the following simple expression for the singular order of g

$$\omega(g) = 4 - 3n_g - m_g, \tag{4.3.2}$$

independent of the order n of perturbation theory.

Before proving (4.3.2), we draw the main conclusions from it. It is evident that there are only finitely many possibilities with $\omega \geq 0$, namely:

$$\omega=4,\quad n_g=m_g=0 \tag{4.3.3}$$
$$\omega=3,\quad n_g=0\ ,\quad m_g=1 \tag{4.3.4}$$
$$\omega=2,\quad n_g=0\ ,\quad m_g=2 \tag{4.3.5}$$
$$\omega=1,\quad n_g=1\ ,\quad m_g=0 \tag{4.3.6}$$
$$\omega=1,\quad n_g=0\ ,\quad m_g=3 \tag{4.3.7}$$
$$\omega=0,\quad n_g=1\ ,\quad m_g=1 \tag{4.3.8}$$
$$\omega=0,\quad n_g=0\ ,\quad m_g=4. \tag{4.3.9}$$

Case (4.3.3) corresponds to the vacuum graphs, the graphs (4.3.4) and (4.3.7) vanish by Furry's theorem (see next section), (4.3.5) is vacuum polarization, (4.3.6) is the self energy, (4.3.8) corresponds to the vertex function and (4.3.9) is photon-photon scattering. Most of these processes have been discussed before and the free parameters have been fixed by physical conditions in the lowest orders, as far as they depend on observable quantities. The same conditions fix the parameters in higher orders. That is to say, in every order n the retarded distributions corresponding to the cases (4.3.3)-(4.3.9) are normalized according to their normalization conditions, which are the same for all n. The resulting n-point distribution T_n is then used without change in higher orders. Then the S-matrix is completely determined, up to unobservable constants. QED is normalizable.

We now turn to the proof of (4.3.2) by induction. We consider a tensor product of two distributions

$$T_r^1(x_1,\ldots,x_r)T_s^2(y_1,\ldots,y_s) \tag{4.3.10}$$

with known singular orders ω_1, ω_2. According to the inductive construction (Sect. 3.2), this has to be normally ordered giving rise to all possible contractions. The contractions are of two types.

1) First we consider l contractions between photon operators. Taking translation invariance into account, the numerical distribution of the contracted expression is of the form

$$t_1(x_1-x_r,\ldots,x_{r-1}-x_r)\prod_{j=1}^{l} D_0^{(+)}(x_{r_j}-y_{s_j})t_2(y_1-y_s,\ldots,y_{s-1}-y_s)$$

$$\stackrel{\text{def}}{=} t(\xi_1,\ldots,\xi_{r-1},\eta_1,\ldots,\eta_{s-1},\eta). \tag{4.3.11}$$

Here, $\{x_{r_j}\}$ is a subset of $\{x_1,\ldots,x_r\}$ and $\{y_{s_j}\}$ is a subset of $\{y_1,\ldots,y_s\}$, and we have introduced relative coordinates

$$\xi_j=x_j-x_r,\quad \eta_j=y_j-y_s,\quad \eta=x_r-y_s. \tag{4.3.12}$$

We compute the Fourier transform (omitting powers of 2π)

$$\hat{t}(p_1,\ldots,p_{r-1},q_1,\ldots,q_{s-1},q)=\int t(\boldsymbol{\xi},\boldsymbol{\eta})e^{ip\xi+iq\eta}d^{4r-4}\xi d^{4s}\eta. \tag{4.3.13}$$

Since products in (4.3.11) go over into convolutions, we get

$$\hat{t}(\cdots) = \int \prod_j d\kappa_j\, \delta(q - \sum_{j=1}^{l} \kappa_j)$$

$$\times\, \hat{t}_1(\ldots, p_i - \kappa_{r(i)}, \ldots) \prod_j \hat{D}_0^{(+)}(\kappa_j)\, \hat{t}_2(\ldots, q_i + \kappa_{s(i)} \ldots). \qquad (4.3.14)$$

Here, $r(i)$ and $s(i)$ are indices of two coordinates x and y in (4.3.11) which are joined by a contraction. Applying this to a test function $\varphi \in S(\mathbb{R}^{\mathbf{m}}), m = 4(r+s-1)$, we obviously have

$$\langle \hat{t}, \varphi \rangle = \int d^{4r-4}p' d^{4s-4}q'\, \hat{t}_1(\boldsymbol{p}')\hat{t}_2(\boldsymbol{q}')\psi(\boldsymbol{p}', \boldsymbol{q}'), \qquad (4.3.15)$$

with

$$\psi = \int d^4q \prod_j d\kappa_j\, \delta\Big(q - \sum_j \kappa_j\Big)\varphi(\ldots, p_i' + \kappa_{r(i)} \ldots, \ldots, q_i' - \kappa_{s(i)} \ldots, q)$$

$$\times \prod_{j=1}^{l} \hat{D}_0^{(+)}(\kappa_j). \qquad (4.3.16)$$

In order to determine the singular order of $\hat{t}$ in p-space (3.2.12, 16), we have to consider the scaled distribution

$$\Big\langle \hat{t}\Big(\frac{p}{\delta}\Big), \varphi \Big\rangle = \delta^m \langle \hat{t}(p), \varphi(\delta p) \rangle$$

$$= \delta^m \int d^{4r-4}p'\, d^{4s-4}q'\, \hat{t}_1(p')\hat{t}_2(q')\psi_\delta(p', q'), \qquad (4.3.17)$$

where

$$\psi_\delta(p', q') = \int d^4q \prod_{j=1}^{l} d\kappa_j\, \delta\Big(q - \sum_j \kappa_j\Big)$$

$$\varphi(\ldots, \delta(p_i' + \kappa_{r(i)}), \ldots; \ldots, \delta(q_i' - \kappa_{s(i)}), \ldots, \delta q) \prod_j \hat{D}_0^{(+)}(\kappa_j). \qquad (4.3.18)$$

We introduce scaled variables $\tilde{\kappa}_j = \delta \kappa_j$, $\tilde{q} = \delta q$ and note that

$$\hat{D}_0^{(+)}\Big(\frac{\tilde{\kappa}}{\delta}\Big) = \delta^1\Big(\frac{\tilde{\kappa}^2}{\delta^2}\Big)\Theta\Big(\frac{\tilde{\kappa}^0}{\delta}\Big) = \delta^2 \hat{D}_0^{(+)}(\tilde{\kappa}). \qquad (4.3.19)$$

To avoid confusion, we recall that $\delta^1(\cdot)$ is a 1-dimensional δ-distribution, whereas δ^2 is the square of the scaling parameter. This implies

$$\psi_\delta(p', q') = \frac{\delta^{2l}}{\delta^{4l}} \int d^4\tilde{q} \prod_{j=1}^{l} d\tilde{\kappa}_j\, \delta\Big(\tilde{q} - \sum_j \tilde{\kappa}_j\Big)$$

$$\times\, \varphi(\ldots \delta p_i' + \tilde{\kappa}_{r(i)}, \ldots; \ldots \delta q_i' - \tilde{\kappa}_{s(i)} \ldots, \tilde{q}) \prod_j \hat{D}_0^{(+)}(\tilde{\kappa}_j)$$

$$= \frac{1}{\delta^{2l}}\psi(\delta p', \delta q'). \tag{4.3.20}$$

Using scaled variables $\delta p' = \tilde{p}$, $\delta q' = \tilde{q}$, again, we find

$$\left\langle \hat{t}\Big(\frac{p}{\delta}\Big), \varphi \right\rangle = \frac{\delta^4}{\delta^{2l}} \int d^{4r-4}\tilde{p}\, d^{4s-4}\tilde{q}\, \hat{t}_1\Big(\frac{\tilde{p}}{\delta}\Big)\hat{t}_2\Big(\frac{\tilde{q}}{\delta}\Big)\psi(\tilde{p},\tilde{q}). \tag{4.3.21}$$

By the induction hypothesis, $\hat{t}_1$ and $\hat{t}_2$ have singular orders ω_1, ω_2 with power counting functions $\rho_1(\delta)$, $\rho_2(\delta)$, respectively. Then the following limit exists:

$$\lim_{\delta\to 0} \delta^{2l-4}\rho_1(\delta)\rho_2(\delta)\left\langle \hat{t}\Big(\frac{p}{\delta}\Big), \varphi \right\rangle = \langle \hat{t}_0(p), \varphi\rangle, \tag{4.3.22}$$

hence, the singular order of $\hat{t}(p)$ is equal to

$$\omega = \omega_1 + \omega_2 + 2l - 4. \tag{4.3.23}$$

It remains to check that (4.3.23) is satisfied by (4.3.2). Substituting

$$\omega_j = 4 - 3n_j - m_j, \quad j = 1, 2 \tag{4.3.24}$$

into (4.3.23), we find

$$\omega = 4 - 3(n_1 + n_2) - (m_1 + m_2 - 2l).$$

Since the last bracket is just the number of photon operators after the l contractions, (4.3.2) is proven in this case.

2) We now consider l contractions between fermionic operators $\overline{\psi}(x)$ and $\psi(y)$. The proof in this case goes through as 1) with only the following modifications: In (4.3.14) one has to substitute

$$\hat{D}_0^{(+)}(k) \longrightarrow \hat{S}^{(+)}(k) = (\not k + m)\delta(k^2 - m^2)\Theta(k^0). \tag{4.3.25}$$

Then instead of (4.3.19), we have the following scaling limit

$$\lim_{\delta\to 0} \frac{1}{\delta}\hat{S}^{(+)}\Big(\frac{\tilde{\kappa}}{\delta}\Big) = \hat{S}_0^{(+)}(\tilde{\kappa}). \tag{4.3.26}$$

This leads to

$$\lim_{\delta\to 0} \delta^{3l-4}\rho_1(\delta)\rho_2(\delta)\left\langle \hat{t}\Big(\frac{p}{\delta}\Big), \varphi \right\rangle = \langle \hat{t}_0(p), \varphi\rangle. \tag{4.3.27}$$

Hence, the singular order of $\hat{t}$ is

$$\omega = \omega_1 + \omega_2 + 3l - 4. \tag{4.3.28}$$

Inserting for ω_1 and ω_2 the induction assumption (4.3.24), we get

$$\omega = 4 - 3(n_1 + n_2 - l) - (m_1 + m_2).$$

This is again (4.3.2) after l contractions of Fermi operators which completes the proof in this case. The proof can be easily carried through for a mixture of photonic and fermionic contractions. The results remain also valid if two contracted field operators belong to the same vertex (Problem 4.2).

Finally we have to convince ourselves that the singular order is not changed by distribution splitting. But this was already established in the general theory in Sect. 3.2 (3.2.45). This completes the proof of normalizability (4.3.2). We see that the simple method of power counting in momentum space has a rigorous meaning in the causal theory.

4.4 Discrete Symmetries

To discuss the discrete symmetries of the S-matrix, we have first to define the corresponding transformation operators in Fock space. For charge conjugation this has already been done in Sect. 2.2. The corresponding unitary operator $\mathbf{U}_C$ is defined by (2.2.44, 45)

$$\mathbf{U}_C\psi(f)\mathbf{U}_C^{-1} = \psi(f_C)^+, \tag{4.4.1}$$

$$\mathbf{U}_C\psi(f)^+\mathbf{U}_C^{-1} = \psi(f_C), \tag{4.4.2}$$

where

$$f_C = C\gamma^0 f^* \tag{4.4.3}$$

is the charge-conjugate Dirac spinor. This implies the following transformation laws for the field operators (2.2.53, 54)

$$\mathbf{U}_C\psi(\boldsymbol{x})\mathbf{U}_C^{-1} = C\overline{\psi}(\boldsymbol{x})^T \tag{4.4.4}$$

$$\mathbf{U}_C\overline{\psi}(\boldsymbol{x})\mathbf{U}_C^{-1} = \psi(\boldsymbol{x})^T C. \tag{4.4.5}$$

Note that

$$\mathbf{U}_C^2 = 1. \tag{4.4.6}$$

The adjoint in (4.4.1) is necessary in order to have invariance of the anticommutation relations:

$$\mathbf{U}_C\{\psi(f), \psi(g)^+\}\mathbf{U}_C^{-1} = (f, g)$$
$$= \{\psi(f_C)^+, \psi(g_C)\} = (g_C, f_C). \tag{4.4.7}$$

The discussion of space-reflection runs along similar lines. We define a unitary operator by

$$\mathbf{U}_P\psi(f)\mathbf{U}_P^{-1} = \eta\psi(f_P), \tag{4.4.8}$$

where

$$f_P(\boldsymbol{x}) = \gamma^0 f(-\boldsymbol{x}) \tag{4.4.9}$$

is the space-reflected Dirac spinor (1.5.69) and η is a phase factor, to be determined below. The corresponding transformation of the field operator follows from

$$\int d^3x\, f_a^*(\boldsymbol{x})(\mathbf{U}_P\psi(\boldsymbol{x})\mathbf{U}_P^{-1})_a = \eta\int d^3x\,(\gamma^0 f(-\boldsymbol{x}))_a^*\psi_a(\boldsymbol{x})$$
$$= \eta\int d^3x\, f_a^*(\boldsymbol{x})(\gamma^0\psi(-\boldsymbol{x}))_a.$$

This implies

$$\mathbf{U}_P\psi(\boldsymbol{x})\mathbf{U}_P^{-1} = \eta\gamma^0\psi(-\boldsymbol{x}), \tag{4.4.10}$$

in agreement with (1.5.69), up to the factor η. The adjoint operators transform as follows

$$\mathbf{U}_P\overline{\psi}(f)\mathbf{U}_P^{-1} = \eta^*\overline{\psi}(f), \tag{4.4.11}$$
$$\mathbf{U}_P\overline{\psi}(\boldsymbol{x})\mathbf{U}_P^{-1} = \eta^*\overline{\psi}(-\boldsymbol{x})\gamma^0. \tag{4.4.12}$$

The phase η is now determined by the requirement that $\mathbf{U}_P$ commutes with charge conjugation $\mathbf{U}_C$. This was not true in the one-particle theory (1.5.74). Comparing

$$\mathbf{U}_C\mathbf{U}_P\psi(f)\mathbf{U}_P^{-1}\mathbf{U}_C^{-1} = \mathbf{U}_C\eta\psi(f_P)\mathbf{U}_C^{-1}$$
$$= \eta\psi(U_C f_P)^+,$$

with

$$\mathbf{U}_P\mathbf{U}_C\psi(f)\mathbf{U}_C^{-1}\mathbf{U}_P^{-1} = \mathbf{U}_P\psi(f_C)^+\mathbf{U}_P^{-1}$$
$$= \eta^*\psi(U_P f_C)^+ = -\eta^*\psi(U_C U_P f))^+,$$

by (1.5.74), it follows $\eta = \pm i$. We choose

$$\eta = i. \tag{4.4.13}$$

Since the phase factor is purely imaginary, $\mathbf{U}_P$ is not an involution:

$$\mathbf{U}_P^2\psi(f)\mathbf{U}_P^{-2} = -\psi(f), \tag{4.4.14}$$

so that $\mathbf{U}_P^2$ anticommutes with all spinor fields

$$\{\mathbf{U}_P^2, \psi(f)\} = 0, \quad \{\mathbf{U}_P^2, \psi(f)^+\} = 0, \tag{4.4.15}$$

for all test functions f. As in Theorem 1.2 (2.1.44), this implies that

$$\mathbf{U}_P^2 = \begin{cases} 1, & \text{on sectors with an even number of fermions} \\ -1 & \text{on sectors with an odd number of fermions} \end{cases} \tag{4.4.16}$$
$$\stackrel{\text{def}}{=} U_V.$$

This unitary operator is called valency operator. It is easy to check that the anticommutation relations are preserved under space-reflection.

For time-reversal it is convenient to define the symmetry transformation directly on the field operators

$$\mathbf{V}_T\psi(x)\mathbf{V}_T^{-1} = \eta\gamma^5 C\psi(x_T), \tag{4.4.17}$$

$$\mathbf{V}_T\overline{\psi}(x)\mathbf{V}_T^{-1} = \eta'\overline{\psi}(x_T)C^{-1}\gamma^5, \tag{4.4.18}$$

where $x_T = (-x^0, \boldsymbol{x})$. Up to the phase factors η, η', which will be soon fixed, the right-hand sides agree with the one-particle theory (1.5.66). However, in order to preserve the anticommutation relations, it is now necessary that $\mathbf{V}_T$ is an antiunitary operator:

$$\mathbf{V}_T\{\psi(x), \overline{\psi}(y)\}\mathbf{V}_T^{-1} = \eta\eta'\Big[\gamma^5 C\psi(x_T)\overline{\psi}(y_T)C^{-1}\gamma^5$$
$$+\left(\overline{\psi}(y_T)C^{-1}\gamma^5\right)\left(\gamma^5 C\psi(x_T)\right)\Big]$$
$$= \eta\eta'\gamma^5 C\frac{1}{i}S(x_T - y_T)C^{-1}\gamma^5 = -\eta\eta'\frac{1}{i}S(x-y)^*, \tag{4.4.19}$$

where (2.3.53) has been used. If $\eta' = \eta^{-1}$, this is just the complex conjugate of the original anticommutator. Hence, $\mathbf{V}_T, \mathbf{V}_T^{-1}$ on the left-hand side must be antiunitary.

To further restrict the phase factor η, we consider the product

$$\mathbf{U}_C\mathbf{V}_T\psi(x)\mathbf{V}_T^{-1}\mathbf{U}_C^{-1} = \mathbf{U}_C\eta\gamma^5\psi(-x^0, \boldsymbol{x})\mathbf{U}_C$$
$$= \eta\gamma^5 CC\overline{\psi}(-x^0, \boldsymbol{x})^T = -\eta\gamma^5\overline{\psi}(-x^0, \boldsymbol{x})^T, \tag{4.4.20}$$

and compare it with

$$\mathbf{V}_T\mathbf{U}_C\psi(x)\mathbf{U}_C^{-1}\mathbf{V}_T^{-1} = \mathbf{V}_T C\overline{\psi}(x)^T\mathbf{V}_T^{-1} = C^*\eta^{-1}\gamma^{5T}C^{-1T}\overline{\psi}(-x^0, \boldsymbol{x})^T$$
$$= -\eta^{-1}C\gamma^{5T}C^{-1}\overline{\psi}(-x^0, \boldsymbol{x})^T = -\eta^{-1}\gamma^5\overline{\psi}(-x^0, \boldsymbol{x})^T. \tag{4.4.21}$$

The two transformations commute if $\eta^2 = 1$. We choose

$$\eta = 1. \tag{4.4.22}$$

There remains to investigate the products

$$\mathbf{V}_T\mathbf{U}_P\psi(x)\mathbf{U}_P^{-1}\mathbf{V}_T^{-1} = \mathbf{V}_T i\gamma^0\psi(x^0, -\boldsymbol{x})\mathbf{V}_T^{-1}$$
$$= -i\gamma^{0*}\gamma^5 C\psi(-x) = -i\gamma^5 C\gamma^0\psi(-x),$$

and

$$\mathbf{U}_P\mathbf{V}_T\psi(x)\mathbf{V}_T^{-1}\mathbf{U}_P^{-1} = \mathbf{U}_P\gamma^5 C\psi(-x^0, \boldsymbol{x})\mathbf{U}_P$$
$$= \gamma^5 Ci\gamma^0\psi(-x).$$

Since the two results differ in sign, it follows as above (4.4.14, 15) that

$$\mathbf{V}_T\mathbf{U}_P = U_V\mathbf{U}_P\mathbf{V}_T. \tag{4.4.23}$$

Similarly one finds

$$\mathbf{V}_T^2 = U_V. \tag{4.4.24}$$

The three symmetry transformations $\mathbf{U}_C, \mathbf{U}_P$ and $\mathbf{V}_T$ generate a multiplicative group G_{16} of 16 elements. In fact, since $\mathbf{U}_C$ commutes with all other transformations, the cyclic group $C_2 = \{1, \mathbf{U}_C\}$ is an invariant subgroup. Consequently, G_{16} is the direct product of C_2 times the group G_8 generated by $\mathbf{U}_P$ and $\mathbf{V}_T$

$$G_8 = \{[1], [\mathbf{U}_P, \mathbf{U}_P^{-1}], [\mathbf{V}_T, \mathbf{V}_T^{-1}], [U_V], [\mathbf{U}_P\mathbf{V}_T, \mathbf{V}_T\mathbf{U}_P]\}. \tag{4.4.25}$$

The square brackets denote the classes of conjugated elements. G_8 is isomorphic to the dihedral group D_4 (Problem 4.3).

Next we have to study the transformation properties of the bilinear expression

$$j^\mu(x) = :\overline{\psi}(x)\gamma^\mu\psi(x):, \tag{4.4.26}$$

appearing in the first order S-matrix, for example

$$\mathbf{V}_T j^\mu(x)\mathbf{V}_T^{-1} = :\overline{\psi}(x_T)C^{-1}\gamma^5\mathbf{V}_T\gamma^\mu\mathbf{V}_T^{-1}\gamma^5 C\psi(x_T):$$

$$= -g^{\mu\mu} :\psi(x_T)C^{-1}\gamma^{\mu T}C\psi(x_T): = g^{\mu\mu} :\overline{\psi}(x_T)\gamma^\mu\psi(x_T):$$

$$= g^{\mu\mu} j^\mu(x_T), \tag{4.4.27}$$

without summation over μ. Similarly one finds

$$\mathbf{U}_P j^\mu(x)\mathbf{U}_P^{-1} = g^{\mu\mu} j^\mu(x_P) \tag{4.4.28}$$

$$\mathbf{U}_C j^\mu(x)\mathbf{U}_C^{-1} = -j^\mu(x). \tag{4.4.29}$$

The same transformation properties hold for the classical electromagnetic potentials (1.5.63). Consequently, the symmetry transformations must be extended to the photon sector as follows

$$\mathbf{U}_P A^\mu(x)\mathbf{U}_P^{-1} = g^{\mu\mu} A^\mu(x_P) \tag{4.4.30}$$

$$\mathbf{U}_C A^\mu(x)\mathbf{U}_C^{-1} = -A^\mu(x) \tag{4.4.31}$$

$$\mathbf{V}_T A^\mu(x)\mathbf{V}_T^{-1} = g^{\mu\mu} A^\mu(x_T). \tag{4.4.32}$$

Then the first order

$$T_1(x) = iej_\mu(x)A^\mu(x) \tag{4.4.33}$$

is P- and C-invariant, because all signs cancel out. For T-invariance we have

$$\mathbf{V}_T T_1(x)\mathbf{V}_T^{-1} = -T_1(x_T) = \tilde{T}_1(x_T), \tag{4.4.34}$$

because $\mathbf{V}_T$ is antiunitary.

We are now ready to prove the same symmetries for the n-point distributions by induction. Let us assume that all m-point distributions with $m \leq n-1$ are C-invariant. Then all products of them are also C-invariant and so are

$$\mathbf{U}_C A'_n(x_1, \ldots, x_n)\mathbf{U}_C^{-1} = \mathbf{U}_C \sum \tilde{T}_{n_1}(X)T_{n-n_1}(Y, x_n)\mathbf{U}_C^{-1} = A'_n(x_1, \ldots, x_n)$$

$$\mathbf{U}_C R'_n(x_1,\dots,x_n)\mathbf{U}_C^{-1} = \mathbf{U}_C \sum T_{n-n_1}(Y,x_n)\tilde{T}_{n_1}(X)\mathbf{U}_C^{-1} = R'_n(x_1,\dots,x_n)$$

$$\mathbf{U}_C D_n(x_1,\dots,x_n)\mathbf{U}_C^{-1} = \mathbf{U}_C(R'_n(x_1,\dots,x_n) - A'_n(x_1,\dots,x_n))\mathbf{U}_C^{-1}$$
$$= D_n(x_1,\dots,x_n). \tag{4.4.35}$$

The next step in the inductive construction is the distribution splitting of D_n. If R_n is an arbitrary splitting solution, then $\mathbf{U}_C R_n \mathbf{U}_C^{-1}$ is also one, and so is the symmetric combination

$$\bar{R}_n = \tfrac{1}{2}[R_n + \mathbf{U}_C R_n \mathbf{U}_C^{-1}]. \tag{4.4.36}$$

This retarded solution is C-invariant

$$\mathbf{U}_C \bar{R}_n \mathbf{U}_C^{-1} = \bar{R}_n,$$

due to (4.4.6). Then the difference $T_n = \bar{R}_n - R'_n$ is also C-invariant

$$\mathbf{U}_C T_n \mathbf{U}_C^{-1} = T_n. \tag{4.4.37}$$

This finishes the inductive proof.

We always write T_n in normally ordered form. Then, there can be no compensation between terms with different operator factors. Consequently, if all terms with the same operator factor are taken together, the sum must be separately C-invariant. As an application we consider a graph with an odd number of photon operators and all Fermi operators contracted, for example the three-point function

$$T_3(x_1,x_2,x_3) = t(x_1,x_2,x_3) \; : A(x_1)A(x_2)A(x_3) : . \tag{4.4.38}$$

Since

$$: A_1A_2 := A_1A_2 - \overline{A_1A_2},$$

it follows

$$U_C : A_1A_2 : U_C^{-1} = A_1A_2 - \overline{A_1A_2} =: A_1A_2 : . \tag{4.4.39}$$

Similarly, from

$$: A_1A_2A_3 := A_1A_2A_3 - : \overline{A_1A_2}A_3 : - : \overline{A_1A_2A_3} : - : A_1\overline{A_2A_3} :$$

we conclude that

$$U_C : A_1A_2A_3 : U_C^{-1} = - : A_1A_2A_3 : . \tag{4.4.40}$$

Obviously, all normally ordered products of an even number of photon operators are invariant under charge conjugation, an odd number of photon operators changes sign. Since T_3 (4.4.38) is C-invariant, it must be zero. The same is obviously true for higher n-point functions containing an odd number of photon operators and no fermion operator. This is Furry's theorem for full QED.

The inductive proof of perturbative P-invariance

$$\mathbf{U}_P T_n(x_1,\ldots,x_n)\mathbf{U}_P^{-1} = T_n(x_{1P},\ldots,x_{nP}) \tag{4.4.41}$$

runs along similar lines. Assuming (4.4.41) to hold for all $m < n$, it follows that R'_n, A'_n and D_n are P-invariant. If $D_n = R_n - A_n$ is an arbitrary causal splitting, one can construct a P-invariant retarded solution as follows:

$$\bar{R}_n(X) = \frac{1}{4}\Big[R_n(X) + \mathbf{U}_P R_n(X_P)\mathbf{U}_P^{-1} + \mathbf{U}_P^2 R_n(X) U_P^{-2} + \mathbf{U}_P^3 R_n(X_P)\mathbf{U}_P^{-3}\Big], \tag{4.4.42}$$

where $X_P = (x_{1P},\ldots,x_{nP})$. In fact,

$$\mathbf{U}_P \bar{R}_n(x_1,\ldots,x_n)\mathbf{U}_P^{-1} = R_n(x_{1P},\ldots,x_{nP}), \tag{4.4.43}$$

because the terms on the right-hand side of (4.4.42) are transformed in cyclic order by the parity transformation. Using this retarded solution in $T_n = \bar{R}_n - R'_n$, we arrive at a P-invariant T_n (4.4.41). In this proof no special property of the theory is used.

The discussion of T-invariance is slightly different. It is our aim to prove

$$\mathbf{V}_T T_n(x_1,\ldots,x_n)\mathbf{V}_T^{-1} = \tilde{T}_n(x_{1T},\ldots x_{nT}). \tag{4.4.44}$$

The inductive construction of

$$T_n = R_n - R'_n = A_n - A'_n, \tag{4.4.45}$$

where

$$A'_n = \sum_{P_2} \tilde{T}_{n_1}(X) T_{n-n_1}(Y, x_n) \tag{4.4.46}$$

$$R'_n = \sum_{P_2} T_{n-n_1}(Y, x_n)\tilde{T}_{n_1}(X), \tag{4.4.47}$$

can be carried through for the n-point distribution $\tilde{T}_n$ of the inverse S-matrix in the same way; one must only exchange retarded and advanced distributions:

$$\tilde{T}_n = \tilde{R}_n - \tilde{R}'_n = \tilde{A}_n - \tilde{A}'_n, \tag{4.4.48}$$

where

$$\tilde{A}'_n = \sum_{P_2} \tilde{T}_{n-n_1}(Y, x_n) T_{n_1}(X) \tag{4.4.49}$$

$$\tilde{R}'_n = \sum_{P_2} T_{n_1}(X)\tilde{T}_{n-n_1}(Y, x_n), \tag{4.4.50}$$

and $\tilde{R}_n, \tilde{A}_n$ are given by the splitting of $\tilde{D}_n = \tilde{R}'_n - \tilde{A}'_n$. Only the last quantities (4.4.49, 50) are used in the following. Assuming now (4.4.44) to hold for all $n < m$ and the analogous relation for $\tilde{T}_m$, it follows

$$\mathbf{V}_T R'_n(x_1,\dots,x_n)\mathbf{V}_T^{-1} = \sum_{P_2} \tilde{T}_{n-n_1}(Y_T, x_{nT})T_{n_1}(X_T)$$

$$= \tilde{A}'_n(x_{1T},\dots,x_{nT}), \tag{4.4.51}$$

and

$$\mathbf{V}_T A'_n(x_1,\dots,x_n)\mathbf{V}_T^{-1} = \tilde{R}'_n(x_{1T},\dots,x_{nT}). \tag{4.4.52}$$

On the other hand, the retarded distribution (3.1.36)

$$R_n = \sum_{P_2^0} T_{n-n_1}(Y, x_n)\tilde{T}_{n_1}(X) = R'_n + T_n$$

is equal to

$$= -\sum_{P_2^0} T_{n_1}(X)\tilde{T}_{n-n_1}(Y, x_n) = -\tilde{R}'_n - \tilde{T}_n, \tag{4.4.53}$$

in virtue of (3.1.13). Similarly, we have

$$A_n = A'_n + T_n = -\tilde{A}'_n - \tilde{T}_n, \tag{4.4.54}$$

so that

$$D_n = R'_n - A'_n = R_n - A_n = -\tilde{R}'_n + \tilde{A}'_n. \tag{4.4.55}$$

This implies T-invariance of the D-distribution

$$\mathbf{V}_T D_n(x_1,\dots,x_n)\mathbf{V}_T^{-1} = D_n(x_{1T},\dots,x_{nT}). \tag{4.4.56}$$

We have achieved our goal (4.4.44), if we can find a splitting $D_n = \bar{R}_n - \bar{A}_n$ satisfying

$$\mathbf{V}_T \bar{R}_n(x_1,\dots,x_n)\mathbf{V}_T^{-1} = -\bar{A}_n(x_{1T},\dots,x_{nT}) \tag{4.4.57}$$

$$\mathbf{V}_T \bar{A}_n(x_1,\dots,x_n)\mathbf{V}_T^{-1} = -\bar{R}_n(x_{1T},\dots,x_{nT}). \tag{4.4.58}$$

In fact, for $T_n = \bar{R}_n - R'_n$ we then obtain

$$\mathbf{V}_T T_n(x_1,\dots,x_n)\mathbf{V}_T^{-1} = -\bar{A}_n(x_{1T},\dots,x_{nT}) - \tilde{A}'_n(x_{1T},\dots,x_{nT})$$

$$= \tilde{T}_n(x_{1T},\dots,x_{nT}), \tag{4.4.59}$$

by (4.4.51) and (4.4.54). If $D_n = R_n - A_n$ is an arbitrary splitting, the desired solution is given by

$$\bar{R}_n(x_1,\dots,x_n) = \frac{1}{4}\Big[R_n(x_1,\dots,x_n) - \mathbf{V}_T A_n(x_{1T},\dots,x_{nT})\mathbf{V}_T^{-1}$$

$$+\mathbf{V}_T^2 R_n(x_1,\dots,x_n)\mathbf{V}_T^{-2} - \mathbf{V}_T^3 A_n(x_{1T},\dots,x_{nT})\mathbf{V}_T^{-3}\Big] \tag{4.4.60}$$

$$\bar{A}_n(x_1,\dots,x_n) = \frac{1}{4}\Big[A_n(x_1,\dots,x_n) - \mathbf{V}_T R_n(x_{1T},\dots,x_{nT})\mathbf{V}_T^{-1}$$

$$+\mathbf{V}_T^2 A_n(x_1,\dots,x_n)\mathbf{V}_T^{-2} - \mathbf{V}_T^3 R_n(x_{1T},\dots,x_{nT})\mathbf{V}_T^{-3}\Big]. \tag{4.4.61}$$

This completes the discussion of T-invariance, because (4.4.44) is the perturbative form of

$$\mathbf{V}_T S(g) \mathbf{V}_T^{-1} = S^{-1}(g_T), \tag{4.4.62}$$

where $g_T(x) = g(-x^0, \boldsymbol{x})$.

Finally we have to show how the inductive step must be modified, in order to get an n-point distribution which is simultaneously C-, P- and T-invariant. In this case one has to sum over all 16 elements of the symmetry group G_{16} (4.4.25) to obtain the symmetrical retarded and advanced distributions

$$\bar{R}_n(X) = \frac{1}{16} \sum_{i=1}^{16} U_i R_i(X_i) U_i^{-1} \tag{4.4.63}$$

$$\bar{A}_n(X) = \frac{1}{16} \sum_{i=1}^{16} U_i A_i(X_i) U_i^{-1}. \tag{4.4.64}$$

Here U_i runs through the 16 group elements in G_{16}, R_i is equal to $-A_n$ if U_i contains $\mathbf{V}_T$ or $\mathbf{V}_T^3 = \mathbf{V}_T^{-1}$ and $R_i = R_n$ otherwise. $X = (x_1, \ldots, x_n)$ and X_i is the argument after the appropriate parity and time-reversal transformations, according to U_i. A_i is equal to $-R_n$ if U_i contains $\mathbf{V}_T$ or $\mathbf{V}_T^{-1}$ and $A_i = A_n$ otherwise. It is easy to verify that $\bar{R}_n$ then satisfies the four transformation laws (4.4.37, 43, 57, 58) simultaneously. This retarded distribution provides an n-point distribution T_n which is symmetric under all three discrete symmetries. Its construction is completely general, it can be carried out in every field theory whatsoever. In particular, T_1 may contain terms with an odd number of Fermi fields. Such terms occur if one considers interacting fields (Sect. 4.9).

4.5 Poincaré Covariance

The next symmetry property of the S-matrix to be investigated is Poincaré covariance. Using the known transformation laws (2.2.65) and (2.11.39) of the free fields (cf. also (1.3.22)), we find that the one-point function is Poincaré invariant

$$U(a, \Lambda) T_1(x) U(a, \Lambda)^{-1} = T_1(\Lambda x + a). \tag{4.5.1}$$

Here $U(a, \Lambda)$ is the pseudo-unitary representation of the Poincaré group in Fock space,

$$U(a, \Lambda)^{-1} = U(a, \Lambda)^K, \tag{4.5.2}$$

which was defined by (2.2.58), (2.11.37) and the equations following there. The conjugation K was introduced in (2.11.43). It agrees with the adjoint apart from the scalar photon sector. We are going to show by induction that

$$U(a, \Lambda) T_n(x) U(a, \Lambda)^{-1} = T_n(\Lambda x + a). \tag{4.5.3}$$

This is the perturbative form of the Poincaré covariance of the S-matrix

$$U(a,\Lambda)S(g)U(a,\Lambda)^{-1}=S(g'), \tag{4.5.4}$$

$$\text{with}\quad g'(x)=g(\Lambda^{-1}(x-a)).$$

As we have discussed at the end of Sect. 2.11, this is not yet the physical Poincaré invariance, because $U(a,\Lambda)$ is not unitary on the entire Fock space. The independence of the measuring results from the reference system requires invariance of S-matrix elements between states in the physical subspace $\mathcal{F}_{\rm phys}$, and this only for $g=1$. By our previous results (see (2.11.64)), this is ensured by Poincaré covariance (4.5.4) together with gauge invariance.

By now we are familiar with the form of the inductive proof. Assuming that all m-point functions with $m\le n-1$ are invariant, then all products of them are also invariant and so are R'_n, A'_n and D_n. The process of distribution splitting does not affect the operator parts of the terms in D_n; we have only to concentrate on the splitting of numerical distributions. Poincaré invariance can only be destroyed by the undetermined local terms in the case of non-trivial splitting. The splitting of a Poincaré covariant distribution $\hat{d}_n(p)$ in momentum space obviously leads to a covariant retarded distribution $\hat{r}(p)$, if the central solution (3.2.61)

$$\hat{r}(p)=\frac{i}{2\pi}\int\limits_{-\infty}^{+\infty}dt\,\frac{\hat{d}_n(tp)}{(t-i0)^{\omega+1}(1-t+i0)} \tag{4.5.5}$$

can be applied. It then follows that T_n is also covariant.

However, even if the central solution does not exist, as in case of massless fermions, a covariant splitting solution can always be constructed by suitable normalization. In the following we want to demonstrate this by a direct method which shows an interesting connection of the problem with cohomology theory. A reader not interested in this material may skip the rest of the section without harm for the further understanding.

First we consider the splitting of a Lorentz invariant causal distribution $d(x)$, $x=(x_1,\cdots x_n)$, $x_j\in\mathbf{R}^4$, satisfying

$$d(x')=d(x), \tag{4.5.6}$$

where

$$x'=\Lambda x\stackrel{\rm def}{=}(\Lambda x_1,\ldots,\Lambda x_n) \tag{4.5.7}$$

for arbitrary proper Lorentz transformations $\Lambda\in L_+^\uparrow$. Translations must not be considered because the numerical distributions in QED are always translation invariant, and this property is explicitly preserved in the splitting procedure. Let $r(x)$ be an arbitrary retarded part of the r.h.s. of (5.5.6). Then $r(\Lambda x)$ is a retarded part of the l.h.s. of (5.5.6). Both distributions agree with $d(x)$ on $\Gamma_n^+(x_n)\setminus(x_n,\ldots,x_n)$ and vanish on the complement $\Gamma_n^+(x_n)^C$. Consequently, their difference can only have a point support

$$r(\Lambda x) = r(x) + \sum_u c_u(\Lambda) D^u \delta(x). \tag{4.5.8}$$

Here $\delta(x)$ stands for

$$\delta(x) = \delta^{(n-1)}(x_1, x_n, \ldots, x_{n-1} - x_n). \tag{4.5.9}$$

The sum runs over all multi-indices u with $|u| \leq \omega$, where ω is the singular order of $d(x)$ at the origin. It is our aim to show that the splitting solution can be redefined (or renormalized)

$$\tilde{r}(x) = r(x) + \sum_u b_u D^u \delta(x) \tag{4.5.10}$$

so that it becomes $L_+^\uparrow$ invariant

$$\tilde{r}(\Lambda x) = \tilde{r}(x). \tag{4.5.11}$$

Since the l.h.s. in (4.5.8) depends (weakly) continuous on Λ, $c_u(\Lambda)$ are continuous functions on $L_+^\uparrow$. Consider first the case $|u| = 0$. Applying (4.5.8) to

$$r([\Lambda_1 \Lambda_2] x) = r(\Lambda_1 [\Lambda_2 x]), \tag{4.5.12}$$

we find

$$c_0(\Lambda_1 \Lambda_2) = c_0(\Lambda_1) + c_0(\Lambda_2). \tag{4.5.13}$$

Writing $c_0(\Lambda) = \log \gamma(\Lambda)$ we see that

$$\gamma(\Lambda_1 \Lambda_2) = \gamma(\Lambda_1)\gamma(\Lambda_2) \tag{4.5.14}$$

is a one-dimensional continuous representation of $L_+^\uparrow$, hence $\gamma(\Lambda) \equiv 1$ and $c_0(\Lambda) \equiv 0$: There is no breaking of $L_+^\uparrow$ invariance for $|u| = 0$.

Next we consider $|u| = 1$. We write $x_u = x_\mu^i$, where $\mu \in \{0, 1, 2, 3\}$ is the Lorentz index and $i \in \{1 \cdots n\}$ the number of the argument

$$r(\Lambda x) = r(x) + c_\mu^i(\Lambda) \partial_i^\mu \delta(x). \tag{4.5.15}$$

Applying this to a product $\Lambda = \Lambda_1 \Lambda_2$, we now obtain

$$c_\mu^i(\Lambda_1 \Lambda_2) = c_\mu^i(\Lambda_2) + (\Lambda_2^{-1})_\mu{}^\nu c_\nu^i(\Lambda_1). \tag{4.5.16}$$

This is n times (for any i) the same equation for the four-vector $c_\mu(\Lambda)$. We shall omit the indices and use matrix notation in the following. Introducing the function $f(\Lambda) = c(\Lambda^{-1})$ and interchanging Λ_1 and Λ_2 in (4.5.16) we arrive at

$$f(\Lambda_1 \Lambda_2) = f(\Lambda_1) + \Lambda_1 f(\Lambda_2). \tag{4.5.17}$$

This equation establishes a connection to homological algebra. We give the relevant definitions following Guichardet (*A. Guichardet, Cohomologie des groupes topologiques et des algèbres de Lie (Paris 1980), p. 16, 181*).

Let G be a group, E an abelian group, and Φ an action of G on E which makes E a G-module E_Φ, that means

$$\Phi(g_1g_2)a = \Phi(g_1)\Phi(g_2)a, \tag{4.5.18}$$

$$\Phi(g)(a_1 + a_2) = \Phi(g)a_1 + \Phi(g)a_2. \tag{4.5.19}$$

These equations must hold for all $g, g_i \in G$, $a, a_i \in E$. g_1g_2 means the group multiplication in G, and $a_1 + a_2$ that in E. The (continuous) mappings $f(g_1, \cdots g_n)$ of G^n into E are called n-cochains and form the n-th cochain group $C^n(G, E_\Phi)$, $C^0(G, E_\Phi) = E$ by definition. The coboundary operator $d^n : C^n(G, E_\Phi) \longrightarrow C^{n+1}(G, E_\Phi)$ is defined by

$$(d^n f)(g_1, \cdots g_{n+1}) = \Phi(g_1)f(g_2, \cdots g_{n+1})$$

$$+\sum_{i=1}^{n}(-)^i f(g_1, \cdots \widehat{g_i}, g_ig_{i+1}, g_{i+2}, \cdots g_{n+1}) + (-)^{n+1} f(g_1, \cdots g_n), \tag{4.5.20}$$

where the argument with hat must be omitted. It obviously satisfies

$$d^{n+1} \circ d^n = 0. \tag{4.5.21}$$

For $n = 0$ we have

$$(d^0 a)(g) = \Phi(g)a - a \tag{4.5.22}$$

and for $n = 1$

$$(d^1 f)(g_1, g_2) = \Phi(g_1)f(g_2) - f(g_1g_2) + f(g_1). \tag{4.5.23}$$

One calls

$$Z^n(G, E_\Phi) \stackrel{\text{def}}{=} \operatorname{Ker} d^n,\ B^n(G, E_\Phi) \stackrel{\text{def}}{=} \operatorname{Im} d^{n-1},\ H^n(G, E_\Phi)$$

$$\stackrel{\text{def}}{=} Z^n(G, E_\Phi)/B^n(G, E_\Phi)$$

the n-th groups of cocycles, coboundaries, and cohomology, respectively. Because of (4.5.21), every coboundary is a cocycle.

We now apply these notions to the function $f(\Lambda)$ in (4.5.17). There we have $G = L_+^\uparrow$, $E = \mathbf{R}^4$, and the action $\Phi(\Lambda)$ is the defining representation: $\Phi(\Lambda)a \stackrel{\text{def}}{=} \Lambda a$. $f(\Lambda)$ is a 1-cochain, and (4.5.17) simply reads: $d^1 f = 0$. The solutions of (4.5.17) are, therefore, the 1-cocycles. In particular, the 1-coboundaries

$$f(\Lambda) = [\Lambda - 1]a \tag{4.5.24}$$

solve (4.5.17). We will soon see (Proposition 5.1), that all (continuous) solutions of (4.5.17) are given by (4.5.24), where a varies arbitrarily in $\mathbf{R}^4$. This means that every 1-cocycle $\in Z^1(L_+^\uparrow, \mathbf{R}^4)$ is a 1-coboundary $\in B^1(L_+^\uparrow, \mathbf{R}^4)$, or: $H^1(L_+^\uparrow, \mathbf{R}^4)$ is trivial. Inserting (4.5.24) into (4.5.15) we get

$$r(\Lambda x) = r(x) + [\Lambda^{-1} - 1]_\mu^{\ \nu} a_\nu^i \partial_i^\mu \delta(x) = r(x) + a_\mu^i [\Lambda - 1]^\mu_{\ \nu} \partial_i^\nu \delta(x). \tag{4.5.25}$$

Therefore,

$$\tilde{r}(x) = r(x) - a^i_\mu \partial^\mu_i \delta(x) \tag{4.5.26}$$

is a Lorentz invariant splitting solution.

Proposition 5.1. Every continuous function f from the proper Lorentz group $L^\uparrow_+$ into $\mathbf{R}^4$

$$L^\uparrow_+ \ni \Lambda \longrightarrow f(\Lambda) \in \mathbf{R}^4 \tag{4.5.27}$$

which satisfies

$$f(\Lambda_1 \Lambda_2) = f(\Lambda_1) + \Lambda_1 f(\Lambda_2) \tag{4.5.28}$$

for all $\Lambda_1, \Lambda_2 \in L^\uparrow_+$ is of the form

$$f(\Lambda) = (\Lambda - 1)a, \tag{4.5.29}$$

where $a \in \mathbf{R}^4$ is independent of Λ.

A proof of this result is already contained in the classical paper by E. Wigner (*Ann. Math. 40, 149 (1939) p. 174*) on the unitary representations of the inhomogeneous Lorentz group. We prove the corresponding result for arbitrary spinor representations of the quantum mechanical Lorentz group $SL(2, \mathbf{C})$. It contains Proposition 5.1 as a special case.

Proposition 5.2. Let E_Φ be a $SL(2, \mathbf{C})$-module which carries a finite - dimensional representation $\Phi(A)$, $A \in SL(2, \mathbf{C})$. Then $H^1(SL(2, \mathbf{C}), E_\Phi)$ is trivial, i.e. any (continuous) mapping $f(A)$ from $SL(2, \mathbf{C})$ into E_Φ, satisfying

$$f(A_1 A_2) = f(A_1) + \Phi(A_1) f(A_2), \tag{4.5.30}$$

is of the form

$$f(A) = \Phi(A)a - a, \tag{4.5.31}$$

where $a \in E_\Phi$ is independent of A.

Proof. We first consider the spin $\frac{1}{2}$ (self-)representation $\Phi_1(A)$ of $SL(2, \mathbf{C})$. A rotation around the 1-axis is represented by

$$A_1 = \cos\frac{\alpha}{2} + i\sigma_1 \sin\frac{\alpha}{2} \tag{4.5.32}$$

and a boost in the 1-direction by

$$B_1 = \mathrm{ch}\frac{\chi}{2} - \sigma_1 \mathrm{sh}\frac{\chi}{2}, \tag{4.5.33}$$

where σ_1, σ_2, σ_3 denote the Pauli matrices. These transformations generate a two-parameter abelian subgroup G_1

$$A_1(\alpha, \chi) = \cos\left(\frac{\alpha}{2} + i\frac{\chi}{2}\right) + i\sigma_1 \sin\left(\frac{\alpha}{2} + i\frac{\chi}{2}\right), \tag{4.5.34}$$

satisfying

$$A_1(\alpha_1+\alpha_2,\, \chi_1+\chi_2) = A_1(\alpha_1,\, \chi_1)A_1(\alpha_2,\, \chi_2). \tag{4.5.35}$$

The eigenvalues of $A_1(\alpha, \chi)$ are equal to

$$\lambda_\pm(\alpha, \chi) = \exp \pm i\left(\frac{\alpha}{2} + i\frac{\chi}{2}\right). \tag{4.5.36}$$

Let us now consider (4.5.30) for arbitrary rotations $R \in SU(2)$

$$f(R_1R_2) = f(R_1) + \Phi_1(R_1)f(R_2). \tag{4.5.37}$$

We integrate this equation with the left invariant Haar measure dR_2 of the (compact) subgroup $SU(2)$, which we assume to be normalized $\int dR_2 = 1$. Writing

$$-a_1 \stackrel{\text{def}}{=} \int f(R_2)dR_2 = \int f(R_1R_2)dR_2, \quad a_1 \in \mathbf{C}^2,$$

we get

$$-a_1 = f(R_1) - \Phi_1(R_1)a_1,$$

or

$$f(R_1) = (\Phi_1(R_1) - 1)a_1 \tag{4.5.38}$$

for arbitrary rotations R_1.

On the other hand, two transformations $A_1,\, A_1' \in G_1$ (4.5.34) satisfy

$$f(A_1A_1') = f(A_1) + \Phi_1(A_1)f(A_1') = f(A_1') + \Phi_1(A_1')f(A_1),$$

thus

$$(\Phi_1(A_1') - 1)f(A_1) = (\Phi_1(A_1) - 1)f(A_1').$$

Here, $(\Phi_1(A_1') - 1)$has an inverse iff $\lambda = 1$ is not eigenvalue of $\Phi_1(A_1')$. Due to (4.5.36) this is the case if $A_1' \neq 1$. Then we find

$$f(A_1) = (\Phi_1(A_1') - 1)^{-1}(\Phi_1(A_1) - 1)f(A_1')$$

$$= (\Phi_1(A_1) - 1)(\Phi_1(A_1') - 1)^{-1}f(A_1') \stackrel{\text{def}}{=} (\Phi_1(A_1) - 1)a, \tag{4.5.39}$$

where $a \in \mathbf{C}^2$ is the same for all $A_1 \in G_1$. If we specialize this to rotations $A_1 \in G_1$ (4.5.32) around the 1-axis ($\chi = 0$) and compare with (4.5.38), we get

$$(\Phi_1(R_1) - 1)a_1 = (\Phi_1(R_1) - 1)a.$$

This implies $a_1 = a$, because the inverse exists if $R_1 \neq 1$. It easily follows from (4.5.30) that (4.5.31) is also true for a product A_1A_2, if it holds for A_1 and A_2 separately. Since rotations $\in SU(2)$ and G_1 generate the entire $SL(2, \mathbf{C})$, this proves the proposition for the spin $\frac{1}{2}$ representation Φ_1.

The same proof applies to the complex conjugated representation Φ_1^*. According to spinor calculus (see Sect. 1.1) any finite-dimensional representation Φ is obtained from direct products

$$\Phi = \Phi_1 \times ... \times \Phi_1 \times \Phi_1^* \times ... \times \Phi_1^*.$$

Using the general result (*Guichardet*, loc. cit.)

$$H^1(G, E_1 \times E_2) = H^1(G, E_1) \times H^1(G, E_2), \tag{4.5.40}$$

we conclude that $H^1(SL(2, \mathbf{C}), E_\Phi)$ is trivial for any finite-dimensional (not necessarily irreducible) representation Φ. The representations with integer spin are equivalent to ordinary representations of $L_+^\uparrow$. □

Now we consider the general case $|u| = p$. Then (4.5.8) reads

$$r(\Lambda x) = r(x) + \left(\sum_{1 \leq i_1 \cdots \leq i_p \leq n} c_{\mu_1 \cdots \mu_p}^{i_1 \cdots i_p} \partial_{i_1}^{\mu_1} \cdots \partial_{i_p}^{\mu_p} \delta(x) \right). \tag{4.5.41}$$

Applying this again to $\Lambda = \Lambda_1 \Lambda_2$ gives

$$c_{\mu_1 \cdots \mu_p}^{i_1 \cdots i_p}(\Lambda_1 \Lambda_2) = c_{\mu_1 \cdots \mu_p}^{i_1 \cdots i_p}(\Lambda_2) + (\Lambda_2^{-1})_{\mu_1}^{\nu_1} \cdots (\Lambda_2^{-1})_{\mu_p}^{\nu_p} c_{\nu_1 \cdots \nu_p}^{i_1 \cdots i_p}(\Lambda_1). \tag{4.5.42}$$

This is the same equation for every $\{i_1 \cdots i_p\}$; so we can omit these indices. Using tensor-notation instead of writing Lorentz-indices $\{\mu_1 \cdots \mu_p\}$ and passing again to $f(\Lambda) \stackrel{\text{def}}{=} c(\Lambda^{-1})$ we get

$$f(\Lambda_1 \Lambda_2) = f(\Lambda_1) + \Lambda_1^{\times p} f(\Lambda_2). \tag{4.5.43}$$

This equation is of the form (4.5.30) with $E_\Phi = (\mathbf{R}^4)^{\times p}$ and $\Phi(\Lambda) = \Lambda^{\times p}$. Therefore, Proposition 5.2 can be applied and leads to

$$f(\Lambda) = (\Lambda^{\times p} - 1)a. \tag{4.5.44}$$

It then follows that the redefined splitting solution

$$\tilde{r}(x) \stackrel{\text{def}}{=} r(x) - \left(\sum_{1 \leq i_1 \cdots \leq i_p \leq n} a_{i_1 \cdots i_p}^{\mu_1 \cdots \mu_p} \partial_{i_1}^{\mu_1} \cdots \partial_{i_p}^{\mu_p} \delta(x) \right) \tag{4.5.45}$$

is indeed Lorentz invariant. We thus have shown that every Lorentz invariant causal distribution can be split into Lorentz invariant retarded and advanced parts.

Now we consider Lorentz covariant causal distributions $d(x)$, that means

$$d(\Lambda x) = D(\Lambda) d(x), \tag{4.5.46}$$

where $D(\Lambda) = \{D(\Lambda)^A{}_B\}$ is a finite dimensional representation of $L_+^\uparrow$ and $d(x) = \{d^A(x)\}$ is an element of the representation space V_D. Any retarded part r of d will fulfill the equation

$$r^A(\Lambda x) = \sum_{B=1}^{\dim V_D} D^A{}_B r^B(x) + \sum_{p=0}^{\omega} \left(\sum_{1 \leq i_1 \cdots \leq i_p \leq n} c_{\mu_1 \cdots \mu_p}^{A, i_1 \cdots i_p} \partial_{i_1}^{\mu_1} \cdots \partial_{i_p}^{\mu_p} \delta(x) \right). \tag{4.5.47}$$

To condense the notation we write

$$c^p \stackrel{\text{def}}{=} c^{A,i_1\cdots i_p}_{\mu_1\cdots\mu_p}. \tag{4.5.48}$$

$$\Lambda^{\times p}c^p \stackrel{\text{def}}{=} \Lambda_{\mu_1}^{\nu_1}\cdots\Lambda_{\mu_p}^{\nu_p}c^{A,i_1\cdots i_p}_{\nu_1\cdots\nu_p}, \tag{4.5.49}$$

$$D(\Lambda)c^p \stackrel{\text{def}}{=} \sum_B D^A_B c^{B,i_1\cdots i_p}_{\mu_1\cdots\mu_p}. \tag{4.5.50}$$

Applying (4.5.47) again to a product $\Lambda = \Lambda_1\Lambda_2$ we find

$$c^p(\Lambda_1\Lambda_2) = D(\Lambda_1)c^p(\Lambda_2) + (\Lambda_2^{-1})^{\times p}c^p(\Lambda_1). \tag{4.5.51}$$

Defining $f(\Lambda) \stackrel{\text{def}}{=} D(\Lambda)c^p(\Lambda^{-1})$ we find

$$f(\Lambda_1\Lambda_2) = f(\Lambda_1) + D(\Lambda_1)\Lambda_1^{\times p}f(\Lambda_2). \tag{4.5.52}$$

Once more we can apply Proposition 5.2. This time we have $E_\Phi = \mathbf{R}^{\dim V_D} \times (\mathbf{R}^4)^{\times p}$ and $\Phi(\Lambda)a = D(\Lambda) \times \Lambda^{\times p}a$. So we get

$$f(\Lambda) = [D(\Lambda)\Lambda^{\times p} - 1]a. \tag{4.5.53}$$

The redefined splitting solution

$$\tilde{r}^A(x) \stackrel{\text{def}}{=} r^A(x) - \sum_{p=0}^{\omega}\left(\sum_{1\le i_1\cdots\le i_p\le n} a^{A,i_1\cdots i_p}_{\mu_1\cdots\mu_p}\partial^{\mu_1}_{i_1}\cdots\ \partial^{\mu_p}_{i_p}\delta(x)\right) \tag{4.5.54}$$

is, therefore, indeed Lorentz covariant: $\tilde{r}(\Lambda x) = D(\Lambda)\tilde{r}(x)$. Thus, the existence of a covariant splitting solution is proven in all cases.

4.6 Gauge Invariance and Ward Identities

We now come to the most significant property of QED, namely its gauge structure. Since in the recursive construction the S-matrix

$$S(g) = \sum_n \frac{1}{n!}\int T_n(x_1,\dots x_n)g(x_1)\dots g(x_n)d^4x_1\dots d^4x_n \tag{4.6.1}$$

has been expressed in terms of the free fields, especially $A_\mu(x)$, which changes under gauge transformations, there arises the question whether (4.6.1) is gauge invariant. We only consider gauge transformations by means of a C-number gauge function $\Lambda(x)$

$$A_\mu(x) \longrightarrow A_\mu(x) - \frac{1}{e}\partial_\mu\Lambda(x), \tag{4.6.2}$$

where $\Lambda(x)$ vanishes at infinity and satisfies the wave equation, so that the transformed field (4.6.2) also fulfills the wave equation. We do not consider a transformation of the spinor fields $\psi(x)$ (as in Sect. 1.5), because $\psi(x)$ is a *free* Dirac field which is not gauge invariant. As before, to prove gauge invariance of S (4.6.1), we must show that the first order $T_1(x)$ is gauge invariant and that the inductive procedure does not destroy it.

Let us write the first order S-matrix as follows

$$S_1(g) = \int T_1(x)g(x)d^4x \quad , \quad \text{with} \tag{4.6.3}$$

$$T_1(x) = ij^\mu(x)A_\mu(x) \tag{4.6.4}$$

$$j^\mu(x) = e : \overline{\psi}(x)\gamma^\mu\psi(x) : . \tag{4.6.5}$$

A gauge transformation in (4.6.3) now yields

$$S_1'(g) = i\int d^4x\, j^\mu(x)\Big(A_\mu(x) - \frac{1}{e}\partial_\mu\Lambda(x)\Big)g(x)$$

$$= S_1(g) + \frac{i}{e}\int(\partial_\mu j^\mu(x))\Lambda(x)g(x)d^4x + \frac{i}{e}\int j^\mu(x)\Lambda(x)\partial_\mu g(x)d^4x. \tag{4.6.6}$$

The free current is conserved

$$\partial_\mu j^\mu(x) = 0, \tag{4.6.7}$$

in virtue of the Dirac equation. Therefore, the second term in (4.6.6) vanishes. The last term goes to zero in the adiabatic limit $g(x) \to 1$. This is the gauge invariance of $S_1(g)$.

We now proceed to consider the n-point function T_n, normally ordered with respect to the photon operators:

$$T_n(x_1,\ldots,x_n) = \sum_{l=0}^{n}\sum_{1\le k_1<\ldots<k_l\le n} t^{\mu_1\ldots\mu_l}_{k_1\ldots k_l}(x_1,\ldots,x_n)$$

$$\times : A_{\mu_1}(x_{k_1})\ldots A_{\mu_l}(x_{k_l}) : . \tag{4.6.8}$$

By the same reasoning as in (4.6.6), we conclude that gauge invariance of T_n may be expressed by

$$\frac{\partial}{\partial x^{\mu_j}_{k_j}} t^{\mu_1\ldots\mu_l}_{k_1\ldots k_l}(x_1,\ldots,x_n) = 0, \tag{4.6.9}$$

for all $1\le l\le n$, all $1\le j\le l$, all $1\le k_1<\ldots<k_l\le n$ and all $(x_1,\ldots,x_n)\in \mathbb{R}^{4n}$. The t's in (4.6.8) contain the Fermi operators: $t^{\mu_1\ldots\mu_l}_{k_1\ldots k_l}$ is the sum of all graphs of order n with l external photon lines at the vertices $x_{k_1},\ldots x_{k_l}$ and no other external photon lines, the external fermions being arbitrary. It follows from (4.6.8) that $t^{\mu_1\ldots\mu_l}_{k_1\ldots k_l}$ is symmetrical in $(x_{k_1},\mu_1)\ldots,(x_{k_l},\mu_l)$.

It is our aim to prove gauge invariance (4.6.9) by induction on n. Let us assume that (4.6.9) holds for all m-point distributions with $m\le n-1$.

Going from $n-1$ to n according to the inductive construction, we have first to form

$$R'_n(x_1,\ldots,x_n)=\sum_X T_{n-n_1}(Y,x_n)\tilde{T}_{n_1}(X), \tag{4.6.10}$$

where $\tilde{T}_{n_1}$ comes from the perturbation expansion of $S(g)^{-1}$. Each term in (4.6.10) is a product of T_m's with $m\leq n-1$ and disjoint arguments. In virtue of the induction assumption, each term is gauge invariant, because the normal ordering in the photon operators does not affect it. The same is true for

$$A'_n(x_1,\ldots,x_n)=\sum_X \tilde{T}_{n_1}(X)T_{n-n_1}(Y,x_n), \tag{4.6.11}$$

and for

$$D_n=R'_n-A'_n. \tag{4.6.12}$$

This distribution has causal support with respect to x_n:

$$\operatorname{supp} D_n\subseteq \Gamma^n_+(x_n)\cup\Gamma^n_-(x_n), \tag{4.6.13}$$

$$\Gamma^n_\pm(x_n)=\{(x_1,\ldots,x_n)\mid x_j\in V^\pm(x_n),\ \forall j=1,\ldots n-1\}, \tag{4.6.14}$$

where $V^\pm(x)$ is the closed forward or backward cone of x, respectively. The essential step in the inductive construction is the splitting of $D_n=R_n-A_n$ into a retarded and advanced part. There remains to prove that gauge invariance is preserved under this operation. The final step $T_n=R_n-R'_n$ does not affect gauge invariance.

Our starting point is the relation

$$\partial_\nu d^\nu \stackrel{\text{def}}{=} \frac{\partial}{\partial x_n^\nu}d^{\mu_1\ldots\mu_{l-1}\nu}_{k_1\ldots k_l}(x_1,\ldots x_{n-1},x_n)=0, \tag{4.6.15}$$

where we have introduced a shorthand notation and have taken $x_{k_j}=x_n,\mu_j=\nu$ for simplicity. Since the retarded part is equal to

$$r^\nu(x_1,\ldots,x_n)=\begin{cases} d^\nu(x_1,\ldots,x_n) & \text{on}\quad \Gamma^n_+(x_n)\setminus(x_n,\ldots x_n)\\ 0 & \text{on}\quad \Gamma^n_+(x_n)^C, \end{cases} \tag{4.6.16}$$

where C denotes the complement, we conclude from (4.6.15) that $\partial_\nu r^\nu$ can only have a point support:

$$\operatorname{supp}\partial_\nu r^\nu(x_1,\ldots,x_n)\subseteq(x_n,\ldots x_n). \tag{4.6.17}$$

We decompose

$$d^\nu=\sum_{f=0}^n d^\nu_f\quad,\quad r^\nu=\sum_{f=0}^n r^\nu_f \tag{4.6.18}$$

into contributions of all graphs of order n with $2\cdot f$ external fermion lines and l external photon lines. According to (4.6.15), we must separately have

$$\partial_\nu d^\nu_f(x_1,\ldots,x_n)=0\quad,\quad \forall 0\leq f\leq n. \tag{4.6.19}$$

It is important to note that there are no terms with derivative ∂_ν acting on a Fermi field operator for the following reason: This derivative comes from an external photon operator $A^\nu(x_n)$. If there is also a Fermi operator $\overline{\psi}(x_n)$, the vertex x_n is connected with the rest of the graph by a single fermion line, represented by a Fermi propagation function $S_{\rm ret}(x_n - x_j)$, $S_{\rm av}$ or S_F. Such a contribution can be reduced by means of the Dirac equation

$$\frac{\partial}{\partial x_n^\nu}\Big(\overline{\psi}(x_n)\gamma^\nu S_{\substack{\rm ret\\ \rm av}}(x_n - x_j)\Big) = i\overline{\psi}(x_n)\delta^4(x_n - x_j),$$

$$\frac{\partial}{\partial x_n^\nu}\Big(S_{\substack{\rm ret\\ \rm av}}(x_j - x_n)\gamma^\nu\psi(x_n)\Big) = -i\delta^4(x_j - x_n)\psi(x_n). \tag{4.6.20}$$

Next we carry out the normal product decomposition in (4.6.19) and split all numerical distributions. Now (4.6.17) must also hold for every r_f^ν separately. Using a well-known theorem on distributions with point-like support, we arrive at

$$\partial_\nu r_f^\nu(x_1,\ldots,x_n) = \sum_g :\overline{\psi}(x_{i_1})\ldots\overline{\psi}(x_{i_f})\Big(\sum_{|a|\le\omega(g)+1} K_a^g D^a\delta(x_1 - x_n)\ldots$$

$$\ldots\delta(x_{n-1} - x_n)\Big)\psi(x_{j_1})\ldots\psi(x_{j_f}):, \tag{4.6.21}$$

where the sum runs over all graphs g with $2f$ external fermions and l photons. $\omega(g)$ is the singular order of the graph g which is given by the previous expression (4.3.2)

$$\omega(g) = 4 - 3f - l. \tag{4.6.22}$$

In (4.6.21) we have used the fact that the derivative increases ω by 1. The r.h.s. of (4.6.21) is usually called an anomaly, because it violates gauge invariance. This anomaly goes over from the retarded to the t-distribution without change, since R' is anomaly-free.

For $\omega < -1$, the inner sum in (4.6.21) contains no term, hence the expression vanishes, which proves the desired divergence relation in this case. There remains to investigate the possible cases of $\omega \ge -1$. In virtue of (4.6.22), there are only the following four cases of this kind:

$$\begin{array}{lllll} (f,l) = & (0,2) & (0,4) & (1,1) & (1,2)\\ \omega\ \ = & 2 & 0 & 0 & -1\\ \text{case}: & \text{I} & \text{II} & \text{III} & \text{IV}. \end{array} \tag{4.6.23}$$

These cases must now be examined. We need the following

Lemma 6.1. In the normal product expansion of the causal distribution $D_n(X,x)$, only *connected graphs* contribute. (This is true in any causal quantum field theory!)

Proof. A disconnected graph factorizes

$$D(X,x) =: D^1(X_1)D^2(X_2) : \tag{4.6.24}$$

where

$$X_1 \cup X_2 = (X,x), \quad X_1 \cap X_2 = \emptyset.$$

Then the supports of D^1 and D^2 are completely independent and the support of D is the direct product of them. This is in contradiction to the causal support $\operatorname{supp} D(X,x) \subseteq (\Gamma_+(x) \cup \Gamma_-(x))$. □

According to this lemma, we have only to consider connected diagrams. However, combinations of field operators which differ from each other in d_f^ν may agree in $\partial_\nu d_f^\nu$, in virtue of the identities (4.6.20).

Case I: $(f,l) = (0,2), \omega = 2$

This is vacuum polarization. Since $f = 0$, Eq. (4.6.21) does not contain any field operator, so that we have to deal with numerical distributions only. We write down the anomaly relation (4.6.21) for the t-distribution, assuming the external photon operators at x_1 and x_2 for convenience:

$$\partial_{1\nu}\Pi^{\nu\mu}(x_1,x_2;x_3,\dots x_n) = \Big(\sum_{ijk} K_{ijk}\partial_i^\alpha\partial_{j\alpha}\partial_k^\mu + \sum_k L_k\partial_k^\mu\Big)\delta^{n-1}, \tag{4.6.25}$$

where $\delta^{n-1} = \delta(x_1 - x_n)\cdot \dots \delta(x_{n-1} - x_n)$. Different from (4.6.21), x_n is an inner vertex here. The r.h.s. is the most general Lorentz covariant local distribution with $\omega = 3$. We now claim:

Proposition 6.2. The anomaly (4.6.25) can be restricted to the following form

$$\partial_{1\nu}\Pi^{\nu\mu} = \Big[K_1\sum_{i=3}^n \partial_i^\alpha\partial_{i\alpha}\partial_1^\mu + K_2\sum_{i=3}^n \partial_i^\alpha\partial_{1\alpha}\partial_i^\mu$$

$$+K_3\partial_1^\alpha\partial_{2\alpha}\partial_2^\mu + K_4\partial_2^\alpha\partial_{2\alpha}\partial_1^\mu + (K_3+K_4)\partial_1^\alpha\partial_{1\alpha}\partial_1^\mu + K_5\partial_1^\alpha\partial_{1\alpha}\partial_2^\mu + K_6\partial_1^\alpha\partial_{2\alpha}\partial_1^\mu$$

$$+K_7\partial_1^\mu\Big]\delta^{n-1}. \tag{4.6.26}$$

Proof. Calculating the divergence of (4.6.25) with respect to x_2, the result must be symmetric in x_1, x_2:

$$(\partial_{2\mu}\partial_{1\nu}\Pi^{\nu\mu})(x_1,x_2;\dots) = (\partial_{2\mu}\partial_{1\nu}\Pi^{\mu\nu})(x_2,x_1;\dots). \tag{4.6.27}$$

This implies $L_k = 0$ for $k \neq 1$ and $K_{ijk} = 0$ for $i,j,k > 2$. Furthermore, in virtue of (4.6.27), only the following K_{ijk} can be $\neq 0$: (i) $K_{ij1}, K_{i1k}, K_{1jk}$, (ii) $K_{mpk}, K_{mjp}, K_{imp}$, (iii) K_{lmp}, where $l,m,p \leq 2$, $i,j,k > 2$.

We now use the property that (4.6.25) is symmetric in all inner vertices $x_3, \dots x_n$. This symmetry property holds for the t-distribution, it is not true for the retarded one, where the splitting vertex plays a special rôle. It allows us to express case (ii), depending on one internal index > 2 only, by derivatives with respect to x_1, x_2, for example:

$$\sum_{k=3}^{n} K_{mnk}\partial_k^\mu \delta^{n-1} = K_{mn}\sum_{3}^{n} \partial_k^\mu \delta^{n-1} = K_{mn}(-\partial_1^\mu - \partial_2^\mu)\delta^{n-1}, \qquad (4.6.28)$$

because

$$\sum_{k=1}^{n} \partial_k^\mu \delta(x_1 - x_n)\cdot \dots \delta(x_{n-1} - x_n) = 0. \qquad (4.6.29)$$

This reduces this case (ii) to case (iii). In case (i) the symmetry in the inner vertices implies

$$K_{ij1} = K_{\pi i \pi j 1}, \quad K_{i1k} = K_{\pi i 1 \pi k} \qquad (4.6.30)$$

for all permutations π. This means that the diagonal elements $i = j$, $i = k$ are equal and all off-diagonal elements also. By redefining these constants, the latter sum over $i \neq j$ can be transformed into an independent summation over $i, j = 3, \dots n$, which, by (4.6.29), is also reduced to case (iii). There remains the summation over diagonal terms, leading to the first two terms in (4.6.26). Finally, in case (iii) we have $K_{111} = K_{122} + K_{221}$, $K_{222} = 0$ due to (4.6.27). Then we arrive at the remaining terms in (4.6.26). □

In the anomaly (4.6.26) the derivative $\partial_{1\nu}$ can now be taken out

$$\partial_{1\nu}\Pi^{\nu\mu} = \partial_{1\nu}\Bigg[g^{\mu\nu}K_1\sum_{i=3}^{n}\partial_i^\alpha\partial_{i\alpha} + K_2\sum_{i=3}^{n}\partial_i^\nu\partial_i^\mu$$

$$+K_3(\partial_1^\nu\partial_1^\mu + \partial_2^\nu\partial_2^\mu) + K_4 g^{\mu\nu}(\Box_1 + \Box_2) + K_5\partial_1^\nu\partial_2^\mu + K_6\partial_2^\nu\partial_1^\mu + K_7 g^{\mu\nu}\Bigg]\delta^{n-1}. \qquad (4.6.31)$$

The square bracket is a polynomial of degree $\omega(\Pi^{\mu\nu}) = 2$ and has the symmetry properties of $\Pi^{\mu\nu}$. Therefore it can be transformed away by renormalization of $\Pi^{\nu\mu}$ which completes the proof of gauge invariance in case I. □

The renormalized vacuum polarization tensor in n-th order now satisfies a so-called Ward-Takahashi identity

$$\partial_{1\nu}\Pi^{\nu\mu}(x_1, x_2; x_3, \dots x_n) = 0 = \partial_{2\mu}\Pi^{\nu\mu}(x_1, x_2; x_3, \dots x_n). \qquad (4.6.32)$$

Case II: $(f,l)=(0,4), \omega=0$

This is photon-photon scattering where we have again to deal with one numerical distribution only. As in the proof of Prop.6.2 (4.6.27), the anomaly can be expressed by derivatives with respect to the external coordinates $x_1,\dots x_4$, using the symmetry in the inner vertices:

$$\partial_{1\nu} t^{\nu\mu\alpha\beta}(x_1,x_2,x_3,x_4;x_5,\dots x_n)$$
$$=\sum_{k=1}^{4}\Big(K_{k1}\partial_k^\mu g^{\alpha\beta}+K_{k2}\partial_k^\alpha g^{\mu\beta}+K_{k3}\partial_k^\beta g^{\mu\alpha}\Big)\delta^{n-1}. \tag{4.6.33}$$

Again, x_n is an inner vertex here. Since $\partial_{1\nu}\partial_{2\mu}\partial_{3\alpha}\partial_{4\beta}t^{\nu\mu\alpha\beta}$ must be symmetric in $x_1,\dots x_4$, it follows that $K_{11}=K_{12}=K_{13}=K$ and all other $K_{kj}=0$. Then

$$\partial_{1\nu} t^{\nu\mu\alpha\beta}=K\partial_{1\nu}\Big(g^{\nu\mu}g^{\alpha\beta}+g^{\nu\alpha}g^{\mu\beta}+g^{\nu\beta}g^{\mu\alpha}\Big)\delta^{n-1}, \tag{4.6.34}$$

and this anomaly can again be transformed away by renormalization of $t^{\nu\mu\alpha\beta}$.

We so obtain the following Ward-Takahashi identities

$$\partial_{1\nu} t^{\nu\mu\alpha\beta}=0=\dots=\partial_{4\beta} t^{\nu\mu\alpha\beta}. \tag{4.6.35}$$

Case III: $(f,l)=(1,1), \omega=0$

This is the most interesting case, because different classes of diagrams must be combined. We have one external photon operator that is now attached to x_n, which is the differentiation variable in agreement with (4.6.21). The decomposition of (4.6.21) leads to the following classes of diagrams: (a) the vertex function with external Fermi operators $\overline{\psi}(x_i)\psi(x_j), 1\le i\ne j\le n-1$, (b) taking (4.6.20) into account, we have to include reducible diagrams containing the self-energy Σ, (c) there is an additional class of reducible diagrams containing the vacuum polarization tensor $\Pi^{\mu\nu}$ (Fig. 9). The diagrams (b) and (c) have external Fermi operators $\overline{\psi}(x_i)\psi(x_i)$.

We write down the anomaly relation for the t-distribution

$$\sum_{i\ne j}^{n-1}\partial_{n\nu}\Big[:\overline{\psi}(x_i)\Lambda^\nu(x_n,x_i,x_j;\dots)\psi(x_j):$$
$$+:\overline{\psi}(x_n)\gamma^\nu S_F(x_n-x_i)\Sigma(x_i,x_j;\dots)\psi(x_j):$$
$$+:\overline{\psi}(x_i)\Sigma(x_i,x_j;\dots)S_F(x_j-x_n)\gamma^\nu\psi(x_n):$$
$$+:\overline{\psi}(x_i)\Pi^{\nu\mu}(x_n,x_j;\dots)D_F(x_j-x_i)\gamma_\mu\psi(x_i):\Big]$$
$$=\sum_{i\ne j}^{n-1}:\overline{\psi}(x_i)\Big(K_0\mathbf{1}+K_1\partial\!\!\!/_i+K_1'\partial\!\!\!/_j+K_n\partial\!\!\!/_n\Big)\delta^{n-1}\psi(x_j):. \tag{4.6.36}$$

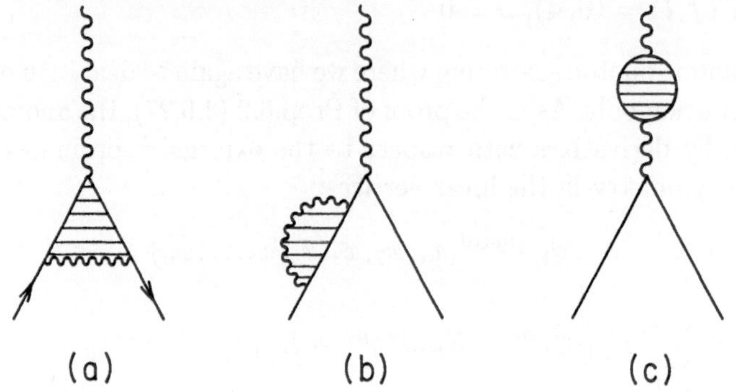

Fig. 9. Diagrams contributing to the 3-point Ward identity

Here we have used the fact that the l.h.s. is symmetrical in $i, j = 1, \dots n-1$, so must be the anomaly on the r.h.s. We test this relation in the variables $x_1, \dots x_{n-1}$ with translation invariant test functions:

$$\sum_{i\neq j}^{n-1} \int \Big[: \overline{\psi}(x_i) \partial_{n\nu} \Lambda^\nu (x_n, x_i, x_j; \dots) \psi(x_j) :$$

$$+ i : \overline{\psi}(x_n) \delta(x_n - x_i) \Sigma(x_i, x_j; \dots) \psi(x_j) : - i : \overline{\psi}(x_i) \Sigma(x_i, x_j; \dots) \delta(x_j - x_n)$$

$$\times \psi(x_n) : \Big] \varphi(x_1 - x_n, \dots, x_{n-1} - x_n)\, dx_1 \dots dx_{n-1}$$

$$= \sum_{i\neq j}^{n-1} \int : \overline{\psi}(x_i) \Big(K_0 \mathbf{1} + K_1 \not\partial_i + K_1' \not\partial_j + K_n \not\partial_n \Big) \delta^{n-1} \psi(x_j) :$$

$$\times \varphi(x_1 - x_n, \dots, x_{n-1} - x_n)\, dx_1 \dots dx_{n-1}, \tag{4.6.37}$$

where (4.6.20, 21) and the Ward identity (4.6.32) in order $(n-1)$ have been taken into account.

Proposition 6.3. The anomaly in (4.6.37) can be restricted to

$$K \sum : \overline{\psi}(x_i) \not\partial_n \delta^{n-1} \psi(x_j) : . \tag{4.6.38}$$

Proof. The term with $\not\partial_i \delta$ can be transformed by means of the Dirac equation as follows

$$\sum_i \int : \overline{\psi}(x_i) \not\partial_i \delta^{n-1} \psi(x_j) : \varphi(x_1 - x_n, \dots, x_{n-1} - x_n)\, dx_1 \dots dx_n =$$

$$= -\sum_i :\overline{\psi}(x_n)\gamma^\mu\psi(x_n): \, (\partial_{i\mu}\varphi)(0,\dots 0)$$

$$+im(n-1) :\overline{\psi}(x_n)\psi(x_n): \, \varphi(0,\dots 0).$$

Translation invariance of the test functions implies

$$(\partial_1 + \ldots + \partial_{n-1})\varphi(x_1 - x_n, \ldots, x_{n-1} - x_n) = -(n-1)\partial_n\varphi.$$

Hence, since

$$\sum_{i\neq j} = \sum_{i,j=1}^{n-1} - \sum_{i=j=1}^{n-1},$$

all $\partial\!\!\!/\delta$ terms can be substituted by $\partial\!\!\!/_n\delta$ plus the mass terms. The latter may be included in K_0 in (4.6.37). Then the anomaly assumes the form

$$\sum_{i\neq j} :\overline{\psi}(x_i)(K_0'\mathbf{1} + K\partial\!\!\!/_n)\delta^{n-1}\psi(x_j):. \tag{4.6.39}$$

Now we use charge conjugation invariance (4.4.37)

$$U_cT_n(x_1,\ldots,x_n)U_c^{-1} = T_n(x_1,\ldots,x_n). \tag{4.6.40}$$

Since (4.6.36) is multiplied by the photon operator $A(x_n)$, which changes sign under charge conjugation, the anomaly (4.6.39) must also change sign if the C-conjugated term of (4.6.39) is enclosed. This implies $K_0' = 0$, which completes the proof of the proposition. □

The remaining anomaly (4.6.38) can be transformed away by renormalization of the vertex function Λ^ν. Then we arrive at

$$\sum_{i\neq j}^{n-1} \int :\overline{\psi}(x_i)\Big[\partial_{n\nu}\Lambda^\nu(x_n, x_i, x_j;\ldots)$$

$$+i\delta(x_n - x_i)\Sigma(x_i, x_j;\ldots) - i\Sigma(x_i, x_j;\ldots)\delta(x_j - x_n)\Big]\psi(x_j):$$

$$\times\, \varphi(x_1 - x_n, \ldots, x_{n-1} - x_n)\, dx_1 \ldots dx_{n-1} = 0. \tag{4.6.41}$$

To get rid of the Dirac field operators, we consider the one-electron states

$$\Phi_l = (2\pi)^{3/2} \int \varphi_l(\boldsymbol{k}) b^+(\boldsymbol{k})\Omega\, d^3k, \quad \varphi_l \in C_0^\infty(\mathbb{R}^3). \tag{4.6.42}$$

Then

$$\psi(x)\Phi_l = \int d^3k\, \varphi_l(\boldsymbol{k}) u(\boldsymbol{k}) e^{-ikx}\Omega = f_l(x)\Omega, \tag{4.6.43}$$

where $f_l(x)$ is a smooth solution of the Dirac equation (also called regular wave packet, see *M. Reed, B. Simon, Modern Methods of Mathematical Physics, vol. III, Academic Press 1975, p. 42*). Abbreviating the square bracket in (4.6.41) by $t(\ldots)$, the equation is now equivalent to

$$\int \left[\sum_{i \neq j}^{n-1} \bar{f}_l(x_i) t(x_n, x_i, x_j; \ldots) g_m(x_j) \right]$$

$$\times \varphi(x_1 - x_n, \ldots x_{n-1} - x_n) dx_1 \ldots dx_{n-1} = 0 \qquad (4.6.44)$$

for arbitrary smooth solutions f_l, g_m.

If there are two external arguments x_i, x_j in $x_1, \ldots x_{n-1} \in \operatorname{supp} t$ which are different from the rest, then there exist neighbourhoods U_i of x_i and U_j of x_j and smooth solutions f_l that do not vanish in U_i and similarly solutions g_m which do not vanish in U_j. Since these solutions can be freely varied in their dependence on the spatial variables $\boldsymbol{x}_i, \boldsymbol{x}_j$, (4.6.44) now implies that the restriction of $t(x_n, x_i, x_j; \ldots)$ to $U_i \times U_j$ is 0.

Suppose next that one external argument, say x_1, coincides with an inner variable. Then it is possible to construct terms $t \neq 0$ with all required symmetries which, nevertheless, give 0 in (4.6.44), for example

$$t_{12} = \sum_{\pi \in S_{n-3}} \Big[\not{\partial}_1^\nu \delta(x_1 - x_{\pi 3}) + \not{\partial}_2^\nu \delta(x_2 - x_{\pi 3})$$

$$+ \frac{i}{2} m \delta(x_1 - x_{\pi 3}) - \frac{i}{2} m \delta(x_2 - x_{\pi 3}) \Big] t_0(x_1 - x_n, \ldots x_{n-1} - x_n), \qquad (4.6.45)$$

where $x_i = x_1, x_j = x_2$ has been assumed and t_0 is symmetrical in all arguments; but x_1 and x_2 can be substituted by inner variables, taking the δ-distributions into account. Here the square bracket has the form of normalization terms for the self energy Σ. Hence, the first term in the sum (4.6.45) corresponds to renormalization of the second order $\Sigma_2(x_3 - x_1)$ in the diagrams of Fig. 10. But the sum of these three terms is 0 by the Ward identity of order $n-2$. Consequently, t_0 in (4.6.45) must vanish. Obviously, all possible semi-local terms correspond to renormalization in lower orders and, therefore, vanish in virtue of the lower order Ward identities.

Finally, t can be a local distribution of the anomalous type (4.6.38) discussed above. Since such anomalies have already been removed by renormalization of the vertex function, it follows that $t = 0$. This leads to the following Ward identity ($i = 1, j = 2$)

$$\partial_{n\nu} \Lambda^\nu(x_1 - x_n, x_2 - x_n; x_3 - x_n, \ldots)$$

$$+ i\delta(x_1 - x_n)\Sigma(x_1 - x_2; x_3 - x_2 \ldots) - i\delta(x_2 - x_n)\Sigma(x_1 - x_2; x_3 - x_2, \ldots) = 0. \qquad (4.6.46)$$

Here translation invariance has again been used. In accordance with our previous conventions, the splitting vertex of Λ^ν is the photon vertex x_n, whereas in $\Sigma(x_1, x_2; \ldots)$ it is the second argument x_2.

We now calculate the Fourier transform in the difference variables $y_j = x_j - x_n$. We must pay attention in the second term, because the different arguments $x_j - x_2 = y_j - y_2$ appear. We arrive at the most important Ward identity

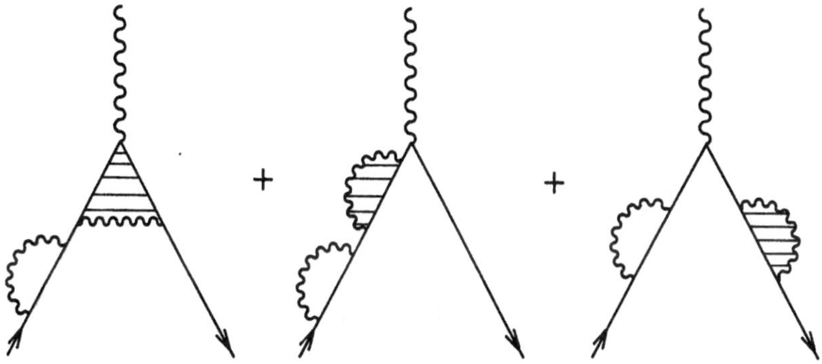

Fig. 10. Diagrams giving rise to semi-local normalization terms

$$(p_1 + \ldots + p_{n-1})_\nu \hat{\Lambda}^\nu(p_1, p_2; p_3 \ldots p_{n-1})$$

$$+\frac{1}{(2\pi)^2}\Big[\hat{\Sigma}(-p_2 - p_3 - \ldots - p_{n-1}; p_3, \ldots p_{n-1}) - \hat{\Sigma}(p_1; p_3, \ldots p_{n-1})\Big] = 0. \tag{4.6.47}$$

For $n = 3$ it assumes the simple form

$$(p_1 + p_2)_\nu \tilde{\Lambda}^\nu(p_1, p_2) + \frac{i}{(2\pi)^2}\Big[\hat{\Sigma}(-p_2) - \hat{\Sigma}(p_1)\Big] = 0.$$

However, in the third order calculation the Fourier transformation was defined with a different sign in the second argument (3.8.23). With the substitutions

$$p_1 \to p, \quad -p_2 \to q, \quad \tilde{\Lambda}^\nu(p, -q) \to \hat{\Lambda}^\nu(p, q)$$

we then arrive at the desired Ward identity

$$(p - q)_\nu \hat{\Lambda}^\nu(p, q) = \frac{1}{(2\pi)^2}\Big[\hat{\Sigma}(p) - \hat{\Sigma}(q)\Big] = 0. \tag{4.6.48}$$

It has repeatedly been used before. A gauge invariant renormalization is now given by

$$\hat{\Lambda}^\nu \to \hat{\Lambda}^\nu + C\gamma^\nu$$

$$\hat{\Sigma}(p) \to \hat{\Sigma}(p) - (2\pi)^2 C \not{p} + C_0, \tag{4.6.49}$$

because it drops out in (4.6.48).

Case IV: $(f, l) = (1, 2), \omega = -1$

In this case we have two external Fermi and two photon operators. If the former have coordinates x_i, x_j and the latter x_n and x_{n-1}, the anomaly relation for the t-distribution is of the following form:

$$\sum_{i\neq j} \partial_{n\nu}\Big[:\overline{\psi}(x_i) t_4^{\nu\mu}(\ldots)\psi(x_j): + \text{reducible terms}\Big]$$

$$= \sum_{i\neq j} K_{ij} :\overline{\psi}(x_i)\gamma^\mu\psi(x_j): \delta^{n-1}. \tag{4.6.50}$$

Here $t_4^{\nu\mu}$ is the contribution of the irreducible diagrams, the reducible terms may have other arguments in the spinor operators, similar to (4.6.38). It is not necessary to write these terms down in detail. Since $\omega = -1$, we have no possibility to remove the anomaly by renormalization. Consequently, gauge invariance can only be true if the anomaly vanishes in virtue of a certain symmetry. In fact, adding the C-conjugated equations, the l.h.s. is even under charge conjugation, because it is multiplied by two photon operators in the total n-point distribution T_n. The anomalous terms on the r.h.s. add up to one local term with support $x_i = \ldots = x_j = x_n$. The latter is odd under charge conjugation, hence, the factor K in front must vanish. This completes the proof of gauge invariance.

4.7 Unitarity

The most important property of the S-matrix, regarding its physical interpretation, is unitarity. We cannot expect unitarity to hold on the entire Fock space in QED, because the scalar potential $A^0(x)$ is not hermitian contrary to the vector potential (see Sect. 2.11). On the other hand, the transition probabilities for all possible final states must add up to one in every scattering process. This is guaranteed by unitarity on the physical subspace $\mathcal{F}_{\mathbf{phys}}$ which has been introduced in Sect. 2.11. This physical unitarity reads as follows

$$\lim_{g\to 1} PS(g)^+ PS(g)P = P, \tag{4.7.1}$$

where P stands for the projection operator onto $\mathcal{F}_{\mathbf{phys}}$. It is our aim now to prove a perturbative version of (4.7.1).

The physical Fock space $\mathcal{F}_{\mathbf{phys}}$ differs from the entire $\mathcal{F}$ only in the photon sector, which in $\mathcal{F}_{\mathbf{phys}}$ is generated by transverse photon operators

$$A_\perp^\mu(x) = PA^\mu(x)P$$

$$= (2\pi)^{-3/2} \int \frac{d^3k}{\sqrt{2\omega}} \sum_{n=1}^{2} \varepsilon_n^\mu(\boldsymbol{k})\Big[a_n(\boldsymbol{k})e^{-ikx} + a_n(\boldsymbol{k})^+ e^{ikx}\Big]. \tag{4.7.2}$$

Here

$$\varepsilon_n^\mu(\boldsymbol{k}) = (0, \boldsymbol{\varepsilon}_n) \quad , \quad \boldsymbol{k} \cdot \boldsymbol{\varepsilon}_n(\boldsymbol{k}) = 0 \quad , \quad n = 1, 2$$

$$\boldsymbol{\varepsilon}_n \cdot \boldsymbol{\varepsilon}_m = \delta_{nm} \tag{4.7.3}$$

are two transverse orthonormal and real polarization vectors and (cf.(3.12.42))

$$a_n(\boldsymbol{k}) = \varepsilon_n^\nu(\boldsymbol{k}) a^\nu(\boldsymbol{k}). \tag{4.7.4}$$

In order to learn what is going on, let us investigate the lowest order terms of the expression

$$PS(g)^+PS(g)P = P + \int P\big[T_1(x) + T_1(x)^+\big]Pg(x)dx$$

$$+ \int PT_1(x_1)^+PT_1(x_2)Pg(x_1)g(x_2)dx_1dx_2$$

$$+\frac{1}{2}\int P\big[T_2(x_1,x_2)^+ + T_2(x_1,x_2)\big]Pg(x_1)g(x_2)dx_1dx_2 + \dots \tag{4.7.5}$$

For (4.7.1) to be valid, all terms on the right side except P must vanish. The first order term vanishes indeed, because

$$T_1(x) = ie : \overline{\psi}(x)\gamma^\mu\psi(x) : A_\mu(x) \tag{4.7.6}$$

is skew-adjoint on $\mathcal{F}_{\rm phys}$. Concerning the second order terms, we concentrate on the electron scattering graph (3.4.5)

$$T_2^{(1)}(x_1,x_2) = -ie^2 : \overline{\psi}(x_1)\gamma^\mu\psi(x_1)\overline{\psi}(x_2)\gamma_\mu\psi(x_2) : D_0^F(x_1 - x_2). \tag{4.7.7}$$

It contributes

$$\frac{1}{2}\big[T_2^{(1)}(x_1,x_2)^+ + T_2^{(1)}(x_1,x_2)\big] = e^2 : \overline{\psi}(x_1)\gamma^\mu\psi(x_1)$$

$$\times\, \overline{\psi}(x_2)\gamma_\mu\psi(x_2) : \operatorname{Im} D_0^F(x_1 - x_2) \tag{4.7.8}$$

to (4.7.5) with

$$\operatorname{Im} D_0^F(x) = (2\pi)^{-4}\pi \int d^4k\, e^{-ikx}\delta(k^2). \tag{4.7.9}$$

The term (4.7.8) should be compensated by some contribution from the $PT_1^+PT_1P$ term. As usual, the product is normally ordered here by means of Wick's theorem 3.1

$$PT_1(x_1)^+PT_1(x_2)P = e^2 : \overline{\psi}(x_1)\gamma_\mu\psi(x_1)\overline{\psi}(x_2)\gamma_\nu\psi(x_2) :$$

$$\times \big[: A_\perp^\mu(x_1)A_\perp^\nu(x_2) : + \overline{A_\perp^\mu(x_1)A_\perp^\nu(x_2)}\big] + \dots \tag{4.7.10}$$

The pairing of the transverse field operators (4.7.2) is given by

$$\overline{A_\perp^\mu(x_1)A_\perp^\nu(x_2)} = (2\pi)^{-3}\int \frac{d^3k}{2\omega}\, e^{-ik(x_1-x_2)} \sum_{n=1}^{2} \varepsilon_n^\mu(\boldsymbol{k})\varepsilon_n^\nu(\boldsymbol{k}). \qquad (4.7.11)$$

Here we use the relation (3.12.44)

$$\sum_{n=1}^{2} \varepsilon_n^\mu(\boldsymbol{k})\varepsilon_n^\nu(\boldsymbol{k}) = -g^{\mu\nu} + \eta^\mu\eta^\nu$$

$$-\frac{1}{\boldsymbol{k}^2}\Big[k_0^2\eta^\mu\eta^\nu + k^\mu k^\nu - k_0(k^\mu\eta^\nu + k^\nu\eta^\mu)\Big], \qquad (4.7.12)$$

where

$$\eta^\mu = (1,0,0,0) \quad , \quad \hat{k}^\mu = \frac{k^\mu - (k^\nu\eta_\nu)\eta^\mu}{|\boldsymbol{k}|} \qquad (4.7.13)$$

are time-like and longitudinal polarization vectors.

From $-g^{\mu\nu}$ we get the covariant distribution

$$d_0^{(+)}(x) \stackrel{\text{def}}{=} -(2\pi)^{-3}\int \frac{d^3k}{2\omega} e^{-ikx}\Big|_{k_0=\omega}$$

$$= -(2\pi)^{-3}\int d^4k\, \delta(k_0^2 - \boldsymbol{k}^2)\Theta(k_0)e^{-ikx}. \qquad (4.7.14)$$

It is important to note that only the part symmetric in x_1, x_2 contributes to (4.7.5). Therefore, instead of (4.7.14), we must consider

$$\frac{1}{2}\big[d_0^{(+)}(x) + d_0^{(+)}(-x)\big] = -\frac{1}{2}(2\pi)^{-3}\int d^4k\, \delta(k^2)e^{-ikx}. \qquad (4.7.15)$$

This is just the negative of $\operatorname{Im} D_0^F(x)$ (4.7.9), consequently this term compensates (4.7.8), as it should. There remains to discuss the remaining non-covariant terms in (4.7.12).

The distribution multiplied by $\eta^\mu\eta^\nu$ in (4.7.11) is equal to

$$\frac{1}{2}(2\pi)^{-3}\int d^4k\, \delta(k^2)e^{-ikx}\Big(1 - \frac{k_0^2}{\boldsymbol{k}^2}\Big) = 0. \qquad (4.7.16)$$

The other terms in (4.7.12) contain at least one k^μ, which leads to a gradient in x-space. Since in (4.7.10) the distribution is multiplied by conserved current densities, the whole expression is a divergence. Then, by partial integration in (4.7.5), the derivative is shifted to the test function g. For $g = 1$ there is no contribution. Hence unitarity is satisfied in the lowest orders as a consequence of current conservation.

In order to prove the same result in all orders, we must again go through the inductive construction. Instead of the simple current conservation we will have to use gauge invariance. The foregoing discussion of the pairing of transverse photon operators may be summarized in the equation

$$\overline{A^\mu(x_1)A^\nu(x_2)} = \overline{A^\mu_\perp(x_1)A^\nu_\perp(x_2)} + \mathrm{grad}\,, \tag{4.7.17}$$

where grad stands for the gradient terms found above. This relation implies the following basic

Lemma 7.1. For disjoint sets X_1, X_2 of points, we have

$$PT_{n_1}(X_1)T_{n_2}(X_2)P = PT_{n_1}(X_1)PT_{n_2}(X_2)P + \mathrm{div}\quad, \tag{4.7.18}$$

where div stands for operator-valued distributions which are a divergence.

Proof. The left side of (4.7.18) and the product written down on the right side are normally ordered by means of Wick's theorem. In the normal products all emission and absorption operators for scalar and longitudinal photons are annihilated by the projection operators P. There survive only transversal photon operators, hence the resulting normal products are the same on both sides. The two results differ only because the pairings of the photon operators are different (4.7.17). Taking gauge invariance into account, the gradient terms in (4.7.17) give rise to divergence terms in (4.7.18). □

In the same way the following slightly more general lemma is proven:

Lemma 7.2. If all X_n, $n = 1, \ldots r$, are disjoint, then

$$PT_{n_1}(X_1)T_{n_2}(X_2)\ldots T_{n_r}(X_r)P = PT_{n_1}(X_1)PT_{n_2}(X_2)\ldots T_{n_r}(X_r)P + \mathrm{div}\,. \tag{4.7.19}$$

Before proving the physical unitarity (4.7.1), we have to show another related symmetry property, namely pseudo-unitarity:

$$S(g)^{-1} = S(g)^K. \tag{4.7.20}$$

The conjugation K has already appeared in (4.5.2). Since the radiation field is self-conjugate

$$A^\mu(x)^K = A^\mu(x), \tag{4.7.21}$$

we obviously have

$$\tilde{T}_1(x)^K = -T_1(x)^K = T_1(x). \tag{4.7.22}$$

We prove inductively

$$\tilde{T}_m(X)^K = T_m(X) \tag{4.7.23}$$

for $m = 1, 2, \ldots$, which is the perturbative version of (4.7.20). Let us assume that (4.7.23) holds for all $m \leq n-1$. As usual we then form $R'_n(x_1, \ldots x_n)$ and in addition

$$R''_n(x_1, \ldots x_n) = \sum_{P_2} T_{n_1}(X)\tilde{T}_{n-n_1}(Y, x_n). \tag{4.7.24}$$

Taking the conjugate (K) of R'_n and using the induction hypothesis (4.7.23), we have

$$R'^K_n = \sum_{P_2} T_{n_1}(X)\tilde{T}_{n-n_1}(Y, x_n) = R''_n. \tag{4.7.25}$$

The advanced distributions

$$A'_n(x_1, \ldots x_n) = \sum_{P_2} \tilde{T}_{n_1}(X)T_{n-n_1}(Y, x_n),$$

$$A''_n(x_1, \ldots x_n) = \sum_{P_2} \tilde{T}_{n-n_1}(Y, x_n)T_{n_1}(X)$$

are handled similarly:

$$A'^K_n = A''_n. \tag{4.7.26}$$

The total retarded distribution (3.1.36)

$$R_n(x_1, \ldots x_n) = \sum_{P_2^0} T_{n-n_1}(Y, x_n)\tilde{T}_{n_1}(X)$$

$$= R'_n + T_n \tag{4.7.27}$$

can also be written as follows

$$R_n = -\sum_{P_2^0} T_{n_1}(X)\tilde{T}_{n-n_1}(Y, x_n), \tag{4.7.28}$$

taking (3.1.11) into account. This implies

$$R_n = R'_n + T_n = -R''_n - \tilde{T}_n \tag{4.7.29}$$

and similarly for A_n:

$$A_n = A'_n + T_n = -A''_n - \tilde{T}_n. \tag{4.7.30}$$

In the inductive procedure we have to form

$$D_n = R'_n - A'_n = R_n - A_n = -R''_n + A''_n. \tag{4.7.31}$$

From (4.7.25) and (4.7.26) we conclude that D_n is skew-conjugate

$$D_n^K = -D_n. \tag{4.7.32}$$

Since this property is independent of the argument $(x_1, \ldots, x_n)$ of the distributions, it is possible to maintain it in the distribution splitting, provided the local terms are chosen appropriately. In fact, if R_n is an arbitrary retarded solution, we form

$$\bar{R}_n = \tfrac{1}{2}[R_n - R_n^K], \tag{4.7.33}$$

then we have

$$\bar{R}_n^K = -\bar{R}_n. \tag{4.7.34}$$

It now follows from (4.7.29) that

$$\tilde{T}_n = -R_n - R_n''$$

and by (4.7.25, 34) and (4.7.29)

$$\tilde{T}_n^K = R_n - R_n' = T_n. \tag{4.7.35}$$

This proves the induction assumption (4.7.23) for $m = n$ and completes the proof of pseudo-unitarity.

We finally come to the physical unitarity (4.7.1). It involves the inverse of

$$PS(g)P = \sum_n \frac{1}{n!} \int d^4x_1 \dots d^4x_n \, PT_n(x_1, \dots, x_n)Pg(x_1)\dots g(x_n) \tag{4.7.36}$$

on the physical subspace $P\mathcal{F}$. In virtue of (3.1.4–6), this inverse has the following perturbation expansion

$$(PS(g)P)^{-1} = \sum_n \frac{1}{n!} \int d^4x_1 \dots d^4x_n \, \tilde{T}_n^P(x_1, \dots, x_n)g(x_1)\dots g(x_n) \tag{4.7.37}$$

with

$$\tilde{T}_n^P(X) = \sum_{r=1}^n (-)^r \sum_{P_r} PT_{n_1}(X_1)P \dots PT_{n_r}(X_r)P, \tag{4.7.38}$$

according to (3.1.6). Applying Lemma 7.2, we arrive at

$$\begin{aligned} \tilde{T}_n^P(X) &= \sum_{r=1}^n (-)^r \sum_{P_r} PT_{n_1}(X_1)\dots T_{n_r}(X_r)P - \mathrm{div} \\ &= P\tilde{T}_n(X)P - \mathrm{div} \\ &= PT_n(X)^K P - \mathrm{div}\,, \end{aligned} \tag{4.7.39}$$

where (4.7.35) was used in the last equality. The conjugation K agrees with the adjoint if it occurs between projection operators P (cf. (2.11.48)), hence

$$\tilde{T}_n^P(X) = PT_n(X)^+ P - \mathrm{div}\,. \tag{4.7.40}$$

This is the desired perturbative version of the physical unitarity (4.7.1), because the divergence terms give no contribution if integrated with $g = 1$.

Our discussion of the simulation of charged particles by the adiabatic limit in Sect. 3.12 has consequences for the structure of the physical subspace $\mathcal{H}_{\text{phys}}$ of QED. Since the adiabatic switching gives rise to unusual asymptotic electromagnetic fields, the latter must be included in $\mathcal{H}_{\text{phys}}$. We therefore define the photonic sector in $\mathcal{H}_{\text{phys}}$ as follows: For $|\boldsymbol{k}| \geq \omega_0$ it contains transversal photons only, but for $|\boldsymbol{k}| < \omega_0$ all four polarizations are included in $\mathcal{H}_{\text{phys}}$. ω_0 is the minimal energy resolution of the photon detectors. In fact, we are forced to modify $\mathcal{H}_{\text{phys}}$ in this way if we want to maintain unitarity of the S-matrix in the sense (4.7.1):

$$\lim_{g\to 1} PS(g)^+ PS(g)P = P, \tag{4.7.41}$$

where P is now the projection operator on the new $\mathcal{H}_{\text{phys}}$. To see this let us investigate (4.7.41) in forth order of perturbation theory.

The forth order terms of (4.7.41) are equal to

$$\frac{1}{4!}\int P[T_4^+ + T_4]Pg(x_1)g(x_2)g(x_3)g(x_4) + \frac{1}{3!1!}\int [PT_3^+ PT_1 P$$

$$+PT_1^+ PT_3 P]g(x_1)\cdots g(x_4) + \frac{1}{(2!)^2}\int PT_2^+ PT_2 Pg(x_1)\cdots g(x_4). \tag{4.7.42}$$

Again let us consider electron scattering on an external potential and take the matrix elements of (4.7.42) between the same initial and final state Φ_i. From the last term we get

$$(\Phi_i, T_2^+ PT_2\Phi_i) = \sum_{\Phi\in\mathcal{H}_{\text{phys}}} (\Phi_i, T_2^+\Phi)(\Phi, T_2\Phi_i)$$

$$= \sum_{\Phi\in\mathcal{H}_{\text{phys}}} |(\Phi, T_2\Phi_i)|^2. \tag{4.7.43}$$

Here the sum over a complete set of states Φ contains the integral over soft photons of bremsstrahlung that diverges in the limit $g \to 1$. This contribution is just the inclusive electron cross section. We therefore look for the same infrared compensation as before (Sect. 3.11). From the second term in (4.7.42) we have

$$\sum_{\Phi}[(\Phi_i, T_3^+\Phi)(\Phi, T_1\Phi_i) + (\Phi_i, T_1^+\Phi)(\Phi, T_3\Phi_i)]$$

$$= 2\text{Re}\sum_{\Phi}(\Phi, T_1\Phi_i)^*(\Phi, T_3\Phi_i). \tag{4.7.44}$$

This contains exactly the contributions coming from the third order vertex and self-energy diagrams in the computation of the cross section $d\sigma^{V+S}/d\Omega$ (3.11.24) (3.12.7), because

$$\sum_{\Phi}|(\Phi, (T_1+T_3)\Phi_i)|^2 = \sum_{\Phi}\Big\{|(\Phi, T_1\Phi_i)|^2 + |(\Phi, T_3\Phi_i)|^2\Big\} + (4.7.44). \tag{4.7.45}$$

Consequently, the same compensation that guarantees finiteness and uniqueness of the cross-section takes place in the unitarity relation (4.7.42). Furthermore, the dependence on the switching function must drop out on the left-hand side of (4.7.41). As emphasized above, this requires summation over four polarizations of soft bremsstrahlung for Φ. These states must therefore be included in $\mathcal{H}_{\text{phys}}$. For frequencies $|\boldsymbol{k}| > \omega_0 > 0$ summation over four or two polarizations gives the same result in the adiabatic limit. This is a consequence of current conservation: Replacing ε^μ in (3.11.27) by k^μ we obtain

$$\frac{pk}{p(k-\varepsilon k_2)} - \frac{qk}{q(k-\varepsilon k_2)} = O(\varepsilon),$$

and this vanishes in the adiabatic limit $\varepsilon \to 0$.

It is clear from the above discussion that the strong unitarity relation (4.7.41) implies infrared cancellations in all orders of perturbation theory. To prove this relation in general is, therefore, an important aim. It is not hard to prove the above weaker form of perturbative unitarity (4.7.39) again. We need the modified contraction of the photon field operator in the physical subspace

$$A^\mu_{\text{phys}}(x) = PA^\mu(x)P, \tag{4.7.46}$$

that is now equal to

$$\overline{A^\mu_{\text{phys}}(x)A^\nu_{\text{phys}}(y)} = (2\pi)^{-3}\int \frac{d^3k}{2\omega} e^{-ik(x-y)}$$

$$\times \left\{ \sum_{n=1}^{2} \varepsilon^\mu_n(\boldsymbol{k})\varepsilon^\nu_n(\boldsymbol{k}) + \Theta(\omega_0 - |\boldsymbol{k}|) \sum_{n=3}^{4} \varepsilon^\mu_n(\boldsymbol{k})\varepsilon^\nu_n(\boldsymbol{k}) \right\}, \tag{4.7.47}$$

where ε^μ_n are four orthogonal, real polarization vectors. Using the completeness of these vectors in Minkowski space (3.12.44), one gets

$$= (2\pi)^{-3}\int d^4k\, \delta(k^2)\Theta(k^0) e^{-ik(x-y)}$$

$$\times \left\{ -g^{\mu\nu} - \Theta(|\boldsymbol{k}| - \omega_0)\frac{1}{\boldsymbol{k}^2}\Big[k^\mu k^\nu - k_0(k^\mu\eta^\nu + k^\nu\eta^\mu)\Big] \right\}, \tag{4.7.48}$$

where $\eta^\mu = (1,0,0,0)$ is the time-like polarization vector. Since the non-covariant terms give rise to a gradient in x-space, one arrives at

$$\overline{A^\mu_{\text{phys}}(x)A^\nu_{\text{phys}}(y)} = \overline{A^\mu(x)A^\nu(y)} + \text{grad}. \tag{4.7.49}$$

The relation (4.7.39) now follows in exactly the same way as before. However, in contrast to (4.7.40) we cannot substitute the conjugation K by the adjoint here, because the physical subspace $P\mathcal{F} = \mathcal{H}_{\text{phys}}$ now contains a little bit of scalar photons.

4.8 Renormalization Group

We have observed at the end of Sect. 3.13 (3.13.16), that the value of the physical charge or fine-structure constant $\alpha = e^2/4\pi$ is determined by the normalization of the vacuum polarization tensor. To study a possible change of normalization, we therefore start from the photon propagator (2-legs distribution) with all vacuum polarization insertions included, as we have discussed it in (3.6.33):

$$T_{AA}(x_1, x_2) = -i : A_\mu(x_1)\Pi^{\mu\nu}(x_1 - x_2)A_\nu(x_2) : \tag{4.8.1}$$

$$\hat{\Pi}^{\mu\nu}(k) = \frac{g^{\mu\nu} - \frac{k^\mu k^\nu}{k^2}}{(k^2 + i0)\left(1 - \frac{\Pi(k)}{k^2}\right)}. \tag{4.8.2}$$

Here $\Pi(k)$ is the scalar vacuum polarization function in arbitrary order of perturbation theory (with all internal momenta $= 0$, due to the adiabatic limit). It is normalized at $k^2 = 0$

$$\Pi(k)|_{k^2=0} = 0, \quad \frac{\Pi(k)}{k^2}\bigg|_{k^2=0} = 0. \tag{4.8.3}$$

A (finite) renormalization of $\Pi(k)$ is given by

$$\frac{\Pi(k)}{k^2} \longrightarrow \frac{\Pi(k)}{k^2} + C. \tag{4.8.4}$$

Writing $C = -\Pi(\lambda)/\lambda^2$, with $\lambda^2 < 0$, we see that this corresponds to normalization (4.8.3) at the space-like point $k = \lambda$, which is the so-called momentum (MOM) subtraction scheme. However, we retain the mass dependence in Π which we want to control. When $\hat{\Pi}^{\mu\nu}(k)$ (4.8.2) is used for an inner photon contraction, it gets multiplied by e^2. Consequently, the change of normalization (4.8.4) can be compensated by a change $\alpha \to \alpha_\lambda$ according to

$$\frac{\alpha}{1 - \frac{\Pi(k,\alpha)}{k^2}} = \frac{\alpha_\lambda}{1 - \frac{\Pi(k,\alpha_\lambda)}{k^2} + \frac{\Pi(\lambda,\alpha_\lambda)}{\lambda^2}}. \tag{4.8.5}$$

Then the physical charge remains unchanged, as it must be. The additive renormalization (4.8.4) is converted into a multiplicative renormalization by summing up all vacuum polarization bubbles, as in (3.6.27). The denominator in (4.8.5) is usually denoted by $Z_3(k^2/\lambda^2, m^2/\lambda^2, \alpha_\lambda)$, where m is the electron mass, and α_λ is used as coupling constant instead of α (running coupling constant). Since Z_3 is dimensionless, it can only depend on dimensionless variables. For two normalization points λ, λ' we now have

$$\frac{\alpha}{1 - \frac{\Pi(k)}{k^2}} = \frac{\alpha_\lambda}{Z_3\left(\frac{k^2}{\lambda^2}, \frac{m^2}{\lambda^2}, \alpha_\lambda\right)} = \frac{\alpha_{\lambda'}}{Z_3\left(\frac{k^2}{\lambda'^2}, \frac{m^2}{\lambda'^2}, \alpha_{\lambda'}\right)}. \tag{4.8.6}$$

If $k^2 = \lambda'^2$, we find

$$\alpha_{\lambda'} = \frac{\alpha_\lambda}{Z_3\left(\frac{\lambda'^2}{\lambda^2}, \frac{m^2}{\lambda^2}, \alpha_\lambda\right)} \tag{4.8.7}$$

in virtue of (4.8.5). Substituting back into (4.8.6) gives

$$\begin{aligned} Z_3\left(\frac{k^2}{\lambda^2}, \frac{m^2}{\lambda^2}, \alpha_\lambda\right) &= Z_3\left(\frac{\lambda'^2}{\lambda^2}, \frac{m^2}{\lambda^2}, \alpha_\lambda\right) Z_3\left(\frac{k^2}{\lambda'^2}, \frac{m^2}{\lambda'^2}, \alpha_{\lambda'}\right) \\ &= \frac{\alpha_\lambda}{\alpha_{\lambda'}} Z_3\left(\frac{k^2}{\lambda'^2}, \frac{m^2}{\lambda'^2}, \alpha_{\lambda'}\right). \end{aligned} \tag{4.8.8}$$

This relation (4.8.8) is the starting point for the renormalization group. We introduce the variables

$$\frac{k^2}{\lambda^2} = x, \quad \frac{m^2}{\lambda^2} = y, \quad \frac{\lambda'^2}{\lambda^2} = t, \tag{4.8.9}$$

$$\frac{\alpha_\lambda}{Z_3\left(\frac{k^2}{\lambda^2}, \frac{m^2}{\lambda^2}, \alpha_\lambda\right)} = S. \tag{4.8.10}$$

Taking (4.8.7) into account, (4.8.8) assumes the following form

$$S(x, y, \alpha_\lambda) = S\left(\frac{x}{t}, \frac{y}{t}, S(t, y, \alpha_\lambda)\right). \tag{4.8.11}$$

For varying t, this describes a one-parameter group of renormalization transformations. If $t = x$, we find the condition

$$S\left(1, \frac{y}{x}, \alpha_\lambda\right) = \alpha_\lambda. \tag{4.8.12}$$

It has become customary to differentiate (4.8.11) with respect to various parameters, but only differentiation in t explores the group action. We therefore differentiate at $t = 1$, taking (4.8.12) into account:

$$\left[x\frac{\partial}{\partial x} + y\frac{\partial}{\partial y} - \psi(y, \alpha_\lambda)\frac{\partial}{\partial \alpha_\lambda}\right] S(x, y, \alpha_\lambda) = 0, \tag{4.8.13}$$

where

$$\psi(y, \alpha_\lambda) = \left.\frac{\partial S(x, y, \alpha_\lambda)}{\partial x}\right|_{x=1}. \tag{4.8.14}$$

We integrate this first order equation by the method of characteristics. Since the equation is linear in S, the characteristic system simply becomes

$$\frac{dx}{d\tau} = x, \quad \frac{dy}{d\tau} = y, \quad \frac{d\alpha_\lambda}{d\tau} = -\psi(y, \alpha_\lambda), \quad \frac{dS}{d\tau} = 0. \tag{4.8.15}$$

If we take y instead of τ as parameter along the characteristics, we find the integrals

$$\log\frac{x}{y} = \text{const.}, \quad S(y) = \text{const.} \tag{4.8.16}$$

and there remains the equation

$$y\frac{\partial \alpha_\lambda}{\partial y} = -\psi(y, \alpha_\lambda) \tag{4.8.17}$$

to be solved. Let the solution be implicitly given by $F_1(y, \alpha_\lambda) = \text{const.}$, then the solution of (4.8.13) is given by

$$S(x, y, \alpha_\lambda) = S_0\Big(\frac{y}{x}, F_1(y, \alpha_\lambda)\Big), \tag{4.8.18}$$

where S_0 is a differentiable function of two variables with the property that (4.8.18) can be solved for F_1:

$$F_1(y, \alpha_\lambda) = F_2\Big(\frac{y}{x}, S(x, y, \alpha_\lambda)\Big).$$

For $x = 1$ the right-hand side becomes identical with the left, hence $F_1 = F_2$, or

$$F_1(y, \alpha_\lambda) = F_1\Big(\frac{y}{x}, S(x, y, \alpha_\lambda)\Big). \tag{4.8.19}$$

We have still to verify that $S(x, y, \alpha_\lambda)$, defined implicitly by (4.8.19), actually satisfies (4.8.11). For this purpose we substitute

$$x \to \frac{x}{t}, \quad y \to \frac{y}{t}, \quad \alpha_\lambda \to S(t, y, \alpha_\lambda),$$

then we get

$$F_1\Big(\frac{y}{t}, S(t, y, \alpha_\lambda)\Big) = F_1(y, \alpha_\lambda)$$

$$= F_1\Big(\frac{y}{x}, S\Big(\frac{x}{t}, \frac{y}{t}, S(t, y, \alpha_\lambda)\Big)\Big) = F_1\Big(\frac{y}{x}, S(x, y, \alpha_\lambda)\Big).$$

Assuming that F_1 can be uniquely solved for the second argument in a certain neighbourhood of $t = 1$, this implies (4.8.11). Similarly (4.8.14) is verified.

The function F_1 in (4.8.19) is given as the solution of (4.8.17) which we write in the form

$$\frac{d\log y}{d\alpha_\lambda} = -\frac{1}{\psi(y, \alpha_\lambda)}. \tag{4.8.20}$$

We now assume that ψ can be expanded around $y = 0$, with $\psi(0, \alpha_\lambda) \neq 0$, so that

$$-\frac{1}{\psi(y, \alpha_\lambda)} = -\frac{1}{\psi(0, \alpha_\lambda)} + O(y, \alpha_\lambda), \tag{4.8.21}$$

where $O(y, \alpha_\lambda)$ vanishes linearly for $y \to 0$. This behaviour can be verified in perturbation theory, as will be shown below. Integration of (4.8.20) now leads to

$$\log y = -\int_{\alpha_1}^{\alpha_\lambda} \frac{d\alpha'}{\psi(0, \alpha')} + O_1(y, \alpha_\lambda) + \text{const.},$$

hence

$$F_1(y, \alpha_\lambda) = \log y + \int_{\alpha_1}^{\alpha_\lambda} \frac{d\alpha'}{\psi(0, \alpha')} - O_1(y, \alpha_\lambda). \tag{4.8.22}$$

All functions $O_j(y, \alpha_\lambda), j = 1, 2, \ldots$ vanish linearly in y. Using (4.8.22) in (4.8.19) we arrive at

$$\log y + \int_{\alpha_1}^{\alpha_\lambda} \frac{d\alpha'}{\psi(0, \alpha')} = \log \frac{y}{x} + \int_{\alpha_1}^{S(x,y,\alpha_\lambda)} \frac{d\alpha'}{\psi(0, \alpha')} - O_2(\frac{y}{x}, S) + O_2(y, \alpha_\lambda).$$

Here the singular $\log y$ drops out and the correction terms are $O_3(y, \alpha_\lambda)$ if $x \neq 0$. Then we obtain the Gell-Mann and Low equation

$$\log x = \int_{\alpha_\lambda}^{S(x,y,\alpha_\lambda)} \frac{d\alpha'}{\psi(0, \alpha')} + O_3(y, \alpha_\lambda), \tag{4.8.23}$$

where one usually puts $y = 0$ in the asymptotic region $m^2 \ll -\lambda^2$. But it is not allowed to neglect the mass correction $O(y, \alpha_\lambda)$ in the differential equation (4.8.20) already.

The more important renormalization group equation is the Callan–Symanzik equation. It involves the original fine-structure constant α instead of α_λ, in the so-called on-shell normalization scheme. The two are related by (4.8.5)

$$\alpha_\lambda = \alpha\Big(1 + \frac{\Pi(\lambda)}{\lambda^2}\Big), \tag{4.8.24}$$

where $\Pi(\lambda)$ depends on α and y but not on x. We wish to derive a differential equation for $S(x, y, \alpha_\lambda(\alpha, y)) = \tilde{S}(x, y, \alpha)$. Differentiating (4.8.23) with respect to x with constant y and α, we get

$$\frac{1}{x} = \frac{1}{\psi(0, \tilde{S})} \frac{\partial \tilde{S}(x, y, \alpha)}{\partial x} + O_4(y, \alpha),$$

and differentiating with respect to α for constant x, y gives

$$0 = \frac{1}{\psi(0, \tilde{S})} \frac{\partial \tilde{S}}{\partial \alpha} - \frac{1}{\psi(0, \alpha_\lambda)} \frac{\partial \alpha_\lambda}{\partial \alpha} + O_5(y, \alpha).$$

Eliminating $\psi(0, \tilde{S})$ from these two equations, we arrive at

$$\Big[x \frac{\partial}{\partial x} - \frac{\beta(\alpha)}{2} \frac{\partial}{\partial \alpha}\Big] \tilde{S}(x, y, \alpha) = O_6(y, \alpha), \tag{4.8.25}$$

where

$$\beta(\alpha) = 2 \frac{\psi(0, \alpha_\lambda)}{\frac{\partial \alpha_\lambda}{\partial \alpha}} \tag{4.8.26}$$

is the so-called β-function in the on-shell scheme. The analogous equation for

$$\tilde{S}^{-1} \stackrel{\text{def}}{=} S_1(x,y,\alpha) \tag{4.8.27}$$

is the Callan–Symanzik equation. It is obtained from (4.8.25) by multiplication with $-2\tilde{S}^{-2}$:

$$2x\frac{\partial S_1}{\partial x} - \beta(\alpha)\frac{\partial S_1}{\partial \alpha} = O(y,\alpha). \tag{4.8.28}$$

There remains to study (4.8.21) in the vicinity of $y=0$. Taking (4.8.14) and (4.8.10) into account, we have to expand

$$S = \frac{\alpha}{1-\frac{\Pi(k)}{k^2}} = \alpha\Big(1+\frac{\Pi(k)}{k^2}+\dots\Big). \tag{4.8.29}$$

Here we use our previous result (3.6.22, 29) for the vacuum polarization function

$$\frac{\Pi(k)}{-k^2} = \frac{\alpha}{2\pi}\frac{\xi}{(1-\xi)^2}\left[\frac{1+\xi}{1-\xi}\Big(\xi-4+\frac{1}{\xi}\Big)\log\xi+\frac{5}{3}\xi-\frac{5}{3\xi}-\frac{22}{3}\right]$$

$$\stackrel{\text{def}}{=} f(\xi), \tag{4.8.30}$$

where

$$\frac{-k^2}{m^2} = \frac{(1-\xi)^2}{\xi}. \tag{4.8.31}$$

Since

$$\xi = 1-\frac{k^2}{2m^2}+\frac{k^2}{2m^2}\sqrt{1-\frac{4m^2}{k^2}}$$

$$\approx -\frac{m^2}{k^2} = -\frac{y}{x}, \tag{4.8.32}$$

the limit $y\to 0$ corresponds to $\xi \to 0$ and

$$\frac{\partial}{\partial x} = \frac{y}{x^2}\frac{\partial}{\partial \xi}. \tag{4.8.33}$$

The function $f(\xi)$ (4.8.30) behaves in the vicinity of $\xi=0$ as follows

$$\frac{3\pi}{\alpha}f(\xi) = (1-6\xi^2)\log\xi - \frac{5}{3}-\frac{32}{3}\xi+\dots,$$

so that

$$\frac{3\pi}{\alpha}f'(\xi) = \frac{1}{\xi}(1-6\xi^2) - 12\xi\log\xi - \frac{32}{3}+\dots.$$

Taking (4.8.33) into account, the function which enters into (4.8.21) is given by

$$\xi f'(\xi) = \frac{\alpha}{3\pi}\Big(1-\frac{32}{3}\xi\Big) = \frac{\alpha}{3\pi}+O(y). \tag{4.8.34}$$

This is the desired behaviour in (4.8.21).

We now wish to discuss an important application of the Callan–Symanzik equation (4.8.28) in the asymptotic region, where the correction $O(y,\alpha)$ can be neglected. We substitute the perturbation series

$$S_1(x,\alpha) = \alpha^{-1}\Big[1 + \sum_{m=1}^{\infty} \alpha^m a_m(x)\Big] \tag{4.8.35}$$

$$\beta(\alpha) = \sum_{n=1}^{\infty} b_n \alpha^{n+1} \tag{4.8.36}$$

into (4.8.28). These series are actually power series in $e^2 = 4\pi\alpha$ and not in e, because sucessive orders differ by an internal photon line, which adds two vertices or an additional loop. We now solve (4.8.28) order by order in α:

$$\begin{aligned}
\alpha^0 : &\qquad 2xa_1' = -b_1,\\
a_1(x) &= -\frac{b_1}{2}\log x + \text{const.}\\
\alpha^1 : &\qquad 2xa_2' = -b_2,\\
a_2(x) &= -\frac{b_2}{2}\log x + \text{const.}
\end{aligned} \tag{4.8.37}$$

$$\begin{aligned}
\alpha^2 : &\qquad 2xa_3' = -b_3 + b_1 a_2\\
a_3' &= -\frac{b_1 b_2}{4}\frac{\log x}{x} + \frac{\text{const.}}{x},\\
a_3(x) &= -\frac{b_1}{2}\frac{b_2}{2}\frac{(\log x)^2}{2} + O(\log x).
\end{aligned} \tag{4.8.38}$$

In order α^n, the leading order in $\log x$ follows from

$$2xa_{n+1}' = (n-1)b_1 a_n .$$

By induction we arrive at the following result

$$a_{n+1}(x) = -\left(\frac{b_1}{2}\right)^{n-1}\frac{b_2}{2}\frac{(\log x)^n}{n} + O(\log x)^{n-1}. \tag{4.8.39}$$

Consequently, the leading logarithm in the vacuum polarization function (4.8.34) in arbitrary order is determined by the two lowest orders b_1, b_2 of the β-function (4.8.35) alone. The result (4.8.36) is in accordance with (4.8.30): In fact, the leading logarithm in (4.8.30) for $\xi \to 0$ comes from the term with $\xi^{-1}\log\xi$, it enables us to identify

$$a_1 = \frac{\log\xi}{3\pi} = -\frac{\log x}{3\pi} + \text{const.} \tag{4.8.40}$$

Comparing this with (4.8.36), we find

$$b_1 = \frac{2}{3\pi}. \tag{4.8.41}$$

The second coefficient of the β-function can be found from the forth order vacuum polarization function (*E. de Rafael, J.L. Rosner, Ann. Phys. 82, 369 (1974)*), it is equal to

$$b_2 = \frac{1}{2\pi^2}. \tag{4.8.42}$$

4.9 Interacting Fields and Operator Products

Until now we have exclusively worked with free fields throughout, which describe or rather simulate the asymptotic states in scattering experiments. Interacting fields, on the other hand, are the are the basic objects in the Lagrangian approach to quantum field theory. It is, therefore, a natural question to ask how they can be rigorously defined in the causal theory. This problem is investigated in this section. Surprisingly enough, we will find that the properties of the interacting fields differ from what one naively assumes. This gives an explanation why the basic expressions in the Lagrangian theory are ill-defined.

Generalizing the idea of causality, we consider the S-matrix as a functional of some additional classical sources

$$S(\underline{g}) = S(g_0, \boldsymbol{g}) = S(g_0, g_1, \dots, g_r). \tag{4.9.1}$$

In QED we deal with

$$S(\underline{g}) = S(g, g_\psi, g_{\overline{\psi}}, g_A, g_j, g_{A\!\!\!/\psi}, g_{\overline{\psi}A\!\!\!/}), \tag{4.9.2}$$

where the first order is given by

$$S_1(\underline{g}) = \int d^4x\, i\Big[:\overline{\psi}(x)A\!\!\!/(x)\psi(x): g(x) + g_\psi(x)\psi(x) + \overline{\psi}(x)g_{\overline{\psi}}(x) + A(x)g_A(x)$$

$$+g_{A\!\!\!/\psi}(x)A\!\!\!/(x)\psi(x) + \overline{\psi}(x)A\!\!\!/(x)g_{\overline{\psi}A\!\!\!/}(x) + :\overline{\psi}\gamma\psi:(x)\, g_j(x)\Big]. \tag{4.9.3}$$

The Lorentz indices are omitted. $g_A(x)$ may be interpreted as an external current density and $g_j(x)$ as an external electromagnetic potential.

The formal perturbation series is given by

$$S(\underline{g}) = 1 + \sum_{n=1}^{\infty} \frac{1}{n!} \int d^4x_1 \dots d^4x_n \sum_{i_1,\dots,i_n} T^{(n)}_{i_1,\dots,i_n}(x_1,\dots,x_n)$$

$$\times\, g_{i_1}(x_1) \dots g_{i_n}(x_n). \tag{4.9.4}$$

Again it can be inductively constructed by means of the causality condition

$$S(\underline{g}_1 + \underline{g}_2) = S(\underline{g}_2)\, S(\underline{g}_1) \quad \text{if} \quad \operatorname{supp} \underline{g}_1 < \operatorname{supp} \underline{g}_2. \tag{4.9.5}$$

This condition implies the following commutation relation

$$\Big[S(\underline{g}_1),\, S(\underline{g}_2)\Big]_- = 0 \quad \text{if} \quad \operatorname{supp} \underline{g}_1 \sim \operatorname{supp} \underline{g}_2. \tag{4.9.6}$$

The symbol $\sim$ means that the two supports are space-like separated. The first order of Eq. (4.9.6) is in conflict with anticommutation of the free Fermi fields: $\{\psi(x),\, \psi(y)\}_+ = 0$ if $x \sim y$. To remedy the conflict, we consider g_ψ, $g_{\overline{\psi}}$, $g_{A\!\!\!/\psi}$, $g_{\overline{\psi}A\!\!\!/}$ as anticommuting C-numbers (Grassmann variables). We explain in Appendix D how the S-matrix is constructed as a functional of Grassmann variables. The n-point distributions $T^{(n)}$ in (4.9.4) are symmetric in the normal (bosonic) variables but antisymmetric with respect to the Grassmann variables, for example

$$T^{(2)}_{\psi\overline{\psi}}(x_1, x_2) = -T^{(2)}_{\overline{\psi}\psi}(x_2, x_1).$$

From $T^{(n)}_{i_1,\dots,i_n}$ the various distributions

$$\tilde{T}^{(n)}_{i_1,\dots,i_n},\ A^{(n)\prime}_{i_1,\dots,i_n},\ R^{(n)\prime},\ A^{(n)}_{i_1,\dots,i_n},\ R^{(n)} \tag{4.9.7}$$

can be constructed as in Sect 3.1.

Now we are ready to define the interacting fields. The idea to define interacting fields as functional derivatives of the S-matrix goes back to Bogoliubov (*N.N. Bogoliubov, D.V. Shirkov, Introduction to the Theory of Quantized Fields, New York 1959*). The general definition is

$$\Phi_{m,\text{int}}(g_0, x) \stackrel{\text{def}}{=} S^{-1}(\underline{g}) \frac{\delta S(\underline{g})}{\delta g_m(x)}\bigg|_{\boldsymbol{g}\equiv\boldsymbol{0}}. \tag{4.9.8}$$

Since we put $\boldsymbol{g} \equiv \boldsymbol{0}$, these functions are auxiliary. However, if in addition to the original interaction some external (C-number) source is present, one may take the corresponding $g_m(x)$ equal to this source. In QED the following interacting fields appear:

$$\psi_{\text{int}}(g, x) = S^{-1}(\underline{g}) \frac{\delta S(\underline{g})}{i\delta g_\psi(x)}\bigg|_{\boldsymbol{g}=\boldsymbol{0}}, \quad \overline{\psi}_{\text{int}}(g, x) = S^{-1}(\underline{g}) \frac{\delta S(\underline{g})}{i\delta g_{\overline{\psi}}(x)}\bigg|_{\boldsymbol{g}=\boldsymbol{0}}, \tag{4.9.9}$$

$$A_{\text{int}}(g, x) = S^{-1}(\underline{g}) \frac{\delta S(\underline{g})}{i\delta g_A(x)}\bigg|_{\boldsymbol{g}=\boldsymbol{0}}, \tag{4.9.10}$$

$$j_{\text{int}}(g, x) = S^{-1}(\underline{g}) \frac{\delta S(\underline{g})}{i\delta g_j(x)}\bigg|_{\boldsymbol{g}=\boldsymbol{0}}, \tag{4.9.11}$$

$$(A\!\!\!/\psi)_{\text{int}}(g, x) = S^{-1}(\underline{g}) \frac{\delta S(\underline{g})}{i\delta g_{A\!\!\!/\psi}(x)}\bigg|_{\boldsymbol{g}=\boldsymbol{0}},$$

$$(\overline{\psi}A\!\!\!/)_{\text{int}}(g, x) = S^{-1}(\underline{g}) \frac{\delta S(\underline{g})}{i\delta g_{\overline{\psi}A\!\!\!/}(x)}\bigg|_{\boldsymbol{g}=\boldsymbol{0}}. \tag{4.9.12}$$

The usefulness of these definitions has to be shown by investigating the properties of these interacting fields, which we are now going to list.

1) Since the functional derivative is a distributional one, the interacting fields are operator-valued distributions, acting in the Fock space of the free fields. Accordingly, the field operators (4.9.8) have to be smeared out in the argument x with some test function $f(x) \in S(\mathbf{R^4})$ to get an operator. By definition of the functional derivative we have

$$\Phi_{m,\text{int}}(g_0,f)=\lim_{\varepsilon\to 0}\frac{1}{\varepsilon}S^{-1}(g_0,\mathbf{0})\Big(S(g_0,\varepsilon f e_m)-S(g_0,\mathbf{0})\Big), \tag{4.9.13}$$

where e_m denotes the unit vector in the direction m:

$$S(g_0,\varepsilon f e_m)=S(g_0,g_1=0,\ldots,g_m=\varepsilon f,\ldots,g_r=0).$$

For example, the spinor field (4.9.9) is given by

$$\psi_{\text{int}}(g,f)=\lim_{\varepsilon\to 0}\frac{1}{\varepsilon}S^{-1}(g,\mathbf{0})\Big(S(g,\varepsilon f e_\psi)-S(g,\mathbf{0})\Big). \tag{4.9.14}$$

The test function $f(x)$ must be of Grassmann type as g_ψ above (see Appendix D).

2) The interacting fields fulfill the requirement of microcausality:

$$\Big[\Phi_{m,\text{int}}(g_0,x),\,\Phi_{n,\text{int}}(g_0,y)\Big]_{\mp}=0 \quad\text{if}\quad x\sim y. \tag{4.9.15}$$

where the commutator or anticommutator corresponds to Bose or Fermi fields, respectively.

Proof. The precise meaning of (4.9.15) is

$$\int dx\,dy\,\Big[S^{-1}(\underline{g})\frac{\delta S(\underline{g})}{\delta g_m(x)}\Big|_{\underline{g}=0},\,S^{-1}(\underline{g})\frac{\delta S(\underline{g})}{\delta g_n(y)}\Big|_{\underline{g}=0}\Big]_{\mp}f_m(x)f_n(y)=0$$
$$\text{if}\quad \operatorname{supp} f_m\sim\operatorname{supp} f_n. \tag{4.9.16}$$

Since $f_m(x)$ and $f_n(y)$ are anticommuting in case of Fermi fields, this is in any case equal to

$$=\frac{\partial}{\partial\varepsilon}\frac{\partial}{\partial\delta}\Big[S^{-1}(g_0,\mathbf{0})S(g_0,\varepsilon f_m e_m),\,S^{-1}(g_0,\mathbf{0})S(g_0,\delta f_n e_n)\Big]_-\Big|_{\varepsilon=\delta=0}. \tag{4.9.17}$$

On the other hand, it follows from (4.9.5) that

$$\Big[S^{-1}(\underline{g})S(\underline{g}+\underline{h}_1),\,S^{-1}(\underline{g})S(\underline{g}+\underline{h}_2)\Big]_-=0\quad\text{if}\quad\operatorname{supp}\underline{h}_1\sim\operatorname{supp}\underline{h}_2. \tag{4.9.18}$$

This implies the commutator (4.9.17) to vanish. □

Remark. Equation (4.9.18) requires some comment. Assume the causality condition

$$S(h_1+h_2)=S(h_2)S(h_1)\quad\text{if}\quad\operatorname{supp}h_2>\operatorname{supp}h_1,$$

which means that in some Lorentz frame the two supports can be separated by a plane $t=$ const. Let $g(x)$ be a test function. Then there exist two approximating sequences with the following properties:

$$\lim_{n\to\infty}[g_{1,n}(x)+g_{2,n}(x)]=g(x)\quad\text{almost everywhere,}$$

$$\operatorname{supp} g_{2,n} > \operatorname{supp} g_{1,n}, \quad \operatorname{supp} g_{1,n} \supset \operatorname{supp} h_1, \quad \operatorname{supp} h_2 \subset g_{2,n}.$$

If $S(g)$ is continuous with respect to the test functions, we may conclude

$$S^{-1}(g)S(g+h_1+h_2) = \lim_{n\to\infty}\Big[S^{-1}(g)S(g_{1,n}+h_1+g_{2,n}+h_2)\Big]$$

$$= \lim_{n\to\infty}\Big[S^{-1}(g)S(g_{2,n}+h_2)S(g_{1,n}+h_1)\Big]$$

$$= \lim_{n\to\infty} S^{-1}(g)S(g_{2,n}+h_2)S(g_{1,n})S^{-1}(g_{1,n})S^{-1}(g_{2,n})S(g_{2,n})S(g_{1,n}+h_1)$$

$$= \lim_{n\to\infty} S^{-1}(g)S(g_{1,n}+g_{2,n}+h_2)S^{-1}(g_{1,n}+g_{2,n})S(g_{1,n}+g_{2,n}+h_1)$$

$$= S^{-1}(g)S(g+h_2)S^{-1}(g)S(g+h_1).$$

This implies

$$\Big[S^{-1}(g)S(g+h_1),\, S^{-1}(g)S(g+h_2)\Big]_- = 0 \quad \text{if} \quad \operatorname{supp} h_1 \sim \operatorname{supp} h_2.$$

In perturbation theory, the final equation follows without any continuity assumption. We take this point of view in the following.

3) The perturbation series for the interacting fields is given by the advanced distributions of the corresponding expansion of the S-matrix:

$$\Phi_{m,\text{int}}(g_0,x) = \Phi_m(x) + \sum_{n=1}^{\infty}\frac{1}{n!}\int d^4x_1\dots d^4x_n\,\varphi_m^{(n)}(x_1,\dots,x_n;x)$$
$$\times\, g_0(x_1)\dots g_0(x_n)$$
$$\text{with}\quad \varphi_m^{(n)}(x_1,\dots,x_n;x) = \frac{1}{i}A^{(n+1)}_{0,\dots,0,m}(x_1,\dots,x_n;x). \tag{4.9.19}$$

As in (4.9.4) the subscripts indicate which kind of vertex i_j is present at x_j. The zeros in the subscripts will be omitted in the following, for example: $T_{0,\dots,0}(x_1,\dots,x_n) = T(X)$ and $A_{0,\dots,0,m}(x_1,\dots,x_n,x) = A_m(X,x)$.

Proof. The perturbative expressions to be inserted into (4.9.8) are

$$S^{-1}(g_0,\mathbf{0}) = 1 + \sum_{n=1}^{\infty}\frac{1}{n!}\tilde{T}(x_1,\dots,x_n)g_0(x_1)\dots g_0(x_n),$$

$$S(g_0,0,\dots g_m,\dots 0)|_{\text{linear in } g_m} = \sum_{n=0}^{\infty}\frac{1}{(n+1)!}\sum_I\int d^4x_1\dots d^4x_{n+1}$$
$$T^{(n+1)}_{i_1\dots i_{n+1}}(x_1,\dots,x_{n+1})g_{i_1}(x_1)\dots g_{i_{n+1}}(x_{n+1}) \tag{4.9.20}$$
$$= \sum_{n=0}^{\infty}\frac{1}{n!}\int d^4x_1\dots d^4x_{n+1}\,T_m^{(n+1)}(x_1,\dots x_{n+1})g_0(x_1)\dots g_0(x_n)g_m(x_{n+1}).$$

The sum in (4.9.20) runs over the $n+1$ permutations I of $(i_1, \ldots, i_{n+1})$ and all terms give the same contribution because there is only one possible Grassmann function g_m. The resulting functional derivative

$$\frac{\delta S(g)}{i\delta g_m(x)}\bigg|_{g=0} = \frac{1}{i}\sum_{n=0}^{\infty}\frac{1}{n!}\int d^4x_1 \ldots d^4x_n \, T_m^{(n+1)}(x_1, \ldots, x_n; x) g_0(x_1) \ldots g_0(x_n) \tag{4.9.21}$$

must be substituted into (4.9.8). By multiplication of the series, we arrive at

$$\varphi_m^{(n)}(X, x) = \frac{1}{i}\sum_{I_1 \cup I_2 = X} \tilde{T}(I_1) T_m(I_2, x) = \frac{1}{i} A_m^{(n+1)}(X, x),$$

where I_1 or I_2 in the partition of X may be the empty set. This is just the advanced distribution (3.1.35).

We give explicit results in lowest orders for the various interacting fields appearing in QED.

$$\psi_{\rm int}(g, x) = \psi(x) + \sum_{n=1}^{\infty}\frac{1}{n!}\int d^4x_1 \ldots d^4x_n \, \psi^{(n)}(x_1, \ldots, x_n; x) g(x_1) \ldots g(x_n)$$

$$\psi^{(1)}(x_1; x) = e S^{\rm ret}(x - x_1)\gamma^\nu \psi(x_1) A_\nu(x_1) \tag{4.9.22}$$

$$\psi^{(2)}(x_1, x_2; x) = e^2 \Big[S^{\rm ret}(x - x_1)\gamma^\mu S^{\rm ret}(x_1 - x_2)\gamma^\nu \psi(x_2) : A_\mu(x_1) A_\nu(x_2) :$$

$$- S^{\rm ret}(x - x_1)\gamma^\mu : \psi(x_1)\overline{\psi}(x_2)\gamma_\mu \psi(x_2) : D^{\rm ret}(x_1 - x_2)$$

$$+ S^{\rm ret}(x - x_1)\Sigma_{\rm ret}(x_1 - x_2)\psi(x_2)\Big] + [x_1 \longleftrightarrow x_2],$$

$$\overline{\psi}_{\rm int}(g, x) = \overline{\psi_{\rm int}(g, x)}.$$

$$A^\mu_{\rm int}(g, x) = A^\mu(x) + \sum_{n=1}^{\infty}\frac{1}{n!}\int d^4x_1 \ldots d^4x_n \, A^{\mu(n)}(x_1, \ldots, x_n; x) g(x_1) \ldots g(x_n)$$

$$A^{\mu(1)}(x_1; x) = -e D_0^{\rm av}(x_1 - x) : \overline{\psi}(x_1)\gamma^\mu \psi(x_1) :, \tag{4.9.23}$$

$$A^{\mu(2)}(x_1, x_2; x) = e^2 \Big[: \overline{\psi}(x_1)\Big(\gamma^\mu S^{\rm ret}(x_1 - x_2)\gamma^\nu D_0^{\rm av}(x_1 - x) A_\nu(x_2)$$

$$+ \gamma^\nu S^{\rm av}(x_1 - x_2)\gamma^\mu D_0^{\rm av}(x_2 - x) A_\nu(x_1)\Big)\psi(x_2) :$$

$$+ D_0^{\rm av}(x_1 - x)\Pi_{\rm av}^{\mu\nu}(x_2 - x_1) A_\nu(x_2)\Big] + [x_1 \longleftrightarrow x_2]. \tag{4.9.24}$$

$$j^\mu_{\rm int}(g, x) = j^\mu(x) + \sum_{n=1}^{\infty}\frac{1}{n!}\int d^4x_1 \ldots d^4x_n \, j^{\mu(n)}(x_1, \ldots, x_n; x) g(x_1) \ldots g(x_n)$$

$$\text{where} \quad j^\mu(x) =: \psi(x)\gamma^\mu \psi(x) : \tag{4.9.25}$$

$$j^{\mu(1)} = e\Big[: \overline{\psi}(x_1)\gamma^\nu S^{\rm av}(x_1-x)\gamma^\mu\psi(x) :$$

$$+ : \overline{\psi}(x)\gamma^\mu S^{\rm ret}(x-x_1)\gamma^\nu\psi(x_1) : \Big] A_\nu(x_1) + e\Pi^{\mu\nu}_{\rm av}(x_1-x)A_\nu(x_1). \quad (4.9.26)$$

$$(A\!\!\!/\psi)_{\rm int}(g,x) = A\!\!\!/(x)\psi(x) + \sum_{n=1}^\infty \frac{1}{n!}\int d^4x_1\ldots d^4x_n\,(A\!\!\!/\psi)^{(n)}(x_1,\ldots,x_n;x)$$

$$\times\, g(x_1)\ldots g(x_n),$$

$$(A\!\!\!/\psi)^{(1)} = -e\Big[: \overline{\psi}(x_1)\gamma_\mu\psi(x_1)D_0^{\rm av}(x_1-x)\gamma^\mu\psi(x) :$$

$$+ : A\!\!\!/(x)S^{\rm ret}(x-x_1)A\!\!\!/(x_1) : \psi(x_1) + \Sigma_{\rm ret}(x-x_1)\psi(x_1)\Big], \quad (4.9.27)$$

$$(\overline{\psi}A\!\!\!/)_{\rm int}(g,x) = \overline{(A\!\!\!/\psi)_{\rm int}(g,x)}.$$

Here

$$\Sigma_{\rm ret}(y) = (2\pi)^{-2}\int d^4p\,\hat{\Sigma}_{\rm ret}(p)\,e^{-ipy}$$

is the retarded form of the self-energy function

$$\hat{\Sigma}_{\rm ret}(p) = (2\pi)^{-4}\Big\{\Big[1-\frac{m^2}{p^2}\Big]\Big[\log\Big|1-\frac{p^2}{m^2}\Big| - i\pi\,{\rm sgn}\,p_0\Theta(p^2-m^2)\Big]$$

$$\times\Big[m-\frac{p\!\!\!/}{4}\Big(1+\frac{m^2}{p^2}\Big)\Big] + \frac{m^2}{4}\frac{p\!\!\!/}{p^2} + \frac{p\!\!\!/}{8} - \frac{3}{8}m\Big\}. \quad (4.9.28)$$

The normalization of (4.9.28) agrees with that of Sect. 3.13 (3.13.19), we will return to this point in the following section (see Remark 10.3). In (4.9.24) the advanced vacuum polarization tensor appears (see (3.6.18))

$$\Pi^{\mu\nu}_{\rm av}(y) = (2\pi)^{-6}\int d^4p\,\Big(-g^{\mu\nu}+\frac{p^\mu p^\nu}{p^2}\Big)\hat{\Pi}_{\rm av}(p)e^{-ipy}, \quad (4.9.29)$$

$$\hat{\Pi}_{\rm av}(p) = \frac{1}{3}p^4\int_{4m^2}^\infty ds\,\frac{s+2m^2}{s^2(p^2-s-ip_00)}\sqrt{1-\frac{4m^2}{s}}.$$

4) If the n-th order term of the perturbation expansion (4.9.19) is normally ordered by means of Wick's theorem, as it is usually done in case of the S-matrix, the resulting terms may again be represented by Feynman graphs. These graphs are amputated at the vertex x. In fact, the external lines representing the external potential $g_m(x)$, which interacts at x in an ordinary S-matrix graph, have to be cut off (Fig. 11), because $g_m(x)$ disappears by the functional differentiation (4.9.21). The singular order ω of an amputated graph agrees with the singular order of the S-matrix graph, because the functional differentiation does not affect the distributional kernel. We therefore have the usual formula

$$\omega = 4 - 3n - m, \quad (4.9.30)$$

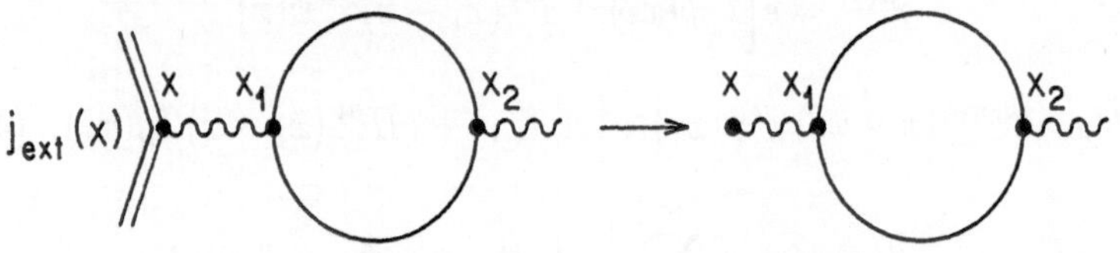

Fig. 11. Amputated graph representing the last term in (4.9.24)

where n is the number of external fermions divided by two and m is the number of external photons with the amputated lines included.

5) The definitions of the interacting fields (4.9.11–12) may be considered as definitions of operator products. In contrast to other such definitions, these products are well-defined quantities without any regularization. Let us illustrate this for the current density (4.9.11)

$$j^\mu_{\rm int}(g,x)=\Big(\overline{\psi}\gamma^\mu\psi\Big)_{\rm int}(g,x).$$

We may compare this with the naive operator product

$$j^\mu_{\rm naive}(g,x)=\overline{\psi}_{\rm int}(g,x)\gamma^\mu\psi_{\rm int}(g,x). \tag{4.9.31}$$

Inserting the lowest orders of the perturbation series, we formally obtain

$$j^\mu_{\rm naive}(g,x)=j^\mu(x)-i\,{\rm tr}\,[\gamma^\mu S^{(-)}(0)]+\int d^4x_1 j^{(1)\mu}_{\rm naive}(x_1;x)g(x_1), \tag{4.9.32}$$

with

$$j^{(1)\mu}_{\rm naive}(x_1;x)=e\Big\{\Big[:\overline{\psi}(x_1)\gamma^\nu S^{\rm av}(x_1-x)\gamma^\mu\psi(x):$$

$$+:\overline{\psi}(x)\gamma^\mu S^{\rm ret}(x-x_1)\gamma^\nu\psi(x_1):\Big]A_\nu(x_1)$$

$$-i\,{\rm tr}\Big[\gamma^\mu S^{\rm ret}(x-x_1)\gamma^\nu S^{(-)}(x_1-x)+\gamma^\nu S^{\rm av}(x_1-x)\gamma^\mu S^{(-)}(x-x_1)\Big]A_\nu(x_1)\Big\}. \tag{4.9.33}$$

The lowest order already contains an infinite constant. While the terms in (4.9.33) coming from tree graphs are correct, the last line is only an unrenormalized form of the correct vacuum polarization expression in (4.9.26). A naive normal product instead of (4.9.31) fails even more, because in first order one term is completely lacking.

The infinite counterterms needed in (4.9.31) become finite if one starts with ultraviolet cut-off fields, as for example in a lattice approach. However, the general structure of the counterterms is very complicated and there sum completely unknown. This might explain why all attempts to construct QED in four dimensions as a continuum limit of a lattice version have failed until now. It seems to be rather hopeless to construct physical QED in a rigorous way by the Lagrangian approach.

6) In this number we study Poincaré covariance. Let $\Pi = (a, \Lambda)$ be an element of the Poincaré group $\mathcal{P}_+^\uparrow$. Then the scalar, spinor and vector test functions in (4.9.3) transform as follows:

$$(\Pi g)(x) = g(\Pi^{-1}x), \quad (\Pi g_A)(x) = \Lambda g_A(\Pi^{-1}x), \quad (\Pi g_j)(x) = \Lambda g_j(\Pi^{-1}x)$$

$$(\Pi g_\psi)(x) = g_\psi(\Pi^{-1}x)\overline{S(\Lambda)}, \quad (\Pi g_{\overline{\psi}})(x) = S(\Lambda)g_{\overline{\psi}}(\Pi^{-1}x)$$

$$(\Pi g_{\not{A}\psi})(x) = g_{\not{A}\psi}(\Pi^{-1}x)\overline{S(\Lambda)}, \quad (\Pi g_{\overline{\psi}\not{A}})(x) = S(\Lambda)g_{\overline{\psi}\not{A}}(\Pi^{-1}x), \tag{4.9.34}$$

where $\overline{S} = \gamma^0 S^+ \gamma^0 = S^{-1}$. The Poincaré covariance of the S-matrix (4.5.4)

$$U(\Pi)S(\underline{g})U(\Pi)^{-1} = S(\Pi\underline{g}) \tag{4.9.35}$$

now implies the covariance of the interacting fields

$$\Phi_{m,\text{int}}(g, x) = S^{-1}(\underline{g})\frac{\delta S(\underline{g})}{\delta g_m(x)}\Big|_{g=0}.$$

We find

$$U(\Pi)\Phi_{m,\text{int}}(g,x)U(\Pi)^{-1} = U(\Pi)S^{-1}(\underline{g})U(\Pi)^{-1}U(\Pi)\frac{\delta S(\underline{g})}{\delta g_m(x)}U(\Pi)^{-1}\Big|_{g=0}$$

$$= S^{-1}(\Pi\underline{g})\frac{\delta}{\delta g_m(x)}S(\Pi\underline{g})\Big|_{g=0} = U_m(\Pi)\, S^{-1}(\Pi\underline{g})\frac{\delta}{\delta \Pi g_m(\Pi x)}S(\Pi\underline{g})\Big|_{g=0}$$

$$= U_m(\Pi)\,\Phi_{m,\text{int}}(\Pi g, \Pi x), \tag{4.9.36}$$

where $U_m(\Pi)$ is the appropriate scalar, vector or spinor representation. The explicit transformation laws are the following

$$U(\Pi)\psi_{\text{int}}(g,x)U(\Pi)^{-1} = S(\Lambda)\psi_{\text{int}}(\Pi g, \Pi x),$$

$$U(\Pi)\overline{\psi}_{\text{int}}(g,x)U(\Pi)^{-1} = \overline{\psi}_{\text{int}}(\Pi g, \Pi x)\overline{S(\Lambda)}, \tag{4.9.37}$$

$$U(\Pi)A_{\text{int}}(g,x)U(\Pi)^{-1} = \Lambda\, A_{\text{int}}(\Pi g, \Pi x),$$

$$U(\Pi)j_{\text{int}}(g,x)U(\Pi)^{-1} = \Lambda\, j_{\text{int}}(\Pi g, \Pi x). \tag{4.9.38}$$

At first sight it may be surprising that the interacting fields transform in the same way as the free fields. However, assuming the test function $g(x)$ to have compact support, the interacting fields agree with free fields if $x < \operatorname{supp} g$. Therefore the transformation properties must be the same.

7) It is pretty clear from the foregoing numbers that the interacting fields are quite different from Wightman fields (*R.F. Streater, A.S. Wightman, PCT, Spin and Statistics and all that, Benjamin, New York 1964*). The most important difference is the appearance of the test function $g(x)$. One might expect that in the adiabatic limit $g \to 1$, some Wightman-type field operators are obtained. However, the adiabatic limit does not exist in the strong sense. To show this, let us look at the first order contribution (4.9.23) to the interacting radiation field, for example,

$$A^{\mu(1)}(x) = -e \int dx_1 \, D_0^{\mathrm{av}}(x_1 - x) g(x_1) \, : \overline{\psi}(x_1)\gamma^\mu \psi(x_1) :$$

$$= -\frac{e}{4\pi} \int d^3x_1 \, \frac{g(x^0 - |\boldsymbol{x}_1 - \boldsymbol{x}|, \boldsymbol{x}_1)}{|\boldsymbol{x} - \boldsymbol{x}_1|} \, : \overline{\psi}\gamma^\mu \psi : (x^0 - |\boldsymbol{x}_1 - \boldsymbol{x}|, \boldsymbol{x}_1). \quad (4.9.39)$$

Applying this to the vacuum, the integral becomes meaningless in the limit $g \to 1$. There is a "volume-divergence" for $|\boldsymbol{x}_1| \to \infty$. Smearing out in x does not help. Even if we consider a theory with massive photons, where D_0^{av} is replaced by D_m^{av}, the adiabatic limit in (4.9.39) still does not exist, due to the form of $D_m^{\mathrm{av}}(x)$ in x-space (2.3.18, 21).

On the other hand, the adiabatic limit does exist in the weak sense, i.e. in the sense of bilinear forms or expectation values. This was already pointed out by Epstein and Glaser (loc.cit.). There is some care required in QED, because of infrared problems. The latter occur if the initial or final states contain charged fermions. But the vacuum expectation values presumably exist

$$W_{m_1,\dots m_p}(x_1, \dots x_p) = \lim_{g_0 \to 1} (\Omega, \Phi_{m_1}(g_0, x_1) \dots \Phi_{m_p}(g_0, x_p) \, \Omega). \quad (4.9.40)$$

The product of field operators in this equation is defined by perturbative expansion

$$(\Omega, \dots \Omega) = \sum_{n=0}^{\infty} \frac{1}{n!} \int W^{(n)}_{m_1,\dots,m_p}(y_1, \dots, y_n; x_1, \dots, x_p)$$
$$\times \, g_0(y_1) \dots g_0(y_n) \, dy_1 \dots dy_n \quad (4.9.41)$$

with

$$W^{|Y|}_{m_1,\dots,m_p}(Y; x_1, \dots, x_p) = (-i)^p \sum_{\cup I_j = Y} \Big(\Omega, \, A^{|I_1|+1}_{0,0,\dots;m_1}(I_1; x_1)$$
$$\dots A^{|I_p|+1}_{0,\dots,0;m_p}(I_p, x_p) \, | \, \Omega\Big). \quad (4.9.42)$$

Again, the I_j in the partition of Y may be empty.

The vacuum expectation values (4.9.40) are expected to have the essential properties of Wightman functions. Then by the general reconstruction theorem of Wightman, it is possible to construct the corresponding Wightman fields. We do not follow this line of investigation for the following reason: The Wightman fields live in their own Hilbert space which has to be also constructed. This Hilbert space presumably does not have a Fock structure with a particle interpretation, in any case one has to work hard to find that out. It seems to be much simpler for actual calculations to work with the interacting fields in free Fock space and to perform the adiabatic limit only very late in observable quantities.

4.10 Field Equations

The properties (1–7) of the interacting fields which we have found in the last section would not be sufficient to justify the introduction of those objects. This justification comes from the fact that the field operators obey field equations.

In order to see which field equations we would like to have, we first write down the equations of classical electrodynamics, i.e. the coupled Maxwell-Dirac equations

$$(i\partial\!\!\!/ - m)\psi_{\rm cl}(x) = -eA\!\!\!/_{\rm cl}(x)\psi_{\rm cl}(x), \tag{4.10.1}$$

$$\square\, A_{\rm cl}(x) = -ej_{\rm cl}(x) \stackrel{\rm def}{=} -e\overline{\psi}_{\rm cl}(x)\gamma\psi_{\rm cl}(x). \tag{4.10.2}$$

The subscript "cl" means that these quantities are classical C-number fields. The Dirac equation (4.10.1) implies current conservation

$$\partial \cdot j_{\rm cl}(x) = 0, \tag{4.10.3}$$

which is a necessary condition for the current density to be the source of the Maxwell field. Equation (4.10.2) is equivalent to the inhomogeneous Maxwell's equations if the Lorentz gauge condition

$$\partial \cdot A_{\rm cl}(x) = 0 \tag{4.10.4}$$

is assumed. Since (4.10.2) and (4.10.3) imply

$$\square\, \partial \cdot A_{\rm cl}(x) = 0, \tag{4.10.5}$$

the Lorentz condition is only a boundary condition and not a dynamical equation in the strict sense.

Now we turn to QED. The field operators introduced above have been constructed by means of adiabatic switching of the interaction by a test function $g(x)$. This requires the substitution $e \to e{\cdot}g(x)$ in the field equations. We therefore expect the following equations to hold:

$$(i\partial\!\!\!/_x - m)\psi_{\rm int}(g,x) = -eg(x)(A\!\!\!/\psi)_{\rm int}(g,x). \tag{4.10.6}$$

This is the quantum version of the Dirac equation, which differs from the classical equation (4.10.1) because the field operator on the right-hand side is the operator (4.9.27) instead of the naive product. The wave equation

$$\square_x A_{\text{int}}(g,x) = -eg(x) j_{\text{int}}(g,x) \tag{4.10.7}$$

and the current conservation

$$\partial_x \cdot j_{\text{int}}(g,x) = 0 \tag{4.10.8}$$

have the same form as the classical equations (4.10.2, 3). This is not the case for the Lorentz condition. It is well-known that this equation cannot hold on the entire Fock space $\mathcal{F}$. Let $\mathcal{F}_{\text{phys}}$ be the physical subspace which contains only transverse photons, then we will show that

$$\partial \cdot A_{\text{int}}(x)|_{\mathcal{F}_{\text{phys}}} = 0. \tag{4.10.9}$$

Note that there is no switching function g in the argument of A_{int}. This means that we have to perform the adiabatic limit $g \to 1$ in the following manner

$$\partial_\mu A^\mu_{\text{int}}(x) \stackrel{\text{def}}{=} \lim_{g\to 1}\Big[\partial_\mu A^\mu_{\text{int}}(g,x)\Big]. \tag{4.10.10}$$

Consequently, according to our discussion of the adiabatic limit in no. 7) of the last section, the Lorentz condition (4.10.9) only holds in the sense of bilinear forms. This is similar to the physical unitarity of the S-matrix (see Sect. 4.7).

It is our aim now to prove the field equation (4.10.6-9). The Dirac and the wave equations are both of the following form:

$$D_x O_{1,\text{int}}(g,x) = g(x) O_{2,\text{int}}(g,x). \tag{4.10.11}$$

D_x is the same linear differential operator which appears in the free field equations. In QFT, at least in the standard model of electro-weak and strong interactions, there are only two kinds of operators D_x

$$D^B_x = \square_x + \mu^2 \quad , \quad D^F_x = i\partial\!\!\!/_x - m \tag{4.10.12}$$

for Bose and Fermi fields, respectively. $O_{1,\text{int}}$ and $O_{2,\text{int}}$ are interacting fields defined by functional differentiation of a causal S-matrix $S(g, g_1, g_2)$

$$O_{1,2,\text{int}} = S^{-1}(g) \frac{\delta S}{i\delta g_{1,2}(x)}\bigg|_{g_1=g_2=0}.$$

$S(g, g_1, g_2)$ is recursively constructed from its first order

$$S_1 = \int d^4x\, [T(x)g(x) + io_1(x)g_1(x) + io_2(x)g_2(x)], \tag{4.10.13}$$

where $o_{1,2}(x)$ are free fields and we assume that $o_1(x)$ is an elementary free field operator, like $\psi(x)$ or $A(x)$, and not a composite operator like $(A\!\!\!/\psi)(x)$ or $j(x)$.

We want to write down the perturbative version of (4.10.11) utilizing (4.9.19). On the left side we get

$$D_x O_{1,\mathrm{int}}(g,x)=\sum_{n=0}^{\infty}\frac{1}{n!}\int d^4x_1\ldots d^4x_n\left[D_x\frac{1}{i}A_1^{(n+1)}(x_1,\ldots,x_n;x)\right]$$
$$\times g(x_1)\ldots g(x_n). \tag{4.10.14}$$

The product on the right side of (4.10.11) can be written as follows

$$g(x)O_{2,\mathrm{int}}(g,x)=\sum_{n=0}^{\infty}\frac{1}{n!}\int d^4x_1\ldots d^4x_n\,d^4x_{n+1}\frac{1}{i}A_2^{(n+1)}(x_1,\ldots,x_n;x)$$
$$\times\delta(x-x_{n+1})g(x_1)\ldots g(x_{n+1}).$$

By symmetrization of the integrand, this is equal to

$$\sum_{n=0}^{\infty}\frac{1}{(n+1)!}\int d^4x_1\ldots d^4x_{n+1}\left[\sum_{j=1}^{n+1}\frac{1}{i}A_2^{(n+1)}(x_1,\ldots,\hat{x}_j,\ldots x_{n+1};x)\delta(x-x_j)\right]$$
$$\times g(x_1)\ldots g(x_{n+1}),$$

where the hat always means that the correponding argument is lacking. Changing the summation index from $n+1$ into n and comparing with (4.10.14), the differential equation becomes the following identity between advanced distributions

$$D_xA_1^{(n+1)}(X;x)=\sum_{x_j\in X}A_2^{(n)}(X\setminus x_j;x)\delta(x_j-x)\stackrel{\mathrm{def}}{=}(A_2^{(n)}\otimes_s\delta)(X;x). \tag{4.10.15}$$

The right side is a symmetrical direct product.

It is our aim now to prove that if (4.10.15) is satisfied for $n=0,1$, then the equation holds for all n, provided in each step of the inductive construction an appropriate solution of the splitting problem is chosen. In addition, the same equation is true if the advanced distributions $A_{1,2}$ in (4.10.15) are replaced by any of the other distributions

$$F\in\left\{A,R,A',R',D,T,\tilde{T}\right\}, \tag{4.10.16}$$

while the induction assumption for $n=0,1$ must only be verified for the advanced distributions. We divide the proof into various steps.

1) First we have to check the start of the induction. For $n=0$ the right side of Eq. (4.10.15) is zero. On the left side we have

$$A_1^{(1)}(x)=R_1^{(1)}(x)=T_1^{(1)}(x)=-\tilde{T}_1^{(1)}(x)=io_1(x),$$
$$\text{or}\quad A_1'^{(1)}(x)=R_1'^{(1)}(x)=D_1^{(1)}(x)=0.$$

Equation (4.10.15), therefore, is the free field equation

$$D_xo_1(x)=0 \tag{4.10.17}$$

which is trivially fulfilled.

2) For $n = 1$ Eq. (4.10.15) reads as follows

$$D_x A_1^{(2)}(x_1, x) = io(x)\delta(x - x_1), \tag{4.10.18}$$

which is obviously fulfilled for the Dirac and wave equations by the explicit expressions (4.9.22) and (4.9.23). Regarding the other fields in (4.10.16), the following expressions appear on the right side of (4.10.15)

$$A_2^{(1)}(x) = R_2^{(1)}(x) = T_2^{(1)}(x) = -\tilde{T}_2^{(1)}(x) = io_2(x), \tag{4.10.19}$$

$$\text{or} \quad A_2^{\prime(1)}(x) = R_2^{\prime(1)}(x) = D_2^{(1)}(x) = 0.$$

On the left side we get

$$A_1^{\prime(2)}(x_1, x) = \tilde{T}_1^{(1)}(x_1) T_1^{(1)}(x), \quad R_1^{\prime(2)}(x_1, x) = T_1^{(1)}(x) \tilde{T}_1^{(1)}(x_1)$$

$$D_1^{(2)} = R_1^{\prime(2)} - A_1^{\prime(2)}, \quad T_1^{(2)} = A_1^{\prime(2)} - A_1^{(2)}, \quad R_1^{(2)} = T_1^{(2)} + R_1^{\prime(2)},$$

and

$$\tilde{T}_1^{(2)}(x_1, x) = -T_1^{(2)}(x_1, x) + T_1^{(1)}(x_1) T_1^{(1)}(x) + T_1^{(1)}(x) T_1^{(1)}(x_1).$$

Applying D_x, we find with the aid of (4.10.17, 18) that (4.10.15) is satisfied.

3) Now we turn to the inductive step. Let us assume that (4.10.15) holds for $n = 0, \ldots, m-1$. We consider

$$A_1^{\prime(m+1)}(X, x) = \sum_{P_2} \tilde{T}(U) T_1(V, x), \tag{4.10.20}$$

where the sum runs over all partitions

$$P_2 : X = U \cup V, \quad U \neq \emptyset. \tag{4.10.21}$$

By the induction assumption we have

$$D_x T_1(V, x) = (T_2 \otimes_s \delta)(V, x),$$

and therefore

$$D_x A_1^{\prime(m+1)}(X, x) = \sum_{P_2} \tilde{T}(U)(T_2 \otimes_s \delta)(V, x)$$

$$= \sum_{P_2} \tilde{T}(U) \sum_{x_i \in V} T_2(V \setminus x_i, x) \delta(x - x_i). \tag{4.10.22}$$

On the other hand we consider

$$(A_2^{\prime(m)} \otimes_s \delta)(X, x) = \sum_{x_i \in X} A_2^{\prime(m)}(X \setminus x_i, x) \delta(x_i - x) =$$

$$= \sum_{x_i \in X} \sum_{P_i} \tilde{T}(U_i) T(V_i, x) \delta(x_i - x), \tag{4.10.23}$$

where P_i are the following partitions

$$P_i : X \setminus x_i = U_i \cup V_i, \quad U_i \neq \emptyset. \tag{4.10.24}$$

The sum (4.10.23) is equal to (4.10.22), because both are the sum of all terms of the following form

$$\tilde{T}(A) T_2(B, x) \delta(x_j - x),$$

with

$$A \cup B = X \setminus x_j \quad , \quad A \neq \emptyset \tag{4.10.25}$$

for some x_j. From (4.10.22) we conclude that

$$D_x A_1^{\prime(m+1)}(X, x) = (A_2^{\prime(m)} \otimes_s \delta)(X, x). \tag{4.10.26}$$

The analogous equation follows for R' and, therefore, also for the causal distributions D_1 D_2

$$D_x D_1^{(m+1)}(X, x) = (D_2^{(m)} \otimes_s \delta)(X, x). \tag{4.10.27}$$

4) In order to prove (4.10.15), there remains to show that the differential equation (4.10.27) is preserved under distribution splitting. If at least one point of X does not coincide with x, then we have either

$$A_1^{(m+1)}(X, x) = -D_1^{(m+1)}(X, x), \quad (A_2^{(m)} \otimes_s \delta)(X, x) = -(D_2^{(m)} \otimes_s \delta)(X, x),$$

$$\text{or} \quad A_1^{(m+1)}(X, x) = 0, \quad (A_2^{(m)} \otimes_s \delta)(X, x) = 0.$$

In both cases (4.10.15) holds. Only the set of points

$$\{(x_1, x_2, \ldots x_m, x)\} \mid x_1 = x_2 = \ldots x_m = x\} \tag{4.10.28}$$

requires a special effort.

We need the following

Lemma 10.1. In the normal product expansion (2.1.16) of the causal distribution $D^{(n)}(X, x)$, only *connected graphs* contribute. (This is true in any causal quantum field theory!)

Proof. A disconnected graph factorizes

$$D(X, x) =: D_1(X_1) D_2(X_2) : \quad , \quad \text{where}$$

$$X_1 \cup X_2 = (X, x), \quad X_1 \cap X_2 = \emptyset.$$

Then the supports of D_1 and D_2 are completely independent and the support of D is the direct product of them. This is in contradiction to the causal support $\operatorname{supp} D(X, x) \subseteq (\Gamma_+(x) \cup \Gamma_-(x))$. □

We write Eq. (4.10.27) in the form

$$D_x D_1^{(m+1)}(X,x) = \sum_{x_j \in X} D_2^{(m)}(X \setminus x_j, x_j)\delta(x_j - x). \tag{4.10.29}$$

In order to derive the general form of $D_1^{(m+1)}$, we start from the definitions (3.1.37) and (3.1.31, 32)

$$D_1^{(m+1)}(X,x) = \sum_{P_2} \left[T(X \setminus Y, x), \tilde{T}(Y)\right]. \tag{4.10.30}$$

As usual this is normally ordered by means of Wick's theorem. By Lemma 6.1 (4.6.24), only connected graphs contribute. Since o_1 is an elementary free field operator, the Lemma implies that, in the graphs contributing to $D_1^{(m+1)}(X,x)$, x is contracted with exactly one vertex x_j. If x is already contracted in $T(X \setminus Y, x)$, a Feynman propagator $B^F(x - x_j)$ is present. Otherwise we get distributions $B^{(+)}(x - x_j)$ or $B^{(-)}(x - x_j)$ which are the usual (anti)-commutators of the positive and negative frequency parts of the free fields. Replacing $B^{(-)}$ by $B^{\rm ret} - B^{\rm av} - B^{(+)}$ and B^F by $\pm(B^{(+)} + B^{\rm av})$ (according to the different conventions of the Feynman propagator for Bose and Fermi fields), and taking the symmetry of $D_1^{(m+1)}(X,x)$ in $x_1, \ldots x_m$ into account, we shall obtain

$$D_1^{(m+1)}(X,x) = \sum_{x_j \in X} [B^{(+)}(x-x_j)g_+(X \setminus x_j, x_j) - B^{\rm ret}(x-x_j)g_{\rm av}(X \setminus x_j, x_j)$$

$$+ B^{\rm av}(x - x_j)g_{\rm ret}(X \setminus x_j, x_j)]. \tag{4.10.31}$$

$B^{\rm ret}$ and $B^{\rm av}$ are the usual retarded and advanced distributions defined by the causal decomposition of

$$B = B^{(+)} + B^{(-)} = B^{\rm ret} - B^{\rm av},$$

$$\text{supp}\, B^{\rm ret} \subset \bar{V}_+, \quad \text{supp}\, B^{\rm av} \subset \bar{V}_-. \tag{4.10.32}$$

Note the relations

$$D_x B^{\rm ret}(x - a) = \delta(x - a) = D_x B^{\rm av}(x - a). \tag{4.10.33}$$

In Eq. (4.10.31) the dependence on x is now explicitly worked out. Next we study the support properties. Since $D_1^{(m+1)}(X,x)$ has a causal support with respect to x, the first term on the right side of (4.10.31) containing $B^{(+)}$ must vanish. Indeed, $B^{(+)}(x - x_j)$ does not vanish for x_j which are space-like with respect to x, such that $g_+(X \setminus x_j, x_j)$ must vanish in this case; but g_+ does not depend on x at all, hence, $g_+ \equiv 0$. In the remaining two terms

$$D_1^{(m+1)}(X,x) = \sum_{x_j \in X} [B^{\rm av}(x-x_j)g_{\rm ret}(X \setminus x_j, x_j) - B^{\rm ret}(x-x_j)g_{\rm av}(X \setminus x_j, x_j)], \tag{4.10.34}$$

we must have

$$\operatorname{supp} g_{\text{ret}}(X \setminus x_j, x_j) \subset \Gamma_+^{(m-1)}(x_j), \quad \operatorname{supp} g_{\text{av}}(X \setminus x_j, x_j) \subset \Gamma_-^{(m-1)}(x_j), \tag{4.10.35}$$

in order to have the correct causal support of $D_1^{(m+1)}(X,x)$. We want to emphasize that the individual terms in (4.10.34) are all well-defined distributions, because they are direct products in the relative coordinates (see also Remark 10.2 below).

We are now ready to substitute (4.10.34) into (4.10.29). Using (4.10.33), we shall obtain

$$0 = \sum_{x_j \in X} \delta(x - x_j)[D_2^{(m)}(X \setminus x_j, x_j) - g_{\text{ret}}(X \setminus x_j, x_j) + g_{\text{av}}(X \setminus x_j, x_j)]. \tag{4.10.36}$$

Since the square bracket in (4.10.36) has to vanish, this together with (4.10.35) implies that g_{ret} and g_{av} form a solution of the splitting problem for $D_2^{(m)}$. Therefore, we may write

$$g_{\text{ret}} \stackrel{\text{def}}{=} R_2^{(m)} \quad , \quad g_{\text{av}} \stackrel{\text{def}}{=} A_2^{(m)}.$$

The relations (4.10.32) and (4.10.35) imply

$$\operatorname{supp} R_2^{(m)}(X \setminus x_j, x_j) B^{\text{av}}(x - x_j) \subset \Gamma_+^{(m)}(x),$$

$$\operatorname{supp} A_2^{(m)}(X \setminus x_j, x_j) B^{\text{ret}}(x - x_j) \subset \Gamma_-^{(m)}(x). \tag{4.10.37}$$

This proves that the decomposition (4.10.34) is a causal splitting of $D_1^{(m+1)}$. Let us call it the 'natural' splitting. Applying D_x to

$$A_1^{(m+1)}(X,x) = \sum_{x_j \in X} B^{\text{ret}}(x - x_j) A_2^{(m)}(X \setminus x_j, x_j), \tag{4.10.38}$$

we arrive at the desired differential equation (4.10.15)

$$D_x A_1^{(m+1)}(X,x) = (A_2^{(m)} \otimes_s \delta)(X,x). \tag{4.10.39}$$

Remark 10.1. The natural splitting of $D_1^{(m+1)}(X,x)$ determines the splitting of $D_2^{(m)}$ uniquely according to (4.10.39). Another solution $\tilde{A}_2^{(m)}$ of the splitting problem for $D_2^{(m)}$ differs from $A_2^{(m)}$ by a contact-term polynomial $P_2^{(m)}$

$$\tilde{A}_2^{(m)} = A_2^{(m)} + P_2^{(m)}.$$

But using this in (4.10.39), the differential equation is no longer true

$$D_x A_1^{(m+1)}(X,x) = (\tilde{A}_2^{(m)} \otimes_s \delta)(X,x) - (P_2^{(m)} \otimes_s \delta)(X,x). \tag{4.10.40}$$

Remark 10.2. The natural splitting (4.10.34) is for $\omega(D_1^{(m+1)}) \geq 0$ *non-trivial*, that means, it cannot be achieved by multiplication with $\Theta(x_1^0 - x^0) \cdot \ldots \cdot \Theta(x_n^0 - x^0)$. Nevertheless, if $D_1^{(m+1)}$ is computed by the inductive construction, it naturally arises in the form (4.10.34), where the individual terms are well-defined distributions. This is due to the fact that we have considered *elementary* field operators like $\psi(x)$ or $A(x)$ which give rise to the propagators $B^{\rm av}(x - x_j)$ and $B^{\rm ret}(x - x_j)$. For composed operators like $j(x)$ or $(A\!\!\!/\psi)(x)$, the natural splitting is impossible: Although also in this case, a decomposition similar to (4.10.34) can formally be written down, the individual terms do not exist.

Remark 10.3. In our previous formulas (4.9.22), (4.9.24) the natural splitting was taken. In general, one could add some contact-term polynomial, but this is unnatural because, in general, it ruins the field equations. If the distributions of lower order appearing in the calculation of $D_1^{(m+1)}(X, x)$ (4.10.30) and $D_2^{(m)}$ are normalized according to our previous conventions, the field operators are completely fixed. The field equations can thus be partly regarded as normalization conditions for the field operators. It is easy to see (Problem 4.6) that the natural splitting (4.10.34) is in agreement with the usual normalization of the corresponding S-matrix graph (see no.4) in the previous section).

We now come to the last step of the proof:

5) From (4.10.26) and (4.10.40) we obtain the differential equation for $T = A - A'$:

$$D_x T_1^{(n+1)}(X, x) = (T_2^{(n)} \otimes_s \delta)(X, x). \tag{4.10.41}$$

There remains to prove the analogous equation for the $\tilde{T}$-distributions. Since $\tilde{T}$ is a sum of direct products of T-distributions, the proof is the similar to the one for A' in 3) (4.10.20–26). This completes the proof of the Dirac and wave equations. □

The last field equations to be discussed are current conservation and the Lorentz condition. These properties of the interacting fields are consequences of gauge invariance. Let us write the m-th order of the S-matrix as follows

$$T^{(m)}(x_1, \ldots x_m) = \sum_{l=0}^{m} \sum_{1 \leq k_1 < \ldots < k_l \leq m} t^{\mu_1 \ldots \mu_l}_{k_1 \ldots k_l}(x_1, \ldots x_m)$$

$$\times : A_{\mu_1}(x_{k_1}) \ldots A_{\mu_l}(x_{k_l}) :, \tag{4.10.42}$$

where the t's contain the Fermi operators. Then gauge invariance is expressed by (4.6.9)

$$\frac{\partial}{\partial x^{\mu_j}_{k_j}} t^{\mu_1 \ldots \mu_l}_{k_1 \ldots k_l}(x_1, \ldots x_m) = 0, \tag{4.10.43}$$

for all $1 \leq l \leq m$, $1 \leq j \leq l$, $1 \leq k_1 < \ldots k_l \leq m$ and all $x_1, \ldots x_m$ in Minkowski space. The same divergence relations hold for the other m-point functions $A', R', D, A, R, \tilde{T}$.

To prove current conservation we consider the advanced function $A^{(n+1)}$ and look at those terms in the corresponding normal product expansion (4.10.42) which have an external photon at the vertex x

$$A^{(n+1)}(x_1, \ldots, x_n; x) = \ldots + \, : A^{(n+1)\mu}(x_1, \ldots, x_n; x) A_\mu(x) : + \ldots . \tag{4.10.44}$$

The divergence relation then reads

$$\frac{\partial}{\partial x^\mu} A^{(n+1)\mu}(x_1, \ldots, x_n; x) = 0. \tag{4.10.45}$$

The interacting current (4.9.25) is given by

$$j^\mu_{\text{int}}(g, x) = j^\mu(x) +$$

$$+ \frac{1}{i} \sum_{n=0}^{\infty} A_j^{(n+1)\mu}(x_1, \ldots, x_n; x) g(x_1) \ldots g(x_n). \tag{4.10.46}$$

According to Sect. 4.9, the distribution $A_j^{(n+1)}$ is constructed by using in the inductive construction at the vertex x the amputated 1-point function $T_j^{(1)\mu}(x) = i : \overline{\psi}(x) \gamma^\mu \psi(x) :$, but at all other vertices x_k the full 1-point function $T^{(1)}(x_k) = T_j^{(1)\mu}(x_k) A_\mu(x_k)$ is used. Graphically speaking, the external photon line representing $A_\mu(x)$ in the term on the right side of (4.10.44) is amputated, hence

$$A_j^{(n+1)\mu}(x_1, \ldots, x_n; x) = A^{(n+1)\mu}(x_1, \ldots, x_n; x). \tag{4.10.47}$$

With (4.10.45) this implies current conservation

$$\partial_\mu j^\mu_{\text{int}}(g, x) = 0. \tag{4.10.48}$$

Now we turn to the Lorentz condition. The interacting photon field (4.9.23) is given by

$$A^\mu_{\text{int}}(g, x) = A^\mu(x) + \frac{1}{i} \sum_{n=0}^{\infty} A_A^{(n+1)\mu}(x_1, \ldots, x_n; x)$$

$$\times g(x_1) \ldots g(x_n). \tag{4.10.49}$$

The advanced distribution $A_A^{(n+1)\mu}$ consists of those graphs which have two external fermion lines $\overline{\psi}(x) \gamma^\mu \psi(x)$ at the vertex x, these fermion lines being amputated (Fig. 12).

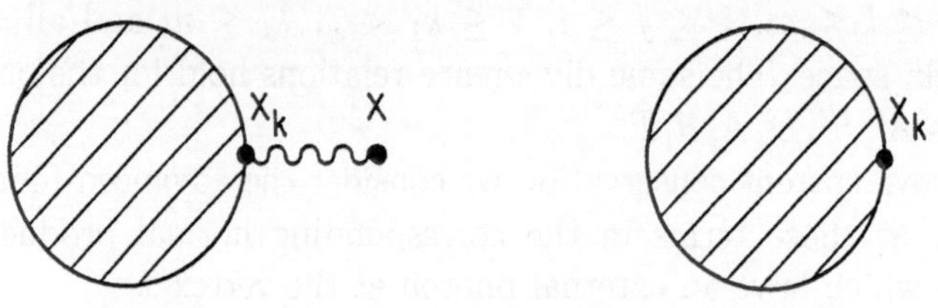

Fig. 12. Once and twice amputated graphs appearing in (4.10.50)

Consequently, $A_A^{(n+1)}$ is of the following form

$$A_{A\mu}^{(n+1)}(x_1,\ldots,x_n;x)=\sum_{k=1}^{n}\tilde{A}_k^{\nu}(x_1,\ldots,x_n)ig_{\mu\nu}D_0^{\mathrm{av}}(x_k-x), \tag{4.10.50}$$

where $\tilde{A}_k^{\nu}$ is advanced relative to x_k (compare (4.10.34)). Splitting off the propagator $g_{\mu\nu}D_0^{\mathrm{av}}(x_k-x)$ in (4.10.50) is the same as amputating an external photon line $A_\nu(x_k)$, which brings us back to the advanced distribution $A_j^{(n)\nu}$ (4.10.47) of the interacting current. Hence,

$$\tilde{A}_k^{\nu}(x_1,\ldots,x_n)=A_j^{(n)\nu}(x_1,\ldots\hat{x}_k,\ldots x_n;x_k)$$

$$=A^{(n)\nu}(x_1,\ldots\hat{x}_k,\ldots x_n;x_k). \tag{4.10.51}$$

Using (4.10.45) we arrive at

$$\frac{\partial}{\partial x_k^{\nu}}\tilde{A}_k^{\nu}(x_1,\ldots,x_n)=0 \tag{4.10.52}$$

for all $k=1,\ldots n$.

Now we are ready to consider the Lorentz condition

$$\frac{\partial}{\partial x^\mu}A_{\mathrm{int}}^{\mu}(g,x)=\partial_\mu A^\mu(x)$$

$$+\frac{\partial}{\partial x_\mu}\sum_{n=0}^{\infty}\sum_{k=1}^{n}\tilde{A}_k^{\nu}(x_1,\ldots,x_n)g_{\mu\nu}D_0^{\mathrm{av}}(x_k-x)g(x_1)\ldots g(x_n)$$

$$=\partial_\mu A^\mu(x)$$

$$-\sum_{n=0}^{\infty}\sum_{k=1}^{n}\tilde{A}_k^{\nu}(x_1,\ldots,x_n)\left[\frac{\partial}{\partial x_k^{\nu}}D_0^{\mathrm{av}}(x_k-x)\right]g(x_1)\ldots g(x_n).$$

Integrating by parts and using (4.10.52), we shall obtain

$$= \partial_\mu A^\mu(x) + \sum_{n=0}^{\infty} \sum_k \tilde{A}_k^\nu(x_1, \ldots, x_n)$$

$$\times D_0^{\mathrm{av}}(x_k - x) g(x_1) \ldots \left[\frac{\partial g(x_k)}{\partial x_k^\nu}\right] \ldots g(x_n). \tag{4.10.53}$$

There are no boundary terms because $g(x)$ vanishes at infinity. Considering now the adiabatic limit $g \to 1$, we have

$$\lim_{g\to 1} \frac{\partial g(x)}{\partial x} = 0,$$

consequently,

$$\partial_\mu A^\mu_{\mathrm{int}}(x) = \partial_\mu A^\mu(x),$$

which vanishes on the physical subspace

$$\partial_\mu A^\mu_{\mathrm{int}}(x)\Big|_{\mathcal{F}_{\mathrm{phys}}} = 0. \tag{4.10.54}$$

As discussed above (4.10.9), the Lorentz condition only holds in the sense of bilinear forms.

The fact that the adiabatic limit is necessary for the Lorentz condition to hold has a simple physical explanation: Multiplication of the interaction by $g(x)$ destroys the local gauge invariance of the theory. Then relations which are connected with special gauges are only satisfied after restoring gauge invariance by taking the limit $g(x) \to 1$. Finally we remark that in contrast to the free term $\partial_\mu A^\mu(x)$, the divergence condition for the difference

$$\partial_\mu \Big[A^\mu_{\mathrm{int}}(x) - A^\mu(x)\Big] = 0$$

holds, quite surprisingly, on the entire Fock space.

4.11 Problems

4.1. Repeate the calculation of $(\Phi, R_2'(x_1, x_2)\Phi)$ (4.2.10) for an electron-photon state (4.2.14).

4.2. Determine the singular order ω (4.3.2) (a) after a double fermionic contraction, (b) after a photonic and fermionic contraction at the same vertex. Show that (4.3.2) still holds.

4.3. The dihedral group D_4 is generated by the following rotations in 3-space: $\pi/2$-rotation around the z-axis, π-rotation around the x-axis. It has 8 elements that fall into 5 classes of conjugated elements. Determine these classes and show that D_4 is isomorphic to G_8 (4.4.25).

4.4. Can a photon decay in vacuum into three photons? Hint: Transform the amplitude by means of crossing symmetry to photon-photon scattering and use the result of gauge invariance (4.6.35). It implies that

$$t^{\nu\mu\alpha\beta}(k_1,k_2,k_3,k_4) = Ck_1^\nu k_2^\mu k_3^\alpha k_4^\beta + \dots \tag{4.11.1}$$

is the beginning of an expasion of the scattering amplitude.

4.5. Verify the results (4.9.22–27) for the interacting fields.

4.6. Check the Dirac equation (4.10.6) and the wave equation (4.10.7) for the interacting fields (4.9.22–23) in lowest orders.

5. Other Electromagnetic Couplings

In this chapter we are concerned with other electrodynamic theories which show new features, not present in four-dimensional spinor QED. The new feature in scalar QED is the derivative coupling between the scalar field and the vector potential. It has important consequences for gauge invariance, in particular, the interaction gets modified by gauge invariance. Such derivative couplings also occur in Yang-Mills theories, and the same implications of gauge invariance appear here. This is briefly discussed in the Epilogue. The axial anomalies, that we describe next, provide an important example where gauge invariance partially breaks down. These anomalies also occur in the standard electro-weak theory and must be compensated there. This implies the lepton-quark duality, which means that to each lepton family (e, μ, τ) there must exist a corresponding quark family. Finally, if QED in two space plus one time dimensions is considered, the vacuum polarization shows unusual behaviour. This has the surprising consequence that the photon becomes massive. These examples illustrate how rich QED really is.

5.1 Scalar QED: Basic Properties

To couple a charged scalar field $\varphi(x)$ to the vector potential $A^\mu(x)$, we need a derivative in the coupling, in order to form a Lorentz scalar. This brings in a new element, which is of particular interest for non-Abelian gauge theories where such derivative couplings also appear.

Our starting point is

$$T_1(x) = e\,A^\mu(x)\;:[\varphi^+(x)\partial_\mu\varphi(x) - \partial_\mu\varphi^+(x)\varphi(x)]: \tag{5.1.1}$$

$$\stackrel{\text{def}}{=} e\,A^\mu(x)\;:\varphi^+(x)\overset{\leftrightarrow}{\partial}_\mu\varphi(x):.$$

In contrast to the lagrangian approach, we have no e^2-vertex in $T_1(x)$! The latter will appear in T_2 as a normalization term (see (5.1.24)). We quantize the free radiation field $A^\mu(x)$ in the same way as before (Sect. 2.11). The free massive charged scalar field $\varphi(x)$ is quantized in the usual manner (see (2.1.79)). The contractions appearing in Wick's theorem are given by

$$\overline{A_\mu(x)A_\nu(y)} = [A_\mu^{(-)}(x),\, A_\nu^{(+)}(y)] = ig_{\mu\nu}D_0^{(+)}(x-y) \tag{5.1.2}$$

$$\overbrace{\varphi(x)^+\varphi(y)} = [\varphi^{+(-)}(x), \varphi^{(+)}(y)] = -iD_m^{(+)}(x-y) \tag{5.1.3}$$

$$\overbrace{\varphi(x)\varphi^+(y)} = [\varphi^{(-)}(x), \varphi^{+(+)}(y)] = -iD_m^{(+)}(x-y) \tag{5.1.4}$$

$$\overbrace{(\partial_\mu\varphi^+)(x)\varphi(y)} = -i(\partial_\mu D_m^{(+)})(x-y),$$

$$\overbrace{\varphi^+(x)(\partial_\mu\varphi)(y)} = i(\partial_\mu D_m^{(+)})(x-y) \tag{5.1.5}$$

$$\overbrace{(\partial_\mu\varphi^+)(x)(\partial_\nu\varphi)(y)} = i(\partial_\mu\partial_\nu D_m^{(+)})(x-y),$$

$$\overbrace{(\partial_\mu\varphi)(x)\varphi^+(y)} = -i\partial_\mu(D_m^{(+)})(x-y),\ldots \tag{5.1.6}$$

where $(+)$ and $(-)$ stand for the emission and absorption parts of the field operators. We use the usual definitions of the commutation functions (Sect. 2.3). The building-stones of the diagrams are the two vertices defined by (5.1.1) and shown in Fig. 13a,b. The dashes on the particle lines represent the derivatives in (5.1.1). We note the free current conservation

$$(\partial_\mu j_\mu)(x) = 0, \quad \text{where} \quad j_\mu(x) =: \varphi^+(x)\overleftrightarrow{\partial}_\mu\varphi(x) :, \tag{5.1.7}$$

which holds because $\varphi(x)$ and the adjoint $\varphi^+(x)$ are solutions of the Klein-Gordon equation.

Having specified $T_1(x)$ and the basic contractions, the inductive construction goes through as described in Sect. 3.1. We first discuss some important general properties of the S-matrix in scalar QED.

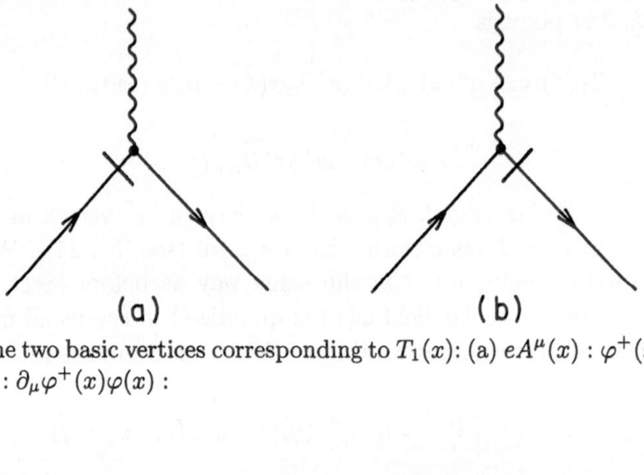

Fig. 13. The two basic vertices corresponding to $T_1(x)$: (a) $eA^\mu(x) : \varphi^+(x)\partial_\mu\varphi(x) :$, (b) $eA^\mu(x) : \partial_\mu\varphi^+(x)\varphi(x) :$

(a) Normalizability. This property can be proven as in Sect 4.3, using the scaling definition of the singular order. We consider a graph with $2s$ external scalar particles and b external photons and we are going to prove that

$$\omega = 4 - b - 2s - d, \tag{5.1.8}$$

where d is the number of derivatives on the external scalar field operators ($d \le 2s$).

In the inductive construction of T_n from the $T_m, m \le n-1$, one must consider tensor products of two distributions

$$T_r^1(x_1, \dots x_r) T_v^2(y_1, \dots y_v)$$

with singular orders, say, ω_1 and ω_2 which obey (5.1.8) as induction hypothesis. This product is normally ordered and we assume that l photon contractions arise in this process. Then, taking translation invariance into account, the numerical part of the contracted expression is of the form

$$t_1(x_1 - x_r, \dots x_{r-1} - x_r) \prod_{j=1}^{l} D_0^{(+)}(x_{r_j} - y_{v_j}) t_2(y_1 - y_v, \dots y_{v-1} - y_v)$$

$$\stackrel{\text{def}}{=} t(\xi_1, \dots \xi_{r-1}, \eta_1, \dots \eta_{v-1}, \eta). \tag{5.1.9}$$

Here, $\{x_{r_j}\}$ is a subset of $\{x_1, \dots x_r\}$ and $\{y_{v_j}\}$ is a subset of $\{y_1, \dots y_v\}$, and we have introduced relative coordinates

$$\xi_j = x_j - x_r, \quad \eta_j = y_j - y_v, \quad \eta = x_r - y_v.$$

The contraction function is given by

$$D_0^{(+)}(x) = \frac{i}{(2\pi)^3} \int d^4p\, \delta^1(p^2) \Theta(p^0) e^{-ipx}. \tag{5.1.10}$$

We compute the Fourier transform (omitting powers of 2π)

$$\hat{t}(p_1, \dots p_{r-1}, q_1, \dots q_{v-1}, q) = \int t(\xi, \eta) e^{ip\xi + iq\eta} d^{4r-4}\xi\, d^{4v}\eta.$$

Since products go over into convolutions, we get

$$\hat{t}(\cdots) = \int \prod_j d\kappa_j\, \delta\Big(q - \sum_{j=1}^{l} \kappa_j\Big)$$

$$\times \hat{t}_1(\dots p_i - \kappa_{r(i)} \dots) \prod_j \hat{D}_0^{+}(\kappa_j)\, \hat{t}_2(\dots q_k + \kappa_{v(k)} \dots).$$

Here, $r(i) = v(k)$ if and only if x_i and y_k are joined by a contraction. Applying this to a test function $\varphi \in \mathcal{S}(\mathbf{R}^{4(r+v-1)})$, we obviously have

$$\langle \hat{t}, \varphi \rangle = \int d^{4r-4}p' \, d^{4v-4}q' \, \hat{t}_1(p')\hat{t}_2(q')\psi(p', q'),$$

with

$$\psi = \int d^4q \prod_j d\kappa_j \, \delta\Big(q - \sum_j \kappa_j\Big)\varphi(\ldots p'_i + \kappa_{r(i)} \ldots, \ldots q'_k - \kappa_{v(k)} \ldots q)$$

$$\times \prod_{j=1}^{l} \hat{D}_0^{(+)}(\kappa_j). \tag{5.1.11}$$

In order to determine the singular order of $\hat{t}$ in p-space, we have to consider the scaled distribution

$$\Big\langle \hat{t}\Big(\frac{p}{\delta}\Big), \varphi \Big\rangle = \delta^m \langle \hat{t}(p), \varphi(\delta p) \rangle$$

$$= \delta^m \int d^{4r-4}p' \, d^{4v-4}q' \, \hat{t}_1(p')\hat{t}_2(q')\psi_\delta(p', q'), \tag{5.1.12}$$

where

$$\psi_\delta(p', q') = \int d^4q \prod_j d\kappa_j \, \delta\Big(q - \sum_j \kappa_j\Big)$$

$$\varphi(\ldots \delta(p'_i + \kappa_{r(i)}) \ldots, \ldots \delta(q'_k - \kappa_{v(k)}) \ldots \delta q) \prod_j \hat{D}_0^{(+)}(\kappa_j) \tag{5.1.13}$$

and $m = 4(r + v - 1)$. We introduce scaled variables $\tilde{\kappa}_j = \delta \kappa_j$, $\tilde{q} = \delta q$ and note that

$$\hat{D}_0^{(+)}\Big(\frac{\tilde{\kappa}}{\delta}\Big) = \delta^1\Big(\frac{\tilde{\kappa}^2}{\delta^2}\Big)\Theta\Big(\frac{\tilde{\kappa}^0}{\delta}\Big) = \delta^2 \hat{D}_0^{(+)}(\tilde{\kappa}). \tag{5.1.14}$$

This implies

$$\psi_\delta(p', q') = \frac{\delta^{2l}}{\delta^{4l}} \int d^4\tilde{q} \prod_j d\tilde{\kappa}_j \, \delta\Big(\tilde{q} - \sum_j \tilde{\kappa}_j\Big)$$

$$\times \varphi(\ldots \delta p'_i + \tilde{\kappa}_{r(i)} \ldots, \ldots \delta q'_k - \tilde{\kappa}_{v(k)} \ldots, \tilde{q}) \prod_j \hat{D}_0^{(+)}(\tilde{\kappa}_j)$$

$$= \frac{1}{\delta^{2l}}\psi(\delta p', \delta q').$$

Using again scaled variables $\delta p' = \tilde{p}$, $\delta q' = \tilde{q}$, we find

$$\Big\langle \hat{t}\Big(\frac{p}{\delta}\Big), \varphi \Big\rangle = \frac{\delta^4}{\delta^{2l}} \int d^{4r-4}\tilde{p} \, d^{4v-4}\tilde{q} \, \hat{t}_1\Big(\frac{\tilde{p}}{\delta}\Big)\hat{t}_2\Big(\frac{\tilde{q}}{\delta}\Big)\psi(\tilde{p}, \tilde{q}).$$

By the induction hypothesis, $\hat{t}_1$ and $\hat{t}_2$ have singular orders ω_1, ω_2 with power counting functions $\rho_1(\delta)$, $\rho_2(\delta)$, respectively. Then the following limit exists:

$$\lim_{\delta \to 0} \delta^{2l-4} \rho_1(\delta)\rho_2(\delta) \Big\langle \hat{t}\Big(\frac{p}{\delta}\Big), \varphi \Big\rangle = \langle \hat{t}_0(p), \varphi \rangle,$$

hence, the singular order of $\hat{t}(p)$ is

$$\omega = \omega_1 + \omega_2 + 2l - 4.$$

It remains to check that this result satisfies (5.1.8). Substituting

$$\omega_j = 4 - b_j - 2s_j - d_j, \quad j = 1, 2 \tag{5.1.15}$$

we find

$$\omega = 4 - (b_1 + b_2 - 2l) - 2(s_1 + s_2) - (d_1 + d_2).$$

Since the first bracket is just the number of photon operators after the l photon contractions, (5.1.8) is proven in this case.

Let us now consider l scalar contractions. Then instead of (5.1.13) we have

$$\psi_\delta(p', q') = \int d^4q \prod_{j=1}^{l} d\kappa_j\, \delta(q - \sum_j \kappa_j)\varphi(\ldots \delta \cdot (p'_i + \kappa_{r(i)}) \ldots,$$

$$\ldots \delta \cdot (q'_k - \kappa_{v(k)}) \ldots, \delta q) \prod_{j=1}^{l} \kappa_j^{a_j} \hat{D}_m^{(+)}(\kappa_j). \tag{5.1.16}$$

Here $a_j = 0, 1, 2$ according to the number of derivatives on the contracted scalar field operators. We use the scaled variables $\tilde{\kappa}_j = \delta\kappa_j$, $\tilde{q} = \delta q$ and

$$\lim_{\delta \to 0} \frac{1}{\delta^2} \hat{D}_m^{(+)}\Big(\frac{\tilde{\kappa}}{\delta}\Big) = \hat{D}_0^{(+)}(\tilde{\kappa}). \tag{5.1.17}$$

Let

$$\sum_{j=1}^{l} a_j = a, \tag{5.1.18}$$

then for $\delta \to 0$, ψ_δ (5.1.16) behaves as

$$\psi_\delta(p', q') \longrightarrow \frac{\delta^{2l-a}}{\delta^{4l}} \int d^4\tilde{q} \prod_j d\tilde{\kappa}_j\, \delta(\tilde{q} - \sum_j \tilde{\kappa}_j)$$

$$\times \varphi(\ldots \delta p'_i + \tilde{\kappa}_{r(i)} \ldots, \ldots \delta q'_i - \tilde{\kappa}_{v(i)} \ldots, \tilde{q}) \prod_j \partial^{a_j} \hat{D}_0^{(+)}(\tilde{\kappa}_j)$$

$$= \frac{1}{\delta^{2l+a}} \psi(\delta p', \delta q'). \tag{5.1.19}$$

Then (5.1.12) becomes

$$\Big\langle \hat{t}\Big(\frac{p}{\delta}\Big), \varphi \Big\rangle \longrightarrow \frac{\delta^4}{\delta^{2l+a}} \int d^{4r-4}\tilde{p}\, d^{4v-4}\tilde{q}\, \hat{t}_1\Big(\frac{\tilde{p}}{\delta}\Big) \hat{t}_2\Big(\frac{\tilde{q}}{\delta}\Big) \psi(\tilde{p}, \tilde{q}). \tag{5.1.20}$$

Consequently, the limit

$$\lim_{\delta\to 0}\delta^{2l+a-4}\rho_1(\delta)\rho_2(\delta)\langle\hat{t}(\frac{p}{\delta}),\varphi\rangle=\langle\hat{t}_0(p),\varphi\rangle \tag{5.1.21}$$

exists, which implies that the singular order is equal to

$$\omega=\omega_1+\omega_2+2l+a-4. \tag{5.1.22}$$

Substituting here the induction hypothesis (5.1.15) we find

$$\omega=4-(b_1+b_2)-2(s_1+s_2-l)-(d_1+d_2-a). \tag{5.1.23}$$

This is in agreement with (5.1.8) because $2l$ scalar operators with a derivatives have disappeared. The same reasonings go through for simultaneous photon and scalar contractions.

The final step of the inductive construction is the splitting of the causal distribution into a retarded and advanced part. In this process the singular order is not changed. Hence, (5.1.8) is true in general. As in spinor QED this implies that there are only finitely many cases with non-negative ω that require normalization. This is the normalizability of scalar QED.

The result (5.1.8) has the surprising consequence that the Compton graphs in Fig. 14a,b have $\omega=0$! Then even these tree graphs have free normalization terms

$$T_2(x_1,x_2)=\ldots-iC'_{\rm Com}e^2\delta(x_1-x_2):\varphi^+(x_1)\varphi(x_2)::A^\mu(x_1)A_\mu(x_2):$$

$$-iC_{\rm Com}e^2\delta(x_1-x_2):\varphi^+(x_2)\varphi(x_1)::A^\mu(x_2)A_\mu(x_1):. \tag{5.1.24}$$

Below the two constants will be fixed by symmetry properties, in particular gauge invariance. In this way the e^2-vertex in Fig. 14c is generated by gauge invariance. In contrast to spinor QED where only loop graphs have non-negative ω, here also certain tree graphs must be carefully normalized. Since this is due to the derivative coupling, this mechanism is of great interest

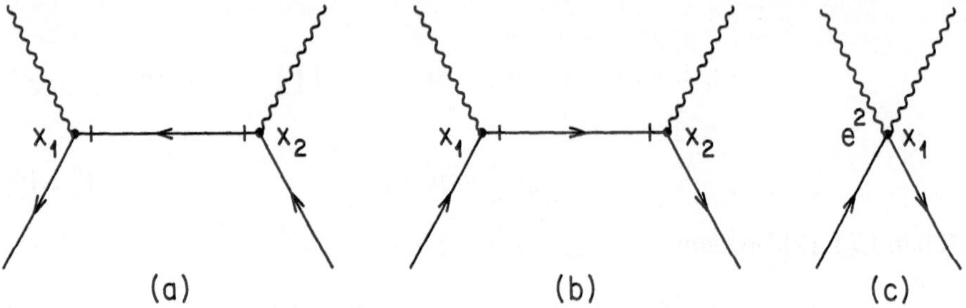

Fig. 14. (a, b) Compton graphs with two derivatives on the inner scalar line. (c) e^2-vertex corresponding to the normalization terms of these graphs

in Yang-Mills theories and quantum gravity. In fact, it turns out that gauge invariant normalization is the best way to find the higher coupling-terms in gauge theories with derivative couplings. At the same time the theory becomes completely transparent because different orders are not mixed by gauge invariance.

Another class of diagrams with $\omega = 0$ is shown in Fig. 15. It has four external boson lines. Here, an undetermined normalization constant λ_n appears again

$$T_n(x_1, \ldots, x_n) = \ldots + e^n \lambda_n \delta(x_1 - x_n) \ldots \delta(x_{n-1} - x_n)$$

$$\sum_{\substack{i_1<i_2, j_1<j_2 \\ i_k \neq j_k (k,l=1,2)}} : \varphi^+(x_{i_1}) \varphi^+(x_{i_2}) \varphi(x_{j_1}) \varphi(x_{j_2}) : . \tag{5.1.25}$$

Since

$$\frac{1}{n!} \int dx_2 \ldots dx_n \, T_n(x_1, \ldots, x_n) = \ldots + e^n \frac{\lambda_n}{4(n-4)!} : \varphi^+(x_1) \varphi^+(x_1)$$

$$\varphi(x_1) \varphi(x_1) :, \tag{5.1.26}$$

the normalization term in (5.1.25) corresponds to a φ^4-vertex (Fig. 15b). Pseudo-unitarity requires λ_n to be purely imaginary (see c) below). Since there are no external photons in Fig. 15, gauge invariance does not further restrict λ_n. The coupling constant

$$\lambda \stackrel{\text{def}}{=} \sum_{l=2}^{\infty} e^{2l} \frac{\operatorname{Im} \lambda_{2l}}{4(2l-4)!} \tag{5.1.27}$$

of the φ^4-theory, that is automatically contained in scalar QED, remains free.

(b) Charge Conjugation Invariance. Let U_φ (U_A) be the unitary charge conjugation operator in the scalar (photon) Fock space, respectively

$$U_\varphi \varphi^{(-)}(x) U_\varphi^{-1} = \varphi^{+(-)}(x), \quad U_\varphi \varphi^{(+)} U_\varphi^{-1} = \varphi^{+(+)}$$

$$U_A A^\mu(x) U_A^{-1} = -A^\mu(x). \tag{5.1.28}$$

We now prove by induction on n that the normalization constants can be chosen in such a way that

$$(U_\varphi \otimes U_A) T_n(x_1, \ldots, x_n) (U_\varphi^{-1} \otimes U_A^{-1}) = T_n(x_1, \ldots, x_n) \tag{5.1.29}$$

holds for all n. This is charge conjugation invariance.

From (5.1.28) we obtain

$$U_\varphi : (\partial_\mu \varphi^+)(x) \varphi(x) : U_\varphi^{-1} =: \varphi^+(x) (\partial_\mu \varphi)(x) :, \quad U_\varphi j_\mu(x) U_\varphi^{-1} = -j_\mu(x), \tag{5.1.30}$$

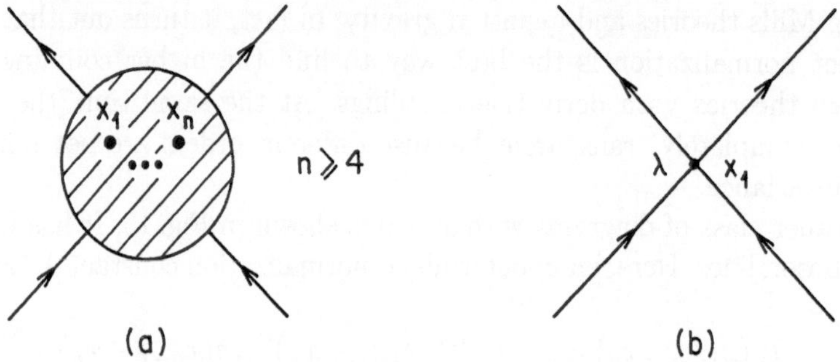

Fig. 15. (a) All graphs of n-th order $\sim$: $\varphi^+\varphi^+\varphi\varphi$. (b) The corresponding normalization term (5.1.25) generates a φ^4-vertex

with j_μ given by (5.1.7). Together with (5.1.28) this proves (5.1.29) for $n = 1$. The inductive step from $n-1$ to n is proven in exactly the same way as in Sect4.4.

To investigate the consequences of C-invariance, we must introduce a notation for a general normally ordered term in the n-point distribution T_n. We write

$$T_n(x_1,\ldots,x_n) = \sum\Big[\,{}^{m_1\ldots m_l}t^{i_1\ldots i_k|i_{k+1}\ldots i_s}_{j_1\ldots j_t|j_{t+1}\ldots j_s}(x_1-x_n,\ldots x_{n-1}-x_n) : \varphi^+(x_{i_1})$$

$$\ldots\varphi^+(x_{i_k})(\partial\varphi^+)(x_{i_{k+1}})\ldots(\partial\varphi^+)(x_{i_s})\varphi(x_{j_1})\ldots\varphi(x_{j_t})(\partial\varphi)(x_{j_{t+1}})\ldots$$

$$(\partial\varphi)(x_{j_s}) :: A(x_{m_1})\ldots A(x_{m_l}) : +\,{}^{m_1\ldots m_l}t^{j_1\ldots j_t|j_{t+1}\ldots j_s}_{i_1\ldots i_k|i_{k+1}\ldots i_s}(x_1-x_n,\ldots x_{n-1}-x_n)$$

$$: \varphi^+(x_{j_1})\ldots\varphi^+(x_{j_t})(\partial\varphi^+)(x_{j_{t+1}})\ldots(\partial\varphi^+)(x_{j_s})\varphi(x_{i_1})\ldots$$

$$\varphi(x_{i_k})(\partial\varphi)(x_{i_{k+1}})\ldots(\partial\varphi)(x_{i_s}) :: A(x_{m_1})\ldots A(x_{m_l}) :\Big]. \tag{5.1.31}$$

We have

$$(-1)^l = (-1)^n \tag{5.1.32}$$

because an even number of photon operators is contracted. The scalar field operators of the two terms in (5.1.31) are exchanged under charge conjugation (5.1.28)

$$(U_\varphi\otimes U_A)T_n(x_1,\ldots,x_n)(U_\varphi^{-1}\otimes U_A^{-1}) = \sum\Big[\,{}^{m}t^{i}_{j} : \varphi^+(x_{j_1})\ldots$$

$$\varphi^+(x_{j_t})(\partial\varphi^+)(x_{j_{t+1}})\ldots(\partial\varphi^+)(x_{j_s})\varphi(x_{i_1})\ldots(\partial\varphi)(x_{i_{k+1}})\ldots : (-1)^l$$

$$: A(x_{m_1})\ldots : +\,{}^{m}t^{j}_{i} : \varphi^+(x_{i_1})\ldots(\partial\varphi^+)(x_{i_{k+1}})\ldots\varphi(x_{j_1})\ldots$$

$$\ldots(\partial\varphi^+)(x_{j_{t+1}})\ldots:(-1)^l:A(x_{m_1})\ldots:\Big]. \tag{5.1.33}$$

Now C-invariance (5.1.29) implies

$$(-1)^n\cdot {}^{m_1\ldots m_l}t^{i_1\ldots i_k|i_{k+1}\ldots i_s}_{j_1\ldots j_t|j_{t+1}\ldots j_s}(x_1-x_n,\ldots x_{n-1}-x_n)$$

$$= {}^{m_1\ldots m_l}t^{j_1\ldots j_t|j_{t+1}\ldots j_s}_{i_1\ldots i_k|i_{k+1}\ldots i_s}(\ldots), \tag{5.1.34}$$

using (5.1.32). In diagrammatic language this means that the numerical distributions of two diagrams with reversed arrows of all scalar lines agree up to a factor $(-1)^n$. For $s=0$ in (5.1.34) we obtain Furry's theorem

$${}^{m_1\ldots m_l}t(x_1-x_n,\ldots x_{n-1}-x_n)=0 \quad \text{for } l \text{ odd}. \tag{5.1.35}$$

For $n=2, s=2, k=t=1$ (5.1.34) implies that the two constants in (5.1.24) must be equal

$$C_{\text{Com}}=C'_{\text{Com}}. \tag{5.1.36}$$

In the next section gauge invariance will fix the constant to $C_{\text{Com}}=-1$. In the adiabatic limit $g(x)=1$, the normalization term (5.1.24) then gives the following contribution

$$\frac{1}{2!}\int d^4x_2\,T_2(x_1,x_2)=\ldots+ie^2:\varphi^+(x_1)\varphi(x_1)::A^\mu(x_1)A_\mu(x_1):. \tag{5.1.37}$$

In diagrammatic language this term corresponds to the e^2-vertex in Fig. 14c. Since the T_n's are constructed inductively, the normalization term (5.1.24) propagates in higher orders. The corresponding terms have the same structure as if they were generated by the e^2-vertex. In this way we arrive at the C-invariant, gauge invariant and, hence, unitary S-matrix of scalar QED.

It is evident from the inductive proof that C-invariance (5.1.29) and (5.1.34) holds also for the various retarded, advanced and causal distributions R'_n, A'_n, D_n, R_n and A_n-distributions in the inductive construction.

(c) Pseudo-Unitarity on the Entire Fock Space. Pseudo-unitarity means

$$T_n^K(X)=\tilde{T}_n(X),\quad n=1,2,\ldots, \tag{5.1.38}$$

where $\tilde{T}_n$ occurs in the perturbation expansion of $S(g)^{-1}$ and ^{K} is the conjugation in Fock space, defined in such a way that $A^\mu(x)$ is self-conjugate (Sect. 2.11). The inductive proof in Sect. 4.7 holds also for scalar QED. It remains only to be checked that (5.1.38) is true for $n=1$ which is obvious from (5.1.1). In the earlier proof, the local normalization terms have to be chosen appropriately, in particular, one must normalize the retarded distributions R_n such that

$$R_n^K=-R_n. \tag{5.1.39}$$

Analogously to (5.1.31), (5.1.33), this can be translated into a condition for the numerical distributions

$$({}^m r^i_j)^* = -({}^m r^j_i) = -(-1)^n({}^m r^i_j), \tag{5.1.40}$$

where (5.1.34) has been used. The retarded distributions are obtained in the process of distribution splitting. It is most convenient to work with the central splitting solution again, that is defined by normalization at $p = 0$. We denote it by a subscript 0. This central solution has all symmetry properties of the original causal distribution, for example it fulfills (5.1.39). Then the normalization constants of the general solution

$$ {}^m r^i_j(x_1 - x_n, \dots x_{n-1} - x_n) = {}^m_0 r^i_j(x_1 - x_n, \dots x_{n-1} - x_n)$$

$$+ \sum_{|a|=0}^{\omega} ({}^m_a C^i_j) D^a(\delta(x_1 - x_n) \dots \delta(x_{n-1} - x_n)) \tag{5.1.41}$$

must be purely imaginary if n is even, or real if n is odd, in order to satisfy (5.1.39) again.

5.2 Scalar QED: Gauge Invariance

We now come to the interesting problem mentioned in the last section, namely gauge invariance of the Compton graphs. The second order causal distribution for Compton scattering is given by

$$D_2^{\rm Com}(x_1, x_2) = ie^2 : A^\mu(x_1)A^\nu(x_2) : \Big\{\Big[- : \varphi^+(x_2)\varphi(x_1) : (\partial_\mu\partial_\nu D_m)(x_1 - x_2)$$

$$+ : \varphi^+(x_2)(\partial_\mu\varphi)(x_1) : (\partial_\nu D_m)(x_1 - x_2) - : (\partial_\nu\varphi^+)(x_2)\varphi(x_1) :$$

$$\times(\partial_\mu D_m)(x_1 - x_2) + : (\partial_\nu\varphi^+)(x_2)(\partial_\mu\varphi)(x_1) : D_m(x_1 - x_2)\Big]$$

$$+\Big[(\partial)\varphi^+ \longleftrightarrow (\partial)\varphi\Big]\Big\}. \tag{5.2.1}$$

The first term in square brackets is represented by Fig. 14b and the second square bracket by Fig. 14a. The diagrams of the other terms are obtained by shifting one or both derivatives from the internal scalar lines to an external line.

The causal splitting of the terms with no or one derivative on $D_m(x_1 - x_2)$ is trivial

$$D_m = D_m^{\rm ret} - D_m^{\rm av}, \quad \partial_\mu D_m = \partial_\mu D_m^{\rm ret} - \partial_\mu D_m^{\rm av}, \tag{5.2.2}$$

but with two derivatives we have $\omega = 0$. Although a splitting solution is trivially obtained, there is an undetermined local term

$$(\partial_\mu\partial_\nu D_m)(x) = [(\partial_\mu\partial_\nu D_m^{\rm ret})(x) + C_{\rm Com} g_{\mu\nu}\delta(x)] - [(\partial_\mu\partial_\nu D_m^{\rm av})(x) +$$

$$+C_{\rm Com}g_{\mu\nu}\delta(x)]. \tag{5.2.3}$$

With $D_m^F = D_m^{\rm ret} - D_m^{(-)}$ we get for $T_2 = R_2 - R_2'$

$$T_2^{\rm Com}(x_1,x_2) = ie^2 :A^\mu(x_1)A^\nu(x_2): \Bigl\{\Bigl[- :\varphi^+(x_2)\varphi(x_1): [\partial_\mu\partial_\nu D_m^F(x_1-x_2)$$

$$+C_{\rm Com}g_{\mu\nu}\delta(x_1-x_2)] + (D_m \longrightarrow D_m^F \quad \text{in (5.2.1)})\Bigr]$$

$$+\Bigl[- :\varphi^+(x_1)\varphi(x_2): [\partial_\mu\partial_\nu D_m^F(x_1-x_2) + C'_{\rm Com}g_{\mu\nu}\delta(x_1-x_2)]$$

$$+(D_m \longrightarrow D_m^F \quad \text{in (5.2.1)})\Bigr]\Bigr\}. \tag{5.2.4}$$

Now C-invariance (5.1.34), that connects the two square brackets, requires $C_{\rm Com} = C'_{\rm Com}$. As in spinor we consider gauge transformations

$$A^\mu(x) \longrightarrow A^\mu(x) + \partial^\mu \Lambda(x) \tag{5.2.5}$$

of the vector field alone, where $\Lambda(x)$ is an arbitrary C-number field. The (free) scalar field φ remains unchanged. The S-matrix should be invariant under this transformation. This means that the divergence of the curly bracket in (5.2.4) with respect to (x_1,μ) and (x_2,ν) must vanish. Due to the symmetry in x_1, x_2, it is sufficient to require the one condition

$$0 = \partial_{x_1\mu}\{\ldots\}^{\mu\nu} = - :\varphi^+(x_2)\varphi(x_1): (\partial^\nu\delta)(x_1-x_2)(1+C_{\rm Com})$$

$$- :\varphi^+(x_1)\varphi(x_2): (\partial^\nu\delta)(x_1-x_2)(1+C_{\rm Com})$$

$$- :\varphi^+(x_2)(\partial^\nu\varphi)(x_1): \delta(x_1-x_2)C_{\rm Com} - - :(\partial^\nu\varphi^+)(x_2)\varphi(x_1): \delta(x_1-x_2)$$

$$- :(\partial^\nu\varphi^+)(x_1)\varphi(x_2): C_{\rm Com}\delta(x_1-x_2) - :\varphi^+(x_1)(\partial^\nu\varphi)(x_2): \delta(x_1-x_2), \tag{5.2.6}$$

where

$$\square\varphi = -m^2\varphi, \quad \square\varphi^+ = -m^2\varphi^+,$$

$$\square D_m^F = \delta - m^2 D_m^F \tag{5.2.6a}$$

has been used. With the distributional identity

$$:\varphi^+(x_2)\varphi(x_1): (\partial^\nu\delta)(x_1-x_2) = - :\varphi^+(x_2)(\partial^\nu\varphi)(x_2): \delta(x_1-x_2)$$

$$+(\partial^\nu\delta)(x_1-x_2) :\varphi^+(x_2)\varphi(x_2): \tag{5.2.7}$$

which is easily proved by smearing out in x_1 and x_2, all field operators in (5.2.6) get the same argument x_2

$$0 = - :\varphi^+(x_2)\varphi(x_2): (\partial^\nu\delta)(x_1-x_2)(1+C_{\rm Com}+1+C_{\rm Com})$$

$$+ :\varphi^+(x_2)(\partial^\nu\varphi)(x_2): \delta(x_1-x_2)(1+C_{\rm Com}-C_{\rm Com}-1)$$

$$+ :(\partial^\nu\varphi^+)(x_2)\varphi(x_2): \delta(x_1-x_2)(1+C_{\rm Com}-1-C_{\rm Com}). \tag{5.2.8}$$

We see that gauge invariance requires $C_{\text{Com}} = -1$. This non-vanishing normalization term may be written as $e^2 : \varphi^+\varphi :: A^\mu A_\mu :$ (5.1.37). Note that all eight Compton diagrams corresponding to the eight terms in (5.2.4) contribute in the gauge invariance proof.

The gauge invariance of $D_2^{\text{Com}}(x_1, x_2)$ may be verified directly

$$\partial_{x_1\mu}\{(5.2.1)\}^{\mu\nu} = 0.$$

How gets it lost (for $C_{\text{Com}} \neq -1$) in the process of distribution splitting? The reason is that $\square D_m = -m^2 D_m$ has no δ-term, in contrast to $\square D_m^F$ in (5.2.6a). If we set $C_{\text{Com}} = 0$ in (5.2.6), gauge invariance is destroyed exactly by those δ-terms. To compensate them, we have to set $C_{\text{Com}} = -1$.

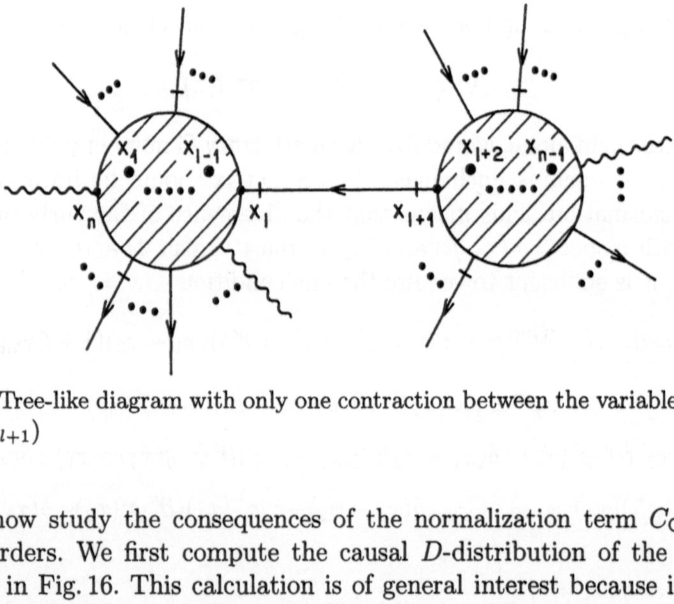

Fig. 16. Tree-like diagram with only one contraction between the variables (X, x_n) and (Y, x_{l+1})

We now study the consequences of the normalization term $C_{\text{Com}} g\delta$ in higher orders. We first compute the causal D-distribution of the tree-like diagram in Fig. 16. This calculation is of general interest because it can be carried through in any causal theory. The important result will be (5.2.22), showing that the D-distribution naturally splits in x-space into retarded and advanced parts. We introduce the notation

$$X = \{x_1, x_2, \ldots x_l\}, \quad Y = \{x_{l+2}, \ldots x_{n-1}\} \tag{5.2.9}$$

and decompose the point sets into disjoint parts

$$P_X : \quad X = X_1 \cup X_2, \quad P_Y : \quad Y = Y_1 \cup Y_2.$$

Here X_1 and Y_1 may be empty but X_2 and Y_2 not. We first compute the left subgraph in Fig. 16

$$R'(X, x_n) = \sum_{P_X} T(X_1, x_n) T^K(X_2)$$

$$= \sum_{P_X, x_l \in X_1} T(X_1, x_n) T^K(X_2) + \sum_{P_X, x_l \in X_2} T(X_1, x_n) T^K(X_2)$$

$$\stackrel{\text{def}}{=} R_l'^{1\mu}(X, x_n)(\partial_\mu \varphi)(x_l) + \ldots + R_l'^{2\mu}(X, x_n)(\partial_\mu \varphi)(x_l) + \ldots . \tag{5.2.10}$$

The subscript l means that we select only those terms necessary to compute the diagram of Fig. 16, that has a contraction between $\partial\varphi(x_l)$ and $\partial\varphi^+(x_{l+1})$. The right subgraph gives

$$R'^K(Y, x_{l+1}) = \sum_{P_Y} T(Y_2) T^K(Y_1, x_{l+1})$$

$$= (R'^K)_l^\nu(Y, x_{l+1})(\partial_\nu \varphi^+)(x_{l+1}) + \ldots \tag{5.2.11}$$

By the induction assumption we have

$$T_l(X, x_n, Y_1, x_{l+1}) = T_l^\mu(X, x_n) i[\partial_\mu \partial_\nu D_m^F(x_l - x_{l+1})$$

$$-g_{\mu\nu}\delta(x_l - x_{l+1})] T_l^\nu(Y_1, x_{l+1}) \tag{5.2.12}$$

$$\Big(T(X, x_n) T^K(Y, x_{l+1})\Big)_l = \Big(T_l(X, x_n) i\partial\partial D_m^{(+)}(x_l - x_{l+1})(T^K)_l(Y, x_{l+1})\Big)_l . \tag{5.2.13}$$

The last subscript l now means that, applying Wick's theorem to $T_l(X, x_n)$ $(T^K)_l(Y, x_{l+1})$, we only take the term without contractions. This implies

$$\Big(T_l(X, x_n)(T^K)_l(Y, x_{l+1}\Big)_l = \Big((T^K)_l(Y, x_{l+1}) T_l(X, x_n)\Big)_l \tag{5.2.14}$$

and we obtain the commutator

$$\Big([T(X, x_n),\, T^K(Y, x_{l+1})]\Big)_l = i\Big(T_l(X, x_n)\partial\partial D_m(x_l - x_{l+1})(T^K)_l(Y, x_{l+1})\Big)_l . \tag{5.2.15}$$

In the same way we get the commutations

$$\Big(\ldots T_{(l)}^{(K)}(\tilde X) T_{(l)}^{(K)}(\tilde Y)\ldots\Big)_l = \Big(\ldots T_{(l)}^{(K)}(\tilde Y) T_{(l)}^{(K)}(\tilde X)\ldots\Big)_l \tag{5.2.16}$$

with $\tilde X = \{X, x_n\}, \{X_1, x_n\}$ or X_2, $\quad \tilde Y = \{Y, x_{l+1}\}, \{Y_1, x_{l+1}\}$ or Y_2, (5.2.17)

if $\{x_l, x_{l+1}\} \not\subset \tilde X \cup \tilde Y$, since there is no contraction between $\tilde X$ and $\tilde Y$ in Fig. 16.

Let us now go over to the causal distribution

$$D(X, Y, x_{l+1}; x_n) = \sum_{P_z} [T(Z_1, x_n),\, T^K(Z_2)], \tag{5.2.18}$$

where P_z is a partition

$$X \cup Y \cup \{x_{l+1}\} = Z_1 \cup Z_2, \quad Z_2 \neq \emptyset. \tag{5.2.19}$$

We have to consider the following different types of partitions P_z (omitting the sums over P_X and P_Y): First we take the cases with x_l and x_{l+1} in the same subset. For $x_l \in X_2$ we have

$$[T(X_1,x_n),\, T^K(X_2,Y,x_{l+1})]_l = -i\Big([T_l(X_1,x_n),\, (T^K)_l(X_2)]_l$$
$$\times[\partial\partial D_m^F(x_l-x_{l+1})^* - g\delta(x_l-x_{l+1})](T^K)_l(Y,x_{l+1})\Big)_l$$
$$= i\Big([A_l'^2(X,x_n) - R_l'^2(X,x_n)][\partial\partial D_m^F(x_l-x_{l+1})^* - g\delta(x_l-x_{l+1})]$$
$$\times(T^K)_l(Y,x_{l+1})\Big)_l \tag{5.2.19a}$$
$$[T(X_1,x_n,Y,x_{l+1}),\, T^K(X_2)]_l =$$

and for $x_l \in X_1$:

$$= i\Big([T_l(X_1,x_n),\, (T^K)_l(X_2)]_l[i\partial\partial D_m^F(x_l-x_{l+1}) - g\delta(x_l-x_{l+1})]T_l(Y,x_{l+1})\Big)_l$$
$$= i\Big([R_l'^1(X,x_n) - A_l'^1(X,x_n)][\partial\partial D_m^F(x_l-x_{l+1}) - g\delta(x_l-x_{l+1})]T_l(Y,x_{l+1})\Big)_l \tag{5.2.19b}$$
$$[T(X,x_n,Y_1,x_{l+1}),\, T^K(Y_2)]_l =$$
$$= i\Big(T_l(X,x_n)[\partial\partial D_m^F(x_l-x_{l+1}) - g\delta(x_l-x_{l+1})][T_l(Y_1,x_{l+1}),\, (T^K)_l(Y_2)]_l\Big)_l$$
$$= i\Big(T_l(X,x_n)[\partial\partial D_m^F(x_l-x_{l+1}) - g\delta(x_l-x_{l+1})]D_l(Y,x_{l+1})\Big)_l. \tag{5.2.19c}$$

For $x_l \in X_2$ we have

$$[T(X_1,x_n,Y_2),\, T^K(X_2,Y_1,x_{l+1})]_l = -i\Big((T_l(X_1,x_n)(T^K)_l(X_2))_l$$
$$\times[\partial\partial D_m^F(x_l-x_{l+1})^* - g\delta(x_l-x_{l+1})](T_l(Y_2)(T^K)_l(Y_1,x_{l+1}))_l\Big)_l$$
$$+i\Big(((T^K)_l(X_2)T_l(X_1,x_n))_l[\partial\partial D_m^F(x_l-x_{l+1})^* - g\delta(x_l-x_{l+1})]$$
$$\times((T^K)_l(Y_1,x_{l+1})T_l(Y_2))_l\Big)_l$$
$$= -i\Big(R_l'^2(X,x_n)[\partial\partial D_m^F(x_l-x_{l+1})^* - g\delta(x_l-x_{l+1})](R'^K)_l(Y,x_{l+1})\Big)_l$$
$$+i\Big(A_l'^2(X,x_n)[\partial\partial D_m^F(x_l-x_{l+1})^* - g\delta(x_l-x_{l+1})](A'^K)_l(Y,x_{l+1})\Big)_l. \tag{5.2.19d}$$

For $x_l \in X_1$ we have

$$[T(X_1,x_n,Y_1,x_{l+1}),\, T^K(X_2,Y_2)]_l = i\Big((T_l(X_1,x_n)(T^K)_l(X_2))_l$$
$$\times[\partial\partial D_m^F(x_l-x_{l+1}) - g\delta(x_l-x_{l+1})](T_l(Y_1,x_{l+1})(T^K)_l(Y_2))_l\Big)_l$$

$$-i\Big(((T^K)_l(X_2)T_l(X_1,x_n))_l[\partial\partial D_m^F(x_l-x_{l+1})-g\delta(x_l-x_{l+1})]$$

$$\times((T^K)_l(Y_2)T_l(Y_1,x_{l+1}))_l\Big)_l$$

$$=i\Big(R_l'^1(X,x_n)[\partial\partial D_m^F(x_l-x_{l+1})-g\delta(x_l-x_{l+1})]R_l'(Y,x_{l+1})\Big)_l$$

$$-i\Big(A_l'^1(X,x_n)[\partial\partial D_m^F(x_l-x_{l+1})-g\delta(x_l-x_{l+1})]A_l'(Y,x_{l+1})\Big)_l. \quad (5.2.19e)$$

We now come to the partitions (5.2.19) with x_l and x_{l+1} in different subsets. The simplest case is (5.2.15). The other cases are:

$$[T(X,x_n,Y_2),\,T^K(Y_1,x_{l+1})]_l$$

$$=i\Big(T_l(X,x_n)[\partial\partial D_m^{(+)}(x_l-x_{l+1})(T_l(Y_2)(T^K)_l(Y_1,x_{l+1}))_l$$

$$+\partial\partial D_m^{(-)}(x_l-x_{l+1})((T^K)_l(Y_1,x_{l+1})T_l(Y_2))_l]\Big)_l$$

$$=i\Big(T_l(X,x_n)[\partial\partial D_m^{(+)}(x_l-x_{l+1})(R'^K)_l(Y,x_{l+1})$$

$$+\partial\partial D_m^{(-)}(x_l-x_{l+1})(A'^K)_l(Y,x_{l+1})]\Big)_l. \quad (5.2.19f)$$

For $x_l\in X_2$ we have also to consider

$$[T(X_1,x_n,Y,x_{l+1}),\,T^K(X_2)]_l$$

$$=-i\Big([(T_l(X_1,x_n)(T^K)_l(X_2))_l\partial\partial D_m^{(-)}(x_l-x_{l+1})$$

$$+((T^K)_l(X_2)T_l(X_1,x_n))_l\partial\partial D_m^{(+)}(x_l-x_{l+1})]T_l(Y,x_{l+1})\Big)_l$$

$$=-i\Big([R_l'^2(X,x_n)\partial\partial D_m^{(-)}(x_l-x_{l+1})$$

$$+A_l'^2(X,x_n)\partial\partial D_m^{(+)}(x_l-x_{l+1})]T_l(Y,x_{l+1})\Big)_l. \quad (5.2.19g)$$

For $x_l\in X_1$ we have

$$[T(X_1,x_n),\,T^K(X_2,Y,x_{l+1})]_l$$

$$=i\Big([(T_l(X_1,x_n)(T^K)_l(X_2))_l\partial\partial D_m^{(+)}(x_l-x_{l+1})$$

$$+((T^K)_l(X_2)T_l(X_1,x_n))_l\partial\partial D_m^{(-)}(x_l-x_{l+1})](T^K)_l(Y,x_{l+1})\Big)_l$$

$$=i\Big([R_l'^1(X,x_n)\partial\partial D_m^{(+)}(x_l-x_{l+1})$$

$$+A_l'^1(X,x_n)\partial\partial D_m^{(-)}(x_l-x_{l+1})](T^K)_l(Y,x_{l+1})\Big)_l. \quad (5.2.19h)$$

Finally for $x_l\in X_2$

$$[T(X_1,x_n,Y_1,x_{l+1}),\,T^K(X_2,Y_2)]_l=$$

$$= -i\Big((T_l(X_1,x_n)(T^K)_l(X_2))_l\partial\partial D_m^{(-)}(x_l-x_{l+1})(T_l(Y_1,x_{l+1})(T^K)_l(Y_2))_l$$

$$+((T^K)_l(X_2)T_l(X_1,x_n))_l\partial\partial D_m^{(+)}(x_l-x_{l+1})((T^K)_l(Y_2)T_l(Y_1,x_{l+1}))_l\Big)_l$$

$$= -i\Big((R_l'^2(X,x_n)\partial\partial D_m^{(-)}(x_l-x_{l+1})R_l'(Y,x_{l+1})$$

$$+A_l'^2(X,x_n)\partial\partial D_m^{(+)}(x_l-x_{l+1})A_l'(Y,x_{l+1}))\Big)_l, \tag{5.2.19i}$$

and for $x_l \in X_1$ we have

$$[T(X_1,x_n,Y_2),\, T^K(X_2,Y_1,x_{l+1})]_l$$

$$= i\Big(((T_l(X_1,x_n)(T^K)_l(X_2))_l\partial\partial D_m^{(+)}(x_l-x_{l+1})(T_l(Y_2)(T^K)_l(Y_1,x_{l+1}))_l$$

$$+((T^K)_l(X_2)T_l(X_1,x_n))_l\partial\partial D_m^{(-)}(x_l-x_{l+1})((T^K)_l(Y_1,x_{l+1})T_l(Y_2))_l)\Big)_l$$

$$= i\Big((R_l'^1(X,x_n)\partial\partial D_m^{(+)}(x_l-x_{l+1})(R'^K)_l(Y,x_{l+1})$$

$$+A_l'^1(X,x_n)(\partial\partial D_m^{(-)})(x_l-x_{l+1})(A'^K)_l(Y,x_{l+1}))\Big)_l. \tag{5.2.19j}$$

Now we substitute (5.2.15) and (5.2.19a–j) into (5.2.18) and make use of

$$D_m = D_m^{(+)} + D_m^{(-)}, \quad D_m^F = D_m^{\mathrm{ret}} - D_m^{(-)} = D_m^{\mathrm{av}} + D_m^{(+)},$$

$$D_m^{F*} = D_m^{\mathrm{ret}} - D_m^{(+)} = D_m^{\mathrm{av}} + D_m^{(-)} \tag{5.2.20}$$

$$T_l = R_l - R_l' = A_l - A_l', \quad (T^K)_l = -R_l - (R'^K)_l = -A_l - (A'^K)_l,$$

$$D_l = R_l - A_l \tag{5.2.21}$$

as well as $R_l'^1 + R_l'^2 = R_l'$ and $A_l'^1 + A_l'^2 = A_l'$ (see (5.2.10)). Then all the terms add up to the desired result

$$D_l(X,Y,x_{l+1};x_n)$$

$$= i\Big(R_l^\mu(X,x_n)[\partial_\mu\partial_\nu D_m^{\mathrm{av}}(x_l-x_{l+1}) - g_{\mu\nu}\delta(x_l-x_{l+1})]R_l^\nu(Y,x_{l+1})\Big)_l$$

$$-i\Big(A_l^\mu(X,x_n)[\partial_\mu\partial_\nu D_m^{\mathrm{ret}}(x_l-x_{l+1}) - g_{\mu\nu}\delta(x_l-x_{l+1})]A_l^\nu(Y,x_{l+1})\Big)_l. \tag{5.2.22}$$

The two terms in (5.2.22) have their supports in $\Gamma^+(x_n)$ and $\Gamma^-(x_n)$, respectively. Hence, this decomposition is a splitting solution (the so-called "natural" splitting). It is easy to check by taking (5.2.10) and (5.2.11) into account, that in $T_n = R - R'$ the D_m^{av} (D_m^{ret}) are changed into D_m^F.

If we would not have used $C_{\mathrm{Com}} = -1$, then $g_{\mu\nu}\delta(x_l - x_{l+1})$ would be everywhere multiplied by C_{Com} or C^*_{Com}. Then the summation of these terms to (5.2.22) would only work if $C_{\mathrm{Com}} = C^*_{\mathrm{Com}}$. Hence, (5.2.22) only holds if the Compton graphs are pseudo-unitary $(R_2^{\mathrm{Com}})^K = -R_2^{\mathrm{Com}}$. If instead of the two derivatives at x_l and x_{l+1} in Fig. 16 there would be only one or no derivative or an internal photon line, then D_l would be given by a formula similar to (5.2.22), but without the $g\delta(x_l - x_{l+1})$-terms. When the Compton graph in Figs. 14a or 14b is part of a loop, the consequences of the normalization term $C_{\rm Com}g\delta$ cannot be formulated in such a simple way as in (5.2.22) for the tree-like diagrams. However, if the loop contains k internal scalar lines with two derivatives, its D-distribution can be decomposed into terms $O(C^0_{\rm Com})$, $O(C^1_{\rm Com})$, $\ldots O(C^k_{\rm Com})$. Since the whole D-distribution has causal support for every value of $C_{\rm Com} \in \mathbf{C}$, the same must be true in every order of $C_{\rm Com}$ separately. Therefore the terms of $O(C^l_{\rm Com})$, $l = 0, 1 \ldots k$, can be split separately. Graphically speaking, the terms $O(C^l_{\rm Com})$ are obtained by replacing l of the k internal scalar lines with two derivatives by vertices of type Fig. 14c.

To complete the proof of gauge invariance, we have to show that gauge invariant normalization is possible in all cases of non-trivial distribution splitting. This can be done as in Sect. 4.6, for details we refer to the paper by M. Dütsch, F. Krahe and G. Scharf (*Nuov. Cim. 106 A, 277 (1993)*).

5.3 Axial Anomalies

Until now it was always possible to remove anomalies, which appeared in the process of distribution splitting, breaking gauge invariance, by appropriate normalization. If we include axial-vector coupling, this is no longer the case. This was first discovered by S. Adler (*Phys. Rev. 177, 2426 (1969)*) and J.S. Bell and R. Jackiw (*Nuovo Cimento A 60, 47 (1969)*) in the framework of usual Feynman perturbation theory. However, the starting point in this approach are the divergent Feynman integrals, which have to be regulated. One then gets the impression that the anomalies are a consequence of this ultraviolet regularization. Therefore, it is urgent to study the problem in the finite causal theory.

We consider QED with pseudovector and pseudoscalar couplings

$$T_1(x) = i\,c_V\,j^\mu_V(x)A_\mu(x) + i\,c_A\,j^\mu_A(x)B_\mu(x)$$
$$+ic_\pi j_\pi(x)\Pi(x). \tag{5.3.1}$$

Here

$$j^\mu_V =: \overline{\psi}\gamma^\mu\psi : \quad , \quad j^\mu_A =: \overline{\psi}\gamma^\mu\gamma^5\psi : \tag{5.3.2a}$$

are the vector and axial-vector currents and

$$j_\pi = i : \overline{\psi}\gamma^5\psi : \tag{5.3.2b}$$

is a pseudoscalar, all being formed from a free massive Dirac field $\psi(x)$ with mass m. The vector-, axialvector- and pseudoscalar vertices defined by (5.3.1) will be abbreviated by V, A and Π in the following. The fields A_μ, B_μ and $\Pi(x)$ play no essential rôle in the following and are, therefore, assumed as classical external fields. Nevertheless, this external field problem will have interesting applications.

From (5.3.2) we have the following divergence relations for the free currents

$$\partial_\mu j_V^\mu = 0, \quad \partial_\mu j_A^\mu = 2mj_\pi. \tag{5.3.3}$$

Our problem is whether similar divergence relations hold in higher orders, in particular for the triangular $VVA-$ and $VV\Pi$-graphs of Fig. 17 which contribute to the 3-point function T_3. To compute the latter, we must first calculate

$$D_3(x_1,x_2,x_3) = T_2(x_1,x_3)T_1^K(x_2) + T_2(x_2,x_3)T_1^K(x_1) + T_1(x_3)T_2^K(x_1,x_2)$$
$$-T_1^K(x_1)T_2(x_2,x_3) - T_1^K(x_2)T_2(x_1,x_3) - T_2^K(x_1,x_2)T_1(x_3), \tag{5.3.4}$$

where we have used pseudo-unitarity to express the T-distributions of the inverse S-matrix. Concerning the triangle graphs of Fig. 17, the 2-point distributions which contribute, come from Compton scattering. For these distributions the usual divergence relations (5.3.3) still hold, so that

$$\frac{\partial}{\partial x_3^{\mu_3}} d_B^{\mu_1\mu_2\mu_3}(x_1,x_2,x_3) = 2m\frac{c_A}{c_\pi} d_\pi^{\mu_1\mu_2}(x_1,x_2,x_3). \tag{5.3.5}$$

Here d_B, d_π are the 3-point distributions corresponding to Fig. 17 without the external fields A, B, Π. The question is whether the same relation remains true after splitting for the retarded distributions. One therefore defines the anomaly by

$$a^{\mu_1\mu_2} = \frac{\partial}{\partial x_3^{\mu_3}} r_B^{\mu_1\mu_2\mu_3}(x_1,x_2,x_3) - 2m\frac{c_A}{c_\pi} r_\pi^{\mu_1\mu_2}(x_1,x_2,x_3). \tag{5.3.6}$$

The t-distributions have the same anomaly because the r'-distributions are anomaly-free.

Since we work with massive Fermi fields, we can perform the splitting in momentum space by means of the central solution

$$\hat{r}_B(p,q) = \frac{i}{2\pi}\int_{-\infty}^{+\infty} \frac{\hat{d}_B(tp,tq)}{(1-t+i0)t^{\omega_B+1}}\,dt \tag{5.3.7}$$

where p, q are assumed to be in the forward cone V^+, and similarly for $\hat{r}_\pi$. ω is the singular order of the d-distributions. The Fourier transformation is carried out in the difference variables $y_1 = x_1 - x_3$, $y_2 = x_2 - x_3$, taking translation invariance into account. From Eq. (5.3.5) we then get

$$i(p_{\mu_3} + q_{\mu_3})\hat{d}_B^{\mu_1\mu_2\mu_3}(p,q) = 2m\frac{c_a}{c_\pi}\hat{d}_\pi^{\mu_1\mu_2}(p,q), \tag{5.3.8}$$

and the anomaly (5.3.6) becomes

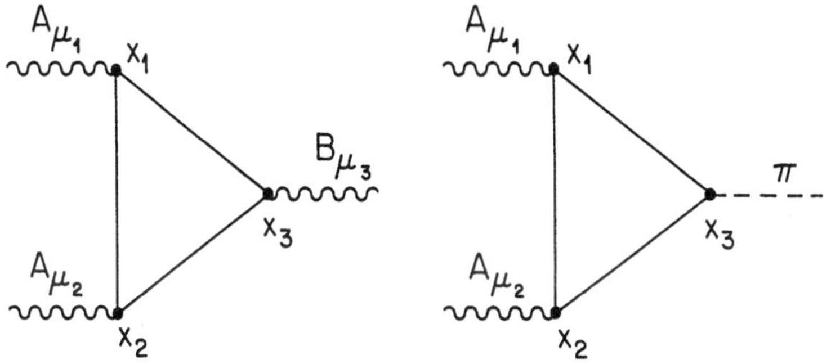

Fig. 17. Axial vector and pseudoscalar triangular graphs

$$\hat{a}^{\mu_1\mu_2}(p,q) = i(p_{\mu_3} + q_{\mu_3})\hat{r}_B^{\mu_1\mu_2\mu_3} - 2m\frac{c_a}{c_\pi}\hat{r}_\pi^{\mu_1\mu_2}. \tag{5.3.9}$$

Substituting (5.3.7) and the analogous equation for $\hat{r}_\pi$ herein, and using (5.3.8), we arrive at the following formula for the anomaly

$$\hat{a}^{\mu_1\mu_2}(p,q) = \frac{i}{\pi}\frac{c_A}{C_\pi}m\int_{-\infty}^{+\infty} dt\, \frac{\hat{d}_\pi^{\mu_1\mu_2}(tp,tq)}{1-t+i0}\left(\frac{1}{t^{\omega_B+2}} - \frac{1}{t^{\omega_\pi+1}}\right). \tag{5.3.10}$$

Hence, the anomaly is due to the fact that $\omega_\pi - \omega_B \neq 1$.

To evaluate (5.3.10) we only need the pseudoscalar d-distribution. From the first tree terms in (5.3.4) we find

$$r'^{\mu_1\mu_2}(y_1,y_2) = c_V^2 c_\pi \mathrm{tr}\,\big[i\gamma_5 S^{(+)}(x_3-x_2)\gamma^{\mu_2}S^{AF}(x_2-x_1)\gamma^{\mu_1}S^{(-)}(x_1-x_3)$$

$$+i\gamma_5 S^{(+)}(x_3-x_2)\gamma^{\mu_2}S^{(-)}(x_2-x_1)\gamma^{\mu_1}S^{F}(x_1-x_3)+$$

$$+i\gamma_5 S^{F}(x_3-x_2)\gamma^{\mu_2}S^{(+)}(x_2-x_1)\gamma^{\mu_1}S^{(-)}(x_1-x_3)\big] + \mathrm{tr}\,\big[x_1\leftrightarrow x_2, \mu_1\leftrightarrow\mu_2\big]. \tag{5.3.11}$$

Here S^{AF} denotes the anti-Feynman propagator which is obtained from the Feynman propagator in momentum space by changing $+i0$ into $-i0$:

$$\hat{S}^{AF}(p) = -(2\pi)^{-2}\frac{\not{p}+m}{p^2-m^2-i0}. \tag{5.3.12}$$

It comes from the conjugation K in (5.3.4). If one replaces $S^{(+)}$ by $S^{(-)}$ and vice versa without changing the arguments, one gets $a'^{\mu_1\mu_2}$. The difference $r' - a'$ gives $d^{\mu_1\mu_2}$.

Expressing the spinor distributions by scalar ones, we see that the terms with three $\not{\partial}$ contain γ_5 plus five other γ-matrices. Then the trace vanishes. In the non-vanishing terms one has at least one factor m instead of $\not{\partial}$.

This lowers ω by one, so that we get $\omega_\pi = 0$, instead of the power-counting estimate 1. But we will see below (see 5.3.34)) that the splitting with $\omega = 0$ or 1 gives the same result. If we replace $i\gamma_5$ by $\gamma^{\mu_3}\gamma_5$ in (5.3.12), we get the r'-distribution for the axial-vector graph: $r'^{\mu_1\mu_2\mu_3}$. Then the terms with three $\not\partial$ contain γ_5 plus six γ^μ matrices and the trace does not vanish. In this case we have the power-counting result $\omega_A = 1$.

After Fourier transformation

$$\hat{r}'^{\mu_1\mu_2}(p,q) = (2\pi)^{-4}\int r'^{\mu_1\mu_2}(y_1,y_2)e^{i(py_1+qy_2)}dy_1\,dy_2 \tag{5.3.13}$$

we shall obtain

$$\hat{r}'^{\mu_1\mu_2}(p,q) = \frac{c_V^2 c_\pi}{(2\pi)^2}\int\Big\{\mathrm{tr}\,\Big[i\gamma_5\hat{S}^{(+)}(-P+k)\gamma^{\mu_2}\hat{S}^{AF}(-p+k)\gamma^{\mu_1}\hat{S}^{(-)}(k)$$
$$+i\gamma_5\hat{S}^{(+)}(-P+k)\gamma^{\mu_2}\hat{S}^{(-)}(-p+k)\gamma^{\mu_1}\hat{S}^F(k)$$
$$+i\gamma_5\hat{S}^F(-P+k)\gamma^{\mu_2}\hat{S}^{(+)}(-p+k)\gamma^{\mu_1}\hat{S}^{(-)}(k)\Big] + \mathrm{tr}\,\big[p\longleftrightarrow q, \mu_1\longleftrightarrow\mu_2\big]\Big\}. \tag{5.3.14}$$

Here we have introduced $P = p + q$. Computing the trace (Problem 5.5) we get

$$\hat{r}'^{\mu_1\mu_2}(p,q) = -\frac{4m^2c_V^2c_\pi}{(2\pi)^6}\varepsilon^{\mu_1\mu_2\alpha\beta}p_\alpha q_\beta$$
$$\times\big\{[I_-(P,p) + I_+(q,-p) + I_+(p,P)] + [p\longleftrightarrow q]\big\}, \tag{5.3.15}$$

where the Lorentz invariant integrals $I_\pm$ are given by

$$I_\pm(p,q) \stackrel{\mathrm{def}}{=} \int d^4k\,\Theta(-k^0)\delta(k^2-m^2)$$
$$\times\,\Theta(k^0-p^0)\delta[(k-p)^2-m^2]\frac{1}{(k-q)^2-m^2\pm i0}. \tag{5.3.16}$$

Owing to the two δ- and Θ-functions, these integrals vanish if p is not in the region $p^2 \geq 4m^2$, $p_0 < 0$. But if p is in this region, one can use $\mathcal{L}_+^\uparrow$ invariance

$$I_\pm(\Lambda p,\Lambda q) = I_\pm(p,q), \quad \forall\Lambda\in\mathcal{L}_+^\uparrow, \tag{5.3.17}$$

to choose $\Lambda p = (-\sqrt{p^2},\mathbf{0})$. Then the integration is done as follows: first we integrate over k^0, using the first δ. In the spatial integration $d^3\boldsymbol{k}$, we use polar coordinates with $\boldsymbol{p}$ as polar axis. Integration over $|\boldsymbol{k}|$ kills the second δ, while the integral over the azimuth φ gives trivially 2π. The remaining integration over $\cos\vartheta = \boldsymbol{k}\cdot\boldsymbol{p}/(|\boldsymbol{k}||\boldsymbol{p}|)$ is elementary. The result in an arbitrary Lorentz system is given by

$$I_\pm(p,q) = \frac{\pi}{4}\Theta(-p^0)\Theta(p^2-4m^2)\frac{1}{\sqrt{N}}\times$$

$$\times \log\left(\frac{-pq+q^2+\sqrt{(1-4m^2/p^2)N}\pm i0}{-pq+q^2-\sqrt{(1-4m^2/p^2)N}\pm i0}\right), \tag{5.3.18}$$

where

$$N = N(p,q) = (pq)^2 - p^2q^2. \tag{5.3.19}$$

We now want to discuss this result. First we notice that

$$\Theta(p^2-4m^2) \neq 0 \Rightarrow \{N \geq 0 \wedge [N=0 \Leftrightarrow q=\lambda p, \text{with} \lambda \in \mathbb{R}]\}. \tag{5.3.20}$$

Let us now assume that $q \neq \lambda p$, then $I_\pm$ can have logarithmic singularities only. Furthermore, $I_\pm$ are continuous at $p^2 = 4m^2$, because the argument of the logarithm is =1 at these points. For $q = \lambda p$, it appears as if $N^{-1/2}$ causes a singularity. However, for $\lambda \neq 1, 0$ we have

$$-pq + q^2 = \lambda(\lambda-1)p^2 \neq 0, \tag{5.3.21}$$

and we can expand the logarithm in powers of

$$\beta \overset{\text{def}}{=} \frac{\sqrt{(1-4m^2/p^2)N}}{\lambda(\lambda-1)p^2}. \tag{5.3.22}$$

This gives

$$I_\pm(p,\lambda p) = \frac{\pi}{2}\Theta(-p^0)\Theta(p^2-4m^2)\sqrt{1-\frac{4m^2}{p^2}}\,\frac{1}{\lambda(\lambda-1)p^2}, \tag{5.3.23}$$

for all $\lambda \neq 0, 1$. In the vicinity of these points the correction terms are $O(\beta^2)$, thus $N^{-1/2}$ cancels. $I_\pm$ is also continuous at $p^2 = 4m^2$. Only the remaining cases $\lambda = 0$ $(q = 0)$ and $\lambda = 1$ $(q = p)$ are difficult to discuss. $I_\pm$ may have pole-like singularities there, but we will not touch these regions when we calculate the anomaly.

The final result is now given by

$$\hat{r}'^{\mu_1\mu_2}(p,q) = \varepsilon^{\mu_1\mu_2\alpha\beta} p_\alpha q_\beta \hat{r}'(p,q), \tag{5.3.24}$$

$$\hat{r}'(p,q) = \frac{mc_V^2 c_\pi}{(2\pi)^5\sqrt{N}}\Big[-\Theta(-p^0)\Theta(p^2-4m^2)\log_1$$

$$-\Theta(-q^0)\Theta(q^2-4m^2)\log_2 - \Theta(-P^0)\Theta(P^2-4m^2)\log_3\Big], \tag{5.3.25}$$

where

$$\log_1 = \log\left(\frac{q^2+pq+\sqrt{(1-4m^2/p^2)N}+i0}{q^2+pq-\sqrt{(1-4m^2/p^2)N}+i0}\right), \tag{5.3.26}$$

$$\log_2 = \log\left(\frac{p^2+pq+\sqrt{(1-4m^2/q^2)N}+i0}{p^2+pq-\sqrt{(1-4m^2/q^2)N}+i0}\right), \tag{5.3.27}$$

$$\log_3 = \log\left(\frac{-pq+\sqrt{(1-4m^2/P^2)N}-i0}{-pq-\sqrt{(1-4m^2/P^2)N}-i0}\right). \tag{5.3.28}$$

The expressions for a' and d are similar:

$$\hat{a}'^{\mu_1\mu_2} = \varepsilon^{\mu_1\mu_2\alpha\beta} p_\alpha q_\beta \hat{a}'(p,q), \tag{5.3.29}$$

$$\hat{a}'(p,q) = \frac{mc_V^2 c_\pi}{(2\pi)^5\sqrt{N}} \Big[-\Theta(p^0)\Theta(p^2-4m^2)\log_1$$

$$-\Theta(q^0)\Theta(q^2-4m^2)\log_2 - \Theta(P^0)\Theta(P^2-4m^2)\log_3 \Big], \tag{5.3.30}$$

$$\hat{d}^{\mu_1\mu_2} = \varepsilon^{\mu_1\mu_2\alpha\beta} p_\alpha q_\beta \hat{d}(p,q), \tag{5.3.31}$$

$$\hat{d}(p,q) = \frac{mc_V^2 c_\pi}{(2\pi)^5\sqrt{N}} \Big[\mathrm{sgn}\,(p^0)\Theta(p^2-4m^2)\log_1$$

$$+\mathrm{sgn}\,(q^0)\Theta(q^2-4m^2)\log_2 + \mathrm{sgn}\,(P^0)\Theta(P^2-4m^2)\log_3 \Big]. \tag{5.3.32}$$

Since the scaling limit in (5.3.32) is equal to

$$\lim_{\lambda\to\infty} \hat{d}^{\mu_1\mu_2}(\lambda p, \lambda q) = \hat{d}^{\mu_1\mu_2}_{m=0}(p,q), \tag{5.3.33}$$

we conclude that $\omega_\pi = 0$. However, the central splitting solution is independent of choosing $\omega = 1$ or $\omega = 0$, respectively. To see this, we calculate the difference

$$\hat{r}^{\mu_1\mu_2}_{\omega=1} - \hat{r}^{\mu_1\mu_2}_{\omega=0} = \frac{i}{2\pi} \int\limits_{-\infty}^{+\infty} dt\, \frac{\hat{d}^{\mu_1\mu_2}(tp,tq)}{1-t+i0} \Big(\frac{1}{t^2} - \frac{1}{t} \Big)$$

$$= \frac{i}{2\pi} \int\limits_{-\infty}^{+\infty} dt\, \frac{\hat{d}^{\mu_1\mu_2}(tp,tq)}{t^2} = 0, \tag{5.3.34}$$

because the denominator is an odd function of t.

To calculate the anomaly, we now insert (5.3.31) into (5.3.10)

$$a^{\mu_1\mu_2}(p,q) = \varepsilon^{\mu_1\mu_2\alpha\beta} p_\alpha q_\beta a(p,q), \tag{5.3.35}$$

where

$$a(p,q) = \frac{im}{\pi} c_V^2 c_A \int\limits_{-\infty}^{+\infty} \frac{dt}{t}\, \hat{d}(tp,tq), \tag{5.3.36}$$

for all $p,q \in V^+$. In (5.3.33) we introduce

$$f_i(p^2,q^2,P^2) = \frac{m}{(2\pi)^5\sqrt{N}} \log_i, \quad i = 1,2,3, \tag{5.3.37}$$

and combine the integrals form $-\infty$ to 0 and from 0 to ∞, taking the sign-functions into account. Substituting $t^2 = \tau$ we get

$$a(p,q)=\frac{i}{\pi}\frac{c_A}{c_\pi}m\int\limits_0^\infty\frac{d\tau}{\tau}\Big[\Theta(\tau p^2-4m^2)f_1(\tau p^2,\tau q^2,\tau P^2)$$

$$+\Theta(\tau q^2-4m^2)f_2(\tau p^2,\tau q^2,\tau P^2)+\Theta(\tau P^2-4m^2)f_3(\tau p^2,\tau q^2,\tau P^2). \quad (5.3.38)$$

Since the anomaly is a polynomial of degree $\omega_B+1=2$, $a(p,q)$ must be a pure number independent of p,q, we can take the limit $p^2\to 0$ and $q^2\to 0$ in (5.3.38), while keeping $P^2>0$. Then only the last term contributes. Substituting $\tau P^2=s$, we obtain

$$a(p,q)=\frac{i}{\pi}\frac{c_A}{c_\pi}m\int\limits_{4m^2}^\infty\frac{ds}{s}\,f_3(0,0,s). \quad (5.3.39)$$

We have for $P^2\geq 4m^2$:

$$f_3(p^2=0,q^2=0,P^2)=\frac{2m}{(2\pi)^5P^2}\log\frac{1-\sqrt{1-4m^2/P^2}}{1+\sqrt{1-4m^2/P^2}}, \quad (5.3.40)$$

which implies

$$a(p,q)=i\frac{4m^2}{(2\pi)^6}c_V^2c_A\int\limits_{4m^2}^\infty\frac{ds}{s^2}\log\frac{1-\sqrt{1-4m^2/s}}{1+\sqrt{1-4m^2/s}}. \quad (5.3.41)$$

Substituting $x=4m^2/s$, we get

$$a(p,q)=\frac{2i}{(2\pi)^6}c_V^2c_A\int\limits_0^1 dx\,\log\frac{1-\sqrt{1-x}}{1+\sqrt{1-x}}, \quad (5.3.42)$$

which shows the mass independence of the anomaly. The further substitution $\sqrt{1-x}=z$ makes the integral elementary and we get

$$a(p,q)=-\frac{2i}{(2\pi)^6}c_V^2c_A. \quad (5.3.43)$$

Summing up, the axial anomaly of central solutions for the triangle graphs is equal to

$$a^{\mu_1\mu_2}(p,q)=-\frac{2i}{(2\pi)^6}c_V^2c_A\varepsilon^{\mu_1\mu_2\alpha\beta}p_\alpha q_\beta. \quad (5.3.44)$$

We have still to investigate whether there exist other splitting solutions which do not have an anomaly while preserving all desired properties of the theory. $\hat{t}^{\mu_1\mu_2}(p,q)$ is a pseudotensor of rank two. The lowest order normalization polynomial with this property is $\sim\varepsilon^{\mu_1\mu_2\alpha\beta}p_\alpha q_\beta$. But this has already $\omega=2$, in contrast to $\omega_\pi=0$. Hence, renormalization of the $VV\Pi$ triangle does not help. There seems to be a better chance with $\hat{t}^{\mu_1\mu_2\mu_3}(p,q)$, which is a pseudotensor of rank three with $\omega_B=1$. The most general normalization polynomial which preserves unitarity and the symmetry in the two V vertices is now given by

$$P^{\mu_1\mu_2\mu_3}(p,q)=C\varepsilon^{\mu_1\mu_2\mu_3\alpha}(p_\alpha-q_\alpha), \tag{5.3.45}$$

where C is a real constant. But this would destroy vector gauge invariance

$$p_{\mu_1}P^{\mu_1\mu_2\mu_3}(p,q)=-C\varepsilon^{\mu_1\mu_2\mu_3\alpha}p_{\mu_1}q_\alpha)\not\equiv 0, \tag{5.3.46}$$

which we do not allow for. That means that the axial anomaly cannot be removed by renormalization. We have to live with it.

An important application of the axial anomaly is the decay of the neutral π^0 meson. This is a pseudoscalar particle which decays into two photons according to Fig. 17b. The fermions going around the triangle are not electrons but quarks. Nevertheless, taking all (coloured) quark states into account, the coupling constant with the photons is again the elementary charge $c_V=e$. When we calculate the 3-point function

$$T_3(x_1,x_2,x_3)=(2\pi)^{-4}\int dp\,dq\,t_\pi^{\mu_1\mu_2}(p,q)e^{-ip(x_1-x_3)-iq(x_2-x_3)}$$

$$\times :A_{\mu_1}(x_1)A_{\mu_2}(x_2):\varPi(x_3) \tag{5.3.47}$$

from the anomaly (5.3.9)

$$a^{\mu_1\mu_2}(p,q)=i(p_{\mu_3}+q_{\mu_3})t_B^{\mu_1\mu_2\mu_3}-\frac{2mc_A}{c_\pi}t_\pi^{\mu_1\mu_2},$$

the first term gives no contribution because it is a divergence. Hence, the $VV\varPi$ diagram can be obtained from the anomaly alone by inserting

$$t_\pi^{\mu_1\mu_2}(p,q)\to\frac{ie^2}{(2\pi)^6m}c_\pi\varepsilon^{\mu_1\mu_2\alpha\beta}p_\alpha q_\beta. \tag{5.3.48}$$

If the anomaly would vanish, π^0 could not decay into two photons.

We evaluate (5.3.47) between initial and final states with sharp four-momenta k_i and k_1,k_2, respectively. With our normalization (5.3.10, 11) we get

$$(\varphi_f,T_3\varphi_i)=(2\pi)^{-17/2}\frac{\varepsilon_{\mu_1}(k_1)\varepsilon_{\mu_2}(k_2)}{\sqrt{8E_i\omega_1\omega_2}}\int dp\,dq$$

$$t_\pi^{\mu_1\mu_2}(p,q)e^{-ip(x_1-x_3)-iq(x_2-x_3)}e^{ik_1x_1+ik_2x_2-ik_ix_3}. \tag{5.3.49}$$

Here $\varepsilon_\mu(k)$ are the transversal polarization vectors of the outgoing photons. The S-matrix element is obtained by integrating over x_1,x_2,x_3 with switching functions $g(x_1)$, $g(x_2)$ and $g(x_3)$. In view of the adiabatic limit, we can safely set $g(x_1)=1=g(x_2)$ which gives two δ-distributions, but we must retain $g(x_3)$. The physical reason for it will become clear below. Then we obtain

$$(\varphi_f,S_3(g)\varphi_i)=(2\pi)^{3/2}\frac{\varepsilon_{\mu_1}(k_1)\varepsilon_{\mu_2}(k_2)}{2\sqrt{2E_i\omega_1\omega_2}}t_\pi^{\mu_1\mu_2}(k_1,k_2)\hat{g}(k_1+k_2-k_i)$$

$$\stackrel{\text{def}}{=} M_{fi}\hat{g}(k_1 + k_2 - k_i). \tag{5.3.50}$$

In order to calculate the decay rate of the pion from the transition amplitude (5.3.50), we must again use wave packets. Let $f_j^*(\boldsymbol{p}, \boldsymbol{q})$ and $f(\boldsymbol{k})$ be square integrable final and initial wave packets, respectively. Then the transition probability is given by

$$|(\Phi_j, S_3(g)\Phi_i)|^2 = \left|\int d^3k_1 d^3k_2 d^3k\, f_j^*(\boldsymbol{k}_1, \boldsymbol{k}_2) M_{fi}\hat{g}(k_1 + k_2 - k) f(\boldsymbol{k})\right|^2. \tag{5.3.51}$$

We sum over a complete set of final states taking the completeness relation

$$\sum_j f_j^*(\boldsymbol{p}, \boldsymbol{q}) f_j(\boldsymbol{p}', \boldsymbol{q}') = \delta(\boldsymbol{p} - \boldsymbol{p}')\delta(\boldsymbol{q} - \boldsymbol{q}') \tag{5.3.52}$$

into account. Then we arrive at

$$\sum_f p_{fi} = \int d^3k_1 d^3k_2 d^3k d^3k' \, |M_{fi}(k_1, k_2)|^2$$

$$\hat{g}^*(k_1 + k_2 - k')\hat{g}(k_1 + k_2 - k) f^*(\boldsymbol{k}) f(\boldsymbol{k}')$$

$$\approx \int d^3k_1 d^3k_2 d^3k \, |M_{fi}|^2 \left|\int d^3k\, \hat{g}(p + q - k) f(\boldsymbol{k})\right|^2, \tag{5.3.53}$$

assuming that M_{fi} is slowly varying compared to $f(\boldsymbol{k})$ which is sharply peaked at the initial momentum k_i of the pion. Then the function inside the absolute square, denoted by F, is also a sharply peaked at $P = k_1 + k_2 = k_i$. Introducing the wave packet in x-space

$$\tilde{f}(x) = (2\pi)^{-3/2} \int d^3k\, f(\boldsymbol{k}) e^{-ikx}, \tag{5.3.54}$$

we get

$$F(P) = (2\pi)^{-2} \int d^4\, g(x) \int d^3k\, f(\boldsymbol{k}) e^{i(P-k)x}$$

$$= (2\pi)^{-1/2} \int d^4x\, e^{iPx} \tilde{f}(x) g(x). \tag{5.3.55}$$

Without $g(x)$ this would give a measure on the mass shell $\sim \delta(P^2 - m^2)$; but then the square $|F(P)|^2$ would be meaningless. Now $|F(P)|^2$ is a positive function, peaked at $P = k_i$ and normalized according to

$$\int d^4P\, |F(P)|^2 = (2\pi)^{-1}(2\pi)^4 \int d^4x |\tilde{f}(x) g(x)|^2.$$

Then it can be approximated by

$$|F(P)|^2 = \delta(P - k_i)(2\pi)^3 \int d^4x |\tilde{f}(x)|^2 |g(x)|^2. \tag{5.3.56}$$

It is now necessary to discuss how the decay rate of the pion is actually measured, because the normal assumption of scattering theory where free particles are coming in from infinity is not applicable here, since the pion is unstable. In the experiment the pion triggers a clock at time $t = 0$. After a time T one looks whether the pion has decayed or not. For the experimenter it only has a chance to decay within the time interval $[0, T]$. Then the switching function $g(x)$ gets quite a physical significance: it is $=1$ for $t \in [0, T]$ and zero outside. This is no contradiction to our previous assumption $g(x_1) = 1 = g(x_2)$, as long as T is large compared to any microscopic time scale. If we neglect the spreading of the wave packet

$$\tilde{f}(x) = f_1(\boldsymbol{x} + \boldsymbol{x}_1 + vt),$$

the integral in (5.3.56) becomes equal to

$$\int d^4x |\tilde{f}(x)|^2 |g(x)|^2 = \int d^3x |f_1(x)|^2 \int_0^T dt = T.$$

This leads to the following general result for the decay rate

$$\Gamma \overset{\text{def}}{=} \frac{1}{T} \sum_f p_{fi} = (2\pi)^3 \int d^3k_1 d^3k_2 \, |M_{fi}(k_1, k_2)|^2 \delta(k_1 + k_2 - k_i). \quad (5.3.57)$$

Let us now evaluate this for the π^0 decay. The transition matrix element follows from (5.3.50) and (5.3.48)

$$M_{fi} = (2\pi)^{-9/2} \frac{\varepsilon_{\mu_1}(k_1) \varepsilon_{\mu_2}(k_2)}{2\sqrt{2E_i \omega_1 \omega_2}} \frac{ie^2 c_\pi}{m} \varepsilon^{\mu_1 \mu_2 \alpha \beta} k_{1\alpha} k_{2\beta}. \quad (5.3.58)$$

The ε-tensor gives rise to a determinant of the four 4-vectors, so that (Problem 5.2)

$$\varepsilon^{\mu_1 \mu_2 \alpha \beta} \varepsilon_{\mu_1} \varepsilon'_{\mu_2} k_{1\alpha} k_{2\beta} = -2\omega \boldsymbol{p} \cdot (\boldsymbol{\varepsilon} \wedge \boldsymbol{\varepsilon}') = \pm 2\omega^2. \quad (5.3.59)$$

Here we caculate in the center-of-mass frame:

$$k_1 = (\omega, \boldsymbol{k}), \quad k_2 = (\omega, -\boldsymbol{k}), \quad E_i = m_\pi. \quad (5.3.60)$$

The summation over the polarizations in (5.3.58) gives a factor of 2, but this cancels against $\frac{1}{2}$ from the Bose statistics of the outgoing photons; since these photons are indistinguishable, the solid angle is only half as big. Then we end up with

$$\Gamma = (2\pi)^{-6} \frac{e^4 c_\pi^2}{2m^2 m_\pi} \int d^3k_1 d^3k_2 \omega^2 \delta(k_1 + k_2 - k_i)$$

$$= (2\pi)^{-6} \frac{e^4 c_\pi^2}{2m^2 m_\pi} \int d^3k_1 \omega^2 \delta(\omega - \tfrac{1}{2} m_\pi) \tfrac{1}{2}$$

$$= (2\pi)^{-6} \frac{e^4 c_\pi^2}{4m^2 m_\pi} 4\pi \int_0^\infty d\omega\, \omega^4 \delta(\omega - \tfrac{1}{2} m_\pi)$$

$$= (2\pi)^{-5} \frac{e^4}{2} \frac{c_\pi^2}{m^2 m_\pi} \left(\frac{m_\pi}{2}\right)^4 = \frac{\alpha^2 c_\pi^2}{m^2} \frac{m_\pi^3}{64\pi^3}. \tag{5.3.61}$$

The coupling constant

$$\frac{c_\pi}{m} = \frac{1}{f_\pi} = (93\text{MeV})^{-1} \tag{5.3.62}$$

can be obtained from the lifetime of the charged pion (*S.B. Treiman, in Lectures in Current Algebras and Its Applications, Princeton University Press 1972*). With $m_\pi = 135$ MeV this gives a decay constant $\Gamma = 7.64$ eV, which corresponds to a lifetime

$$\tau = \frac{\hbar}{\Gamma} = 0.8615 \cdot 10^{-16}\text{s}$$

of the π^0 in agreement with the experimental value $(0.828 \pm 0.057) \cdot 10^{-16}$ s (*Particle Data Group, Phys. Lett. B 239, 1 (1990)*).

Besides the VVA-graph also the AAA-graph with three axialvector vertices has an anomaly. In analogy to (5.3.8) it is defined by

$$a_{AAA}^{\mu_1\mu_2}(p,q) = i(p+q)_{\mu_3} \hat{r}_{AAA}^{\mu_1\mu_2\mu_3}(p,q) - 2m\hat{r}_{AA\Pi}^{\mu_1\mu_2}. \tag{5.3.63}$$

Its calculation by means of (5.3.10) runs along the same lines as before (Problem 5.3). The final result is one third of the previous anomaly

$$a_{AAA}^{\mu_1\mu_2}(p,q) = -\frac{2i}{3(2\pi)^6} \varepsilon^{\mu_1\mu_2\alpha\beta} p_\alpha q_\beta. \tag{5.3.64}$$

We must again investigate whether this anomaly can be removed by different normalization. The free normalization polynomial which can be added to $\hat{t}_{AAA}^{\mu_1\mu_2\mu_3}$ is of the following form

$$P^{mu_1\mu_2\mu_3}(p,q) = C\varepsilon^{mu_1\mu_2\mu_3\alpha}(p_\alpha - q_\alpha) \tag{5.3.65}$$

with real C. There is no restriction by vector gauge invariance here, because we have no vector field. However, $t_{AAA}^{mu_1\mu_2\mu_3}(x_1 - x_3, x_2 - x_3)$ must be symmetric in the three vertices. The permutation of 1 and 3 leads to the condition

$$P^{mu_1\mu_2\mu_3}(p,q) = P^{mu_3\mu_2\mu_1}(-p-q,q),$$

which implies

$$C\varepsilon^{mu_1\mu_2\mu_3\alpha}(p_\alpha - q\alpha) = C\varepsilon^{mu_1\mu_2\mu_3\alpha}(p_\alpha + 2q_\alpha).$$

This is only possible if $C = 0$. Hence, this anomaly cannot be removed by renormalization. In the VVA case this was possible, if vector gauge invariance is violated instead. Here is no such way out. Consequently, the axial anomaly cannot be avoided, but vector gauge invariance can be always achieved. The two triangle graphs discussed here are the only anomalous graphs, all other graphs in arbitrary order are anomaly-free (see *F. Krahe, Nuovo Cimento 106A, 917 (1993)*).

5.4 (2+1)-Dimensional QED: Vacuum Polarization

In two space dimensions, QED shows interesting new features. They have their origin in different properties of the 3-dimensional gamma matrices. Our conventions for the metric tensor and the representation of the gamma matrices are

$$g_{\mu\nu} = \begin{pmatrix} 1 & 0 & 0 \\ 0 & -1 & 0 \\ 0 & 0 & -1 \end{pmatrix} \tag{5.4.1}$$

$$\gamma^0 = \sigma_3, \quad \gamma^1 = i\sigma_1, \quad \gamma^2 = i\sigma_2, \tag{5.4.2}$$

where σ_i are the Pauli matrices. We have

$$\{\gamma^\mu, \gamma^\nu\} = \gamma^\mu\gamma^\nu + \gamma^\nu\gamma^\mu = 2g^{\mu\nu}, \quad \text{and} \tag{5.4.3}$$

$$\gamma^\mu\gamma^\nu = g^{\mu\nu} - i\varepsilon^{\mu\nu\alpha}\gamma_\alpha. \tag{5.4.4}$$

The occurrence of the 3-dimensional ε-tensor is the interesting new feature.

The above relations imply

$$\gamma_\mu\gamma^\mu = 3, \quad \gamma_\mu \not{p} \gamma^\mu = -\not{p}. \tag{5.4.5}$$

$$\varepsilon_{\mu\nu\alpha}\varepsilon^{\beta\gamma\alpha} = \delta_\mu^\beta\delta_\nu^\gamma - \delta_\mu^\gamma\delta_\nu^\beta \tag{5.4.6}$$

$$\varepsilon^{\alpha\beta\gamma}\varepsilon_{\alpha\beta\gamma} = 3 \tag{5.4.7}$$

$$\mathrm{tr}\,(\gamma^\mu) = 0 \tag{5.4.8}$$

$$\mathrm{tr}\,(\gamma^\mu\gamma^\nu) = 2g^{\mu\nu}. \tag{5.4.9}$$

By successive use of (5.4.4), (5.4.8) and (5.4.9), we find

$$\mathrm{tr}\,(\gamma^\mu\gamma^\nu\gamma^\rho) = -i\varepsilon^{\mu\nu\alpha}\mathrm{tr}\,(\gamma_\alpha\gamma^\rho) = -2i\varepsilon^{\mu\nu\rho} \tag{5.4.10}$$

$$\begin{aligned} \mathrm{tr}\,(\gamma^\mu\gamma^\nu\gamma^\rho\gamma^\sigma) &= g^{\mu\nu}g^{\rho\sigma} - \varepsilon^{\mu\nu\alpha}\varepsilon^{\rho\sigma\beta}\mathrm{tr}\,(\gamma_\alpha\gamma_\beta) \\ &= 2(g^{\mu\nu}g^{\rho\sigma} - g^{\mu\rho}g^{\nu\sigma} + g^{\mu\sigma}g^{\nu\rho}). \end{aligned} \tag{5.4.11}$$

All greek indices run from 0 to 2 according to (5.4.2). We note that the parity transformation P corresponds to inversion of *one* of the axes in space, since inversion of both leads to a rotation of angle π.

The theory is defined by the old first order term

$$T_1(x) = ie : \overline{\psi}(x)\gamma^\mu\psi(x) : A_\mu(x) = -\tilde{T}_1(x). \tag{5.4.12}$$

The basic contraction functions are given by

$$D_0^{(+)}(x) = \frac{i}{(2\pi)^2}\int d^3p\,\delta^1(p^2)\Theta(p^0)e^{-ipx} \tag{5.4.13}$$

for photonic contractions and

$$\hat{S}_m^{(+)}(p) = (\not{p} + m)\hat{D}_m^{(+)}, \tag{5.4.14}$$

for fermionic contractions. We first look at the normalizability of the theory by means of our scaling method of Sect. 4.3. If two t-distributions $\hat{t}_1$ and $\hat{t}_2$ in momentum space are contracted by l photon contractions, the new singular order is equal to

$$\omega = \omega_1 + \omega_2 + l - 3, \tag{5.4.15}$$

instead of (4.3.23) (Problem 5.3). In case of l fermionic contractions one finds

$$\omega = \omega_1 + \omega_2 + 2l - 3. \tag{5.4.16}$$

To find the general formula, one has to start the induction by looking at the diagrams of lowest orders. One makes a general ansatz for ω which fits the lowest diagrams and then checks whether it is preserved in the inductive step (5.4.15, 16). This ansatz contains the number b of external boson (photon) lines and the total number f of external electrons and positrons. But this is not enough in QED_3, the order n of perturbation theory explicitly appears: every term $\sim n$ with constant coefficient is obviously preserved under (5.4.15, 16). In this way one finds (Problem 5.4)

$$\omega = 3 - f - \frac{1}{2}b - \frac{1}{2}n. \tag{5.4.17}$$

The negative coefficient of n implies that the theory is super-normalizable (see Sect. 4.3). In fact, there are only a few graphs with $\omega \geq 0$ in the lowest orders, in contrast to QED_4 which has graphs with $\omega \geq 0$ in any order. In QED_3 there appear only three such graphs, namely: for $n = 2$ the vacuum polarization ($f = 0$, $b = 2$) with $\omega = 1$ and the electron self-energy ($f = 2$, $b = 0$) with $\omega = 0$ and the vacuum polarization in fourth order $n = 4$ ($f = 0$, $b = 4$) with $\omega = 0$. The vertex graph $n = 3$, $f = 2$ and $b = 1$ has already $\omega = -1$. The appearance of the order n in (5.4.17) may be surprising, but if $n/2$ were not there, ω could be half-integer which is impossible.

The most interesting graph is the second order vacuum polarization which we now calculate. The first order T_1 is given by (5.4.12). On going from $n = 1$ to $n = 2$ the inductive method proceeds by forming

$$A_2'(x_1, x_2) = \tilde{T}_1(x_1)T_1(x_2) = -T_1(x_1)T_1(x_2)$$

$$R_2'(x_1, x_2) = T_1(x_2)\tilde{T}_1(x_2) = -T_1(x_2)T_1(x_1), \tag{5.4.18}$$

and

$$D_2(x_1, x_2) = R_2' - A_2' = T_1(x_1)T_1(x_2) - T_1(x_2)T_1(x_1). \tag{5.4.19}$$

By using Wick's theorem, the term due to vacuum polarization is obtained by two fermionic contractions

$$D_2^{\rm vp}(x_1, x_2) = -e^2 {\rm tr}\,[\gamma^\mu S^{(-)}(y)\gamma^\nu S^{(+)}(-y)$$

$$-\gamma^\mu S^{(+)}(y)\gamma^\nu S^{(-)}(-y)] \; : A_\mu(x_1)A_\nu(x_2) : \tag{5.4.20}$$

where $y = x_1 - x_2$ and $S^{(+)}$ was defined in (5.4.14):

$$S^{(+)}(x) = \frac{i}{(2\pi)^2} \int \frac{d^2p}{2E} (\not{p} + m) e^{-ipx}. \tag{5.4.21}$$

Above, as usual, $\not{p} = \gamma^\mu p_\mu$, and $px = p^\mu x_\mu$ and $E = \sqrt{\mathbf{p}^2 + m^2}$. Similarly

$$S^{(-)}(x) = \frac{i}{(2\pi)^2} \int \frac{d^2p}{2E} (\not{p} - m) e^{ipx}. \tag{5.4.22}$$

Due to invariance of the trace under cyclic permutations, one may write (5.4.20) in the form:

$$D_2(x_1, x_2) = [P^{\mu\nu}(y) - P^{\nu\mu}(-y)] \, : A_\mu(x_1) A_\nu(x_2) : \tag{5.4.23}$$

where

$$P^{\mu\nu}(y) = e^2 \mathrm{tr}\, [\gamma^\mu S^{(+)}(y) \gamma^\nu S^{(-)}(-y)]. \tag{5.4.24}$$

The numerical distribution occurring in the 2-point function is therefore

$$d^{\mu\nu}(x_1, x_2) = d^{\mu\nu}(y) = P^{\mu\nu}(y) - P^{\nu\mu}(-y). \tag{5.4.25}$$

It has causal support, as may be verified by writing

$$\gamma^\mu [S^{(-)}(y) \gamma^\nu S^{(+)}(-y) - S^{(+)}(y) \gamma^\nu S^{(-)}(-y)]$$
$$= \gamma^\mu [S(y) \gamma^\nu S^{(+)}(-y) - S^{(+)}(y) \gamma^\nu S(-y)]$$

where

$$S(x) = S^{(+)}(x) + S^{(-)}(x) = (i\not{\partial} + m) D(x)$$

and

$$D(x) = \frac{i}{(2\pi)^2} \int d^3p \, \delta(p^2 - m^2) \mathrm{sgn}\, p^0 e^{-ipx}$$

is the (3-dimensional) Jordan-Pauli distribution, which has support in $\bar{V}^+(0) \cup \bar{V}^-(0)$.

The rest of the analysis proceeds in momentum space. The Fourier transform $\hat{P}^{\mu\nu}(k)$ follows from (5.4.24) and (5.4.21, 22)

$$\hat{P}^{\mu\nu}(k) = e^2 (2\pi)^{-1} \mathrm{tr} \int dq_1 \, dq_2 \, \gamma^\mu (\not{q}_1 + m) \gamma^\nu (\not{q}_2 + m)$$

$$\times \frac{i}{2\pi} \Theta(q_1^0) \delta(q_1^2 - m^2) \Big(-\frac{i}{2\pi} \Big) \Theta(-q_2^0) \delta(q_2^2 - m^2) \delta(q_1 - q_2 - k). \tag{5.4.26}$$

Using $p \equiv q_1$ and $q \equiv -q_2$ as new integration variables, we arrive at

$$\hat{P}^{\mu\nu}(k) = -e^2 (2\pi)^{-3} \int dp \, dq \, \mathrm{tr}\, [\gamma^\mu (\not{p} + m) \gamma^\nu (\not{q} - m)]$$

$$\times \Theta(p^0) \delta(p^2 - m^2) \Theta(q^0) \delta(q^2 - m^2) \delta(p + q - k)$$
$$\equiv -e^2 (2\pi)^{-3} J^{\mu\nu}(k), \tag{5.4.27}$$

where

$$J^{\mu\nu}(k) = \int dp\, \delta(p^2 - m^2)\Theta(p^0)\delta[(k-p)^2 - m^2] \times\Theta(k^0 - p^0)j^{\mu\nu}(k,p) \tag{5.4.28}$$

with

$$j^{\mu\nu}(k,p) = \mathrm{tr}\,[\gamma^\mu(\not p + m)\gamma^\nu(\not k - \not p - m)]. \tag{5.4.29}$$

By the identities (5.4.8) and (5.4.9), we find

$$j^{\mu\nu}(k,p) = -2[(m^2 - p^2)g^{\mu\nu} + 2p^\mu p^\nu - (p^\mu k^\nu + p^\nu k^\mu) + g^{\mu\nu} p\cdot k - im\varepsilon^{\mu\nu\alpha}k_\alpha]. \tag{5.4.30}$$

By (5.4.27-30) it follows that $\hat{P}^{\mu\nu}$ is gauge-invariant:

$$k_\mu \hat{P}^{\mu\nu}(k) = 0. \tag{5.4.31}$$

This implies the following tensor structure of $\hat{P}^{\mu\nu}$

$$\hat{P}^{\mu\nu}(k) = \hat{P}^{\mu\nu}_S(k) + \hat{P}^{\mu\nu}_A(k) \tag{5.4.32}$$

with

$$\hat{P}^{\mu\nu}_S(k) = (k^\mu k^\nu - k^2 g^{\mu\nu})\tilde{B}(k^2) \tag{5.4.33}$$

and

$$\hat{P}^{\mu\nu}_A(k) = im\varepsilon^{\mu\nu\alpha}k_\alpha \tilde{\Pi}^{(2)}(k^2). \tag{5.4.34}$$

It follows from (5.4.32-34) that

$$-2k^2\tilde{B}(k^2) = \hat{P}^\mu_\mu(k) = g_{\mu\nu}\hat{P}^{\mu\nu}(k) \tag{5.4.35}$$

$$\tilde{\Pi}^{(2)}(k^2) = -\frac{i}{2m}\frac{k^\alpha}{k^2}\varepsilon_{\mu\nu\alpha}\hat{P}^{\mu\nu}(k). \tag{5.4.36}$$

Note that in three space-time dimensions $\hat{P}^{\mu\nu}$ has an antisymmetric part $\hat{P}^{\mu\nu}_A$. In four dimensions such a term would involve the tensor $\varepsilon^{\mu\nu\alpha\beta}$ and the corresponding term $\varepsilon^{\mu\nu\alpha\beta}k_\alpha k_\beta$ is identically zero.

We now turn to the calculation of $\tilde{B}(k^2)$. By (5.4.35) and (5.4.27–30)

$$-2k^2\tilde{B}(k^2) = \hat{P}^\mu_\mu(k) = -e^2(2\pi)^{-3}I^\mu_\mu(k)$$

$$= -\frac{e^2}{2\pi}\int d^3p\, \delta(p^2 - m^2)\Theta(p^0)\delta\Big((k-p)^2 - m^2\Big)\Theta(k^0 - p^0)i^\mu_\mu(k,p), \tag{5.4.37}$$

where i^μ_μ follows from (5.4.30):

$$i^\mu_\mu(k,p) = -2(3m^2 - p^2 + pk). \tag{5.4.38}$$

In (5.4.37), since $p^2 = m^2 = q^2$, where $q = k - p$, it follows that $k = p + q$ is time-like and there exists a Lorentz frame where $k = (k_0, \mathbf{0})$. In this frame, using $(k-p)^2 = m^2 = k^2 - 2pk + p^2$, which implies $2pk = k^2$, we get

$$\tilde{B}(k_0^2) = -\frac{e^2}{(2\pi)^3}\frac{4m^2+k_0^2}{2k_0^2}\int \frac{d^2p}{2E}\,\delta(k_0^2 - 2k_0E)\Theta(k_0 - E). \qquad (5.4.39)$$

Here,

$$E = \sqrt{\boldsymbol{p}^2 + m^2} = \frac{k_0}{2},$$

which implies

$$|\boldsymbol{p}| = \sqrt{\frac{k_0^2}{4} - m^2}.$$

It follows that

$$\tilde{B}(k_0^2) = -\frac{e^2}{(2\pi)^3}\frac{4m^2+k_0^2}{4k_0^2}2\pi\Theta(k_0^2-4m^2)\int_0^\infty d|\boldsymbol{p}|\,\frac{|\boldsymbol{p}|}{E}\delta\Big(2k_0(\frac{k_0}{2}-E)\Big)\Theta(k_0-E)$$

$$= -\frac{e^2}{(2\pi)^3}2\pi\frac{k_0^2+4m^2}{2k_0^2}\Theta(k_0^2-4m^2)\int_m^\infty dE\,\frac{\delta(E-k_0/2)}{2|k_0|}\Theta(\frac{k_0}{2})$$

$$= -\frac{e^2}{(2\pi)^3}2\pi\left(\frac{k_0^2+4m^2}{2k_0^2}\right)\Theta(k_0^2-4m^2)\frac{1}{2|k_0|}\Theta(k_0)$$

$$= -\frac{e^2}{(2\pi)^3}2\pi\left(\frac{k_0^2+4m^2}{2k_0^2}\right)\Theta(k_0^2-4m^2)\frac{1}{2\sqrt{k_0^2}}\Theta(k_0). \qquad (5.4.40)$$

Hence

$$\tilde{B}(k_0^2) = -\frac{e^2}{4}\frac{k_0^2+4m^2}{k_0^2}\Theta(k_0^2-4m^2)\frac{1}{\sqrt{k_0^2}}\Theta(k_0), \qquad (5.4.41)$$

which implies

$$\tilde{B}(k^2) = -\frac{e^2}{4}\frac{k^2+4m^2}{k^2}\Theta(k^2-4m^2)\frac{1}{\sqrt{k^2}}\Theta(k_0), \qquad (5.4.42)$$

upon returning to an arbitrary Lorentz frame. The calculation of $\tilde{\Pi}^{(2)}(k^2)$ is much simpler and follows the same lines (Problem 5.5). The result is

$$\tilde{\Pi}^{(2)}(k^2) = -\frac{1}{2\sqrt{k^2}}\Theta(k^2-4m^2)\Theta(k_0). \qquad (5.4.43)$$

By (5.4.25) we therefore find

$$d^{\mu\nu}(k) = d_S^{\mu\nu}(k) + d_A^{\mu\nu}(k) \qquad (5.4.44)$$

where

$$d_S^{\mu\nu}(k) = (k^\mu k^\nu - k^2 g^{\mu\nu})B(k^2) \qquad (5.4.45)$$

and

$$d_A^{\mu\nu}(k) = im\varepsilon^{\mu\nu\alpha}k_\alpha \Pi^{(2)}(k^2). \qquad (5.4.46)$$

$B(k^2)$ and $\Pi^{(2)}(k^2)$ follows from (5.4.25), (5.4.32-34), (5.4.42) and (5.4.43):

$$B(k^2) = -\frac{e^2}{4}\Big(1+\frac{4m^2}{k^2}\Big)\Theta(k^2-4m^2)\frac{\operatorname{sgn} k_0}{\sqrt{k^2}} \tag{5.4.47}$$

$$\Pi^{(2)}(k^2) = -\frac{1}{2\sqrt{k^2}}\Theta(k^2-4m^2)\operatorname{sgn} k_0\Big(-\frac{e^2}{2\pi}\Big). \tag{5.4.48}$$

The order of $d_S^{\mu\nu}$ is therefore $\omega = 1$ and that of $d_A^{\mu\nu}$, $\omega = 0$.

Next we have to perform the distribution splitting. We first treat the symmetrical part. The corresponding retarded, central splitting solution, given by (3.2.62), is ($\omega = 1$):

$$r_S^{\mu\nu}(k) = \frac{i}{2\pi}\int\limits_{-\infty}^{+\infty} dt\,\frac{t^2(k^\mu k^\nu - k^2 g^{\mu\nu})B(t^2k^2)}{(t-i0)^2(1-t+i0)}. \tag{5.4.49}$$

This result holds true for $k \in \bar{V}^+$, hence $\operatorname{sgn} k_0 = 1$. Putting (5.4.47) into (5.4.49), we obtain

$$r_S^{\mu\nu} = \frac{i\alpha}{2\pi}(k^\mu k^\nu - k^2 g^{\mu\nu})\int\limits_{-\infty}^{+\infty}\frac{dt}{1+t+i0}\Big(1+\frac{4m^2}{t^2k^2}\Big)\Theta(t^2k^2-4m^2)\frac{\operatorname{sgn}(tk_0)}{|t|\sqrt{k^2}}$$

$$= \frac{i\alpha}{2\pi}(k^\mu k^\nu - k^2 g^{\mu\nu})\frac{\operatorname{sgn} k_0}{\sqrt{k^2}}(J_1+J_2), \tag{5.4.50}$$

where

$$J_1 = -2\int\limits_{\sqrt{\frac{4m^2}{k^2}}}^{\infty}\frac{dt}{t^2-1-i0}, \tag{5.4.51}$$

$$J_2 = \frac{4m^2}{k^2}\int\limits_{\sqrt{\frac{4m^2}{k^2}}}^{\infty}\frac{dt}{t^2(t^2-1-i0)} \tag{5.4.52}$$

and we set

$$\alpha = -\frac{e^2}{4}. \tag{5.4.53}$$

It is clear from (5.4.49, 50) that the symmetric splitting solution preserves gauge invariance. The integral in (5.4.51) can be evaluated:

$$J_1 = \log\bigg(\frac{1-\sqrt{k^2/4m^2}}{1+\sqrt{k^2/4m^2}}\bigg). \tag{5.4.54}$$

J_2 may be calculated from (5.4.51)

$$J_2 = \frac{4m^2}{k^2} J_1 + \frac{8m^2}{k^2}\sqrt{\frac{k^2}{4m^2}}. \tag{5.4.55}$$

We hence obtain from (5.4.50, 54) and (5.4.55)

$$r_S^{\mu\nu}(k) = \frac{i\alpha}{2\pi}(k^\mu k^\nu - k^2 g^{\mu\nu})\Big[\frac{1}{\sqrt{k^2}}\Big(1+\frac{4m^2}{k^2}\Big)$$

$$\times \log\Big(\frac{1-\sqrt{k^2/4m^2}}{1+\sqrt{k^2/4m^2}}\Big)+\frac{4m}{k^2}\Big], \quad k \in \bar{V}^+. \tag{5.4.56}$$

Expanding the logarithm for $k^2 \to 0$, we find

$$R_S^{\mu\nu}(k) \to \frac{i\alpha}{2\pi}(k^\mu k^\nu - k^2 g^{\mu\nu})\Big(-\frac{1}{6m}\Big). \tag{5.4.57}$$

The general splitting solution is now given by

$$\tilde{r}_S^{\mu\nu}(k) = r_S^{\mu\nu}(k) + C_0 g^{\mu\nu} + C_1^\mu k^\nu + C_2^\nu k^\mu, \tag{5.4.58}$$

where we have written the most general covariant polynomial of degree $\omega = 1$. It follows immediately that $C_1^\mu = C_2^\nu = 0$ because these terms are not in conformance with the tensor structure. Finally, because of gauge invariance

$$k_\mu \tilde{r}_S^{\mu\nu} = 0,$$

C_0 must also vanish. Hence, (5.4.56) is the unique solution for the symmetric part.

5.5 (2+1)-Dimensional QED: Mass Generation

We now consider the interesting antisymmetric part of the vacuum polarization tensor. Its central splitting solution follows from (5.4.46, 48) ($\omega = 0$):

$$r_A^{\mu\nu}(k) = -\frac{e^2}{2\pi}\frac{i}{2\pi} im\varepsilon^{\mu\nu\alpha}\int\limits_{-\infty}^{+\infty} dt\, \frac{tk_\alpha \Pi^{(2)}(t^2k^2)}{(t-i0)(1-t+i0)}$$

$$= -\frac{e^2 m}{2(2\pi)}\varepsilon^{\mu\nu\alpha}\frac{k_\alpha}{\sqrt{k^2}}\int\limits_{-\infty}^{+\infty} dt\, \frac{\operatorname{sgn}(tk_0)\Theta(t^2k^2-4m^2)}{|t|(1-t+i0)}$$

$$= -\frac{e^2 m}{2(2\pi)}\varepsilon^{\mu\nu\alpha}\frac{k_\alpha}{\sqrt{k^2}}\log\Big(\frac{1-\sqrt{k^2/4m^2}}{1+\sqrt{k^2/4m^2}}\Big), \tag{5.5.1}$$

upon comparison with (5.4.52). Comparing with (5.4.46), we get

$$r_A^{\mu\nu}(k) = im\varepsilon^{\mu\nu\alpha}k_\alpha \Pi_r^{(2)}(k^2), \tag{5.5.2}$$

where

$$\Pi_r^{(2)}(k^2) = \frac{ie^2}{2(2\pi)\sqrt{k^2}} \log \frac{1-\sqrt{k^2/4m^2}}{1+\sqrt{k^2/4m^2}}. \tag{5.5.3}$$

Expanding again for $k^2 \to 0$, we find

$$\Pi_r^{(2)}(0) = -\frac{ie^2}{4\pi}. \tag{5.5.4}$$

The general splitting solution with $\omega = 0$ is now equal to

$$\tilde{r}_A^{\mu\nu}(k) = r_A^{\mu\nu}(k) + C_0 g^{\mu\nu}. \tag{5.5.5}$$

There is no other constant Lorentz tensor of second rank. However, The constant term does not preserve the antisymmetric structure, hence $C_0 = 0$ and (5.5.1) is, again, the unique solution of the splitting problem.

As in QED$_4$ (3.6.28) we write the photon propagator $\mathcal{D}$ in the form

$$(\mathcal{D}^{-1})_{\mu\nu} = (D^{-1})_{\mu\nu} - i\Pi_{\mu\nu}, \tag{5.5.6}$$

where

$$D_{\mu\nu}(k) = \frac{-i}{k^2+i0}\Big(g_{\mu\nu} - \frac{k_\mu k_\nu}{k^2}\Big) - i\alpha \frac{k_\mu k_\nu}{k^4 + i0k^2} \tag{5.5.7}$$

is the free propagator, $\Pi_{\mu\nu}$ the vacuum polarization tensor and α a gauge parameter (above we have used the so-called Feynman gauge $\alpha = 1$, see (3.6.33)). Equation (5.5.6) follows from the usual proper vacuum polarization insertions (see (3.6.26))

$$\mathcal{D} = D + iD\Pi D + i^2 D\Pi D\Pi D + \ldots = D + iD\Pi\mathcal{D}. \tag{5.5.8}$$

Writing the vacuum polarization tensor in the conventional form

$$\Pi_{\mu\nu}(k) = \Big(g_{\mu\nu} - \frac{k_\mu k_\nu}{k^2}\Big)\Pi^{(1)}(k^2) + im\varepsilon_{\mu\nu\alpha}k^\alpha \Pi^{(2)}(k^2), \tag{5.5.9}$$

we have, from (5.4.57) and (5.5.4),

$$\Pi^{(1)}(k^2) \to Ck^2, \quad k^2 \to 0 \tag{5.5.10}$$

with

$$C = \frac{e^2}{48\pi m} \neq 0 \tag{5.5.11}$$

and

$$\Pi^{(2)}(0) = \frac{e^2}{4\pi m}. \tag{5.5.12}$$

To compute $\mathcal{D}$ from (5.5.6) one uses the following three orthogonal projection operators

$$P_{\mu\nu}^{(1)} = \frac{1}{2}\Big(g_{\mu\nu} - \frac{k_\mu k_\nu}{k^2} + i\varepsilon_{\mu\nu\beta}\frac{k^\beta}{\sqrt{k^2}}\Big)$$

$$P^{(2)}_{\mu\nu} = \frac{1}{2}\left(g_{\mu\nu} - \frac{k_\mu k_\nu}{k^2} - i\varepsilon_{\mu\nu\beta}\frac{k^\beta}{\sqrt{k^2}}\right)$$

$$P^{(3)}_{\mu\nu} = \frac{k_\mu k_\nu}{k^2}, \tag{5.5.13}$$

satisfying

$$P^{(j)2} = P^{(j)}, \qquad P^{(j)}P^{(k)} = 0, \quad j \neq k$$

and

$$\sum_{j=1}^{3} P^{(j)}_{\mu\nu} = g_{\mu\nu}.$$

Then (5.5.7) and (5.5.9) assume the following form

$$D = \frac{-i}{k^2 + i0}\left(P^{(1)} + P^{(2)} + \alpha P^{(3)}\right) \tag{5.5.14}$$

$$\Pi = \Pi^{(1)}(P^{(1)} + P^{(2)}) + m(P^{(1)} - P^{(2)})\sqrt{k^2}\Pi^{(2)}. \tag{5.5.15}$$

Since

$$D^{-1} = ik^2\left(P^{(1)} + P^{(2)} + \frac{1}{\alpha}P^{(3)}\right),$$

one finds for (5.5.6)

$$\begin{aligned}\mathcal{D}^{-1} &= (ik^2 - i\Pi^{(1)} + im\sqrt{k^2}\Pi^{(2)})P^{(1)} \\ &+(ik^2 - i\Pi^{(1)} - im\sqrt{k^2}\Pi^{(2)})P^{(2)} + \frac{i}{\alpha}k^2 P^{(3)}.\end{aligned} \tag{5.5.16}$$

The inverse is given by

$$\begin{aligned}\mathcal{D} &= \frac{1}{ik^2 - i\Pi^{(1)} + im\sqrt{k^2}\Pi^{(2)}}P^{(1)} \\ &+ \frac{1}{ik^2 - i\Pi^{(1)} - im\sqrt{k^2}\Pi^{(2)}}P^{(2)} - \frac{i\alpha}{k^2 + i0}P^{(3)}.\end{aligned} \tag{5.5.17}$$

Substituting the original expressions (5.5.13), one finally gets

$$\begin{aligned}\mathcal{D}_{\mu\nu}(k) &= \frac{-i}{k^2 - \Pi(k^2)}\left[g_{\mu\nu} - \frac{k_\mu k_\nu}{k^2} - im\frac{\Pi^{(2)}}{1 - \Pi^{(1)}/k^2}\varepsilon_{\mu\nu\beta}\frac{k^\beta}{k^2}\right] \\ &-i\alpha\frac{k_\mu k_\nu}{k^4 + i0k^2},\end{aligned} \tag{5.5.18}$$

with

$$\Pi(k^2) = \Pi^{(1)} + \frac{m^2(\Pi^{(2)})^2}{1 - \Pi^{(1)}/k^2}. \tag{5.5.19}$$

By (5.5.10, 12) it follows that, in second-order perturbation theory

$$\Pi(0) = \frac{[e^2/(4\pi)]^2}{1 - e^2/(48\pi m)} \neq 0. \tag{5.5.20}$$

Strictly speaking, we have only calculated the retarded distributions. For the full vacuum polarization tensor we must add $-R_2'$ (5.4.18). But the latter vanishes for $k^2 < 4m^2$ due to (5.4.42, 43), and this is the region we are only interested in.

It follows from (5.5.18) and (5.5.20) that the photon acquires non-perturbatively a (nonzero) "dynamically generated" mass μ of order e^2. This feature depends only on property (5.5.20). The word "nonperturbative" requires, however, a clarification: by that we mean that the nonzero mass $\mu \neq 0$ is a result of summing an infinite number of *selected* graphs, i.e. precisely the vacuum polarization bubbles, contributing to the r.h.s. of (5.5.8). The effect is, therefore, absent in any finite order of perturbation theory! If one treats this problem by the conventional methods of regularizing divergent Feynman integrals, one may get ambiguous results which depend on the method of regularization (*S. Deser, R. Jackiw, S. Templeton, Ann. Phys. (N.Y.) 140, 372 (1982))*. This shows clearly the merit of the causal approach.

For the sake of completeness we also discuss the other second order graph with non-negative ω, which is the self-energy. The corresponding C-number distribution is given by (3.7.2)

$$d(y) = -e^2\gamma^\mu[S^{(-)}(y)D_0^{(+)}(-y) + S^{(+)}(y)D_0^{(+)}(y)]\gamma_\mu$$

$$= -e^2\gamma^\mu[d_-(y) + d_+(y)]\gamma_\mu, \tag{5.5.21}$$

where

$$d_-(y) = S^{(-)}(y)D_0^{(+)}(-y) \tag{5.5.22}$$

$$d_+(y) = S^{(+)}(y)D_0^{(+)}(y). \tag{5.5.23}$$

The Fourier transform of d is

$$\hat{d}_-(p) = -(2\pi)^{-3/2}\int d^3q\,\hat{D}_0^{(-)}(p-q)(\not{q}+m)\hat{D}_m^{(-)}(q) =$$

$$= I_1 + I_2, \tag{5.5.24}$$

where

$$I_1 = (2\pi)^{-3/2}\int d^3q\,\Theta(q^0-p^0)\delta((p-q)^2)\not{q}\Theta(-q^0)\delta(q^2-m^2) \tag{5.5.25}$$

and

$$I_2 = (2\pi)^{-3/2}\int d^3q\,\Theta(q^0-p^0)\delta((p-q)^2)m\Theta(-q^0)\delta(q^2-m^2). \tag{5.5.26}$$

We now take p time-like in the form $p = (p_0, \mathbf{0})$. Then

$$I_1 = \int d^3q\,\Theta(q^0-p^0)\Theta(-q^0)\delta(p^0 - 2p^0q^0 + m^2)\delta(q^2-m^2)$$

$$= \int \frac{d^2q}{2E_q} \delta(p_0^2 + 2p_0 E_q + m^2)\Theta(-E_q - p_0). \qquad (5.5.27)$$

It follows from (5.5.27) that $p_0 < 0$ and

$$E_q = \frac{m^2 + p_0^2}{-2p_0} = \sqrt{m^2 + \boldsymbol{q}^2}, \quad \text{whence}$$

$$|\boldsymbol{q}| = \frac{m^2 - p_0^2}{2p_0} \quad \text{and} \quad m^2 - p_0^2 < 0.$$

Therefore

$$I_1 = \frac{2\pi}{2}\Theta(p_0^2 - m^2)\Theta(-p_0) \int_0^\infty d|\boldsymbol{q}| \frac{|\boldsymbol{q}|}{E_q 2|p_0|}\delta\Big(E_q - \frac{p_0^2 + m^2}{2|p_0|}\Big)$$

$$= \pi\Theta(p_0^2 - m^2)\Theta(-p_0) \int_m^\infty dE \frac{E}{E2|p_0|}\delta\Big(E - \frac{p_0^2 + m^2}{2|p_0|}\Big)$$

$$= \frac{\pi}{2}\Theta(p_0^2 - m^2)\Theta(-p_0)\frac{1}{\sqrt{p_0^2}}. \qquad (5.5.28)$$

Now, $I_2 = I_{2\nu}\gamma^\nu$, and $I_{2\nu} = 0$ if $\nu \neq 0$ by symmetry. For I_{20} we have

$$I_{20} = \int d^3q\, q_0 \Theta(q_0 - p_0)\delta(p_0^2 - 2p_0 q_0 + m^2)\frac{\delta(q_0 + E_q)}{2E_q}$$

$$= -\frac{\pi}{2}\Theta(p_0^2 - m^2)\Theta(-p_0)\frac{p_0^2 + m^2}{2p_0^2}\frac{p_0}{\sqrt{p_0^2}} \qquad (5.5.29)$$

in the same way as above for I_1. For general p, we have thus to replace p_0^2 by p^2 in I_1 and I_2 (except in $\Theta(-p_0)$, which remains as such), and the linear term $\gamma^0 p_0$ in $\gamma^0 I_{20}$, with I_{20} given by (5.5.29), by $\not p$. We therefore obtain

$$\hat{d}_-(p) = (2\pi)^{-3/2}\frac{\pi}{2}\Theta(p^2 - m^2)\Theta(-p_0)\frac{1}{\sqrt{p^2}}\Big[m - \frac{\not p}{2}\Big(1 + \frac{m^2}{p^2}\Big)\Big]. \qquad (5.5.30)$$

d_+ (5.5.23) is calculated in the same way.

It follows now from (5.5.21), (5.5.30) and (5.4.5) that

$$d(p) = e^2\gamma^\mu(\hat{d}_-(p) - \hat{d}_+(p))\gamma_\mu$$

$$= e^2(2\pi)^{-3}\frac{\pi}{2}\Theta(p^2 - m^2)\frac{\operatorname{sgn} p_0}{\sqrt{p^2}}\Big[3m + \frac{\not p}{2}\Big(1 + \frac{m^2}{p^2}\Big)\Big]. \qquad (5.5.31)$$

From (5.5.31) we see that the singular order is $\omega = 0$. Hence, the central splitting solution is therefore given by

$$r(p) = e^2(2\pi)^{-3}\frac{i}{2\pi}\int\limits_{-\infty}^{+\infty}\frac{dt}{(t-i0)(1-t+i0)}\,\Theta(t^2p^2-m^2)$$

$$\times\frac{\mathrm{sgn}\,(tp_0)}{|t|\sqrt{p^2}}\Big[3m+\frac{t\not p}{2}\Big(1+\frac{m^2}{t^2p^2}\Big)\Big]. \tag{5.5.32}$$

Here p must be in the forward cone $\bar{V}^+$. We call J_3, J_4, J_5 the terms corresponding to the summands $3m$, $t\not p/2$, m^2/t^2p^2, respectively, in (5.5.32). We have

$$J_3 = \int\limits_{\sqrt{\frac{m^2}{p^2}}}^{\infty}\frac{dt}{t^2}\Big(\frac{1}{1-t+i0}+\frac{1}{1+t+i0}\Big)$$

$$= J_2 = J_1 + 2\sqrt{\frac{p^2}{m^2}} = \log\frac{1-\sqrt{p^2/m^2}}{1+\sqrt{p^2/m^2}}+2\sqrt{\frac{p^2}{m^2}}. \tag{5.5.33}$$

Now,

$$J_4 = \int\limits_{\sqrt{\frac{m^2}{p^2}}}^{\infty}\frac{dt}{|t|}\Big(\frac{1}{1-t+i0}-\frac{1}{1+t+i0}\Big)$$

$$= J_1 = \log\frac{1-\sqrt{p^2/m^2}}{1+\sqrt{p^2/m^2}} \tag{5.5.34}$$

and

$$J_5 = \int\limits_{\sqrt{\frac{m^2}{p^2}}}^{\infty}\frac{dt}{|t|t^2}\,\frac{2t}{(1+i0)^2-t^2}$$

$$= -2\int\limits_{\sqrt{\frac{m^2}{p^2}}}^{\infty}\frac{dt}{t^2(t^2-1-i0)} = J_3. \tag{5.5.35}$$

It follows from (5.5.32)-(5.5.35) that

$$r(p) = e^2(2\pi)^{-4}\frac{i}{\sqrt{p^2}}\Big[\Big(3m+\frac{\not p}{2}(1+\frac{m^2}{p^2})\Big)\log\frac{1-\sqrt{p^2/m^2}}{1+\sqrt{p^2/m^2}}$$

$$+\sqrt{\frac{p^2}{m^2}}\Big(6m+\not p\frac{m^2}{p^2}\Big)\Big]. \tag{5.5.36}$$

The general solution of the splitting problem is

$$\tilde{r}(p) = r(p) + C_0. \tag{5.5.37}$$

On the other hand, the tensor structure of r is

$$r(p) = r_0(p^2) + \not{p} r_1(p^2) \tag{5.5.38}$$

as seen from (5.5.31), whereby r_0 has order $\omega = -1$ and $(\not{p} r_1(p^2))$ has order $\omega = 0$. But C_0 in (5.5.37) would correspond to r_0, and this is impossible, since $\omega = -1$. Thus $C_0 = 0$, because the distribution splitting must respect the tensor structure. Hence, (5.5.36) is the general solution. The logarithmic sigularity at the mass shell $p^2 = m^2$ is worse than in four dimensions (3.7.34). However, the singularity is integrable, and $r(p)$ remains a well-defined tempered distribution. In any case, $S(g)$ with $g \in S$, is well defined in perturbation theory also in three dimensions. We do not consider the adiabatic limit $g \to 1$ here. The latter is more subtle in three dimensions.

If one goes down to (1+1) dimensions, quantum electrodynamics becomes even soluble, at least for massless fermions. This is the so-called Schwinger model which has extensively been discussed in the literature (see the bibliographical notes). But its properties are not typical for QED in higher dimensions, which is the price for the solubility. A more physical theory is totally scalar QED in four dimensions. Here the "electron" is taken as a charged scalar (massive) particle and the "photon" as a (massless) scalar field as well:

$$T_1(x) = ie : \varphi^+(x)\varphi(x) : A(x). \tag{5.5.39}$$

This theory has much resemblance with spinor QED and scalar QED, because a Furry theorem still holds, so that the graphs have the same structure. One also finds similar infrared properties, for example the same compensation between vertex function and bremsstrahlung. The nice feature is that this 4-dimensional theory is super-normalizable. The reader is know sufficiently trained to verify all this. If he is ambitious, he may try to estimate $S_n(g)$ in a suitable way. We conjecture that totally scalar QED will be the first non-trivial theory in four dimensions where the perturbation series can be completely controlled.

5.6 Problems

5.1. Calculate the trace in (5.3.14). Hint: Use problem 1.9.

5.2. Prove the Eq. (5.3.59).

5.3. Calculte the graph with three axialvector vertices in analogy to the VVA-graph.

5.4. Verify the results (5.4.15, 16) for the singular order ω in (2+1)-dimensional QED.

5.5. Determine the singular orders of all first and second order graphs in (2+1)-dimensional QED to prove the general formula (5.4.17) for ω.

5.6. Verify the result (5.4.43) for the antisymmetric part of the vacuum polarization tensor.

6. Epilogue: Non-Abelian Gauge Theories

It is of course highly desirable to extend the methods developed in this book to non-abelian gauge theories. The main problem that arises in this connection is the meaning of non-abelian gauge invariance *in the quantum theory.* To introduce this concept, we formulate the gauge invariance of QED, discussed in Sect. 4.6, in a slightly different way. In QED we have considered the simple gauge transformation

$$A'^{\mu}(x) = A^{\mu} + \lambda\partial^{\mu}u(x) \tag{6.1}$$

where $u(x)$ is a C-number field satisfying the wave equation

$$\square u(x) = 0. \tag{6.2}$$

This transformation can be pseudo-unitarily implemented in the form

$$A'^{\mu}(x) = e^{-i\lambda Q}A^{\mu}e^{i\lambda Q} = A^{\mu} - \frac{i\lambda}{1!}[Q,\, A^{\mu}] - \frac{\lambda^2}{2!}\Big[Q,\, [Q,\, A^{\mu}]\Big] + \dots \tag{6.3}$$

by means of the charge

$$Q = \int d^3x\, (\partial_\nu A^\nu \overleftrightarrow{\partial}_0 u). \tag{6.4}$$

This is a consequence of the commutation rule

$$[Q,\, A^{\mu}(x)] = i\partial^{\mu}u(x), \tag{6.5}$$

and all higher commutators vanish. Perturbative QED is invariant under this transformation. In fact, since Q commutes with ψ and $\overline{\psi}$, we obtain

$$[Q, T_1] = ie[Q, :\overline{\psi}\gamma^\nu\psi : A_\nu]$$

$$= -e : \overline{\psi}\gamma^\nu\psi : \partial_\nu u = \partial_\nu(-e : \overline{\psi}\gamma^\nu\psi : u) \stackrel{\text{def}}{=} i\partial_\nu T^\nu_{1/1}. \tag{6.6}$$

This is a divergence, consequently, if integrated with a switching function $g(x)$, the commutator vanishes in the adiabatic limit $g \to 1$.

It is not hard to extend the discussion to higher orders. For a fixed x_l we consider from T_n all terms with the external field operator $A_\mu(x_l)$

$$T_n(x_1, ..., x_n) =: t^\mu_l(x_1, ..., x_n)A_\mu(x_l) : + ... \tag{6.7}$$

the dots mean the terms without $A_\mu(x_l)$. Since the commutator acts like a derivative, we find

$$[Q, T_n(x_1, ..., x_n)] = t_l^\mu(x_1, ..., x_n) i\partial_\mu u(x_l) + ... \tag{6.8}$$

This is a divergence if and only if

$$\partial_\mu^l t_l^\mu(x_1, ..., x_n) = 0, \tag{6.9}$$

which is precisely our old condition (4.6.9) of gauge invariance.

We now want to apply these ideas to pure Yang-Mills theories (without matter fields), where all essential points can already be seen. This theory is defined by the following gluon coupling

$$T_1^A(x) = igf_{abc} : A_{\mu a}(x) A_{\nu b}(x) \partial^\nu A_c^\mu(x) :, \tag{6.10}$$

which can also be written in the form

$$= i\frac{g}{2} f_{abc} : A_{\mu a} A_{\nu b} F_c^{\nu\mu} : \tag{6.11}$$

where

$$F_c^{\mu\nu}(x) = \partial^\mu A_c^\nu - \partial^\nu A_c^\mu \tag{6.12}$$

are the *free* asymptotic Yang-Mills fields. f_{abc} are the antisymmetric structure constants of the gauge group, say $SU(N)$, and $A_{\mu a}(x)$ now are the gauge potentials, satisfying free field commutation rules

$$[A_a^\mu(x), \, A_b^\nu(y)] = i\delta_{ab} g^{\mu\nu} D(x-y) \tag{6.13}$$

$$[A_a^{(-)\mu}(x), \, A_b^{(+)\nu}] = i\delta_{ab} g^{\mu\nu} D^{(+)}(x-y), \tag{6.14}$$

where $A^{(\pm)}$ are the emission and absorption parts (negative and positive frequency parts) of A and $D^{(\pm)}$ the (mass zero) Jordan-Pauli distributions. The time dependence of A is given by the wave equation

$$\square A_a^\mu(x) = 0.$$

One may wonder why (6.10) does not contain a quadrilinear term $\sim g^2$. As we shall shortly see, this term is automatically generated in second order by gauge invariance, similarly as in Sect. 5.2 in scalar QED. We now try to apply the same gauge transformation (6.1). In the charge operator Q we must sum over the colour index a

$$Q = \int d^3x \, (\partial_\nu A_a^\nu \overleftrightarrow{\partial}_0 u_a). \tag{6.15}$$

For the sake of completeness we calculate the commutators in detail

$$[Q, \, A_a^{(\pm)\mu}(x)] = \int d^3y \, [\partial_\nu A_b^{(\mp)\nu} \partial_0 u_b - \partial_0 \partial_\nu A_b^{(\mp)\nu} u_b, \, A_a^{(\pm)\mu}(x)]$$

$$= \int d^3y \, i\Big(\partial_y^\mu D^{(\pm)}(y-x) \partial_0 u_a(y) - \partial_0^y \partial_y^\mu D^{(\pm)}(y-x) u_a(y)\Big)$$

$$= i\partial^\mu u_a^{(\pm)}(x), \tag{6.16}$$

because $u_a(x)$ are solutions of the wave equation. For the commutator of Q with $T_1^A(x)$ (6.10) we now obtain

$$[Q,\, T_1^A(x)] = igf_{abc} : \Big\{[Q,\, A_{\mu a}]A_{\nu b}\partial^\nu A_c^\mu$$

$$+A_{\mu a}[Q,\, A_{\nu b}]\partial^\nu A_c^\mu + A_{\mu a}A_{\nu b}[Q,\, \partial^\nu A_c^\mu]\Big\}:$$

$$= -gf_{abc} : \Big\{\partial_\mu u_a A_{\nu b}\partial^\nu A_c^\mu + \partial_\nu(A_{\mu a}u_b\partial^\nu A_c^\mu)\Big\}: . \tag{6.17}$$

Here the last term is a divergence, but the first term spoils gauge invariance. To restore it, this term must be compensated. With a C-number gauge function $u_a(x)$ this is impossible. We therefore consider $u_a(x)$ to be a *free* quantum field that has additional couplings to the gauge potentials. In this way we are led to operator gauge transformations. The fields $u(x)$ occurring in the gauge transformations are called ghost fields. Gauge invariance requires the coupling to those gauge fields.

It turns out that a gauge invariant Yang-Mills theory can only be obtained by using *fermionic* ghost fields. Let $u_a(x)$ and $\tilde{u}_a(x)$ be free massless Fermi fields satisfying the anti-commutation relations

$$\{u_a^{(\pm)}(x),\, \tilde{u}_b^{(\mp)}(y)\} = -i\delta_{ab}D^\mp(x-y), \tag{6.18}$$

all other anti-commutators vanish. We take the same charge operator Q as before (6.15) and find by a similar calculation as above (6.16)

$$\{Q,\, \tilde{u}_a^{(\pm)}(x)\} = -i\partial_\nu A_a^{(\pm)\nu}(x), \tag{6.19}$$

$$\{Q,\, u_b^{(\pm)}\} = 0, \quad \{Q,\, \partial^\mu \tilde{u}_a(x)\} = -i\partial_\nu F_a^{\mu\nu}(x). \tag{6.20}$$

The ghost fields must be coupled to the gauge potentials by derivative couplings

$$T_1^u = igf_{abc} : A_{\mu a}u_b\partial^\mu\tilde{u}_c :, \tag{6.21}$$

as in scalar QED. For the commutator with Q we get

$$[Q,\, T_1^u] = -gf_{abc} : \{\partial_\mu u_a u_b \partial^\mu \tilde{u}_c + A_{\mu a}u_b\partial^\mu\partial^\nu A_{\nu c}\} : . \tag{6.22}$$

Taking the antisymmetry of f_{abc} into account, we see that the first term is a divergence

$$f_{abc}\partial_\mu u_a u_b \partial^\mu \tilde{u}_c = \frac{1}{2} f_{abc}\partial_\mu(u_a u_b \partial^\mu \tilde{u}_c),$$

because $\tilde{u}_a(x)$ fulfills the wave equation. The second term can be transformed as follows

$$[Q,\, T_1^u] = -gf_{abc} : \{\tfrac{1}{2}\partial^\mu(u_a u_b \partial_\mu \tilde{u}_c) + \partial^\nu(A_{\mu a}u_b\partial^\mu A_{\nu c})$$

$$-\partial^\nu u_b A_{\mu a}\partial^\mu A_{\nu c}\} : . \tag{6.23}$$

Interchanging $a \leftrightarrow b$ and $\mu \leftrightarrow \nu$ in the last term, it becomes equal to the first term in (6.17). Hence, the desired compensation can be achieved by taking

$$T_1(x) = T_1^A(x) - T_1^u(x)$$
$$= igf_{abc} : \{A_{\mu a}A_{\nu b}\partial^\nu A_c^\mu - A_{\mu a}u_b\partial^\mu \tilde{u}_c\} : . \tag{6.24}$$

Then the commutator

$$[Q, T_1(x)] = gf_{abc} : \Big\{-\partial_\nu\Big(A_{\mu a}u_b(\partial^\nu A_c^\mu - \partial^\mu A_c^\nu)\Big) + \frac{1}{2}\partial_\nu(u_a u_b \partial^\nu \tilde{u}_c)\Big\} : \tag{6.25}$$

is a sum of divergences of normal products. If this is integrated with a test function $g(x)$, the result vanishes in the adiabatic limit $g \to 1$. We therefore call (6.25) gauge invariance of $T_1(x)$.

Now the Yang-Mills coupling (6.24) bears a resemblance with the trilinear derivative coupling of scalar QED (5.1.1). Consequently, the second order tree graphs with two derivatives on the inner line $\sim \partial_x^2 D^F(x-y)$ have again singular order $\omega = 0$ and, hence, are subject to normalization $\sim \delta(x-y)$. Gauge invariance

$$[Q, T_2(x,y)] = \mathrm{div} \tag{6.26}$$

determines the normalization term uniquely. In this way the four-gluon coupling is generated. Then the theory is gauge invariant in arbitrary order. The inductive proof, however, is much more complicated than in QED. It has been published in the papers "Causal Construction of Yang-Mills theories I–IV" by M. Dütsch, T. Hurth, F. Krahe and G. Scharf (*Nuovo Cim. 106A, 1029 (1993), 107A, 375 (1994), 108A, 679+737 (1995)*). The unitarity on the physical subspace can also be shown, using similar methods as in Sect. 4.7.

From these brief remarks it is clear that QED is still a source of ideas and methods for current research in quantum field theory. It is true that *classical* Yang-Mills theories show very interesting new structures with nice connection to geometry. But the quantum theory lives from the old mother QED.

Appendices

A: The Hydrogen Atom According to the Schrödinger Equation

We consider the radial Schrödinger equation

$$\left[-\frac{1}{r}\frac{d^2}{dr^2}r + \frac{l(l+1)}{r^2} - \frac{2k}{r}\right]\varphi = E\varphi, \tag{A.1}$$

where we substitute

$$r\varphi(r) = u(r). \tag{A.2}$$

The eigenvalue equation

$$\left[-\frac{d^2}{dr^2} + \frac{l(l+1)}{r^2} - \frac{2k}{r}\right]u \stackrel{\text{def}}{=} H_l u = Eu \tag{A.3}$$

can be factorized by introducing raising and lowering operators

$$a_l^+ = -\frac{d}{dr} + \frac{l}{r} - \frac{k}{l} \tag{A.4}$$

$$a_l = \frac{d}{dr} + \frac{l}{r} - \frac{k}{l}. \tag{A.5}$$

a_l^+ is the adjoint of a_l in $L^2(0,\infty)$. In fact, computing the products

$$a_l a_l^+ = -\frac{d^2}{dr^2} + \frac{l(l+1)}{r^2} - \frac{2k}{r} + \frac{k^2}{l^2} = H_l + \frac{k^2}{l^2} \tag{A.6}$$

$$a_l a_l^+ = -\frac{d^2}{dr^2} + \frac{l(l-1)}{r^2} - \frac{2k}{r} + \frac{k^2}{l^2} = H_{l-1} + \frac{k^2}{l^2}, \tag{A.7}$$

we get the Hamiltonian H_l up to a constant. Then the eigenvalue equation (A.3) can be written in two ways

$$H_l u = Eu = \left(a_l^+ a_l - \frac{k^2}{l^2}\right)u \tag{A.8}$$

$$= \left(a_{l+1}a_{l+1}^+ - \frac{k^2}{(l+1)^2}\right)u. \tag{A.9}$$

We now operate on (A.8) with a_l, taking (A.7) into account,

$$(a_l a_l^+) a_l u = \Big(E + \frac{k^2}{l^2}\Big) a_l u = \Big(H_{l-1} + \frac{k^2}{l^2}\Big) a_l u. \tag{A.10}$$

This means, if $u_l(r)$ is eigenvector of H_l, then $a_l u_l$ is eigenvector of H_{l-1} with the same eigenvalue E. Hence, E is independent of l, a_l is called a lowering operator. Assuming the eigenvectors $\|u_l\| = 1$ to be normalized, we have

$$a_l u_l = C_l^{l-1} u_{l-1}, \tag{A.11}$$

where C_l^{l-1} is a normalization constant. Similarly, operating with a_{l+1}^+ on (A.9), we get

$$\begin{aligned}(a_{l+1}^+ a_{l+1}) a_{l+1}^+ u_l &= \Big(E + \frac{k^2}{(l+1)^2}\Big) a_{l+1}^+ u_l \\ &= \Big(H_{l+1} + \frac{k^2}{(l+1)^2}\Big) a_{l+1}^+ u_l. \end{aligned} \tag{A.12}$$

Hence a_{l+1}^+ is a raising operator and

$$a_{l+1}^+ u_l = C_l^{l+1} u_{l+1}. \tag{A.13}$$

Equation (A.8) implies

$$a_l^+ a_l u_l = \Big(E + \frac{k^2}{l^2}\Big) u_l \tag{A.14}$$

and the scalar product

$$(u_l, a_l^+ a_l u_l) = \|a_l u_l\|^2 = E + \frac{k^2}{l^2} \geq 0 \tag{A.15}$$

is non-negative. Consequently, negative eigenvalues $E < 0$ can only correspond to finitely many l-values $l = 0, 1, \dots n-1$. The number n is determined by the condition that further raising is impossible:

$$a_n^+ u_{n-1} = 0 = \Big(-\frac{d}{dr} + \frac{n}{r} - \frac{k}{n}\Big) u_{n-1}(r). \tag{A.16}$$

This simple equation has the following square-integrable solution

$$u_{n-1}(r) = C_n r^n \exp\Big(-\frac{k}{n} r\Big), \tag{A.17}$$

where the normalization constant is equal to

$$C_n = \Big(\frac{n}{2k}\Big)^{-n-\frac{1}{2}} (2n)!^{-\frac{1}{2}}. \tag{A.18}$$

Let us determine the corresponding eigenvalues. Operating with a_n on (A.16) we find

$$0 = a_n a_n^+ u_{n-1} = \left(H_{n-1} + \frac{k^2}{n^2}\right) u_{n-1} = \left(E + \frac{k^2}{n^2}\right) u_{n-1}.$$

This gives

$$E_n = -\frac{k^2}{n^2}, \tag{A.19}$$

which is the well-known discrete spectrum of the hydrogen atom. The normalization constants in (A.11) can now be calculated as follows

$$(C_l^{l-1})^2 = (a_l u_l,\, a_l u_l) = (u_l,\, a_l^+ a_l u_l)$$

$$= E_n + \frac{k^2}{l^2} = k^2 \frac{n^2 - l^2}{n^2 l^2},$$

thus

$$C_l^{l-1} = \frac{k}{nl}\sqrt{n^2 - l^2}. \tag{A.20}$$

The other eigenfunctions follow from (A.17) by successive application of the lowering operators a_l (A.5). The result is of the following form

$$u_{ln}(r) = r^{l+1} L_{ln}(r) \exp -\frac{k}{n} r, \tag{A.21}$$

where $L_{ln}(r)$ is a polynomial of degree $n - l - 1$. The inductive step

$$u_{l-1,n}(r) = \left(C_l^{l-1}\right)^{-1} \left(\frac{d}{dr} + \frac{l}{r} - \frac{k}{l}\right) r^{l+1} L_{ln} \exp -\frac{k}{n} r$$

$$= \frac{nl}{k\sqrt{n^2 - l^2}} r^l \left[\left(2l + 1 - \frac{k}{n} r - \frac{k}{l} r\right) L_{ln} + r L_{ln}'\right] \exp -\frac{k}{n} r$$

$$\stackrel{\text{def}}{=} r^l L_{l-1,n}(r) \exp -\frac{k}{n} r \tag{A.22}$$

implies the recursion relation

$$L_{l-1,n}(r) = \frac{nl}{k\sqrt{n^2 - l^2}} \left[\left(2l + 1 - \frac{k}{n} r - \frac{k}{l} r\right) L_{ln} + r L_{ln}'\right]. \tag{A.23}$$

According to (A.17), the relation gets started by

$$L_{n-1,n} = C_n. \tag{A.24}$$

B: Regularly Varying Functions

Since this theory is not so widely known, we give rather detailed proofs of the basic properties of functions of regular variation, following essentially the presentation by E. Senata (*Lecture Notes in Mathematics 508, Springer-Verlag 1976*).

Definition 1. *A positive function $\rho(t)$ is called regularly varying at $t = 0$, if it is measurable in $(0, t_0]$ for some $t_0 > 0$, and the limit*

$$\lim_{t\to 0} \frac{\rho(at)}{\rho(t)} = a^{\omega} \tag{B.1}$$

exists for every $a > 0$, with some fixed ω, $-\infty < \omega < +\infty$.

The number ω is called the order of ρ. Introducing the function $r(t)$ by

$$\rho(t) = t^{\omega} r(t), \tag{B.2}$$

it follows from (B.1) that

$$\lim_{t\to 0} \frac{r(at)}{r(t)} = 1, \tag{B.3}$$

for every $a > 0$, i.e. we have reduced the order to $\omega = 0$. This suggests the

Definition 2. *A regularly varying function with order $\omega = 0$ is called slowly varying.*

The most important properties of such functions are contained in a theorem of uniform convergence (Theorem 3) and in a representation theorem (Theorem 4):

Theorem 3. Let $r(t)$ be a slowly varying function. Then the relation (B.3) holds uniformly in $a \in [a_1, a_2]$, in every fixed finite interval $0 < a_1 < a_2 < \infty$.

Theorem 4. Let $r(t)$ be defined for $0 < t < t_0$ and slowly varying. Then there exists t_1, $0 < t_1 < t_0$, such that

$$r(t) = \exp\left[\eta(t) + \int_t^{t_1} \frac{h(s)}{s}\, ds\right] \tag{B.4}$$

holds for all $0 < t < t_1$. Here η is bounded and measurable on $(0, t_1]$, and $\eta \to c$ for $t \to 0$ ($|c| < \infty$), and $h(t)$ is continuous on $(0, t_1]$ and $h(t) \to 0$ for $t \to 0$.

Let us consider the function

$$f(x) = \log r\left(e^{-x}\right). \tag{B.5}$$

In virtue of (B.3) we have

$$\log r(at) - \log r(t) \to 0 \quad \text{for} \quad t \to 0. \tag{B.6}$$

Substituting

$$a = e^{-\mu}, \qquad t = e^{-x}, \tag{B.7}$$

this implies for f that

$$f(x+\mu) - f(x) \to 0 \quad \text{for} \quad x \to \infty \tag{B.8}$$

and every μ.

The proof of Theorem 3 is a direct corollary of

Lemma 5. Let $f(x)$ be a real measurable function on $[x_0, \infty)$ which satisfies (B.8). Then (B.8) holds uniformly in μ on every fixed, finite closed interval.

In fact, this implies that (B.3) holds true uniformly in $a \in [a_1, a_2]$.

Proof. We first prove Lemma 5 indirectly for the interval $[0,1]$. If (B.8) does not hold uniformly in $\mu \in [0,1]$, then there exists an $\varepsilon > 0$ and sequences $x_n \to \infty$ and $\mu_n \in [0,1]$ with

$$|f(x_n + \mu_n) - f(x_n)| \geq \varepsilon, \quad \forall n. \tag{B.9}$$

Let us consider the measurable sets

$$U_n = \{\mu \in [0,2] \,|\, |f(x_m + \mu) - f(x_m)| < \frac{\varepsilon}{2}, \\ \forall m \geq n\} \tag{B.10}$$

$$V_n = \{\lambda \in [0,2] \,|\, |f(x_m + \mu_m + \lambda) - f(x_m + \mu_m)| < \frac{\varepsilon}{2}, \\ \forall m \geq n\}. \tag{B.11}$$

Both sets are monotonely increasing towards $[0,2]$ in virtue of (B.8). Consequently, their Lebesgue measure is bigger than 3/2, $m(U_n) > 3/2$, $m(V_n) > 3/2$ for big enough n, $n \geq N$. Now we consider the translated set $V'_N = V_N + \mu_N$, which has $m(V'_N) = m(V_N) > 3/2$. Since $U_N \subset [0,2] \subset [0,3]$ and $V'_N \subset [0,3]$, this implies that U_N and V'_N have a non-void intersection. Consequently, there exists a point $\mu \in U_N \cap V'_N$ and $\mu - \mu_N \in V_N$. Since this μ is in U_N, we have

$$|f(x_N + \mu) - f(x_N)| < \frac{\varepsilon}{2}, \tag{B.12}$$

and

$$|f(x_N + \mu_N + \mu - \mu_N) - f(x_N + \mu_N)| < \frac{\varepsilon}{2},$$

or

$$|f(x_N + \mu) - f(x_N + \mu_N)| < \frac{\varepsilon}{2},$$

because $\mu - \mu_N$ is in V_N. This together with (B.12) implies

$$|f(x_N + \mu_N) - f(x_N)| < \varepsilon$$

in contradiction to (B.9).

To prove the lemma for an arbitrary interval $[a, b]$, we consider

$$\tilde{f}(x) = f((b-a)x). \tag{B.13}$$

Then the difference (B.8) is equal to

$$f(x + \mu) - f(x) = \tilde{f}(y + \nu) - \tilde{f}(y) + f(x + a) - f(x), \tag{B.14}$$

where

$$y = \frac{x + a}{b - a}, \quad \nu = \frac{\mu - a}{b - a}, \tag{B.15}$$

and ν is now in $[0, 1]$. $x \to \infty$ is equivalent to $y \to \infty$. The second difference on the r.h.s. of (B.14) goes to 0 because a is fixed, and the first difference goes to 0 uniformly in $\nu \in [0, 1]$, i.e. in $\mu \in [a, b]$. □

For the proof of the representation Theorem 4 we need

Lemma 6. There exists a fixed constant x_0, such that the function $f(x)$ (B.5) is bounded on every interval $[X, X']$, $X' > X \geq x_0$.

Proof. According to Lemma 5 there exists an X such that

$$|f(x + \mu) - f(x)| < 1 \tag{B.16}$$

for all $x \geq X$ and all $\mu \in [0, 1]$. Choosing $x = X$ and $X + \mu = y \in [X, X + 1]$, it follows that

$$|f(y)| \leq |f(X)| + 1. \tag{B.17}$$

In the same way we find for $x \in [X + 1, X + 2]$

$$|f(x)| \leq |f(X + 1)| + 1 \leq |f(X)| + 2, \quad \text{etc.}$$

thus

$$|f(x)| \leq |f(X)| + k \tag{B.18}$$

for $x \in [X + k - 1, X + k]$, and therefore for all $x \in [X, X + k]$. □

Since $f(x)$ is measurable and bounded, we get

Corollary 7. $f(x)$ is integrable in $[X, X']$ for every X'.

Lemma 8. For $x \geq X$, $f(x)$ as above can be represented as follows:

$$f(x) = g(x) + \int_X^x h(s)\,ds, \tag{B.19}$$

where g and h are measurable and bounded on every interval $[X, X']$ and $g(x) \to g_0$, $|g_0| < \infty$, $h(x) \to 0$ for $x \to \infty$.

Proof. We start for $x \geq X$ from the trivial identity

$$f(x) = \int_x^{x+1} (f(x) - f(s))\,ds + \int_X^x (f(s+1) - f(s))\,ds$$

$$+ \int_X^{X+1} f(s)\,ds$$

$$\stackrel{\text{def}}{=} k(x) + \int_X^x h(s)\,ds + g_0. \tag{B.20}$$

Here $k(x)$ denotes the first and g_0 the last integral. The integrand

$$h(s) = f(s+1) - f(s) \tag{B.21}$$

of the second integral goes to 0 for $s \to \infty$ due to (B.8). Introducing

$$g(x) = k(x) + g_0, \tag{B.22}$$

we see that $g(x)$ is bounded by (B.16) and

$$k(x) = \int_0^1 (f(x) - f(x+\mu))\,d\mu \to 0 \tag{B.23}$$

for $x \to \infty$ by (B.8) again. □

Lemma 9. There exists X^*, such that for all $x \geq X^*$

$$f(x) = g^*(x) + \int_{X^*}^x h^*(s)\,ds, \tag{B.24}$$

where g^* and h^* are as in Lemma 8, however, h^* is continuous.

Proof. Let

$$f^*(x) = \int_X^x h(s)\,ds = \int_X^x (f(s+1) - f(s))\,ds. \tag{B.25}$$

Then (B.19) implies

$$f(x) - f^*(x) = g(x) \to g_0, \tag{B.26}$$

for $x \to \infty$. For all $\mu > 0$ we get

$$f^*(x+\mu) - f^*(x) = \int_x^{x+\mu} (f(s+1) - f(s))\,ds$$

$$= \int_0^{\mu} (f(y+x+1) - f(y+x))\,dy. \tag{B.27}$$

The difference

$$f(y+x+1) - f(y+x) = f(x+y+1) - f(x) - [f(x+y) - f(x)]$$

goes to 0 for $x \to \infty$, uniformly in y by Lemma 5. Hence

$$f^*(x+\mu) - f^*(x) \to 0 \tag{B.28}$$

for $x \to \infty$ for all $\mu > 0$. This follows in the same way for $\mu < 0$.

Lemmas 5–8 can now be applied to $f^*(x)$ which leads to the representation

$$f^*(x) = k^*(x) + \int_{X^*}^x h^*(s)\,ds + g_0^*, \tag{B.29}$$

where $h^*(s) = f^*(s+1) - f^*(s)$ is continuous, because f^* is continuous. From (B.26) we now get

$$f(x) = g(x) + f^*(x) = g(x) + k^*(x) + \int_{X^*}^x h^*(s)\,ds + g_0^*. \tag{B.30}$$

Denoting the sum of the three terms besides the integral by $g^*(x)$, we arrive at (B.24). □

By repeating this argument we can make $h^*(s)$ arbitrarily smooth. All the non-smooth behaviour in f is then accumulated in g^*. But $g^*(x)$ has a finite limit for $x \to \infty$ and, therefore, it can be left out in the definition of the quasi-asymptotics. Consequently, the power-counting functions can be assumed to be arbitrarily smooth.

Finally we have to prove Theorem 4:

Proof. We start from (B.24)

$$f(x) = g^*(x) + \int_{X^*}^{x} h^*(s)\,ds = \log r\left(e^{-x}\right).$$

Using the original variable $t = \exp(-x)$, we get

$$\log r(t) = g^*(-\log t) + \int_{X^*}^{-\log t} h^*(s)\,ds. \tag{B.31}$$

Denoting the first function by $\eta(t)$ and introducing $s = -\log t'$ as integration variable, we arrive at

$$\log r(t) = \eta(t) - \int_{t_1}^{t} \frac{h^*(-\log t')}{t'}\,dt',$$

where $t_1 = \exp(-X^*)$. Writing $h^*(-\log t') = h(t')$, (B.4) follows by exponentiation. □

By Theorem 4, a regularly varying power-counting function can be represented as follows

$$\rho(t) = t^\omega \exp\left\{\eta(t) + \int_{t}^{t_1} \frac{h(s)}{s}\,ds\right\}, \tag{B.32}$$

where $\eta(t) \to c$ and $h(s) \to 0$ if the arguments approach 0. Let $\varepsilon > 0$ be an arbitrarily small number, then there exists $s_0(\varepsilon)$ such that $|h(s)| < \varepsilon$ for $s \leq s_0(\varepsilon)$. Thus

$$\left|\int_{t}^{s_0(\varepsilon)} \frac{h(s)}{s}\,ds\right| \leq -\varepsilon \log t + \text{const.} \tag{B.33}$$

if $t \leq s_0(\varepsilon)$. This leads to the upper estimate

$$\rho(t) \leq C t^{\omega-\varepsilon}. \tag{B.34}$$

From $h(s) > -\varepsilon$, we get the lower estimate

$$\rho(t) \geq C' t^{\omega+\varepsilon}. \tag{B.35}$$

If the order ω is positive or negative, $\rho(\delta)$ shows radically different behaviour for $\delta \to 0$: if $\omega < 0$ we have $\rho(\delta) \to \infty$, but $\rho(\delta) \to 0$ for $\omega > 0$.

C: Spence Functions

In this appendix we follow essentially the paper by *K. Mitchell, Phil. Mag. 40, 351 (1949)*. For complex $z \notin [1, +\infty)$, one defines the Spence function by

$$L(z) = \int_0^z \frac{\log(1-t)}{t} dt, \tag{C.1}$$

where

$$\log z = \log |z| + i \arg z \quad , \quad |\arg z| < \pi. \tag{C.2}$$

It is directly related to the Euler dilogarithm (see *L. Lewin, Dilogarithms and associated functions, Mc Donald, London 1958*). The path of integration in (C.1) is the straight line from 0 to z or some other contour avoiding $(1, +\infty)$. The integral appearing in the applications corresponds to real $z = x$

$$\int_0^x \frac{\log|1-t|}{t} dt = \frac{1}{2}\big(L(x+i0) + L(x-i0)\big)$$

$$\stackrel{\text{def}}{=} L(x). \tag{C.3}$$

This definition is useful because by taking the arithmetic mean in (C.3), the arguments of the logarithms cancel out.

Inserting the logarithmic series

$$\log(1-t) = -\sum_{n=1}^{\infty} \frac{t^n}{n} \quad , \quad |t| < 1,$$

into (C.1), we obtain the power series expansion

$$L(z) = -\sum_{n=1}^{\infty} \frac{z^n}{n^2}, \tag{C.4}$$

which is valid for $|z| \le 1$. It yields the special values

$$L(1) = -\sum_{n=1}^{\infty} \frac{1}{n^2} = -\frac{\pi^2}{6} \tag{C.5}$$

$$L(-1) = \sum_{n=1}^{\infty} (-)^{n+1} \frac{1}{n^2} = \frac{\pi^2}{12}. \tag{C.6}$$

We now turn to identities satisfied by $L(z)$. Unfortunately, there is a great many. For z not real, we write (C.1) in the form

$$L(z) = \left(\int_0^1 + \int_1^z\right) \frac{\log(1-t)}{t} dt \tag{C.7}$$

and integrate the second integral by parts

$$= -\frac{\pi^2}{6} + \log z \log(1-z) + \int_1^z \frac{\log t}{1-t} dt, \tag{C.8}$$

where (C.5) has been used. Introducing the new integration variable $s = 1-t$ in the remaining integral, we find

$$L(z) = -L(1-z) + \log z \log(1-z) - \frac{\pi^2}{6}. \tag{C.9}$$

For real $z = x$, it follows from (C.3) that

$$L(1-x) = -L(x) + \log|x| \log|1-x| - \frac{\pi^2}{6}. \tag{C.10}$$

In order to calculate $L(1/z)$, we proceed for z not real as follows:

$$L\Big(\frac{1}{z}\Big) = L(1) + \int_1^{1/z} \frac{\log(1-t)}{t} dt = L(1) - \int_1^z \frac{\log(1-\frac{1}{s})}{s} ds$$

$$= L(1) - \int_1^z \frac{ds}{s}\big(\log(1-s) - \log(-s)\big) = 2L(1) - L(z) + \int_1^z \frac{\log(-s)}{s} ds.$$

The last integral can be easily carried out

$$L\Big(\frac{1}{z}\Big) = -L(z) - \frac{\pi^2}{3} + \frac{1}{2}\log^2(-z) - \frac{1}{2}\log^2(-1)$$

$$= -L(z) + \frac{1}{2}\log^2(-z) + \frac{\pi^2}{6}. \tag{C.11}$$

Here, $\log^2 z \overset{\text{def}}{=} (\log z)^2$. The relation (C.11) holds also for real $x < 0$. For positive x we must use (C.3). Then the term $\log^2(-z)$ gives a contribution $-\pi^2$:

$$L\Big(\frac{1}{x}\Big) = -L(x) + \frac{1}{2}\log^2|x| + \begin{cases} \frac{\pi^2}{6}, & x < 0 \\ -\frac{\pi^2}{3}, & x > 0. \end{cases} \tag{C.12}$$

From (C.10) and (C.12) one obtains

$$L\Big(\frac{x-1}{x}\Big) = L\Big(1 - \frac{1}{x}\Big) = -L\Big(\frac{1}{x}\Big) - \log|x| \log\Big|1 - \frac{1}{x}\Big| - \frac{\pi^2}{6}$$

$$= L(x) + \frac{1}{2}\log^2|x| - \log|x|\log|x-1| + \begin{cases} -\frac{\pi^2}{3}, & x < 0 \\ \frac{\pi^2}{6}, & x > 0. \end{cases} \tag{C.13}$$

Again from (C.12) we get

$$L\Big(\frac{x}{x-1}\Big)= -L\Big(\frac{x-1}{x}\Big)+\frac{1}{2}\log^2\Big|\frac{x-1}{x}\Big|+\begin{cases}\frac{\pi^2}{6} & \text{(a)}\\ -\frac{\pi^2}{3} & \text{(b)}.\end{cases}$$

The case (a) corresponds to $0 < x < 1$, whereas (b) corresponds to the values $x < 0$ or $x > 1$. We find with the aid of (C.13)

$$L\Big(\frac{x}{x-1}\Big)= -L(x)+\frac{1}{2}\log^2|x-1|+\begin{cases}0, & x<1\\ -\frac{\pi^2}{2}, & x>1.\end{cases} \tag{C.14}$$

Likewise we shall obtain from (C.12) and (C.10)

$$L\Big(\frac{1}{1-x}\Big)= L(x)-\frac{1}{2}\log|1-x|\log\frac{x^2}{|1-x|}+\begin{cases}\frac{\pi^2}{3}, & x>1\\ -\frac{\pi^2}{6}, & x<1.\end{cases} \tag{C.15}$$

The formulas used in the main text are special cases of (C.10) and (C.13–15), using the variable $\xi = -x > 0$:

$$L\Big(\frac{\xi}{1+\xi}\Big)= -L(-\xi)+\frac{1}{2}\log^2(1+\xi) \tag{C.16}$$

$$L\Big(\frac{1}{1+\xi}\Big)= L(-\xi)-\frac{1}{2}\log(1+\xi)\log\frac{\xi^2}{1+\xi}-\frac{\pi^2}{6} \tag{C.17}$$

$$L(1+\xi) = -L(-\xi)+\log\xi\log(1+\xi)-\frac{\pi^2}{6} \tag{C.18}$$

$$L\Big(\frac{1+\xi}{\xi}\Big)= L(-\xi)-\frac{1}{2}\log\xi\log\frac{(1+\xi)^2}{\xi}-\frac{\pi^2}{3}. \tag{C.19}$$

D: Grassmann Test Functions

D.1 Grassmann Algebra

D.1.1 Finite-Dimensional Case. Let E be a n-dimensional vector space over $\mathbf{C}$ and $A^p(E)$ the space of antisymmetric p-forms, that is the set of all p-linear and antisymmetric maps

$$S : \underbrace{E\times\ldots\times E}_{\text{p factors}} \to \mathbf{C}. \tag{D.1}$$

We define the wedge product

$$A^p(E)\times A^q(E)\to A^{p+q}(E):$$

$$S\wedge T(u_1,\ldots,u_{p+q}) \overset{\text{def}}{=} \frac{1}{p!q!}\sum_{\pi\in P_{p+q}}(-1)^\pi S(u_{\pi 1},\ldots,u_{\pi p})$$
$$T(u_{\pi(p+1)},\ldots,u_{\pi(p+q)}), \tag{D.2}$$

where P_{p+q} is the permutation group of $p+q$ elements. Note the relations

$$(R \wedge S) \wedge T = R \wedge (S \wedge T)\,,\; S \wedge T = (-1)^{pq} T \wedge S. \tag{D.3}$$

The Grassmann algebra is the direct sum

$$A(E) \stackrel{\text{def}}{=} \bigoplus_{p=0}^{n} A^p(E). \tag{D.4}$$

Let $(e_j)_{j=1}^n$ be a basis in E, such that

$$u = \sum_{j=1}^{n} u^j e_j \quad , \quad u^j \in \mathbf{C}, \tag{D.5}$$

for any $u \in E$. The 'generators' $\delta^j \in A^1(E)$ of $A(E)$ are defined by

$$\delta^j(u) = u^j \quad , \quad j = 1, 2, \ldots n. \tag{D.6}$$

By (D.3) we have

$$\delta^j \wedge \delta^k = -\delta^k \wedge \delta^j. \tag{D.7}$$

For every $S \in A^p(E)$ there exist expansion coefficients $S_{j_1 \cdots j_p} \in \mathbf{C}$ with

$$S = \frac{1}{p!} \sum_{j_1 \cdots j_p = 1}^{n} S_{j_1 \cdots j_p} \delta^{j_1} \wedge \ldots \wedge \delta^{j_p} = \sum_{j_1 < \ldots < j_p} S_{j_1 \cdots j_p} \delta^{j_1} \wedge \ldots \wedge \delta^{j_p}, \tag{D.8}$$

which must be antisymmetric

$$S_{j_{\pi 1} \cdots\cdots j_{\pi p}} = (-1)^{\pi} S_{j_1 \cdots j_p}. \tag{D.9}$$

It follows from (D.8) that

$$A^q(E) = 0 \quad \text{for} \quad q > n, \quad \text{and} \quad \dim A^p(E) = \binom{n}{p}. \tag{D.10}$$

D.1.2 Infinite-Dimensional Case. Let now E be the space $S(\mathbf{R^4}, \mathbf{C^4})$ of Schwartz test functions of rapid decrease. Physically speaking, they are Dirac spinors over Minkowski space. Then $A^p(E)$ (D.1) is the space of antisymmetric direct products of p tempered distributions with the wedge product defined in the same way as above (D.2). $A(E)$ is the infinite direct sum

$$A(E) = \bigoplus_{p=0}^{\infty} A^p(E). \tag{D.11}$$

The generators $\delta_x^a \in A^1(E)$ are defined by

$$(\delta_x^a, f) \stackrel{\text{def}}{=} f^a(x),\; a = 1, 2, 3, 4,\; \forall f \in E. \tag{D.12}$$

This is a 4-dimensional δ-distribution and a Kronecker delta in the spinor index a. (x, a) corresponds to the index j in the finite dimensional case. Equation (D.7) now reads

$$\delta_x^a \wedge \delta_y^b = -\delta_y^b \wedge \delta_x^a. \tag{D.13}$$

This implies

$$f^a(x) \wedge g^b(y) = -g^b(y) \wedge f^a(x), \tag{D.14}$$

hence, we may consider $g^a(x)$ as "Grassmann-valued" test functions. These are the quantities which multiply Fermi operators in the extended S-matrix.

D.2 The S-Matrix as a Functional of Grassmann Variables

The S-matrix $S(\underline{g})$ (4.9.3) is smeared out with ordinary test functions $g(x)$, $g_A(x)$, $g_j(x)$ which multiply Bose operators together with $g_\psi(x)\ldots$ multiplying Fermi operators. According to (D.14), the Fermi test functions are multiplied with the wedge product. To have a uniform concise notation, we have always dropped the wedge in the main text, and instead, we have considered $g_\psi \ldots$ as "anticommuting C-numbers (Grassmann variables)". We use the same notation of omitting the wedge from now on and also omit the spinor indices.

The S-matrix of the extended theory in n-th order is now given by

$$S_n(\boldsymbol{g}) = \frac{1}{n!} \sum_{i_1 \ldots i_n} \int dx_1 \ldots dx_n \, T^{(n)}_{i_1 \ldots i_n}(x_1, \ldots, x_n) g_{i_1}(x_1) \ldots g_{i_n}(x_n). \tag{D.15}$$

The distribution $T^{(n)}_{i_1 \ldots i_n}$ is an element of $A^p(E)$, where p is the number of indices in $i_1, \ldots, i_n$ which are equal to ψ, $\overline{\psi}$, $A\!\!\!/\psi$, $\overline{\psi}A\!\!\!/$, $(0 \leq p \leq n)$, apart from the fact that these distributions now are operator-valued instead of being complex valued. The test functions g_ψ, $g_{\overline{\psi}}$, $g_{A\!\!\!/\psi}$, $g_{\overline{\psi}A\!\!\!/}$ are Grassmann variables, the remaining arguments g, g_A, g_j are ordinary C-number functions. The distribution $T^{(n)}$ then has a mixed symmetry

$$T^{(n)}_{i_{\pi 1} \ldots i_{\pi n}}(x_{\pi 1}, \ldots x_{\pi n}) = (-1)^{Q(\pi)} T^{(n)}_{i_1 \ldots i_n}(x_1, \ldots, x_n). \tag{D.16}$$

Here, the sign of the permutation π is determined in the following way: (1) Decompose the permutation π in a product of transpositions. (2) For every transposition $i_l \longleftrightarrow i_k$ with i_l and i_k both $\in \{\psi, \overline{\psi}, A\!\!\!/\psi, \overline{\psi}A\!\!\!/\}$ one has a factor -1, for all other transpositions the factor is $+1$. (3) $(\pm 1)^\pi$ is the product of all these factors. – Note that (D.17) vanishes if two Grassmann test functions are the same. Consequently, there is a fixed maximal number of Fermi interaction vertices which may at most appear in any term of the perturbation series.

We will briefly discuss, how the construction of the S-matrix must be modified in the mixed case. First note that, due to (D.16), the set $X =$

$\{(x_1, i_1), (x_2, i_2), \ldots (x_n, i_n)\}$ of pairs (x_l, i_l) must be *ordered* in the fermionic vertices. Now we consider the causality condition in n-th order:

$$S^{(n)}(\boldsymbol{g} + \tilde{\boldsymbol{g}}) = \sum_{k=0}^{n} S^{(k)}(\boldsymbol{g}) S^{(n-k)}(\tilde{\boldsymbol{g}}) \quad \text{for} \quad \operatorname{supp} \boldsymbol{g} \geq \operatorname{supp} \tilde{\boldsymbol{g}}. \tag{D.17}$$

Using (D.15) and permuting (by π^{-1}) all g's to the left of all $\tilde{g}$'s in an arbitrary term of the left side of (D.17), we have

$$\sum_{i_1 \ldots i_n} (-1)^{Q(\pi)} \int dx_1 \ldots dx_n \, T^{(n)}_{i_1 \ldots i_n}(x_1, \ldots x_n) g_{i_{\pi 1}}(x_{\pi 1}) \ldots g_{i_{\pi l}}(x_{\pi l})$$

$$\times \tilde{g}_{i_{\pi(l+1)}}(x_{\pi(l+1)}) \ldots \tilde{g}_{i_{\pi n}}(x_{\pi n})$$

$$= \sum_{i_1 \ldots i_n} \Big((-1)^{Q(\pi)} \Big)^2 \int dx_1 \ldots dx_n \, T^{(n)}_{i_{\pi 1} \ldots i_{\pi n}}(x_{\pi 1}, \ldots x_{\pi n})$$

$$\times g_{i_{\pi 1}}(x_{\pi 1}) \ldots g_{i_{\pi l}}(x_{\pi l}) \tilde{g}_{i_{\pi(l+1)}}(x_{\pi(l+1)}) \ldots \tilde{g}_{i_{\pi n}}(x_{\pi n})$$

$$= \sum_{i_1 \ldots i_n} \int dx_1 \ldots dx_n \, T^{(n)}_{i_1 \ldots i_n}(x_1, \ldots x_n) g_{i_1}(x_1) \ldots g_{i_l}(x_l)$$

$$\tilde{g}_{i_{l+1}}(x_{l+1}) \ldots \tilde{g}_{i_n}(x_n), \tag{D.18}$$

where (D.16) has been used. Comparing this with the $k = l$ summand on the right-hand side of (D.17), we obtain

$$T^{(n)}_{i_1 \ldots i_n}(x_1, \ldots x_n) = T^{(l)}_{i_1 \ldots i_l}(x_1, \ldots x_l) T^{(n-l)}_{i_{l+1} \ldots i_n}(x_{l+1}, \ldots x_n), \tag{D.19}$$

for $\{x_1, \ldots x_l\} \geq \{x_{l+1}, \ldots x_n\}$. Performing an arbitrary permutation $\pi : X \to (P, Q)$, we may generalize (D.19) as follows

$$T(X) = (-1)^{Q(\pi)} T(P, Q) = (-1)^{Q(\pi)} T(P) T(Q), \quad \text{for} \quad P \geq Q. \tag{D.20}$$

Similarly, the relation $S(\boldsymbol{g}) S^{-1}(\boldsymbol{g}) = 1$ leads to

$$\sum_{P_2^0} (-1)^{Q(\pi)} T^{(l)}(Y) \tilde{T}^{(n-l)}(X \setminus Y) = 0, \tag{D.21}$$

where P_2^0 are the partitions $X = Y \cup (X \setminus Y)$, $|X| = n$, $|Y| = l$ with empty sets $X, Y = \emptyset$ allowed, and π is the permutation

$$\pi : \quad X \to \Big(Y, (X \setminus Y) \Big). \tag{D.22}$$

(D.21) implies the general formula

$$\tilde{T}^{(n)}(X) = \sum_{r=1}^{n} (-1)^r \sum_{P_r} (-1)^{Q(\pi)} T^{(n_1)}(X_1) \ldots T^{(n_r)}(X_r), \tag{D.23}$$

where the second sum runs over all partitions P_r of X into r disjoint subsets $X = X_1 \cup X_2 \cup \ldots \cup X_r$, $X_j \neq \emptyset$, $|X_j| = n_j$, and π is the permutation

$$\pi: \quad X \to (X_1, X_2, \ldots X_r). \tag{D.24}$$

Going through the inductive construction of the $T^{(n)}$'s in the pure bosonic case, it is now evident, which modifications must be done fore our mixed fermionic-bosonic case: Every term must be multiplied by $(-1)^{Q(\pi)}$, where π is the permutation which brings the arguments $(x_1, i_1), \ldots (x_n, i_n)$ in that order, as they occur in the considered term. For example, the $R'^{(n)}$-distribution must be defined by

$$R'^{(n)}(X, x_n) = \sum_{P_2} (-1)^{Q(\pi)} T^{(n-l)}(X \setminus Y, x_n) \tilde{T}^{(l)}(Y), \tag{D.25}$$

where $|X| = n - 1$ and the sum runs over all partitions P_2:

$$X = (X \setminus Y) \cup Y, \quad Y \neq \emptyset \quad \text{with}$$

$|Y| = l \geq 1$, and π is the permutation

$$\pi: \quad (X, x_n) \to \Big((X \setminus Y), x_n, Y\Big). \tag{D.26}$$

The most difficult step in the inductive procedure, the distribution splitting, is not affected by the additional factors $(-1)^{Q(\pi)}$.

The result of the modified inductive construction is very simple: Every graph belonging to $T^{(n)}_{i_1 \ldots i_n}(x_1, \ldots, x_n)$ can be obtained by amputation of certain external lines of a graph belonging to $T^{(n)}_{0 \ldots 0}(x_1, \ldots, x_n)$. This was discussed in no. (4) of Sect. 4.9. Consequently, every numerical distribution $t^{(n)}_{i_1 \ldots i_n; k}(x_1, \ldots, x_n)$ appearing in the normal product expansion of $T^{(n)}_{i_1 \ldots n_n}(x_1, \ldots, x_n)$ is equal to a certain numerical distribution appearing in $T^{(n)}_{0 \ldots 0}(x_1, \ldots, x_n)$. By introducing the additional interaction terms

$$L_i(x) = \psi(x), \overline{\psi}(x), A(x), j(x), (A\!\!\!/\psi)(x), (\overline{\psi}A\!\!\!/)(x)$$

in (4.9.3) we do not get any new *numerical* distribution. But the symmetry property of the *operator-valued* distribution $T^{(n)}$ is changed by the amputation, if a Fermi operator is amputated.

Bibliographical Notes

In this section we want to collect some important references which are strongly related to the various chapters in the book. No attempt for completeness is made, the reader will easily find further references in the cited papers and books. Some references were already given in the text at the places where they have been used.

Chapter 0

The history of quantum field theory is intimately connected with that of particle physics. This subject has been fully described by A. Pais in his book *"Inward Bound", Oxford University Press 1986*, which contains several chapters on field theory. Earlier reviews are: *International colloquium on history of particle physics, Paris 1982, J. de Physique 43, Suppl. to No. 12 (1982)* and *"Shelter Island II", edited by R. Jackiw, N.N. Khuri, S. Weinberg and E. Witten, MIT Press 1985* in particular the article by S.S. Schweber (see also *Les Houches 1983, edited by B.S. DeWitt, R. Stora, North-Holland 1984*).

From the large literatur on the Lorentz group we mention the classical books by B.L. van der Waerden *"Group Theory and Quantum Mechanics", Springer-Verlag 1974* and H. Weyl *"The Theory of Groups and Quantum Mechanics", Dover Publ. 1950.* The still fresh original papers on relativity theory are reprinted in *H.A. Lorentz, A. Einstein, H. Minkowski and H. Weyl, "The Principle of Relativity", Dover Publ. 1952.*

The time-dependent approach to scattering theory can be found in *M. Reed, B. Simon, Methods of Modern Mathematical Physics, Vol. III, Scattering Theory, Academic Press 1979.* Volume II of this work is also important, because time-dependent Hamiltonians are treated there.

Chapter 1

The spinor calculus was invented by van der Warden (loc.cit.). The discussion of invariant field equations for arbitrary spin can be found in *H. Umezawa, "Quantum Field Theory", North-Holland 1956* or in the original work by M. Fierz *Helv. Phys. Acta 12 (1939) 3.* From the physical point of view, it would be better to derive the field equations from covariance under the Poincaré group. Since this gives no other equation, the simpler approach with the Lorentz group is justified.

The Dirac equation is discussed in all textbooks on quantum field theory. From the large literature we mention the following ones: *S.S. Schweber, "An Introduction to Relativistic Quantum Field Theory", Harper and Row 1961, J.D. Bjorken, S.D. Drell, "Relativistic Quantum Mechanics", McGraw-Hill 1964*, where the Feynman rules are derived without field quantization, as in the original work of Feynman (see *R.P. Feynman, "Quantum Electrodynamics", W.A. Benjamin 1962)* and *J.D. Bjorken, S.D. Drell, "Relativistic Quantum Field Theory". McGraw-Hill 1965*. More modern books are *V.B. Berestetski, E.M. Lifschitz, L.P. Pitajewski, "Relativistic Quantum Theory", Vol. IV of L.D. Landau, E.M. Lifschitz, "Theoretical Physics", Pergamon Press 1970* and *C. Itzykson, J.B. Zuber, "Quantum Field Theory", McGraw-Hill 1980*. The latter authors have already used the quadratic Dirac equation to treat the hydrogen atom, but have not calculated the wave functions in this way.

Chapter 2

A standard reference on second quantization is *F.A. Berezin, "The Method of Second Quantization", Academic Press 1966*. The theorem of spin and statistics goes back to *W. Pauli, Phys. Rev. 58 (1940) 716*. It has various generalizations to interacting fields (see *R.F. Streater, A.S. Wightman, "PCT, Spin and Statistics, and All That", Benjamin 1964*), which, however, are not needed in the causal theory.

The discussion of the various Lorentz-invariant distributions in x-space is crucial for the causal method. More on this subject can be found in *V.S. Vladimirov, "Methods of the Theory of Functions of Several Complex Variables", Cambridge 1966* or *F. Constantinescu, "Distributionen und ihre Anwendungen in der Physik", Teubner 1974*.

The external field problem of QED was already treated perturbatively by *G. Källen, "Quantum Electrodynamics", Springer-Verlag 1972*. The non-perturbative theory was developed in different directions by *R. Seiler, Comm. Math. Phys. 25 (1972) 127, G. Labonté, Comm. Math. Phys. 36 (1974) 59, Canad. J. Phys. 53 (1975) 1533, J. Bellisard, Comm. Math. Phys. 41 (1975) 235, 46 (1976) 53, P. Bongaarts, S. Ruijsenaars, Ann. Phys. (N.Y.) 101 (1976) 289, W. Hochstenbach, Comm. Math. Phys. 51 (1976) 211, M. Klaus, G. Scharf, Helv. Phys. Acta 50 (1977) 779, 803, S. Ruijsenaars, J. Math. Phys. 18 (1977) 720, Comm. Math. Phys. 52 (1977) 267* and *H.P. Seipp, Helv. Phys. Acta 55 (1982) 1*.

The causal phase of the S-matrix and its connection with vacuum polarization was studied by the author in *Nuov. Cim. A 74 (1983) 302*. The implementation of causality by means of dispersion relations has a long tradition. It started with the paper by *R. de L. Kronig, J. Opt. Soc. Amer. Rev. Sci. Instrum. 12 (1926) 547*. In the 1950s the idea was taken over to potential scattering (*N.N. Khuri, Phys. Rev. 107 (1957) 1148, W. Hunziker, Helv. Phys. Acta 34 (1961) 593, H.M. Nussenzveig, "Causality and Dispersion Re-*

lations", Academic Press 1972) and field theory. The standard reference in the latter context is *N.N. Bogoliubov, D.V. Shirkov, "Introduction to the Theory of Quantized Fields", Wiley-Interscience 1959*, see also *"Dispersion Relations and Their Connection with Causality", edited by E.P. Wigner, Academic Press 1964*. Some mathematical problems concerning the tempered distributions appearing in the dispersion relations have been discussed by *J.G. Taylor, Ann. Phys. (N.Y.) 5 (1958) 391*, see also *H. Bremermann, "Distributions, Complex Variables and Fourier Transformation, Addison-Wesley 1965.*

For quantization of the electromagnetic field, there exist essentially two different methods. (i) Canonical quantization in the Coulomb gauge div $\boldsymbol{A} = 0$ leads to a theory with only physical degrees of freedom, namely the two transversal polarizations. The price one has to pay is the loss of manifest Lorentz covariance and the occurance of the non-local Coulomb interaction (see *Björken and Drell*, loc. cit.). (ii) In the Gupta–Bleuler method (*S. Gupta, Proc. Roy. Soc. A 63 (1950) 681, K. Bleuler, Helv. Phys. Acta 23 (1950) 567*) the theory is manifestly covariant and local, but has unphysical (four) polarizations. In addition, one usually uses a state space with indefinite metric. This we have avoided by introducing a new conjugation in ordinary Fock space (with positive definite metric). In the causal approach we need a local theory, so that we are forced to use method (ii).

Chapter 3

Important older books on QED are *A.I. Akhiezer, V.B. Berestetski, "Quantum Electrodynamics", Wiley 1965*, which contains many explicit results, *I. Bialynicki-Birula, Z. Bialynicka-Birula, "Quantum Electrodynamics", Pergamon Press 1975* and *J.M. Jauch, F. Rohrlich, "The Theory of Photons and Electrons", Springer-Verlag 1975*. The causal approach was advocated by Bogoliubov and Shirkov (loc. cit.). This book inspired many people, in particular Epstein and Glaser, although the method was already invented by E.C.G. Stückelberg and collaborators (see *G. Wanders, Fortschritte der Physik 4 (1956) 611*).

The paper by Epstein and Glaser (*Ann. Inst. Poincaré A 19 (1973) 211*) was almost completely ignored. To our knowledge their method was only used by P. Blanchard, R. Sénéor (*Ann. Inst. Poincaré A 23 (1975) 147*) to study the adiabatic limit in Green's functions and by H.G. Dosch, V.F. Müller (*Fortschr. Phys. 23 (1975) 661*) to include external fields in QED. Obviously, it was not realized that the method is superior to the usual recipes of "infinite counterterms", in particular in gauge theories. Nevertheless, dispersion techniques have been frequently used in perturbative calculations (see *G. Källen*, loc. cit. and *M.D. Scadron, "Advanced Quantum Theory", Springer-Verlag 1991*), but without knowing the causal background.

The notion quasi-asymtotics of distributions was introduced by russian mathematicians (see *V.S. Vladimirov, Y.N. Drozzinov, B.I. Zavialov,*

"Tauberian Theorems for Generalized Functions", Kluwer Acad. Publ. 1988). It involves the scaling limit which tests the behaviour for small x or large p in momentum space. This is a better tool to define the singular order of distributions than the original estimates by Epstein and Glaser.

A review on higher order calculations in QED is *B. Lautrup, A. Peterman, E. de Rafael, Physics Reports 3C (1972) 193.* The complete calculation of (scalar) 1-loop integrals (off mass-shell) were carried out by G. 't Hooft and M. Veltman, *Nucl. Phys. B 153 (1979) 365.* We have checked our results for the vertex function with their results. Although the analytic expressions in terms of Spence functions are different, the results agree due to identities for Spence functions. Some of these identities, but not all, can be found in *L. Lewin, "Dilogarithms and Associated Functions", McDonald 1958.*

The usual method to treat the infrared problem is by assuming a finite photon mass $\mu > 0$ and taking the limit $\mu \to 0$ in the right inclusive cross sections (*D.R. Yennie, S.C. Frautschi, H. Suura, Ann. Phys. 13 (1961) 379*). The problem with this method is that it modifies the interaction at all distances, whereas in the adiabatic switching it is changed in the asymptotic region $x \to \infty$, only. But the method is justified, because the final results agree.

Chapter 4

Stability of the vacuum is an important and not at all trivial property. There are arguments that in Yang-Mills theories there may be no unique stable vacuum, but different so-called ϑ-vacua (see e.g. *K. Huang, "Quarks, Leptons and Gauge Fields", World Scientific 1982*).

Renormalizability is difficult to prove in the conventional method with counterterms, in particular in gauge theories where one must show that the counterterms can be chosen in a gauge invariant manner. The main obstacle are the overlapping divergences which require so much sophistication, that it is hard to understand the proofs completely. The situation was ironically characterized by A. Salam who has also worked on the problem (*Phys. Rev. 86 (1952) 731*) as follows: "The difficulty, as in all this work, is to find a notation which is both concise and intelligible to at least two persons, of whom one may be an author" *P. Matthews, A. Salam, Rev. Mod. Phys. 23 (1951) 314.*

The more modern work was started by N.N. Bogoliubov and O.S. Parasiuk (see *N.N. Bogoliubov, D.V. Shirkov*, loc. cit.) and further developed by K. Hepp (*"Théorie de la Renormalisation", Springer-Verlag 1969*), W. Zimmermann (*Ann. Phys. (N.Y.) 77 (1973) 536*) and J.H. Lowenstein (*Nucl. Phys. B 86 (1975) 77, Comm. Math. Phys. 47 (1976) 53*). This method is therefore called the BPHZ-method (see *V.A. Smirnov, "Renormalization and Asymptotic Expansion", Birkhäuser Verlag 1991*). Later on, other methods were introduced by *C. de Calan, V. Rivasseau, Comm. Math. Phys. 82 (1981) 69, K. Gawedzki, A. Kupiainen, Comm. Math. Phys. 92 (1984) 531, J. Magnen,*

R. Sénéor, Ann. Phys. (N.Y.) 152 (1984) 130, J. Polchinski, Nucl. Phys. B 231 (1984) 269, J.S. Feldman, T.R. Hurd, L. Rosen, J.D. Wright, "QED: A Proof of Renormalizability", Springer-Verlag 1988, G. Keller, C. Kopper, Phys. Lett. B 273 (1991) 323 and *M. Bonini, M. D'Attanasio, G. Marchesini, Nucl. Phys. B 418 (1994) 81.* The continuous activity in this field shows that in all previous approaches something remained to be desired.

The discrete symmetries were already discussed by G. Lüders (*Dansk. Mat. Fys. Medd. 28 (1954) 5*) and W. Pauli (*"Niels Bohr and the Development of Physics", edited by W. Pauli, Pergamon Press 1955*). Thereafter, R. Jost (*Helv. Phys. Acta 30 (1957) 409*) showed that the product of C, T, P is a symmetry in a local quantum field theory under very general conditions. This is the famous CTP-theorem (see *R.F. Streater, A.S. Wightman* loc. cit. and *F.J. Dyson, Phys. Rev. 110 (1958)*).

The existence of a Lorentz covariant splitting solution was first shown by Epstein and Glaser (loc. cit. *p. 249*).

The older proofs of gauge invariance are problematic because they have to face the renormalizability problem at the same time. In the causal theory the issues are clearly separated. The Ward identities go back to *J.C. Ward, Phys. Rev. 78 (1950) 182* and in slightly more general form to *Y. Takahashi, Nuov. Cim. 6 (1957) 371.* Our formulation is still more general because the inner vertices are not integrated out.

Unitarity of the S-matrix is so important that it is often assumed as an axiom (*R. Eden, V. Landshof, D.I. Olive, J. Polkinghorne, "Analytic S-Matrix Theory", Cambridge University Press 1966, D. Iagolnitzer, "The S-Matrix", North-Holland 1978*). In the causal theory it is a simple consequence of the inductive construction in case of massive scalar theories (*Epstein and Glaser* loc. cit.). But in gauge theories this simple unitarity is changed into pseudo-unitarity due to the unphysical degrees of freedom. In addition, unitarity on the physical subspace is important. The latter is a direct consequence of gauge invariance.

The renormalization group was discovered by E.C.G. Stückelberg and A. Petermann (*Helv. Phys. Acta 26 (1953) 499*) and by M. Gell-Mann and F. Low (*Phys. Rev. 95 (1954) 1300*). It was further developed by Bogoliubov and Shirkov (loc. cit), K. Symanzik (*Comm. Math. Phys. 18 (1970) 227* and C.G. Callen (*Phys. Rev. D 2 (1970) 1541*).

The definition of interacting fields by functional derivatives of the S-matrix goes back to Bogoliubov and Shirkov (loc.cit). Epstein and Glaser (loc.cit.) combined it with the causal method. The definition of operator products which is obtained in this way should be compared with the operator product expansion of K. Wilson (*Phys. Rev. D 3 (1971) 1818*) and W. Zimmermann (*Ann. Phys. (N.Y.) 77 (1973) 570*). The field equations for interacting fields in QED were derived in *M. Dütsch, F. Krahe, G. Scharf, Nuovo Cim. 103 A (1990) 871.*

Chapter 5

Scalar QED in the Lagrangian approach can be found in most of the cited textbooks, but as far as we know, gauge invariance in all orders is not discussed. We follow the paper by *M. Dütsch, F. Krahe, G. Scharf, Nuovo Cim. 106 A (1993) 277.*

Axial anomalies in the triangle graphs were discovered by S. Adler (*Phys. Rev. 177 (1969) 2426*) and J.S. Bell, R. Jackiw (*Nuovo Cim. 60 A (1969) 47*). Radiative corrections to these graphs were considered by S. Adler and W.A. Bardeen (*Phys. Rev. 182 (1969) 1517*) and A. Zee (*Phys. Rev. Lett. 29 (1972) 1198*). These authors found that the anomalies are not changed by radiative corrections. In the electro-weak theory (Weinberg-Salam model) all anomalies must cancel. This requirement imposes important constraints on possible models (see *D.J. Gross, R. Jackiw, Phys. Rev. D 6 (1972) 477*), in fact, it provides an argument for the standard family structure of the fermions. Since axial anomalies can be seen in the external field problem, they show up in fermionic determinants (*R. Jackiw in Les Houches 1983, edited by B.S. DeWitt, R. Stora, North-Holland 1984*).

Quantum electrodynamics in two space dimensions was studied by W. Siegel (*Nucl. Phys. B 156 (1979) 135)*), J. Schonfeld (*Nucl. Phys. B 185 (1981) 157*), R. Jackiw and S. Templeton (*Phys. Rev. D 23 (1981) 291*). These authors showed that in this case the photon becomes massive without destroying gauge invariance. The mass term has a topological aspect (*S. Deser, R. Jackiw, S. Templeton, Ann. Phys. (N.Y.) 140 (1982) 372*). Higher order corrections to the topological mass term were investigated by S. Coleman, B. Hill (*Phys. Lett. 159 B (1985) 184*) T. Lee (*Phys. Lett. 171 B (1986) 247*) and C.R. Hagen, P. Panigrahi, S. Ramaswamy (*Phys. Rev. Lett. 61 (1988) 389*). All this work is based on the usual Feynman perturbation theory and, therefore, contains a certain regulator ambiguity (*S. Deser, R. Jackiw, S. Templeton,* loc. cit.). We have followed our work *G. Scharf, W.F. Wreszinski, B.M. Pimentel, J.L. Tomazelli, Ann. Phys. (N.Y.) 231 (1994) 185.* The theory has interesting applications to the quantum Hall effect and high-T_c superconductivity.

(1+1)-dimensional QED was first studied by J. Schwinger (*Phys. Rev. 128 (1962) 2425*), and is therefore called Schwinger model (in the massless case). It was used as laboratory for the development of various tools in field theory, as for example dynamical mass generation, the treatment of anomalies (*R. Jackiw, R. Rajaraman, Phys. Rev. Lett. 54 (1985) 1219*) or the test of quantization methods (*L.D. Fadeev, S.L. Shatashvili, Phys. Lett. B 176 (1986) 225*).

Subject Index

The numbers in this index are related to the equations next to the subject. The first number refers to the chapter, the second one to the section and the third one is the number of the equation. The numbers in italics are page numbers.

Absorption operator (2.1.24) *70*, (2.2.4) *78*
Adiabatic limit (3.11.1) *240*, (4.1.30) *267*, (4.9.39) *322*
- switching (0.3.21) *16*, (3.1.1) *160*, (3.12.1) *248*
Adjoint (0.3.6) *14*, (2.1.25) *70*
- Dirac equation (1.4.1) *36*
- Dirac field (2.2.34) *83*
- spinor (1.2.26) *29*
Advanced function (0.3.33) *18*
- distribution (3.1.53) *167*, (4.9.11) *315*
Advanced part (2.3.20) *89*
Angular momentum ((1.2.38) *31*, (1.6.1) *54*
Annihilation operators (2.1.21) *70*, (2.2.4) *78*
Anomalous magnetic moment (3.10.19) *239*
Anomaly (4.6.21) *292*
Anticommutation relations (1.2.30) *29*, (2.1.40) *73*, (2.2.14) *80*, (2.2.29) *82*
Antisymmetric tensor (0.2.12) *12*, (1.1.26) *24*, (1.3.23) *35*
Antiunitary operator (1.5.54) *51*, (4.4.19) *277*
Asymptotic completeness (0.3.18) *16*
Axial anomaly (5.3.6) *352*, (5.3.10) *353*, (5.3.44) *357*

Bhaba scattering (3.14.2) *262*
Bilinear form (0.2.6) 12
Bilinears (1.3.21) *35*, (1.3.30) *36*
Bohr's magneton (1.5.31) *48*
Boost (0.1.13) *9*, (1.1.19) *23*, (1.4.49) *41*
Bose statistics (2.1.3) *67*, (2.1.41) *73*
Bremsstrahlung (3.11.25) *244*

Callen–Symanzik equation (4.8.28) *312*
Causal connection (0.1.9) *8*, (2.3.19) *89*
- function (2.8.25) *127*
- phase (2.8.63) *133*, (2.9.47) *140*
- support (2.3.19) *89*, (3.1.54) *167*, (3.2.29) *174*
Causality condition (0.3.27) *17*, (2.8.7) *125*, (3.1.23) *163*, (3.1.28) *164*
Central splitting solution (3.2.61) *181*, (3.7.41) *214*, (3.9.43) *235*
Center-of-mass-system (3.4.35) *192*, (3.8.34) *220*
Charge (1.5.22) *47*, (2.2.23) *81*, (6.4) *376*
- conjugate spinor (1.5.52) *51*
- conjugation (1.5.53) *51*, (2.2.44) *84*, (4.4.1) *275*, (5.1.28) *341*
- conservation (2.4.82) *105*, (2.10.8) *141*
- normalization (3.6.35) *207*, (3.13.1) *258*
- operator (2.2.23) *81*, (2.4.79) *104*, (6.4) *376*

Charged scalar field (2.1.78) *78*, (5.1.1) *335*, (5.5.39) *374*
Chiral representation (1.2.28) *29*, (1.3.7) *33*, (1.5.44) *50*
C-invariance (1.5.51) *51*
Classical electrodynamics (3.12.24) *254*, (4.10.1) *323*
Clifford algebra (1.3.1) *32*
C-number (2.1.44) *73*
Coboundary operator (4.5.20) *285*
Cohomology theory (4.5.17) *284*
Commutation functions (2.3.1) *87*
– relations (2.1.40) *73*, (2.11.20) *149*
Completeness relations (2.6.21) *114*, (3.4.21) *190*, (3.5.46) *201*
Compton scattering (3.5.1) *195*, (3.14.6) *262*
– in scalar QED (5.1.24) *340*, (5.2.1) *344*
– wave length (1.2.13) *27*
Confinement (3.11.1) *240*
Conjugation (2.11.43) *152*
Conserved current (3.12.46) *257*, (4.6.7) *290*, (4.10.8) *324*, (4.10.44) *331*
Continuity equation (1.5.20) *46*
Contraction of tensors (0.2.7) *12*, (1.1.27) *24*
– of field operators (3.3.7) *184*, (3.7.17) *210*
Contravariant (0.2.4) *11*, (0.2.13) *13*
Coulomb interaction (3.11.1) *240*
– potential (2.10.25) *144*
Covariance (1.4.55) *42*, (2.2.58) *86*
Covariant (0.2.4) *11*, (0.2.11) *12*
– derivative (1.5.7) *45*
– projection operators (1.4.43) *41*
Creation operators (2.1.14) *68*, (2.2.13) *80*
Cross section (2.6.33) *116*, (3.4.32) *192*, (3.11.30) *245*
Crossed graph (3.5.6) *196*
Crossing symmetry (3.14.7) *262*
Curl (0.2.12) *12*
Current density (1.5.19) *46*, (2.10.2) *141*, (3.12.33) *255*

Darwin term (1.5.41) *48*
Decay rate (5.3.51) *359*, (5.3.57) *360*
Differentiation of tensors (0.2.8) *12*
Differential cross section (2.6.37) *116*, (3.4.38) *193*
Dirac conjugation (1.3.18) *34*
– equation (1.2.25) *29*, (1.4.1) *36*, (4.10.6) *323*
– field (2.2.13) *80*, (2.2.34) *83*
– form factor (3.9.48) *236*
– Fourier transform (1.4.37) *40*
– spinor (1.2.26) *29*
Dispersion relations (2.8.45) *130*, (3.2.60) *180*
– with subtractions (2.8.52) *131*, (3.2.60) *180*, (3.2.65) *183*
Distribution splitting (3.2.30) *175*
Divergence (0.2.10) *12*
Divergences, infrared (3.1.1) *160*, (3.7.34) *213*
– ultraviolet (3.1.68) *170*, (3.2.61) *181*, (3.6.11) *204*, (3.7.20) *210*
Dual space (0.2.1) *11*
– basis (0.2.2) *11*
Dynamics (0.3.7) *14*
Dyson series (0.3.15) *15*, (3.1.67) *170*

Eigenfunctions (1.4.34) *40*
Electromagnetic field (1.5.6) *45*, (2.11.5) *146*
– classical (3.12.24) *254*, (4.10.2) *323*
– quantization (2.11.20) *149*
Electron–photon scattering (3.5.1) *195*
Electron–positron annihilation (3.14.7) *262*
– field (2.2.13) *80*
– scattering (3.14.2) *262*
Electron radius, classical (3.5.42) *201*
Electron scattering (2.4.38) *98*, (2.6.3) *111*, (3.4.1) *186*, (3.11.4) *240*
Emission operator (2.1.14) *68*, (2.2.13) *80*
Energy (1.4.20) *38*
– momentum conservation (3.4.16) *189*, (3.5.15) *198*, (3.12.30) *254*
Epsilon-tensor (1.1.26) *24*, (1.2.9) *27*
Euler–Lagrange equation (1.5.11) *45*
Expectation values (2.1.70) *77*

External field (2.2.6) *79*, (2.9.2) *134*, (5.3.2) *351*

Fermi statistics (2.1.4) *67*, (2.1.41) *73*
– operators (2.1.40) *73*
Feynman diagrams (3.3.16) *185*
– gauge (5.5.7) *369*
– propagator (2.3.41) *92*, (2.6.13) *113*, (2.7.16) *120*, (3.4.5) *187*, (3.5.3) *196*, (3.7.38) *214*
– rules (3.1.1) *160*, (3.3.16) *185*, (3.6.1) *202*, (3.7.21) *210*
Feynman's integration trick (3.11.31) *246*
Fibration (2.11.54) *154*
Field operators (2.1.19) *69*, (2.2.34) *83*
– quantization (2.1.40) *73*, (2.2.13) *80*
Fine structure (1.6.36) *58*, (1.6.41) *59*
– constant (1.6.15) *56*, (3.4.51) *195*, (4.8.24) *311*
Fock representation (2.1.42) *73*, (2.4.1) *93*
– space (2.1.6) *67*, (2.1.52) *75*, (2.11.47) *153*
Form factors (3.9.48) *236*, (3.13.15) *260*
Fourier transformation (1.4.11) *37*, (1.4.37) *40*, (2.3.1) *87*, (2.3.24) *90*
Furry theorem (2.4.67) *102*, (2.8.63) *133*, (4.4.40) *279*, (5.1.35) *343*

Gamma matrices (1.2.28) *29*, (1.3.1) *32*
– trace of (2.6.43) *117*
Gauge (1.2.39) *31*, (2.11.4) *146*
– field (1.5.5) *45*
– invariance (1.5.60) *52*, (2.4.83) *105*, (2.7.34) *122*, (2.10.5) *141*, (2.11.64) *156*, (4.6.9) *290*, (5.2.1) *344*, (6.26) *379*
– transformation (1.5.5) *45*, (1.5.59) *52*, (2.11.54) *154*, (4.6.2) *289*, (6.1) *376*
Gauss theorem (0.2.16) *13*
Gell-Mann Low equation (4.8.23) *311*
Ghost fields (6.18) *378*
Gordon decomposition (1.4.68) *44*
Gradient (0.2.8) *12*
Grassmann (anticommuting) variables (4.9.6) *314*
Gyromagnetic ratio (g) (3.10.19) *239*

Hamiltonian (0.3.2) *14*, (1.4.6) *37*, (1.5.40) *49*, (1.5.57) *51*, (2.2.6) *79*
– second quantized (2.1.66) *76*, (2.2.20) *80*, (2.11.27) *150*
Heisenberg equations (2.1.67) *76*, (2.2.21) *81*, (2.11.26) *150*
– field operators (2.1.67) *76*
Helicity (1.4.25) *39*
Hilbert–Schmidt condition (2.4.60) *101*, (2.5.27) *109*
Hilbert space (1.4.7) *37*, (2.11.50) *153*
Hydrogen atom (1.6.1) *54*, (1.6.69) *61*
– Schrödinger equation (1.6.25) *57*, (1.6.41) *59*, (A1) *381*

Inclusive cross section (3.11.30) *245*
Indefinite scalar product (0.1.8) *8*, (2.11.19) *149*
Infrared divergence (3.1.1) *160*, (3.7.34) *213*
Inhomogeneous Lorentz transformation (0.1.18) *10*
Interacting fields (4.9.8) *315*
Interaction picture (0.3.13) *15*, (2.5.2) *105*
Invariant (0.2.7) *12*, (1.1.33) *25*
– field equations (1.2.10) *27*
Irreducible (1.1.35) *25*, (1.3.5) *32*, (2.1.42) *73*

Jordan–Pauli distribution (2.3.7) *88*, (2.11.18) *148*, (3.2.21) *173*

Klein–Gordon equation (1.2.13) *27*, (1.2.31) *29*, (1.2.42) *31*
Klein–Nishina formula (3.5.41) *201*

Lagrange principle (1.5.9) *45*
Leading logarithms (4.8.39) *313*
Lifting (0.2.4) *11*, (2.1.30) *71*, (2.2.55) *86*, (2.11.36) *151*
Light cone (0.1.1) *7*
Light-like (0.1.9) *8*

Light rays (0.1.1) *7*
Linear form (0.2.1) *11*
Local distributions (3.2.46) *178*, (4.5.8) *284*, (4.6.21) *292*
Loop graphs (3.6.11) *204*, (3.7.20) *210*
Lorentz boost (0.1.13) *9*, (1.1.19) *23*
– condition (1.2.40) *31*, (2.11.1) *146*, (4.10.9) *324*, (4.10.49) *331*
– covariance (1.4.63) *43*, (2.2.58) *86*, (2.11.61) *165*, (3.1.19) *163*
– group (0.1.7) *7*, (1.1.15) *23*
– group, proper (0.1.11) *9*, (1.1.19) *23*
– transformations (0.1.7) *7*, (0.2.6) *12*, (1.1.15) *23*, (1.2.32) *30*
– transformations, special (0.1.13) *9*, (1.1.19) *23*, (1.4.49) *41*
– transformations, inhomogeneous (0.1.18) *10*
– tensor (2.7.35) *122*
Lowering (0.2.4) *11*

Magnetic moment (1.5.31) *48*, (3.10.11) *237*
Majorana representation (1.7.14) *64*
Mandelstam variables (3.4.44) *194*
Mass normalization (3.7.39) *214*, (3.11.22) *243*
– scale (3.2.61) *181*
Maxwell equations (1.2.42) *31*, (1.5.18) *46*, (2.11.5) *146*
Metric tensor (0.1.6) *7*, (0.2.5) *11*
Microcausality (4.9.15) *316*
Minimal coupling (1.5.9) *45*
Minkowski space (0.1.8) *8*
– scalar product (0.1.8) *8*
Moeller scattering (3.4.1) *186*

Noether's theorem (1.5.12) *46*
Non-abelian gauge theories (6.10) *377*
Non-relativistic limit (1.5.28) *47*, (3.4.50) *195*
Norm (2.1.17) *69*, (4.2.1) *268*
Normal order (2.1.36) *72*, (2.2.20) *80*, (2.4.24) *96*, (3.1.3) *161*, (3.3.6) *184*
Normalizability (4.3.1) *271*, (5.1.8) *337*
Normalization conditions (3.2.56) *179*
– polynomial (3.2.54) *179*

One-particle operator (2.1.30) *71*
– sector (2.1.45) *74*
– theory (1.1.1) *21*
Operator-valued distribution (2.1.18) *69*, (2.11.11) *147*, (3.1.1) *160*, (3.2.1) *170*
Operator character (4.2.1) *268*
– products (4.9.31) *320*
Order, singular (3.2.16) *173*

Pair annihilation (2.4.38) *98*, (3.14.7) *262*
– creation (2.4.38) *98*, (2.7.2) *118*, (2.7.43) *124*, (2.8.32) *128*
Parity (0.1.9) *8*, (1.2.17) *28*, (1.2.35) *30*, (1.3.20) *35*, (1.5.62) *52*, (4.4.8) *275*
Partial integration (0.2.18) *13*
Particle number operator (2.1.11) *68*, (2.11.40) *152*
Pauli equation (1.5.30) *48*
– form factor (3.9.48) *236*
Pauli–Jordan function (2.3.7) *88*, (2.11.18) *148*, (3.2.21) *173*
Pauli matrices (1.1.1) *21*
– principle (2.1.41) *73*, (3.4.18) *190*
– Theorem of, (1.3.7) *33*
Perturbation series (2.5.1) *105*, (3.1.1) *160*, (3.1.4) *161*
– first order (3.3.1) *183*
Phase transformations (1.5.2) *44*
Photon (2.11.29) *150*
Photon, transversality (2.11.33) *151*, (3.11.28) *245*, (3.12.45) *257*
Photon–photon scattering (4.3.9) *272*, (4.6.33) *295*
Physical subspace (2.11.31) *150*, (4.7.41) *306*
Pion decay (5.3.31) *356*
Plane wave (1.4.33) *39*, (1.7.2) *62*, (4.2.12) *270*
Poincaré covariance (1.4.55) *42*, (2.2.58) *86*, (2.11.37) *151*, (4.5.1) *282*, (4.9.34) *321*
– group (0.1.18) *10*, (1.4.63) *43*
Polarization potential (2.10.27) *144*
Polarizations (3.5.45) *201*, (3.12.10) *250*

Polarization vectors (2.11.32) *151*, (3.5.9) *197*, (3.12.42) *256*, (4.7.3) *301*
Positron scattering (2.4.38) *98*
– spectrum (2.7.11) *119*
Power counting (3.2.15) *172*
Probability (0.3.8) *15*, (2.6.19) *114*
Proca theory (1.2.42) *31*
Projections (1.4.40) *40*, (2.2.7) *79*
Propagator (0.3.16) *15*
Proper Lorentz group (0.1.11) *9*
– transformations (0.1.13) *9*, (1.1.19) *23*
Pseudo-scalar (1.3.28) *36*, (5.3.1) *351*
Pseudo-unitarity (3.12.11) *251*, (4.7.20) *303*, (5.1.38) *343*
Pseudo-vector (1.3.30) *36*, (5.3.1) *351*

Quantization (2.1.6) *67*, (2.2.13) *80*, (2.11.9) *147*
Quantum mechanics (0.3.9) *15*, (1.4.6) *37*, (1.5.24) *47*
Quasi-asmptotics (3.2.6) *171*

Radiation, classical (3.12.37) *255*
– field (2.11.2) *146*
– gauge (2.11.5) *146*
Radiative corrections (3.11.4) *240*
Rank (0.2.8) *12*, (1.1.4) *21*
Reference system (0.1.2) *7*, (1.2.2) *26*
Regularization (0.0) *4*, (5.3.1) *351*
Regularly varying functions (3.2.20) *173*, (B1) *383*
Relativity (0.1.5) *7*, (2.11.37) *151*
Renormalizable theories (4.3.1) *271*
Renormalization (3.13.1) *258*, (4.6.49) *299*, (4.8.5) *308*
– group (4.8.8) *309*
Representation (1.1.6) *22*, (1.1.35) *25*, (2.1.42) *73*
– complex conjugate (1.1.29) *24*
Rest system (1.4.50) *42*
Retarded distribution (3.1.52) *167*, (3.2.63) *182*, (3.4.4) *187*
– function (0.3.33) *18*
Retarded part (2.3.20) *89*, (3.2.2) *171*
Rotation (0.1.12) *9*, (1.1.4) *21*, (1.2.37) *31*, (1.4.51) *42*
– group (0.1.12) *9*, (1.1.8) *22*, (1.2.37) *31*
Running coupling constant (4.8.6) *308*
Rutherford scattering (2.6.50) *118*, (3.4.52) *195*

Scalar (1.3.21) *35*, (1.5.10) *45*
– field (0.2.8) *12*, (1.2.2) *26*, (2.1.76) *77*
– product (0.1.8) *8*, (1.4.8) *37*, (1.4.56) *42*, (2.1.9) *68*, (2.2.2) *78*
– QED (5.1.1) *335*
– totally scalar QED (5.5.39) *374*
Scaling limit (3.2.27) *174*, (3.2.40) *177*
Scaling transformation (3.2.13) *172*
Scattering cross section (2.6.33) *116*, (3.4.32) *192*
– matrix (S-matrix) (0.3.7) *14*, (1.5.58) *52*, (2.4.7) *94*
– operator (2.4.7) *94*
– theory (0.3.7) *14*, (1.5.58) *52*, (2.4.7) *94*
Schrödinger equation (0.3.9) *15*, (1.5.30) *48*, 1.6.25) *57*
– wave function (1.5.39) *49*
Second quantization (2.1.6) *67*, (2.1.61) *76*
Self-energy (3.7.1) *208*, (3.7.41) *214*, (3.13.18) *261*, (4.3.6) *272*
Singular order (3.2.16) *173*, (4.3.2) *271*
Singularities, infrared (3.9.18) *231*, (3.11.6) *242*
SL(2,C) (1.1.20) *23*, (1.1.31) *24*
Slash (1.4.45) *41*
S-matrix (0.3.7) *14*, (1.5.58) *52*, (3.1.1) *160*, (4.2.1) *268*
SO(3) (0.1.12) *9*, (1.4.52) *42*
Soft photons (3.11.25) *244*
Space-like (0.1.8) *8*, (1.4.59) *42*, (2.3.14) *89*
Space reflection (0.1.9) *8*, (1.2.17) *28*, (1.2.35) *30*, (1.5.62) *52*, (4.4.8) *275*
– time (0.1.1) *7*
Spence function (3.9.6) *229*, (C1) *389*
Spherical harmonics (1.6.4) *55*
Spin (1.2.38) *31*, (1.3.11) *33*, (1.4.52) *42*

Spin-orbit coupling (1.5.41) *49*
Spinor field (1.2.3) *26*
- representation (1.1.15) *23*, (1.1.35) *25*
Spin-statistics theorem (2.2.16) *80*
Splitting of distributions (3.1.65) *169*, (3.2.30) *175*, (3.2.61) *181*
- trivial (3.2.30) *175*, (3.4.3) *187*
Stability (4.1.32) *268*
Standard representation (1.3.10) *33*, (1.4.15) *38*, (1.5.25) *47*
Strong field QED (2.7.11) *119*
Structure constants (6.11) *377*
SU(2) (1.1.5) *21*
SU(N) Yang–Mills theories (6.10) *377*
Subtractions (2.8.52) *131*
Super-normalizable theories (4.3.1) *271*, (5.4.17) *363*
Switching function (0.3.21) *16*, (3.1.1) *160*, (5.3.56) *359*
Symmetries, and conservation laws (1.5.12) *46*
- gauge (1.5.8) *45*
Symplectic form (1.1.24) *24*

Tensor (0.2.1) *11*, (1.3.23) *35*
- field (0.2.8) *12*
- product (1.1.32) *25*, (2.1.5) *67*
Thomson formula (3.5.49) *202*
Time evolution (0.3.4) *14*, (2.2.31) *82*, (2.11.25) *149*
- ordered exponential (0.3.20) *16*, (3.1.67) *170*
- ordered product (0.3.20) *16*, (0.3.36) *19*, (3.1.30) *164*, (3.1.67) *170*
- ordering (0.3.20) *16*, (3.1) *159*, (3.1.30) *164*
Time-like (0.1.8) *8*, (1.4.57) *42*
- reflection (0.1.9) *8*, (1.5.61) *52*, (4.4.17) *276*
- reversal (0.1.9) *8*, (1.5.61) *52*, (4.4.17) *276*
Totally scalar QED (5.5.39) *374*
T-product, n-point function (0.3.20) *16*, (0.3.36) *19*, (3.1.30) *164*, (3.1.68) *170*
Trace formulae (2.6.43) *117*
Transformation properties (1.4.46) *41*
Transition probability (0.3.8) *15*, (2.6.19) *114*, (2.7.23) *121*
Translation invariance (1.4.63) *43*, (3.1.16) *162*
Transversal (2.11.4) *146*, (2.11.32) *151*, (3.12.37) *255*, (4.7.2) *300*
Tree graph (3.4.5) *187*
Triangular graphs (5.3.4) *352*
Two-particle operators (2.1.36) *72*

U(1) gauge invariance (1.5.2) *44*
Ultraviolet divergences (3.1.68) *170*, (3.2.61) *181*, (3.6.11) *204*, (3.7.20) *210*
Unimodular group (1.1.5) *21*
Unitarity (0.3.18) *16*, (1.4.39) *40*, (2.4.46) *99*, (3.1.14) *162*
- physical (4.7.1), (4.7.40) *305*
- pseudo (4.7.20) *303*
Unitary group (1.4.9) *37*
- propagator (0.3.15) *15*
- representation (1.4.63) *43*, (2.2.58) *86*
- time evolution (0.3.1) *14*, (2.1.55) *75*, (2.4.62) *101*

Vacuum (2.1.7) *67*, (2.1.42) *73*, (3.3.2) *183*
- bubbles (3.6.26) *206*, (4.8.2) *308*, (5.5.8) *369*
- graphs (4.1.1), (4.3.3) *272*
- polarization (2.10.21) *143*, (3.6.1) *202*, (3.6.26) *206*, (4.3.5) *272*, (4.6.32) *294*, (4.8.1) *308*, (5.4.56) *368*, (5.5.8) *369*
- polarization potential (2.10.23) *143*
Valency operator (4.4.16) *276*
Vector (0.1.8) *8*, (1.3.22) *35*
- field (9.2.9), (1.2.5) *26*, (1.2.39) *31*, (2.11.1) *146*
- product (1.1.2) *21*
Vertex function (3.8.1) *214*, (3.13.1) *258*, (4.3.8) *272*
- causal distribution (3.8.84) *228*
- retarded distribution (3.9.1) *228*

Ward–Takahashi identities (3.13.10) *259*, (4.6.32) *294*, (4.6.47) *299*

Wave equation (1.2.13) *27*, (2.11.2) *146*, (4.10.7) *324*
– operator (0.3.3) *14*, (1.2.13) *27*
Weyl equation (1.2.15) *27*
Weyl representation (1.2.28) *29*, (1.3.7) *33*, (1.5.44) *50*
Weyl spinor (1.2.15) *27*, (1.2.32) *30*, (1.4.46) *41*
Wick ordering (2.1.36) *72*, (3.3.6) *184*
Wick's theorem (3.3.6) *184*, (3.14.1) *261*

Yang–Mills theories (6.10) *377*

Springer-Verlag and the Environment

We at Springer-Verlag firmly believe that an international science publisher has a special obligation to the environment, and our corporate policies consistently reflect this conviction.

We also expect our business partners – paper mills, printers, packaging manufacturers, etc. – to commit themselves to using environmentally friendly materials and production processes.

The paper in this book is made from low- or no-chlorine pulp and is acid free, in conformance with international standards for paper permanency.